Elementary Particle Physics

Elementary Particle Physics

The Standard Theory

J. Iliopoulos and T.N. Tomaras

OXFORD
UNIVERSITY PRESS

OXFORD
UNIVERSITY PRESS

Great Clarendon Street, Oxford, OX2 6DP,
United Kingdom

Oxford University Press is a department of the University of Oxford.
It furthers the University's objective of excellence in research, scholarship,
and education by publishing worldwide. Oxford is a registered trade mark of
Oxford University Press in the UK and in certain other countries

Published in the United States of America by Oxford University Press
198 Madison Avenue, New York, NY 10016, United States of America

British Library Cataloguing in Publication Data

Data available

Library of Congress Control Number: 2021936417

ISBN 978–0–19–284420–0 (hbk.)
ISBN 978–0–19–284421–7 (pbk.)

DOI: 10.1093/oso/9780192844200.001.0001

Printed and bound by
CPI Group (UK) Ltd, Croydon, CR0 4YY

4

To Alexander, Avra and Jiarui

Contents

1
Introduction

The question of the microscopic composition of matter has preoccupied natural philosophers since at least as far back as the 5th century BC. The *atomic hypothesis*, i.e. the existence of "elementary" constituents, is attributed to Democritus of Abdera (c.460 – c.370 BC). Determining the nature of these constituents and understanding the interactions among them is the subject of a branch of fundamental physics called *the physics of elementary particles*. It is the subject of this book. In the past few decades this field has gone through a "phase transition". At first sight the critical time was somewhere between the late 1960s and the early 1970s when the theory, which became known as *the Standard Model*, was fully formulated. However, this is only part of the story and the term "phase transition" may be misleading. It took several decades of intense effort by experimentalists and theorists, both before and after the "critical point", for the various ingredients of this Model to be discovered and for its main predictions to be computed theoretically and verified experimentally. The agreement has been impressive to the extent that we should no more talk about *the Standard Model*, but rather *the Standard Theory*.

This transition has brought profound changes in our way of thinking and understanding the nature of the fundamental forces. The changes are subtle and a casual observer may miss them. Superficially the Standard Model is not fundamentally different from other models that people have considered for many years. They are all based on relativistic quantum mechanics or, as it is usually called, quantum field theory. This has been the language of elementary particle physics since the early 1930s, when Enrico Fermi introduced the notion of a quantum field associated with every elementary particle. The main new element brought by the Standard Model concerns the nature of the interactions between these various quantum fields. In the old days the interactions were chosen on purely phenomenological grounds, like the various potential functions $V(x)$ used in non-relativistic quantum mechanics. The new vision brought by the Standard Model is based on geometry: the interactions are required to satisfy a certain geometrical principle. In the physicists' jargon this principle is called *gauge invariance*; in mathematics it is a branch of differential geometry. This "geometrisation" of physics is the main legacy of the Standard Model.[1]

It is the purpose of this book to present and explain this modern viewpoint to a readership of well-motivated undergraduate students. It is our impression that, although the Standard Model is well established in elementary particle physics and is

---

[1] Above the entrance of Plato's Academy there was the inscription: "Μηδείς αγεωμέτρητος εισίτω μοι τη θύρα", i.e. "Let no one ignorant of geometry enter my door".

Elementary Particle Physics. John Iliopoulos and Theodore N. Tomaras, Oxford University Press.
© John Iliopoulos and Theodore N. Tomaras (2021). DOI: 10.1093/oso/9780192844200.003.0001

widely used, its underlying principles are not easily found in books that undergraduate students usually read. Our ambition is to show that this theory is more than an efficient way to compute scattering amplitudes at lowest order in the perturbation expansion. The subjects we cover and the way we choose to present them are dictated by this goal. We believe that it is time to introduce undergraduate students to these new concepts and methods. And we mean physics concepts. Mathematics will be introduced only when it is absolutely needed. The emphasis will be on the theoretical aspects, and this choice is partly due to our own competence and partly to limitations of space. A good exposition of experimental techniques would take a second volume.

The plan of the book is as follows: We start with a presentation of Dirac's theory of spontaneous emission in atomic physics. This makes it possible to introduce concepts such as the quantised radiation field, canonical commutation relations, creation and annihilation operators, gauge invariance and gauge fixing, but also transition probabilities and Fermi's golden rule, and all these in a concrete and well-defined physical context. Indeed the problem we were facing was that presenting the Standard Model required lengthy chapters of pure formalism before real physics questions could be addressed.

In the following two chapters we recall some results from classical field theory and we introduce the scattering formalism in non-relativistic quantum mechanics. Chapter 5 presents an elementary introduction to the theory of Lie groups and Lie algebras, including Lorentz and Poincaré. The student who has attended a course on group theory may go very fast through it.

Chapter 6 constitutes the introduction to the physics of elementary particles. It is phenomenological and follows, to a certain extent, the historical evolution of the field during the twentieth century. History is not our primary goal, but we believe that it helps in understanding the birth and development of new ideas. The following two chapters present, in a systematic way, the classical relativistic wave equations for fields of low spin and the attempts to use them in order to build a relativistic one-particle quantum mechanics. We derive the well known result that all these attempts point unambiguously to a system with an infinite number of degrees of freedom, i.e. to quantum field theory.

Going from a classical field theory to its quantum descendant is the object of Chapters 9–12. We decided to do it using Feynman's path integral method. Several reasons made us choose this approach, although it is somewhat unorthodox for an undergraduate textbook. First, we believe that the sum over histories, with its relations to stochastic processes, offers a more profound vision of the quantum world. Second, it is by far the most practical way to obtain the quantum theory of a nonlinear constrained system, such as a Yang–Mills theory. Third, and very important, it offers the only quantisation method that is not restricted to the perturbation expansion. The path integral formulation does not assume that the coupling is weak. In fact, appropriately truncated on a space-time lattice, it becomes suitable for numerical simulations in the strong coupling regime. Non-perturbative results from lattice simulations have already reached a remarkable precision, and their agreement with the observed hadronic spectrum is impressive. Furthermore, in the coming years, with the continuing rise in computing power, the importance of these calculations is predicted to increase accordingly.

A systematic introduction to gauge invariant theories and the phenomenon of spontaneous symmetry breaking are the subjects of Chapters 13 and 14. We want to emphasise the conceptual step involved in the introduction of gauge invariant theories. What we physicists call "gauge fields" are not like any other field. The correct mathematical description is given by differential geometry, but we present a simplified version based on a formulation on a space-time lattice. Lattice field theory is poor man's differential geometry.

A book on particle physics today cannot be limited to the calculation of tree-level Feynman diagrams. The precision of the experiments is such that a meaningful comparison requires us to take into account the effects of higher orders. This cannot be done consistently without some notions from the theory of renormalisation and the renormalisation group. They can be found in Chapter 15. A simple treatment of the infrared divergences associated with massless particles is given in Chapter 17.

Chapters 16 to 21 present the Standard Model. They contain a one-loop calculation of the electron gyromagnetic ratio in quantum electrodynamics, the phenomenology of the weak interactions, the $SU(2) \times U(1)$ electroweak gauge theory and quantum chromodynamics. A discussion of neutrino physics, with the poorly understood phenomenon of neutrino oscillations, can be found in Chapter 20

We end with Chapter 22, which offers a panorama of the comparison of the Standard Model theoretical predictions with experimental measurements, including the most recent results. When the Large Hadron Collider (LHC) started operating in 2008 we were all expecting new physics to be around the corner. Today, more than a decade later, we must admit that no corner has been found. The reasons why we still believe that there must be physics beyond the Standard Model are briefly exposed in the last chapter. Finally, in an appendix we explain the notation we are using and present a collection of some useful formulae.

One of us (J.I.) has recently co-authored a book on quantum field theory.[2] Although the scope and level of this present book are different, there is some overlap in chapters that are common to both.

---

[2]"From Classical to Quantum Fields", by Laurent Baulieu, J.I. and Roland Sénéor, Oxford University Press 2017.

2
Quantisation of the Electromagnetic Field and Spontaneous Photon Emission

2.1 Introduction

The first great success of quantum mechanics was the accurate description of atomic spectra. However, this very success also showed its limitations. Indeed, by solving the Schrödinger equation, we find the eigenstates of the Hamiltonian which correspond to the stationary states. It follows that all levels should describe stable states of the atom. On the other hand, we know experimentally, that only the ground state is stable. All excited states decay by the emission of one or more quanta of radiation – photons. Here we shall analyse the simplest case

$$A^{(n)} \quad \rightarrow \quad A^{(0)} + \gamma \tag{2.1}$$

in which the transition to the ground state is a single-step process accompanied by the emission of one photon. $A^{(n)}$ represents the atom in the n-th excited state and $A^{(0)}$ the same atom in the ground state. The phenomenon is known as *spontaneous emission of radiation*, and it is not described by the Schrödinger equation. It was the need to compute the rate of such decays that prompted Dirac in 1927 to develop and use the quantum description of the electromagnetic field.

In quantum mechanics the time evolution of a physical system is given by the operator $U(t, t_0) = \mathrm{e}^{-\mathrm{i}H(t-t_0)}$, where H is the Hamiltonian. So, we must first find the Hamiltonian which, when applied to the state $A^{(n)}$, can yield the atom in its ground state *and* a photon. In other words, this Hamiltonian should have terms with non-zero matrix elements between states containing different numbers of particles. Since this will turn out to be a central theme in our efforts to describe the phenomena we observe in particle physics experiments, it will be useful, as a warm-up exercise, to start with this problem of atomic physics. It will allow us to introduce, in a well-defined physical context, several concepts that will be essential later.

2.2 The Principle of Canonical Quantisation

In the following we shall follow Dirac and build up a formalism, which will make it possible for particles to be created, or absorbed, as a result of the interaction. This formalism will turn out to be that of a quantised field theory and we will study first the quantum theory of the electromagnetic field.

Elementary Particle Physics. John Iliopoulos and Theodore N. Tomaras, Oxford University Press.
© John Iliopoulos and Theodore N. Tomaras (2021). DOI: 10.1093/oso/9780192844200.003.0002

Let us start with a brief reminder of the principle of canonical quantisation. It is based on the knowledge of the physical system at the classical level. Let us consider, as an example, a system with one degree of freedom. In classical mechanics it is described by a generalised coordinate $q(t)$ and its canonical conjugate momentum $p(t)$. The so-called "canonical quantisation" of this system is given, by definition, by the prescription according to which q and p are promoted to operators, acting in a certain Hilbert space, and satisfying the equal time commutation rule[1]

$$[\hat{q}(t), \hat{p}(t)] = i\hbar \tag{2.2}$$

where we have used the notation $\hat{A}$ to denote the operator corresponding to the classical quantity A.[2] We recover the classical theory in the limit $\hbar \to 0$, in which the operators become commuting variables.

This prescription to pass from a classical to a quantum system generalises easily to n degrees of freedom. The commutation relation (2.2) becomes

$$[\hat{q}_I(t), \hat{p}_J(t)] = i\hbar \delta_{IJ}, \quad [\hat{q}_I(t), \hat{q}_J(t)] = [\hat{p}_I(t), \hat{p}_J(t)] = 0, \quad I, J = 1, \ldots, n \tag{2.3}$$

A special case of a system given by (2.3) consists of $\alpha = 1, 2, \ldots, s$ degrees of freedom living on each site of a space lattice with N points and lattice spacing a. In that case the index I is naturally denoted as $I = \{i, \alpha\}$ with $i = 1, 2, \ldots, N$ labelling the lattice points and $n = sN$. For $s = 1$, in particular, the indices i and j denote both the lattice points and the variable at each point. We have studied such systems in statistical mechanics where we often considered the limit $N \to \infty$. We can also consider an appropriate double limit $N \to \infty$ and $a \to 0$, in which i becomes the label $\boldsymbol{x}$ of points of the spatial continuum and $q_I(t) \to q_{x,\alpha}(t)$, which we shall write conveniently as $q_\alpha(t, \boldsymbol{x})$. We thus obtain a system with s continuous infinities of degrees of freedom. In classical physics such a system is called a *classical field* and the best known example is the electromagnetic field.

If the classical theory is well defined and the canonical variables $q_\alpha(t, \boldsymbol{x})$ and $p_\beta(t, \boldsymbol{x})$ correctly identified, the quantisation of such a system is, in principle, straightforward: The commutation relation (2.3) is replaced by

$$[\hat{q}_\alpha(t, \boldsymbol{x}), \hat{p}_\beta(t, \boldsymbol{y})] = i\hbar \delta_{\alpha\beta} \delta^3(\boldsymbol{x} - \boldsymbol{y}) \tag{2.4}$$

with the remaining equal time commutators of two q's, as well as of two p's equal to zero. We notice that the right-hand side of the relations (2.4) is proportional to the Dirac δ-function, which is not a "function" in the usual sense of the word. In particular, the square, or any power of it, cannot be defined. In mathematics, such generalised functions are called *distributions*. It follows that the operators $\hat{q}$ and $\hat{p}$ must also be

---

[1] In the usual formulation of non-relativistic quantum mechanics, the so-called *Schrödinger representation*, states $|\Psi(t)\rangle$ depend on time, and evolve according to $|\Psi(t)\rangle = U(t,0)|\Psi(0)\rangle = \exp(-iHt)|\Psi(0)\rangle$, while observables A without explicit time dependence are represented by time-independent operators A_S. In the commutation relation (2.2) we used the *Heisenberg representation*, in which the states are fixed $|\Psi(0)\rangle = U^{-1}(t,0)|\Psi(t)\rangle$ and the observables $A_H(t) = U^{-1}(t,0)A_S U(t,0)$ are time-dependent. At $t = 0$ the two coincide: $A_H(0) = A_S$.

[2] In order to simplify the formulae we will often drop the "hat" if there is no danger of confusion.

represented by distributions, and we expect their powers to be ill defined. We say that $\hat{q}$ and $\hat{p}$ are not *operator valued functions* of the space point, but become instead *operator valued distributions*. This leads to some mathematical difficulties, which are common to all quantum field theories and which are only partially mastered. We shall come back to this point quite often in this book.

As promised, we next apply this programme to the electromagnetic field and obtain the quantum descendant of Maxwell's theory. It will be our first example of a quantum field theory and the only one which has a well known classical limit.

2.3 The Quantum Theory of Radiation

2.3.1 Maxwell's theory as a classical field theory

The simplest version of Maxwell's equations takes the form[3]

$$\boldsymbol{\nabla} \cdot \boldsymbol{E} = \rho \, , \quad \boldsymbol{\nabla} \cdot \boldsymbol{B} = 0 \tag{2.5}$$

$$\boldsymbol{\nabla} \wedge \boldsymbol{B} - \frac{\partial \boldsymbol{E}}{\partial t} = \boldsymbol{j} \, , \quad \boldsymbol{\nabla} \wedge \boldsymbol{E} + \frac{\partial \boldsymbol{B}}{\partial t} = 0 \tag{2.6}$$

where $\boldsymbol{E}(t, \boldsymbol{x})$ and $\boldsymbol{B}(t, \boldsymbol{x})$ are the electric and magnetic fields, respectively, while $\rho(t, \boldsymbol{x})$ and $\boldsymbol{j}(t, \boldsymbol{x})$ are the external electric charge and current densities. Consistency of (2.5), (2.6), requires that ρ and $\boldsymbol{j}$ satisfy the continuity condition: $\partial \rho / \partial t + \boldsymbol{\nabla} \cdot \boldsymbol{j} = 0$. According to our recipe, if we want to describe this system as a classical field theory, we must first identify a set of independent variables $q_\alpha(t, \boldsymbol{x})$. These cannot be the six components of $\boldsymbol{E}(t, \boldsymbol{x})$ and $\boldsymbol{B}(t, \boldsymbol{x})$ because they are constrained by the first two equations (2.5). We should solve these constraints and eliminate the redundant variables.

It seems that a first step in this direction was taken by Gauss in 1835, long before Maxwell wrote his equations. It consists of introducing the vector and scalar potentials $\boldsymbol{A}(t, \boldsymbol{x})$ and $\phi(t, \boldsymbol{x})$. It will be convenient to use a compact relativistic notation in which $x = (t, \boldsymbol{x})$ and introduce the four-vectors $j^\mu(x) = (\rho, \boldsymbol{j})$ and $A^\mu(x) = (\phi, \boldsymbol{A})$. We then construct the two-index antisymmetric tensor

$$F_{\mu\nu}(x) = \frac{\partial A_\nu}{\partial x^\mu} - \frac{\partial A_\mu}{\partial x^\nu} \tag{2.7}$$

Since the derivative operator $\partial/\partial x_\mu$ will appear very often, we introduce a short-hand notation for it: $\partial/\partial x_\mu = \partial^\mu$. Similarly, $\partial/\partial x^\mu = \partial_\mu$. The electric and magnetic fields are given in terms of $F_{\mu\nu}$ by

$$F_{0i} = -\partial_0 A^i - \partial_i A^0 = \left(-\frac{\partial \boldsymbol{A}}{\partial t} - \boldsymbol{\nabla} A^0 \right)^i = E^i \, , \quad B^i = \frac{1}{2} \epsilon_{ijk} F_{jk} \tag{2.8}$$

---

[3]We use the symbol $\wedge$ to denote the vector product of two three-dimensional vectors : $(\boldsymbol{a} \wedge \boldsymbol{b})_i = \epsilon_{ijk} a^j b^k$.

where $\epsilon^{ijk}(= -\epsilon_{ijk})$ is the three-index completely antisymmetric tensor, equal to $+1$ if $\{ijk\}$ form an even permutation of $\{123\}$.[4] It is now easy to check that the two inhomogeneous equations (2.6) combine to

$$\partial^\mu F_{\mu\nu}(x) = j_\nu(x) \qquad (2.9)$$

while the two homogeneous ones (2.5) are automatically satisfied. The equation (2.9) follows from a variational principle applied to the action

$$S = \int d^4x \, \mathcal{L} = \int d^4x \left(-\frac{1}{4} F_{\mu\nu} F^{\mu\nu} + A_\mu j^\mu \right) \qquad (2.10)$$

Can we choose the four components of A_μ as independent dynamical variables? The answer is no, because the Lagrangian (2.10) does not contain the time derivative of A_0; in other words, the canonical conjugate momentum of A_0 would be identically zero. We know that this problem is related to the fact that the Lagrangian density in (2.10) is invariant under the transformation $A_\mu(x) \to A_\mu(x) + \partial_\mu \theta(x)$ with $\theta(x)$ an arbitrary function of x. In classical electrodynamics we call this invariance "gauge invariance" and we can use the freedom of choosing a particular function θ to reduce the number of independent variables. In Chapter 13 we will study this problem in a more general context, but here we just recall that, experimentally, an electromagnetic wave in empty space has only transverse degrees of polarisation. Therefore, we can impose the transversality condition $\boldsymbol{\nabla} \cdot \boldsymbol{A}(x) = 0$ (the so-called "Coulomb gauge condition") under which the zero component of the vector potential satisfies[5]

$$\Delta A_0(x) + \rho(x) = 0 \qquad (2.11)$$

where Δ is the Laplacian. This implies the Coulomb law (hence the name of this condition)

$$A_0(t, \boldsymbol{x}) = \frac{1}{4\pi} \int \frac{d^3x'}{|\boldsymbol{x} - \boldsymbol{x}'|} \rho(t, \boldsymbol{x}') \qquad (2.12)$$

which shows that A_0 is entirely given by the external source and it is not an independent dynamical degree of freedom.[6] We are left with the spatial components, which are constrained by the Coulomb condition. The simplest way to solve it and obtain an unconstrained system is to take the three-dimensional Fourier transform

$$\boldsymbol{A}(t, \boldsymbol{x}) = \frac{1}{(2\pi)^3} \int d^3k \, e^{i\boldsymbol{k} \cdot \boldsymbol{x}} \widetilde{\boldsymbol{A}}(t, \boldsymbol{k}) \qquad (2.13)$$

in terms of which the constraint becomes

---

[4]In this formula, raising and lowering the indices of three-dimensional vectors is performed using the Minkowski metric, as we explain in Appendix A.

[5]Another choice is the Lorentz covariant gauge condition of the form $\partial_\mu A^\mu(x) = 0$, the so-called "Lorenz gauge condition", first introduced by the Dane Ludvig Lorenz in 1867.

[6]Unless noted otherwise, we will assume that both the sources and the dynamical fields vanish at infinity.

$$\boldsymbol{k} \cdot \widetilde{\boldsymbol{A}}(t, \boldsymbol{k}) = 0 \tag{2.14}$$

which suggests to choose an orthonormal system of unit vectors $\epsilon^{(3)}(\boldsymbol{k}) = \boldsymbol{k}/|\boldsymbol{k}|$ and $\epsilon^{(\lambda)}(\boldsymbol{k})$, $\lambda = 1, 2$, satisfying

$$\boldsymbol{k} \cdot \epsilon^{(\lambda)}(\boldsymbol{k}) = 0 \;, \quad \epsilon^{(\lambda)}(\boldsymbol{k}) \cdot \epsilon^{(\lambda')}(\boldsymbol{k}) = \delta_{\lambda\lambda'} \;, \quad \epsilon^{(\lambda)}(-\boldsymbol{k}) = (-)^\lambda \epsilon^{(\lambda)}(\boldsymbol{k}) \tag{2.15}$$

i.e. $\epsilon^{(3)}(\boldsymbol{k})$ is parallel to the wave vector and the other two are transverse to it. Because of the gauge condition, in this frame the vector potential has only transverse components

$$\boldsymbol{A}(t, \boldsymbol{x}) = \sum_{\lambda=1}^{2} \frac{1}{(2\pi)^3} \int \mathrm{d}^3 k \; \mathrm{e}^{\mathrm{i}k \cdot x} \widetilde{A}^{(\lambda)}(t, \boldsymbol{k}) \epsilon^{(\lambda)}(\boldsymbol{k}) \tag{2.16}$$

in agreement with the experimental fact that the electromagnetic waves in empty space are transverse.

Let us summarise: Formulating the theory in terms of the vector potential A_μ, and making full use of its gauge invariance, allowed us to show that, out of the six components of $\boldsymbol{E}$ and $\boldsymbol{B}$, only two are really independent, as expected from the known properties of electromagnetic radiation. This result, which, as we shall prove later, follows from general invariance principles of the theory, made it possible to formulate electromagnetism as a dynamical system. For each $\{\boldsymbol{k}, \lambda\}$, $\widetilde{A}^{(\lambda)}(t, \boldsymbol{k})$ is an independent variable. The associated canonical momentum can be computed from the Lagrangian (2.10). We find

$$q_I \to \widetilde{A}^{(\lambda)}(t, \boldsymbol{k}) \;, \quad p_I \to \partial \mathcal{L}/\partial \dot{\widetilde{A}}^{(\lambda)}(t, \boldsymbol{k}) = \dot{\widetilde{A}}^{(\lambda)*}(t, \boldsymbol{k}) \tag{2.17}$$

where we have used the standard notation of mechanics in which "dot" means derivative with respect to time.

In the absence of external sources, the vector potential $\boldsymbol{A}(x)$ in the Coulomb gauge satisfies the wave equation

$$\ddot{\boldsymbol{A}}(x) - \Delta \boldsymbol{A}(x) = 0 \tag{2.18}$$

If $\widetilde{\boldsymbol{A}}(k)$ is the four-dimensional Fourier transform of $\boldsymbol{A}(x)$, the wave equation (2.18) becomes an algebraic equation: $k^2 \widetilde{\boldsymbol{A}}(k) = 0$, which implies that $\widetilde{\boldsymbol{A}}(k) = \boldsymbol{F}(k^2)\,\delta(k^2)$ with $\boldsymbol{F}(k^2)$ arbitrary functions of k^2, provided they are regular at $k^2 = 0$.

The general solution of (2.18) can be expanded in plane waves, i.e. functions of the form $\mathrm{e}^{-\mathrm{i}k \cdot x}$ with $k \cdot x = k_0 t - \boldsymbol{k} \cdot \boldsymbol{x}$, and with the four vector k^μ satisfying $k^2 \equiv k_0^2 - \boldsymbol{k}^2 = 0$. It will be convenient to introduce the notation

$$\mathrm{d}\Omega_m(k) = \frac{\mathrm{d}^3 k}{(2\pi)^3 2E_k} = \frac{\mathrm{d}^4 k}{(2\pi)^4} (2\pi)\delta(k^2 - m^2)\theta(k_0) \;, \quad E_k = \sqrt{\boldsymbol{k}^2 + m^2} \tag{2.19}$$

which is the Lorentz invariant measure on the positive energy branch of the mass hyperboloid given by $k^2 = m^2$. We thus write

$$\boldsymbol{A}(t, \boldsymbol{x}) = \int \mathrm{d}\Omega_0(k) \sum_{\lambda=1}^{2} \boldsymbol{\epsilon}^{(\lambda)}(\boldsymbol{k}) \left[a^{(\lambda)}(\boldsymbol{k}) \mathrm{e}^{-\mathrm{i}k\cdot x} + a^{(\lambda)*}(\boldsymbol{k}) \mathrm{e}^{\mathrm{i}k\cdot x} \right] \qquad (2.20)$$

where $\mathrm{d}\Omega_0(k)$ is the value for $m = 0$ of the expression (2.19). It follows that the integration in (2.20) is over the boundary of the positive energy $k^2 = 0$ cone.

We can invert (2.20) to express the coefficient functions $a^{(\lambda)}(\boldsymbol{k})$ as

$$a^{(\lambda)}(\boldsymbol{k}) = \int \mathrm{d}^3 x \; \mathrm{e}^{-\mathrm{i}k\cdot x} \boldsymbol{\epsilon}^{(\lambda)}(\boldsymbol{k}) \cdot \left[E_k \boldsymbol{A}(t, \boldsymbol{x}) + \mathrm{i}\dot{\boldsymbol{A}}(t, \boldsymbol{x}) \right] \qquad (2.21)$$

In Problem 2.1 we ask the reader to verify that the energy H and momentum $\boldsymbol{P}$ of the field in terms of the coefficient functions are

$$H = \int \mathrm{d}\Omega_0(k) \; E_k \sum_{\lambda=1}^{2} \left[a^{(\lambda)*}(\boldsymbol{k}) a^{(\lambda)}(\boldsymbol{k}) \right], \quad \boldsymbol{P} = \int \mathrm{d}\Omega_0(k) \; \boldsymbol{k} \sum_{\lambda=1}^{2} \left[a^{(\lambda)*}(\boldsymbol{k}) a^{(\lambda)}(\boldsymbol{k}) \right]$$
$$(2.22)$$

2.3.2 Quantum theory of the free electromagnetic field–photons

We are now ready to apply our general quantisation prescription and obtain the corresponding quantum theory. The canonical variables (2.17) are promoted to operators satisfying the canonical commutation relations (2.4).[7]

$$[\widetilde{A}^{(\lambda)}(t, \boldsymbol{k}), \dot{\widetilde{A}}^{(\lambda')\dagger}(t, \boldsymbol{k'})] = \mathrm{i}\delta^{\lambda\lambda'}(2\pi)^3 \delta^3(\boldsymbol{k} - \boldsymbol{k'})$$

$$[\widetilde{A}^{(\lambda)}(t, \boldsymbol{k}), \widetilde{A}^{(\lambda')}(t, \boldsymbol{k'})] = 0 \;, \quad [\dot{\widetilde{A}}^{(\lambda)\dagger}(t, \boldsymbol{k}), \dot{\widetilde{A}}^{(\lambda')\dagger}(t, \boldsymbol{k'})] = 0 \qquad (2.23)$$

where † denotes the Hermitian adjoint of the operator. The presence of the factor $(2\pi)^3$ is due to our convention on the Fourier transform. Here, and throughout this book, we have adopted the system of units we introduce in Appendix A in which $c = \hbar = 1$. Physical units will be restored only when it is necessary.

The relations (2.23) imply for the operators corresponding to the coefficients of the plane wave expansion (2.20) the commutation relations

$$[a^{(\lambda)}(\boldsymbol{k}), a^{(\lambda')\dagger}(\boldsymbol{k'})] = \delta^{\lambda\lambda'} 2E_k (2\pi)^3 \delta^3(\boldsymbol{k} - \boldsymbol{k'})$$

$$[a^{(\lambda)}(\boldsymbol{k}), a^{(\lambda')}(\boldsymbol{k'})] = 0 \;, \quad [a^{(\lambda)\dagger}(\boldsymbol{k}), a^{(\lambda')\dagger}(\boldsymbol{k'})] = 0 \qquad (2.24)$$

These relations are still formal because we have not yet defined the space in which these operators act. In order to do it we remark that, for every fixed value of $\{\boldsymbol{k}, \lambda\}$, the operators $a^{(\lambda)}(\boldsymbol{k})$ and $a^{(\lambda)\dagger}(\boldsymbol{k})$, appropriately rescaled, satisfy the commutation relations of annihilation and creation operators for a harmonic oscillator with frequency $E_k = |\boldsymbol{k}|$. In other words, we can interpret the electromagnetic field in the vacuum as a doubly (one for each value of λ) continuous infinite set of independent harmonic

---

[7]We omit the symbol ∧ on the operators.

oscillators. Following the example of the harmonic oscillator, we define the space of states as follows:

1. First we assume the existence of a state which is annihilated by all annihilation operators $a^{(\lambda)}(\boldsymbol{k})$, for both values of λ and all $\boldsymbol{k}$. We assume this state to be unique and normalised to one. We denote it by $|0\rangle$ and we call it *the vacuum state*.

$$a^{(\lambda)}(\boldsymbol{k})\,|0\rangle = 0 \ , \quad \langle 0|0\rangle = 1 \qquad (2.25)$$

2. Starting from this state we build excited states by applying the creation operators on it. For example, the first excited states are given by

$$|\boldsymbol{k},\lambda\rangle = \ a^{(\lambda)\dagger}(\boldsymbol{k})\,|0\rangle \qquad (2.26)$$

Using the commutation relations (2.24) and the equation (2.25), which defined the vacuum state, we find

$$\langle \boldsymbol{k}',\lambda'|\boldsymbol{k},\lambda\rangle \ = \ \langle 0|\,a^{(\lambda')}(\boldsymbol{k}')a^{(\lambda)\dagger}(\boldsymbol{k})\,|0\rangle = \delta^{\lambda\lambda'}2E_k(2\pi)^3\delta^3(\boldsymbol{k}-\boldsymbol{k}') \qquad (2.27)$$

Of course, we expect the state $|\boldsymbol{k},\lambda\rangle$, since it is a state with fixed momentum, to correspond to a wave function described by a plane wave and, therefore, to be non-normalisable. This is the meaning of the δ function on the r.h.s. of equation (2.27). As we did in quantum mechanics, we can build normalisable states using wave packets. Given a function $\Phi(\boldsymbol{k})$ satisfying

$$\int \mathrm{d}\Omega_0(k)|\Phi(\boldsymbol{k})|^2 = 1 \qquad (2.28)$$

we define the wave packet

$$|\Phi,\lambda\rangle = \int \mathrm{d}\Omega_0(k)\Phi(\boldsymbol{k})\,|\boldsymbol{k},\lambda\rangle \qquad (2.29)$$

which is normalised to one.

In the terminology of the harmonic oscillator, $|\boldsymbol{k},\lambda\rangle$ is the state of one excitation of type $\{\boldsymbol{k},\lambda\}$. We see that, for each value of λ, the space of the one-excitation states is our familiar Hilbert space of square-integrable complex-valued functions, which we denote by $\mathcal{H}_1$.[8]

In a similar way we build "multi-excitation" states by acting on the vacuum with products of creation operators

$$|\boldsymbol{k}_1,\lambda_1;\boldsymbol{k}_2,\lambda_2;\dots;\boldsymbol{k}_n,\lambda_n\rangle = \ a^{(\lambda_n)\dagger}(\boldsymbol{k}_n)\dots a^{(\lambda_2)\dagger}(\boldsymbol{k}_2)a^{(\lambda_1)\dagger}(\boldsymbol{k}_1)\,|0\rangle \qquad (2.30)$$

Again, for each set of values $\{\lambda_i, i = 1,...,n\}$, we denote by $\mathcal{H}_{n,\lambda}$ the Hilbert space of states in the direct product of $\mathcal{H}_1$ with itself n times. The entire space of states is

---

[8]More precisely, the space we use in quantum mechanics is a ray-space. Let $|\psi\rangle$ be a vector in $\mathcal{H}_1$ normalised to 1 $\langle \psi|\psi\rangle = 1$ and $C \in \mathbb{C} \ C \neq 0$. A unit ray associated to $|\psi\rangle$ is the set of all vectors of the form $C\,|\psi\rangle$. In quantum mechanics we have learned that all vectors in this set represent the same physical state because the wave functions $\Psi(x)$ and $C\Psi(x)$ are identified.

the direct sum of all $\mathcal{H}_n$'s, for all n and all sets $\{\lambda\}$. We call this space *the Fock space of states*

$$\mathcal{F} = \sum_{n=0}^{\infty} \sum_{\{\lambda\}} \oplus \mathcal{H}_{n,\lambda} \tag{2.31}$$

where, for notational simplicity, we have defined $\mathcal{H}_0$ as the one-dimensional space spanned by the vacuum state, i.e. the ray-space of complex numbers.

The physical meaning of the states in the Fock space becomes more transparent if we express the Hamiltonian and the momentum operators of the system in terms of a and $a^\dagger$, to obtain the quantum versions of the relations (2.22). It is easy to see that, since creation and annihilation operators do not commute, we end-up with

$$H = \frac{1}{2} \int d\Omega_0(k) \, E_k \sum_{\lambda=1}^{2} \left[a^{(\lambda)\dagger}(\boldsymbol{k}) a^{(\lambda)}(\boldsymbol{k}) + a^{(\lambda)}(\boldsymbol{k}) a^{(\lambda)\dagger}(\boldsymbol{k}) \right] \tag{2.32}$$

and

$$\boldsymbol{P} = \frac{1}{2} \int d\Omega_0(k) \, \boldsymbol{k} \sum_{\lambda=1}^{2} \left[a^{(\lambda)\dagger}(\boldsymbol{k}) a^{(\lambda)}(\boldsymbol{k}) + a^{(\lambda)}(\boldsymbol{k}) a^{(\lambda)\dagger}(\boldsymbol{k}) \right] \tag{2.33}$$

We next use (2.32) and (2.33) to compute the energy and momentum of each state in $\mathcal{F}$. But here we face a subtle problem: Let us start with the energy of the vacuum state. As usual, it is given by the expectation value of the Hamiltonian $\langle 0| H |0\rangle$. Using the commutation relations (2.24), we obtain

$$H = \int d\Omega_0(k) \, E_k \sum_{\lambda=1}^{2} \left(a^{(\lambda)\dagger}(\boldsymbol{k}) a^{(\lambda)}(\boldsymbol{k}) + \frac{1}{2} \left[a^{(\lambda)}(\boldsymbol{k}) \, , \, a^{(\lambda)\dagger}(\boldsymbol{k}) \right] \right) \tag{2.34}$$

The trouble comes from the commutator which, formally, is proportional to $\delta^3(0)$ and is, therefore, meaningless. It is easy to understand the origin of this problem, both physically and mathematically. First with the physics: We recall that the energy of a single harmonic oscillator with frequency ω is expressed as $H = \omega(a^\dagger a + 1/2)$ and its mean value in the ground state is just $\omega/2$, *the zero-point energy* of the harmonic oscillator. It is a quantum mechanical phenomenon and, as such, in physical units it is proportional to $\hbar$. We have noted already that the quantum electromagnetic field in the absence of sources is equivalent to an infinite set of harmonic oscillators satisfying the relativistic dispersion law $\omega = |\boldsymbol{k}|$. So, it is not surprising that the ground state energy is infinite; it is the infinite sum of the zero-point energies. Now with the mathematics: A divergent expression often results from a mathematical mistake; some quantity is ill defined. Indeed, in writing the Lagrangian density, or the Hamiltonian, of the classical electromagnetic field, we used expressions such as those in equation (2.10), which contain the square of the field $F_{\mu\nu}(x)$. This is fine for the classical theory. In the quantum theory, however, we have noted already that the field variables are distributions, and their square, or any higher power, is not well defined. At this stage the problem is a nuisance rather than a catastrophe. Since we understand its origin we can easily guess its solution. In the absence of a gravitational field we never measure the absolute value of the energy of any state. Only energy differences are measurable. In other words, the expression for the energy of a system is always defined up to an arbitrary additive constant. So, the equation (2.34) should be corrected to $H \to H + C$, with C an arbitrary constant. We shall use this freedom to redefine the energy of the electromagnetic field in the absence of sources so that the vacuum energy vanishes. We define a *renormalised* energy operator by imposing the condition

$$\langle 0 | H_{\text{ren}} | 0 \rangle = 0 \qquad (2.35)$$

We see that this amounts to dropping the constant divergent term in (2.34) and we find

$$H_{\text{ren}} = \int d\Omega_0(k) \, E_k \sum_{\lambda=1}^{2} \left(a^{(\lambda)\dagger}(\boldsymbol{k}) a^{(\lambda)}(\boldsymbol{k}) \right) \qquad (2.36)$$

This is an example of a general procedure, called *renormalisation*, which allows us to attribute precise values to mathematically ill defined expressions involving products of distributions. It is based on physical requirements, such as the vanishing of the vacuum energy and it has been at the basis of all the spectacular success of quantum field theory.

This prescription can also be expressed as an ordering prescription for a product of operators. Consider an operator $\mathcal{O}$, which is the product of n creation and m annihilation operators. We shall call *normal-ordered* $\mathcal{O}$, denoted by $: \mathcal{O} :$, the expression in which we write first the n creation operators and then the m annihilation ones:

$$: \mathcal{O} := a^{(\lambda_1)\dagger}(\boldsymbol{k}_1) a^{(\lambda_2)\dagger}(\boldsymbol{k}_2) \dots a^{(\lambda_n)\dagger}(\boldsymbol{k}_n) a^{(\lambda_1')}(\boldsymbol{k}_1') a^{(\lambda_2')}(\boldsymbol{k}_2') \dots a^{(\lambda_m')}(\boldsymbol{k}_m') \qquad (2.37)$$

Obviously, the definition extends to sums and products of such operators. In any product, all creation operators will be written on the left of all annihilation ones. As a result, the vacuum expectation value of a normal-ordered operator vanishes identically. We see also that for the Hamiltonian operator in particular $: H := H_{\text{ren}}$. Similarly, we define the renormalised version of the momentum operator (2.33) with $: \boldsymbol{P} := \boldsymbol{P}_{\text{ren}}$. Now it is straightforward to verify that a state in the Fock space of the form given in (2.30) is an eigenstate of the renormalised energy and momentum operators with eigenvalues $E_1 + E_2 + \dots + E_n$ and $\boldsymbol{k}_1 + \boldsymbol{k}_2 + \dots + \boldsymbol{k}_n$, respectively. We remind ourselves also that the momenta $k_a^\mu = (E_a, \boldsymbol{k}_a)$ satisfy $|\boldsymbol{k}_a| = E_a$, i.e. $k_a^2 = E_a^2 - \boldsymbol{k}_a^2 = 0$. These properties allow us to give a physical interpretation to the multi-excitation states (2.30).

The state $|\boldsymbol{k}, \lambda\rangle$ has momentum $\boldsymbol{k}$ and energy $E = |\boldsymbol{k}|$, thus its 4-momentum satisfies $k^2 = 0$. Therefore, we can interpret it as describing a massless particle which we shall call a *photon*.[9] Similarly, the states (2.30) are energy and momentum eigenstates with eigenvalues corresponding to any number of non-interacting photons. Therefore, the quantised free electromagnetic Maxwell field theory is the theory of free photons.

---

[9] Looking at the transformation properties of the state under Lorentz transformations, we can prove that λ, defined in (2.15), corresponds to the polarisation of the one-photon state, see section 5.6.2.

Two remarks before closing this section: First, we note that, since the creation operators for any values of momentum and polarisation commute, we can apply them in any order in the formula (2.30). We conclude that the multiphoton state is totally symmetric under the interchange of any pair of photons; in other words, photons obey the Bose–Einstein statistics. The second is more general: Following this quantisation procedure, we expressed a quantum field, here the electromagnetic field $A_\mu(x)$, which satisfies a linear differential equation, as an infinite superposition of creation and annihilation operators acting in a certain Fock space. The states in this space are eigenstates of the operators of energy and momentum. This property makes it possible to interpret these states as describing non-interacting particles. This field–particle duality will be generalised and used extensively in this book. It will become clear that at a fundamental level any types of elementary particles are the quanta of a corresponding relativistic field.

2.4   Interaction of Atoms and Radiation

2.4.1   The Hamiltonian

We are now in a position to attack the problem of spontaneous emission, equation (2.1). In order to simplify the discussion we shall make some assumptions and approximations. First, we take the atom to be infinitely heavy and neglect the recoil resulting from the photon emission. Second, we choose a hydrogen-like atom with only one electron in the outer shell. The approximation consists of treating this electron as moving in an effective Coulomb potential and neglect the multibody interactions with the individual electrons of the inner shells. Third, we shall neglect all relativistic corrections for the electron.

The second assumption implies that the Hamiltonian of the atom alone is given by the usual expression

$$H_a = \frac{1}{2m} \boldsymbol{p}^2 + eU_C \tag{2.38}$$

where m is the effective mass of the electron, e its electric charge and U_C the Coulomb potential. To this expression we must add the Hamiltonian H_r of the free electromagnetic field, equation (2.36).

Finally, we must include the term that describes the interaction between the electron in the atom and the electromagnetic field. In classical electrodynamics we have found a prescription, called *minimal substitution*, which amounts to replacing everywhere the momentum p_μ by $p_\mu - eA_\mu$, with A_μ the electromagnetic vector potential. In Chapter 13 we shall give a justification of this prescription based on fundamental symmetry principles, but here we apply it starting from the Hamiltonian (2.38) with the result

$$H_I = -\frac{e}{2m}[\boldsymbol{p} \cdot \boldsymbol{A} + \boldsymbol{A} \cdot \boldsymbol{p}] + \frac{e^2}{2m} \boldsymbol{A}^2 \tag{2.39}$$

So, the total Hamiltonian of the system is

$$H = H_a + H_r + H_I \tag{2.40}$$

where H_a is the hydrogen-like atom Hamiltonian (2.38) with $U_C = A_0$, H_r that of the electromagnetic field in the vacuum (2.36) and H_I is the interaction Hamiltonian given by (2.39).[10]

The quantisation of this system is, in principle, straightforward. The canonical variables are the position of the electron x and its conjugate momentum p, as well as the infinite set of annihilation and creation operators $a^{(\lambda)}(k)$ and $a^{(\lambda)\dagger}(k)$ of the electromagnetic field. In the Coulomb gauge, p and A commute. However, the interaction term H_I couples the electronic and the photonic degrees of freedom and, as a result, we are unable to diagonalise H exactly and find its eigenvalues and eigenstates. In fact, there is a very small number of physical problems for which we have exact solutions and for the vast majority, we have to use some approximation scheme. The most common is a perturbation expansion, which we have studied already in quantum mechanics. We recall the main features of the method here.

2.4.2 Elements of perturbation theory

In quantum mechanics the probability amplitude $\mathcal{A}_{fi}$ for the transition of a system from an initial state $|\Psi_i(t_i)\rangle$ to a final state $|\Psi_f(t_f)\rangle$ is given by

$$\mathcal{A}_{fi} = \langle \Psi_f(t_f)| \, U(t_f, t_i) \, |\Psi_i(t_i)\rangle \tag{2.41}$$

where $U(t_f, t_i) = \exp[-\mathrm{i}H(t_f - t_i)]$ is the evolution operator and H the Hamiltonian of the system. For most interesting cases this expression is only formal, because we do not know how to diagonalise the Hamiltonian and, consequently, we do not know how to compute the evolution operator U.

To proceed, let us start by addressing a general problem in which the Hamiltonian H, is the sum

$$H = H_0 + H_P \tag{2.42}$$

of a part H_0, whose spectrum is supposed to be known

$$H_0 \, |n\rangle_{(0)} = E_n^{(0)} \, |n\rangle_{(0)} \tag{2.43}$$

and a perturbation H_P. In order to simplify the discussion let us first assume that H_0 has a discrete, non-degenerate spectrum with normalisable eigenstates. We shall deal shortly with the case of the spontaneous emission, in which the spectrum of the Hamiltonian for the free electromagnetic field is continuous. Furthermore, we shall assume that the set of vectors $|n\rangle_{(0)}$ form a basis in the entire Hilbert space, so that any state can be written as a linear superposition of them. If H_0 and H_P were not operators but c-number functions, the evolution operator could be written as

---

[10]We see here the physical interpretation of the constant e: it is the electric charge of the particle, which characterises the strength of its coupling with the electromagnetic field. We shall call it *the coupling constant*, a terminology we shall use more generally, when we want to speak about the strength of an interaction.

$U(t) = U_0(t)U_P(t) = \exp[-\mathrm{i}H_0t]\exp[-\mathrm{i}H_Pt]$. However, for operators this is not true because $\exp[A]\exp[B] \neq \exp[A+B]$. Nevertheless, let us define the unitary operator

$$\widetilde{U}_P(t) = U_0^{-1}(t)U(t) \tag{2.44}$$

Both $U(t)$ and $U_0(t)$ satisfy the Schrödinger equation with Hamiltonian H and H_0, respectively, and it is straightforward to derive the equation satisfied by $\widetilde{U}_P(t)$

$$\mathrm{i}\frac{\partial\widetilde{U}_P}{\partial t} = -U_0^{-1}\mathrm{i}\frac{\partial U_0}{\partial t}U_0^{-1}U + U_0^{-1}(H_0 + H_P)U_0\widetilde{U}_P = \widetilde{H}_P\widetilde{U}_P \tag{2.45}$$

where

$$\widetilde{H}_P(t) = U_0^{-1}(t)H_PU_0(t) \tag{2.46}$$

Equation (2.45), together with the initial condition $\widetilde{U}_P(0) = 1$, can be written as the integral equation

$$\widetilde{U}_P(t) = 1 - \mathrm{i}\int_0^t \widetilde{H}_P(t')\widetilde{U}_P(t')\mathrm{d}t' \tag{2.47}$$

It is this equation that we want to solve perturbatively in powers of H_P. Once $\widetilde{U}_P(t)$ is known we can reconstruct the total evolution operator from the definition $U(t) = U_0(t)\widetilde{U}_P(t)$. We find

$$\widetilde{U}_P(t) = 1 - \mathrm{i}\int_0^t \widetilde{H}_P(t_1)\mathrm{d}t_1 - \int_0^t\int_0^{t_1} \widetilde{H}_P(t_1)\widetilde{H}_P(t_2)\mathrm{d}t_1\mathrm{d}t_2 + \dots \tag{2.48}$$

which gives, for $U(t)$,

$$U(t) = \mathrm{e}^{-\mathrm{i}H_0t}\left(1 - \mathrm{i}\int_0^t \mathrm{e}^{\mathrm{i}H_0t_1}H_P\,\mathrm{e}^{-\mathrm{i}H_0t_1}\mathrm{d}t_1 + \dots\right) \tag{2.49}$$

In this formula H_P is the perturbation Hamiltonian in the Schrödinger representation.

2.4.3 The transition probability

Let us assume that at $t = t_i$ the system is in an eigenstate $|i\rangle$ of H_0. Under the influence of the perturbation, at time $t = t_f$ the system will be in some state $|f\rangle$, which, under our assumptions, we can write as a linear superposition of the eigenstates of H_0. We would like to compute the probability to find the system in a particular eigenstate of H_0. It will be given by the square of the amplitude (2.41):

$$\mathcal{P}_{fi} = |\langle f|U(t_f, t_i)|i\rangle|^2 \tag{2.50}$$

We wish to compute this matrix element in perturbation theory using the expansion (2.49). In all interesting cases, the initial and the final states are different and

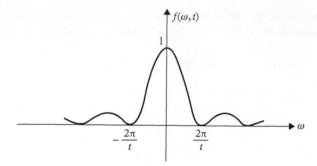

Fig. 2.1 The function $f(\omega, t)$.

orthogonal to each other, so the zeroth order term, the unit operator, drops out. At first order we obtain

$$\mathcal{P}_{fi} = \left| \int_0^t e^{i(E_i - E_f)t_1} \langle f| H_P |i\rangle \, dt_1 \right|^2 \tag{2.51}$$

where we have set $t_i = 0$ and $t_f = t$.

In our problem of spontaneous emission, as well as in most physically interesting problems, the perturbation Hamiltonian is time-independent. In this case the integration over t_1 is trivial and gives

$$\mathcal{P}_{fi} = |\langle f| H_P |i\rangle|^2 f(\omega, t) t^2 \tag{2.52}$$

where $\omega = E_i - E_f$ and

$$f(\omega, t) = \left(\frac{\sin\gamma}{\gamma} \right)^2 \tag{2.53}$$

with $\gamma = \omega t / 2$. This is shown graphically in Figure 2.1. The maximum around $\omega = 0$ gets more and more pronounced as t increases. In fact, we can show that, as a distribution, f satisfies

$$\lim_{t \to \infty} t f(\omega, t) = 2\pi \delta(\omega) \tag{2.54}$$

Contrary to what we have assumed up to now, the unperturbed Hamiltonian $H_0 = H_a + H_r$ of our problem, describing the atom and the free electromagnetic radiation, has a continuous spectrum. We go around this problem by imagining that we quantise the electromagnetic field in a large box of size L with periodic boundary conditions. The spectrum is discrete but, if L is large enough, the energy difference between adjacent levels can be taken to be arbitrarily small, much smaller than the experimental resolution. So, the sensible quantity to compute is the transition probability of the system from the state $|i\rangle$ at $t = t_i$ to any state belonging to a set F of states with energies in the interval $(E_f, E_f + \Delta E)$ at a much later time $t = t_f$, i.e.

$$\mathcal{P}_{i \to F} = \sum_{f \in F} \mathcal{P}_{fi} \tag{2.55}$$

It will prove convenient to replace the sum in (2.55) by an integral, but, in order to do that, we must assume that the matrix element $\langle f| H_P |i \rangle$ for all $f \in F$, depends only on E_f. Let $\rho_F(E_f)dE_f$ be the number of states $f \in F$ with energies between E_f and $E_f + dE_f$. Then the probability (2.55) can be written as

$$\mathcal{P}_{i \to F} = \int_{E_f}^{E_f + \Delta E} \mathrm{d}E_f \rho_F(E_f) |\langle f| H_P |i \rangle|^2 f(\omega, t) t^2 \tag{2.56}$$

Because of the property (2.54), when t tends to infinity, the transition strictly conserves energy, but in such a case both the physical and the mathematical problems are not well defined. Indeed, this calculation shows that the excited state of the atom has a finite lifetime, so we cannot perform infinite time experiments in such a system. On the other hand, equation (2.56) shows that the perturbation expansion will break down for very large times. For a finite time, let us call it Δt, we expect to have an uncertainty in energy of the order of Δt^{-1}. This is a result of classical wave mechanics supplemented with the quantum mechanical condition that relates energy to frequency. Notice also that this result is, to first order in perturbation, independent of the strength of the perturbation.

Thus, let us assume that t is large enough for (2.54) to be approximately valid. A non-zero result for $\mathcal{P}_{i \to F}$ requires that E_i belongs to the interval $(E_f, E_f + \Delta E)$. Finally, dividing by t we obtain for the transition probability per unit time

$$\lambda_{i \to F} = \frac{\mathcal{P}_{i \to F}}{t} = 2\pi \Big[\rho_F(E_f) |\langle f| H_P |i \rangle|^2 \Big]_{E_f = E_i} \tag{2.57}$$

This formula is called *Fermi's golden rule*. The transition probability per unit time is constant, independent of t, with dimensions of energy. The sum of $\lambda_{i \to F}$ over all possible final states F gives what is called the "total decay rate" Γ of the unstable system in the state $|i\rangle$, which in turn leads to the well known "Law of Radioactivity". Indeed, if at time t we have a number $N(t)$ of unstable objects (excited atoms, nuclei, or anything else) with decay probability per unit time given by a constant Γ, in the time interval $(t, t + dt)$ its population will change by $dN(t) = -\Gamma N(t)dt$.[11] Upon integration with initial condition $N(0) = N_0$ we obtain

$$N(t) = N_0 \, \mathrm{e}^{-\Gamma t} \tag{2.58}$$

In the special case at hand, we see that Γ is proportional to the square of the matrix element of the perturbation Hamiltonian, i.e. the square of the coupling constant e, which characterises the strength of the interaction. It is a straightforward exercise in probability theory to show that $\tau = \Gamma^{-1}$, is *the mean lifetime*, or simply *the lifetime* of the excited state, a concept first introduced by Rutherford in 1900.

In non-relativistic quantum mechanics we have established the result that the time evolution of a stationary state with energy E_i is given by $\Psi(t, \boldsymbol{x}) = \Psi(0, \boldsymbol{x})\exp(-\mathrm{i}E_i t)$.

---

[11]If $dN(t)$ objects "succeeded" to decay in the time interval dt out of the total $N(t)$ which "tried" to decay, the probability of success, i.e. decay, per unit time is $dN(t)/(N(t)dt)$.

We can incorporate the result we just obtained by saying that, at first order in perturbation theory, the time evolution is corrected to become

$$\Psi(t, \boldsymbol{x}) = \Psi(0, \boldsymbol{x}) e^{-i(E_i - i\Gamma/2)t} \tag{2.59}$$

It is in this sense that we often say that excited states correspond to complex values of the energy. We shall give a more rigorous justification of this terminology when we discuss resonance phenomena in scattering experiments.

Before closing this general discussion, let us come back to the conditions for the validity of (2.57) and (2.58). We have noted already that, for the first order perturbation theory to be valid, t must be smaller than $\tau = \Gamma^{-1}$. On the other hand, since Γ is given in terms of the matrix elements of the perturbation Hamiltonian, we must have $E_i \gg \Gamma$. Finally, for the delta-function approximation we used in deriving the equation (2.57) to be valid, t must be much larger than $\omega = E_i - E_f$, which is bounded by the experimental resolution ΔE. We conclude that Fermi's golden rule is valid if the following inequalities are satisfied

$$E_i \gg \Gamma \gg \Delta E \tag{2.60}$$

which seem to be intuitively obvious. We often say that they are "obviously" satisfied in real experiments, but we shall see that, at least in one case, namely in the detection of the so-called J/ψ resonance, it was crucial to make sure that they were indeed satisfied.

2.4.4   Application to the problem of spontaneous emission

After all this general discussion, let us return to our problem of spontaneous emission. We start by splitting the Hamiltonian (2.40) into an unperturbed part H_0 and a perturbation: $H = H_0 + H_P$. H_0 must be exactly solvable; in other words we must be able to find its eigenvalues and eigenvectors exactly, so we choose

$$H_0 = H_a + H_r \ , \quad H_P = H_I \tag{2.61}$$

in the notation of the equations (2.40) and (2.39). We denote the eigenvectors of H_0 by $|\mathcal{E}_m, \{n(\boldsymbol{k}, \lambda)\}\rangle$ in an obvious notation in which $\mathcal{E}_m$ is the energy of the m-th level of the atom and $n(\boldsymbol{k}, \lambda)$ the number of photons with momentum $\boldsymbol{k}$ and polarisation λ. In other words, the space of states of H_0 is the direct product of the Hilbert space of the atomic eigenfunctions and the Fock space of the photons.

$$H_a |\mathcal{E}_m, \{n(\boldsymbol{k}, \lambda)\}\rangle = \mathcal{E}_m |\mathcal{E}_m, \{n(\boldsymbol{k}, \lambda)\}\rangle$$

$$H_r |\mathcal{E}_m, \{n(\boldsymbol{k}, \lambda)\}\rangle = \sum_{\lambda, \boldsymbol{k}} n(\boldsymbol{k}, \lambda) |\boldsymbol{k}| \, |\mathcal{E}_m, \{n(\boldsymbol{k}, \lambda)\}\rangle \tag{2.62}$$

Two remarks:

1) When we set up the quantisation prescription for the operator $\boldsymbol{A}$ of the electromagnetic field, we used the Heisenberg representation in which the operator is time

dependent: $A(t, x)$. However, the expressions that appear in the Hamiltonian (2.39) are in the Schrödinger representation. Therefore we must use $A(0, x)$.

2) We shall use the parameter e, the electric charge of the electron, to organise the terms in the perturbation expansion. Terms multiplied by $e^1, e^2, \ldots$ are respectively terms of first order, second order, and so on. It follows that, at first order, we can keep in (2.39) only the term that is linear in A.

We must now compute the matrix element of H_I between the initial and the final states

$$\langle f | H_I | i \rangle = \frac{e}{2m_e} \langle f | \boldsymbol{p} \cdot \boldsymbol{A} + \boldsymbol{A} \cdot \boldsymbol{p} | i \rangle = \int \mathrm{d}^3 x \, \langle \{n_f\} | \, \boldsymbol{A}(0, \boldsymbol{x}) \, | \{n_i\} \rangle \cdot \boldsymbol{j}_{fi}(\boldsymbol{x}) \quad (2.63)$$

where $|\{n_i\}\rangle$ and $|\{n_f\}\rangle$ are the initial and final states of the Fock space of free photons and $\boldsymbol{j}_{fi}(\boldsymbol{x})$ is given by the matrix element of the current operator $\boldsymbol{j}$ between the initial and final atomic wave functions, i.e.

$$\boldsymbol{j}_{fi}(\boldsymbol{x}) = \frac{e}{2 \mathrm{i} \, m_e} \left[\Psi_f^* \boldsymbol{\nabla} \Psi_i - (\boldsymbol{\nabla} \Psi_f^*) \Psi_i \right] \quad (2.64)$$

The expansion (2.20) of the electromagnetic potential $A(0, x)$ in terms of creation and annihilation operators shows that the matrix element (2.63) vanishes unless the number of photons in the final and the initial states differs by ± 1. Therefore this expression contributes only to the processes of emission or absorption of a single photon.

In our case of spontaneous emission we have $n_i = 0$ and $n_f = 1$. The calculation of the matrix element is now straightforward, but for the fact that, in the continuum spectrum, the normalisation of the one photon state contains the factor $[2\omega_k (2\pi)^3 \delta^3(0)]^{-1/2}$; see equation (2.27). It is simpler if we quantise the theory in a cubic box of size L, with arbitrarily large but finite L with periodic boundary conditions. In this case the factor $(2\pi)^3 \delta^3(0)$ is replaced by L^3. The final result is

$$\langle f | H_I | i \rangle = \int \frac{\mathrm{d}^3 x}{\sqrt{2\omega_q L^3}} \mathrm{e}^{-\mathrm{i} \boldsymbol{q} \cdot \boldsymbol{x}} \boldsymbol{\epsilon}^{(\lambda)}(\boldsymbol{q}) \cdot \boldsymbol{j}_{fi}(\boldsymbol{x}) \quad (2.65)$$

where $\boldsymbol{q}$ and λ are the momentum and polarisation of the emitted photon. In principle we have completed the computation because the wave functions of the atom in the initial and the final states are supposed to be known. However, using some reasonable approximations, we can further simplify this expression and make the comparison with experiment easier. First, let us note that the current matrix $\boldsymbol{j}_{fi}(\boldsymbol{x})$ vanishes exponentially fast outside a sphere of roughly the Bohr radius, so in the exponential of (2.65), $\boldsymbol{x}$ is on the order of the Bohr radius which is $1/m_e \alpha$, where $\alpha \equiv e^2/4\pi \simeq 1/137$. Furthermore, $|\mathbf{q}| \simeq \mathcal{E}_i - \mathcal{E}_f$, which is of order $m_e \alpha^2$. So, $\boldsymbol{q} \cdot \boldsymbol{x}$ is of order α and so, to leading order in our computation, we should approximate the exponential by unity. Second, let us replace the current operator (2.64) by the electron electric dipole moment operator $\boldsymbol{d} = e \, \boldsymbol{x}$. This can be done by using the commutation relation

$$[H_a, \boldsymbol{x}] = \left[\frac{\boldsymbol{p}^2}{2m_e}, \boldsymbol{x}\right] = -\frac{\mathrm{i}}{m_e}\boldsymbol{p} \qquad (2.66)$$

so that the spatial integral of the matrix element of the current takes the form

$$\int \mathrm{d}^3 x\, \boldsymbol{j}_{fi}(\boldsymbol{x}) = \frac{e}{m_e}\langle \Psi_f | \boldsymbol{p} | \Psi_i\rangle = \mathrm{i}\langle \Psi_f | [H_a, \boldsymbol{d}] | \Psi_i\rangle = \mathrm{i}(\mathcal{E}_f - \mathcal{E}_i)\boldsymbol{d}_{fi} \qquad (2.67)$$

Let us now compute the density of states ρ_F. Again, it is simpler if we quantise the electromagnetic field in a box. Every component of $\boldsymbol{q}$ takes the values $q_i = (2\pi n_i)L^{-1}$, with integer $n_i = 1, 2, 3, \ldots$. Therefore, the volume in momentum space corresponding to each photon state is $(2\pi)^3/L^3$, and the number of photon states within the volume element $q^2 \Delta E_f \Delta\Omega$ with $q = |\boldsymbol{q}| \simeq \mathcal{E}_i - \mathcal{E}_f$ is

$$\Delta N_F = \frac{L^3 q^2\, \Delta E_f\, \Delta\Omega}{(2\pi)^3} \qquad (2.68)$$

The density of states, up to terms of order $\Delta E_f / q$, is

$$\rho_F(E_f) = \frac{\Delta N_f}{\Delta E_f} = \frac{L^3 q^2 \Delta\Omega}{(2\pi)^3} \qquad (2.69)$$

Putting everything together, we obtain that the transition probability per unit time of an atom in an excited state Ψ_i to the same atom in a state Ψ_f with the simultaneous emission of a photon inside a solid angle $\Delta\Omega_q$ and with polarisation vector $\boldsymbol{\epsilon}^{(\lambda)}(\boldsymbol{q})$ is

$$\Delta\lambda_{i\to F} = \frac{\omega_q^3}{8\pi^2}\left|\boldsymbol{\epsilon}^{(\lambda)}(\boldsymbol{q})\cdot \boldsymbol{d}_{fi}\right|^2 \Delta\Omega_q \qquad (2.70)$$

In order to compute the *intensity* ΔI of the emitted radiation, i.e. the energy emitted per unit time in the solid angle $\Delta\Omega_q$, we multiply (2.70) by ω_q and sum over the two polarisations making use of the relation $\sum_\lambda \epsilon_k^{(\lambda)}(\boldsymbol{q})\,\epsilon_l^{(\lambda)*}(\boldsymbol{q}) = \delta_{kl} - q_k q_l/q^2$.[12] We find

$$\Delta I = \frac{\omega_q^4}{8\pi^2}\sin^2\theta\,|\boldsymbol{d}_{fi}|^2\Delta\Omega_q \qquad (2.71)$$

where θ is the angle between $\boldsymbol{d}_{fi}$ and $\boldsymbol{q}$. In physical units this expression becomes

$$\Delta I = (\omega_q^4/8\pi^2 c^3)\sin^2\theta\,|\boldsymbol{d}_{fi}|^2\,\Delta\Omega_q \qquad (2.72)$$

We observe that ΔI is independent of $\hbar$. Furthermore, it is identical to the expression for the intensity emitted by an oscillating classical dipole. This explains the success of the old Thomson model in describing the phenomenon of atomic radiation.

Formally, we can continue the calculation and compute the higher order terms in the perturbation expansion, but we will not do it here; first, because they present some technical difficulties which we will study in Chapter 15 and second because they are smaller numerically than the relativistic corrections which we have not included. For a qualitative discussion, see Problem 2.4.

---

[12]The left-hand side is a transverse two-index symmetric tensor under rotations, which depends only on the vector $\boldsymbol{q}$ and has trace equal to 2. The right-hand side is the unique expression with these properties.

2.5 Problems

Problem 2.1 Compute the total energy and momentum of the electromagnetic field in terms of the coefficient functions $a^{(\lambda)}(\boldsymbol{k})$ and derive the expressions (2.22).

Problem 2.2 *Emission and absorption of radiation.* In subsection 2.4.4 we computed the probability per unit time for the transition of an atom from the initial state Ψ_i to the final Ψ_f, accompanied by the emission of a photon with polarisation λ and momentum in the solid angle $\Delta\Omega_q$ around $\boldsymbol{q}$ with $|\boldsymbol{q}| = \mathcal{E}_i - \mathcal{E}_f$.

1. Compute the probability per unit time $\Delta\lambda^{\mathrm{em}}$ for the same atomic transition, assuming that there are $N_{q\lambda}$ photons with the above characteristics present in the initial state. Show, in particular, that $\Delta\lambda^{\mathrm{em}}$ is proportional to $N_{q\lambda} + 1$. The 1 corresponds to the spontaneous emission, and the rest defines the *induced* emission. So, $\Delta\lambda^{\mathrm{em}} = \Delta\lambda^{\mathrm{ind}} + \Delta\lambda^{\mathrm{spont}}$.

2. Compute the probability per unit time $\Delta\lambda^{\mathrm{abs}}$ of the inverse atomic process (absorption), i.e. the transition of the atom from Ψ_f to Ψ_i with the absorption of one of the $N_{q\lambda}$ present initially. Show that they satisfy the relations (*A. Einstein, 1916*)

$$\Delta\lambda^{\mathrm{abs}} = \Delta\lambda^{\mathrm{ind}} = N_{q\lambda}\,\Delta\lambda^{\mathrm{spont}}, \qquad \frac{\Delta\lambda^{\mathrm{em}}}{\Delta\lambda^{\mathrm{abs}}} = \frac{N_{q\lambda} + 1}{N_{q\lambda}}$$

Problem 2.3 The purpose of this problem is to study the properties of the electromagnetic field in a cavity with the help of the formalism of coherent states.

I. Introduction of the set of coherent states. Consider a one-dimensional harmonic oscillator with frequency ω and let a and $a^\dagger$ be the annihilation and creation operators, respectively, and $|n\rangle$, $n = 0, 1, \ldots$ the eigenvectors of the operator $N = a^\dagger a : N |n\rangle = n |n\rangle$.

1. Show that to every complex number z, there corresponds a normalised state $|z\rangle$, eigenstate of the operator a with eigenvalue z. Construct explicitly $|z\rangle$ as superposition of the states $|n\rangle$. We shall call the states $|z\rangle$ *coherent states*. Find the coherent state which corresponds to the complex number $z = 0$.

2. Show that $|z\rangle$ can be written as : $|z\rangle = \exp[-\frac{|z|^2}{2}] \exp(za^\dagger) |0\rangle$.

3. If $|z_1\rangle$ and $|z_2\rangle$ are two coherent states, find the value of the scalar product $\langle z_1 | z_2 \rangle$.

4. Give the physical meaning of the real number $|z|^2$.

5. Consider the operator $D(z) = \exp[za^\dagger - z^*a]$, where z^* is the complex conjugate of z. Show that D is unitary and satisfies the relations:
$D^{-1}(z)aD(z) = a + z\mathbf{1}$ and $D(z_1)|z_2\rangle = |z_1 + z_2\rangle$, where $\mathbf{1}$ is the unit operator. Because of these properties we call D *the translation operator*.

II. Physical meaning of the coherent states.

1. Assuming that at $t = 0$ the oscillator is in the coherent state $|\Psi(t = 0)\rangle = |z\rangle$, find its state at a later time t.

2. Study the evolution of a classical harmonic oscillator which, at $t = 0$, is characterised by $q(0)$ and $p(0)$. Find $q(t)$ and $p(t)$ and compare with the evolution of the coherent state.

3. Consider the harmonic oscillator subject to a time-dependent external force. We write the Hamiltonian as:

$$H = \frac{p^2}{2m} + \frac{1}{2}m\omega^2 q^2 - qF(t) = H_0 + H_1$$

where $H_0 = \omega(a^\dagger a + 1/2)$ and $H_1 = -(2m\omega)^{-1/2}(a^\dagger + a)F(t)$.

Show that the evolution operator of the system in the interaction representation, $U_I(t)$ equals, up to a phase, the translation operator $D(z(t))$ and compute $z(t)$. Assuming that at $t = 0$ the system is in the fundamental state $|0\rangle$, find its state $|\Psi(t)\rangle$ at a later time t.

III. Coherent states of the electromagnetic field. Consider the electromagnetic field quantised in a cubic box of dimension L.

1. Define the coherent states a such an electromagnetic field.

2. If $|\Psi(0)\rangle$ is the coherent state describing the state of the field at $t = 0$, find its state $|\Psi(t)\rangle$ at a later time t.

Problem 2.4 *The $\mathcal{O}(e^2)$ terms in the perturbation expansion (2.49).* In the notation of section 2.4 we consider an initial state $|\mathcal{E}_i, n_i = 0\rangle$ and a final state $|\mathcal{E}_f, n_f\rangle$.

1. Write the expression of the matrix element.

2. Show that the possible values of n_f are 0 or 2 and give the physical meaning for each value.

3. Are there any measurable effects corresponding to transitions with $n_f = 0$?

Hint: Think of the case in which the unperturbed Hamiltonian has a degenerate energy level.

Problem 2.5 *"Atom" and photons in a cavity.*

1. Consider a free two-state "atom" at rest: a system with just two states, the ground state $|g\rangle$ with energy set to zero and an excited state $|e\rangle$ with energy E.

We define the operators b and $b^\dagger$ which act in the two-dimensional space spanned by the vectors $|g\rangle$ and $|e\rangle$ as follows:

$$b\,|g\rangle = 0 \;, \quad b\,|e\rangle = |g\rangle \;, \quad b^\dagger\,|g\rangle = |e\rangle \;, \quad b^\dagger\,|e\rangle = 0$$

Prove that b and $b^\dagger$ satisfy the anti-commutation relation $\{b, b^\dagger\} \equiv bb^\dagger + b^\dagger b = 1$ and express the Hamiltonian which describes this atom in terms of them.

2. Consider a "cavity" with electromagnetic radiation, photons, which can only have one value of energy, the energy E.

Using the corresponding creation and annihilation operators $a^\dagger$ and a, write the Hamiltonian for these free photons.

3. Describe the energy eigenstates and the corresponding spectrum of the whole system of the "atom", together with the radiation, ignoring their interaction. Show that with the exception of the ground state, all energy eigenstates are doubly degenerate.

4. Allow the atom and the photons to interact, i.e. introduce an interaction term in the Hamiltonian, which describes the fundamental processes

$$|e; n\rangle \rightarrow |g; n + 1\rangle \;, \quad |g; n + 1\rangle \rightarrow |e; n\rangle \;, \quad \text{for any non-negative } n$$

where the second entry n is the number of photons in the state. *Assume that the interaction can be treated as a small perturbation.* Show that the simplest interaction term which describes the above processes is

$$H_I = \lambda a^\dagger b + \lambda^* a b^\dagger$$

with λ a constant which is assumed to satisfy $|\lambda| \ll E$.

5. Compute to leading order in the perturbation H_I the probabilities of the transitions:

$$|e; n\rangle \rightarrow |g; n+1\rangle$$

of induced $(n \neq 0)$ or spontaneous $(n = 0)$ photon emission by the excited atom.

6. Use the standard *perturbation theory of degenerate states* to compute the energies of the first two excited states, after the degeneracy is lifted by the perturbation. Show that the two new states have energies $E \pm |\lambda|$. What are the corresponding eigenstates?

7. *Rabi oscillations.* Assume that the cavity at time $t = 0$ is in the state $\psi(0) = |e; 0\rangle$. Take for simplicity $\lambda \in R$ and compute the probability that the cavity will be, at time t, in the state $|g; 1\rangle$. Show that the cavity oscillates between the states $|e; 0\rangle$ and $|g; 1\rangle$ and compute the period of oscillation.

3

Elements of Classical Field Theory

3.1 Introduction

Although the natural framework to describe the interactions among elementary particles is the quantum theory, we will start here by recalling some elements of the classical
theory of fields. There is a good reason for that. As we alluded to in the first chapter,
in order to obtain a quantum theory we start from the corresponding classical theory
to which we apply the quantisation prescription. This applies to any physical system,
no matter whether it has a finite or an infinite number of degrees of freedom. It follows
that the knowledge of the classical system is essential in the formulation of the corresponding quantum system. Since the quantum theory of fields will be the language of
elementary particle physics, it is essential to understand the corresponding classical
field theory.

3.2 Lagrangian and Hamiltonian Mechanics

The high level of conceptualisation of classical mechanics has, since the 19th century,
played an essential part in the development of physical theories. We shall give here
a very brief review of the main results with no proofs, essentially in order to fix
terminology and notations.

The Lagrangian. Let us consider a system with N degrees of freedom and let
$q_a(t)$, $a = 1, \ldots, N$, denote the corresponding generalised coordinates. We will assume
that they determine a point $\boldsymbol{q}$ in an N-dimensional differentiable manifold $\mathcal{M}$, for
example the N-dimensional real space $\mathbb{R}^N$.[1] We shall call $\mathcal{M}$ *the configuration space*
of the system. Since $\mathcal{M}$ is differentiable, we can consider at every point $\boldsymbol{q}$ the set of N
tangent vectors $\dot{q}_a(t) = \mathrm{d}q_a(t)/\mathrm{d}t$ of curves passing through $\boldsymbol{q}$. Together with q_a they
span a $2N$-dimensional space, which we shall call $\mathcal{T}(\mathcal{M})$.[2]

---

[1]In the simple case of N unconstrained real variables we just write the corresponding dynamical
equations, for example Newton's equations. However, during the 18th and 19th centuries people
realised that there were problems for which this simple formulation is not straightforward. A typical
example is a problem with constraints, such as a particle subject to an external force but constrained
to move on a given surface. The more abstract Lagrangian and Hamiltonian formulations of classical
mechanics were developed to make possible also the description of such dynamical systems.

[2]If the configuration space is the N-dimensional real space $\mathbb{R}^N$, the space of the tangent vectors is
again $\mathbb{R}^N$ and $\mathcal{T}(\mathcal{M}) = \mathbb{R}^N \times \mathbb{R}^N$. However, for a general differentiable manifold $\mathcal{M}$, this construction
cannot be done globally and we must consider the tangent space $\mathcal{T}_q(\mathcal{M})$ built above the point $\boldsymbol{q}$. We
define $\mathcal{T}(\mathcal{M})$ as the union over all $\boldsymbol{q}$ of these tangent spaces and we call it *the tangent space* of the
manifold $\mathcal{M}$. It is obvious that $\mathcal{T}(\mathcal{M})$ is a $2N$-dimensional vector space because, given any two points
$\boldsymbol{q_1}$ and $\boldsymbol{q_2}$, in order to expand a vector of $\mathcal{T}_{q_1}(\mathcal{M})$ in terms of the basic vectors of $\mathcal{T}_{q_2}(\mathcal{M})$ we need

Elementary Particle Physics. John Iliopoulos and Theodore N. Tomaras, Oxford University Press.
© John Iliopoulos and Theodore N. Tomaras (2021). DOI: 10.1093/oso/9780192844200.003.0003

A *Lagrangian* L is a real function of the $2N$ variables q_a and $\dot{q}_a$ and, possibly, the time t, i.e. $L(q_a, \dot{q}_a, t) : \mathcal{T}(\mathcal{M}) \times \mathbb{R} \to \mathbb{R}$. An important mathematical tool, which was developed for functional analysis problems, of the kind we shall deal with in this book, is *the calculus of variations*. For the simple case of $\mathcal{M} = \mathbb{R}^N$ it derives the following well-known theorem:

- Consider $\boldsymbol{q} \in \mathbb{R}^N$ and let $\gamma = \{t, \boldsymbol{q} \mid \boldsymbol{q} = \boldsymbol{q}(t), t_0 \leq t \leq t_1\}$ be a curve in $\mathbb{R}^N \times \mathbb{R}$ such that $\boldsymbol{q}(t_0) = \boldsymbol{q}_0$ and $\boldsymbol{q}(t_1) = \boldsymbol{q}_1$, and let the Lagrangian $L : \mathbb{R}^N \times \mathbb{R}^N \times \mathbb{R} \to \mathbb{R}$ be a sufficiently regular function of $2N + 1$ variables. We can prove that the curve γ is extremal for the *action* functional defined by $S[\gamma] = \int_{t_0}^{t_1} L(\boldsymbol{q}, \dot{\boldsymbol{q}}, t) \mathrm{d}t$ in the space of the curves joining $(t_0, \boldsymbol{q}_0)$ to $(t_1, \boldsymbol{q}_1)$ if and only if the Euler–Lagrange equations

$$\frac{\mathrm{d}}{\mathrm{d}t}\left(\frac{\partial L}{\partial \dot{\boldsymbol{q}}}\right) - \frac{\partial L}{\partial \boldsymbol{q}} = 0 \tag{3.1}$$

are satisfied along γ.

The Least Action Principle. We can now formulate the *Principle of Least Action* which links Newton's equations to the Euler–Lagrange equations:

- A dynamical system is called *natural* if $L = T - V$, with T the kinetic energy and V the potential energy of the system. It is easy to prove that, for a natural system, the extrema of the functional $S[\gamma]$ are given by the solutions of Newton's equations.

The Hamiltonian. The connection between the Lagrangian and the Hamiltonian formulations is given by a *Legendre transformation*. We define the N quantities $p_a(t)$, usually called *conjugate momenta* to $q_a(t)$, and the *Hamiltonian* $H(\boldsymbol{q}, \boldsymbol{p}, t)$ of the system by

$$p_a(t) = \frac{\partial L}{\partial \dot{q}_a} \quad \text{and} \quad H(\boldsymbol{q}, \boldsymbol{p}, t) = \boldsymbol{p}\,\dot{\boldsymbol{q}} - L(\boldsymbol{q}, \dot{\boldsymbol{q}}, t) \tag{3.2}$$

respectively. In writing equation (3.2) we assume that we can invert the definition of $\boldsymbol{p}$ and express $\dot{\boldsymbol{q}}$ in terms of $\boldsymbol{p}$ and $\boldsymbol{q}$. Then we can prove the following theorem:

- The Euler–Lagrange system (3.1) of N second order differential equations is equivalent to the Hamilton system of $2N$ first order equations

$$\dot{\boldsymbol{p}} = -\frac{\partial H}{\partial \boldsymbol{q}} \quad , \quad \dot{\boldsymbol{q}} = \frac{\partial H}{\partial \boldsymbol{p}} \tag{3.3}$$

As a result of these equations the time evolution of any quantity $f(\boldsymbol{p}, \boldsymbol{q}, t)$ is obtained from its equation of motion

$$\frac{\mathrm{d}f}{\mathrm{d}t} = \frac{\partial f}{\partial t} + \frac{\partial f}{\partial \boldsymbol{q}}\frac{\partial H}{\partial \boldsymbol{p}} + \frac{\partial f}{\partial \boldsymbol{p}}\left(-\frac{\partial H}{\partial \boldsymbol{q}}\right) = \frac{\partial f}{\partial t} + \{H, f\} \tag{3.4}$$

where, by definition, the *Poisson bracket* of two functions f and g is given by

$$\{f, g\} = \frac{\partial f}{\partial p_a}\frac{\partial g}{\partial q_a} - \frac{\partial f}{\partial q_a}\frac{\partial g}{\partial p_a} \tag{3.5}$$

also the basic vectors of $\mathcal{M}$. In addition, we can endow $\mathcal{T}(\mathcal{M})$ with the structure of a fibre bundle, *the tangent bundle*, but we will not use it in this book.

- It follows that if f does not depend explicitly on time, the statement "f is a constant of the motion" is equivalent to $\{H, f\} = 0$. An important corollary of this property is the fact that a Hamiltonian, which does not depend explicitly on time, is conserved, i.e.

$$\frac{\partial H}{\partial t} = 0 \quad \text{implies that} \quad \frac{\mathrm{d}H}{\mathrm{d}t} = 0 \tag{3.6}$$

- Clearly q_a and p_a span a $2N$-dimensional space which we will denote by $\mathcal{T}^*(\mathcal{M})$. It is the *cotangent space* of $\mathcal{M}$.[3] In physics texts it is usually called *the phase space* of the system. So, the Hamiltonian is a real function $H : \mathcal{T}^*(\mathcal{M}) \times \mathbb{R} \to \mathbb{R}$.

Noether's theorem. Between 1915 and 1918 A.E. Noether proved a mathematical theorem which profoundly influenced all aspects of the physical sciences. In rather loose terms it established a connection between continuous symmetries and conservation laws. We will give here a more precise formulation, still avoiding most technical details.

- To each one parameter group of diffeomorphisms of the configuration manifold, which preserves the Lagrangian, there corresponds a prime integral of the equations of motion, i.e. a conserved quantity.

Let us consider a differentiable map $h : \mathcal{M} \to \mathcal{M}$ and let $T(h) : \mathcal{T}(\mathcal{M}) \to \mathcal{T}(\mathcal{M})$ be the induced map on the tangent space. A Lagrangian system $(\mathcal{M}, L)$ is invariant under h, if its Lagrangian function remains unchanged under the action of h on its variables, i.e. if $\forall v \in \mathcal{T}(\mathcal{M})$

$$L(T(h)v) = L(v) \tag{3.7}$$

Noether's theorem asserts that:

Theorem: If the Lagrangian system $(\mathcal{M}, L)$ is invariant under the one parameter group of diffeomorphisms $h_s : \mathcal{M} \to \mathcal{M}, s \in \mathbb{R}$,[4] then the system of the Euler–Lagrange equations admits a prime integral $I : \mathcal{T}(\mathcal{M}) \to \mathbb{R}$, which, furthermore, in a local coordinate system is given by

$$I(\boldsymbol{q}, \dot{\boldsymbol{q}}) = \frac{\partial L}{\partial \dot{\boldsymbol{q}}} \left. \frac{\mathrm{d}h_s(\boldsymbol{q})}{\mathrm{d}s} \right|_{s=0} \tag{3.8}$$

Applications of Noether's theorem. We present some important consequences of this theorem. See also Problem 3.1. Let us consider a system of N particles[5] with Lagrangian

$$L = \sum_{a=1}^{N} \frac{1}{2} m_a \dot{\boldsymbol{q}}_a^2 - V(\boldsymbol{q}_1, \dots, \boldsymbol{q}_N), \ \boldsymbol{q}_a \in \mathbb{R}^3, \ a = 1, \dots, N$$

- We assume that the Lagrangian is invariant under translations

---

[3]We can again endow it with the structure of a fibre bundle, *the cotangent bundle.*

[4]In fact, we can prove Noether's theorem without assuming the invariance of the Lagrangian expressed by equation (3.7). A more global version, in which only the action S is invariant, suffices. The difference is important because it allows to consider transformations under which the Lagrangian changes by the time derivative of a function. We will need this more general version later in this chapter.

[5]In this case the variables q_a are just the position vectors r_a of the particles.

$$h_s : \boldsymbol{q}_a \to \boldsymbol{q}_a + s\,\boldsymbol{c}, \quad \boldsymbol{c} \in \mathbb{R}^3, \quad a = 1, \ldots, N \tag{3.9}$$

The corresponding conserved quantity is the *total momentum* of the system.

$$0 = \frac{\mathrm{d}\boldsymbol{P}}{\mathrm{d}t} \cdot \boldsymbol{c} \quad \text{from which} \quad \frac{\mathrm{d}\boldsymbol{P}}{\mathrm{d}t} = 0 \tag{3.10}$$

We see that the invariance of the Lagrangian under translations implies the *conservation of the total momentum*.

• The Lagrangian is invariant under rotations

$$\boldsymbol{q}_a \to R(\boldsymbol{n}, \theta)\boldsymbol{q}_a, \quad a = 1, \ldots, N \tag{3.11}$$

where $R(\boldsymbol{n}, \theta)$ is a rotation with axis of rotation and direction determined by $\boldsymbol{n}$ and angle θ. In this case, what follows is the *conservation of the total angular momentum*

$$\frac{\mathrm{d}\boldsymbol{L}}{\mathrm{d}t} = 0 \quad \text{with} \quad \boldsymbol{L} = \sum_{a=1}^{N} \boldsymbol{q}_a \wedge \boldsymbol{p}_a \tag{3.12}$$

• The conservation of the Hamiltonian expressed in equation (3.6) is also a consequence of Noether's theorem. It is valid whenever the Lagrangian does not depend explicitly on time, in which case it is invariant under time translations.

3.3 Classical Field Theory

Although in the discussion of the previous section we used mainly the example of a system of point particles moving in space, nothing in the formalism depends on this. We can consider a general dynamical system with $q_a(t)$ $a = 1, \ldots, N$, the corresponding generalised coordinates. In Chapter 2 we indicated a formal way to take the large N limit and obtain a classical field theory starting from a system defined on the points of a spatial lattice. In this book we shall be interested in Lorentz invariant field theories, so we are led to formulate a classical field theory based on a set of postulates, which extend those we used in the formulation of classical mechanics.

The Minkowski space. The base space of the theory is the four-dimensional Minkowski space $\mathbb{M}^4$ with the metric $\eta_{\mu\nu} = \eta^{\mu\nu} = \text{diag}(1, -1, -1, -1)$, which is left invariant by the ten parameter group of the Poincaré transformations. A point in this space will be denoted by $x = (x^0, \boldsymbol{x})$.

The fields. The dynamical variables are real or complex valued functions of x, i.e. $\phi : \mathbb{M}^4 \to \mathbb{R}$ or $\mathbb{C}$. Unless otherwise stated, they are taken to be C^∞ and to vanish at infinity. In addition, we will assume that the fields $\phi(x)$ may transform non-trivially under a group of transformations, for example Lorentz transformations, whose properties we will study in Chapter 5.

The set of fields forms an infinite-dimensional functional space $\mathcal{M}$, which is the configuration space of our dynamical system. It is a differentiable manifold, so at every point ϕ we can consider the tangent vectors $\partial_\mu \phi$. This way we can build the tangent space at each point ϕ. The union of the tangent spaces for all ϕ forms the tangent space of $\mathcal{M}$ and, by analogy with what we did in classical mechanics, we build the space $\mathcal{T}(\mathcal{M})$ as the functional space build out of ϕ and $\partial_\mu \phi$.

Obviously, we can generalise this construction by considering a family of fields $\{\phi_a(x)\}_{a \in I}$ indexed by I.

The Lagrangian density. The Lagrangian density is a real function of ϕ, $\partial_\mu \phi$ and, possibly, the space-time point x, that is $\mathcal{L} : \mathcal{T}(\mathcal{M}) \times M^4 \to \mathbb{R}$. We will always choose $\mathcal{L}$ to be a Lorentz scalar. The action S is a functional of the fields given by

$$S[\phi] = \int_\Omega \mathcal{L}(\phi(x), \partial_\mu \phi(x), x) \, \mathrm{d}^4 x. \tag{3.13}$$

where Ω is a regular region of space-time over which the Lagrangian density is being integrated. We will usually take Ω to be the entire Minkowski space M^4.

The principle of least action and the Euler–Lagrange equations. Only very few changes are necessary in order to adapt the variational calculus of mechanics to the case of a field theory. We define the *functional derivative*, $\delta F[\phi]/\delta\phi(x)$, of a functional $F[\phi]$ by the limit, when it exists, of

$$\frac{\delta F[\phi]}{\delta \phi(x)} = \lim_{\varepsilon \to 0} \frac{F[\phi + \varepsilon \delta_x] - F[\phi]}{\varepsilon} \tag{3.14}$$

where δ_x is the Dirac measure at the point x, $\delta_x(y) = \delta(y - x)$. As an example, for $F = \int \phi^n(y) \, \mathrm{d}^4 y$, we obtain

$$\frac{\delta F[\phi]}{\delta \phi(x)} = \lim_{\varepsilon \to 0} \frac{1}{\varepsilon} \left(\int \mathrm{d}^4 y \, [(\phi(y) + \varepsilon \delta_x(y))^n - \phi^n(y)] \right)$$

$$= n \int \mathrm{d}^4 y \, \phi^{n-1}(y) \delta(y - x) = n \, \phi^{n-1}(x) \tag{3.15}$$

On the other hand, the change in the Lagrangian, induced by an infinitesimal shift $\delta\phi_a(x)$ of the fields, $\phi_a(x) \to \phi_a(x) + \delta\phi_a(x)$, is given by

$$\delta\mathcal{L} = \frac{\partial \mathcal{L}}{\partial \phi_a} \delta\phi_a + \frac{\partial \mathcal{L}}{\partial(\partial_\mu \phi_a)} \delta\partial_\mu \phi_a$$

$$= \frac{\partial \mathcal{L}}{\partial \phi_a} \delta\phi_a + \partial_\mu \left(\frac{\partial \mathcal{L}}{\partial(\partial_\mu \phi_a)} \delta\phi_a \right) - \partial_\mu \left(\frac{\partial \mathcal{L}}{\partial(\partial_\mu \phi_a)} \right) \delta\phi_a$$

$$= \left[\frac{\partial \mathcal{L}}{\partial \phi_a} - \partial_\mu \left(\frac{\partial \mathcal{L}}{\partial(\partial_\mu \phi_a)} \right) \right] \delta\phi_a + \partial_\mu \left(\frac{\partial \mathcal{L}}{\partial(\partial_\mu \phi_a)} \delta\phi_a \right) \tag{3.16}$$

This relation is very important and can be used in various ways. Let us first consider the induced variation of the action as defined in equation (3.13). The last term in (3.16) is a 4-divergence. It gives no contribution upon integration in Ω, because of the vanishing on the boundary $\partial\Omega$ of the fields and of their infinitesimal variations. Thus, the first result derived from (3.16) is that the requirement that the action $S[\phi]$ be stationary under an arbitrary variation $\phi_a(x) \to \phi_a(x) + \delta\phi_a(x)$ with ϕ_a and $\delta\phi_a$ vanishing on $\partial\Omega$, implies the equations of motion for the fields ϕ_a:

$$\frac{\partial \mathcal{L}}{\partial \phi_a(x)} - \partial_\mu \left(\frac{\partial \mathcal{L}}{\partial(\partial_\mu \phi_a(x))} \right) = 0 \tag{3.17}$$

These are the Euler–Lagrange equations of motion of the classical fields.

It should be noted that, under the conditions we imposed for the derivation of the Euler–Lagrange equations and, in particular, the vanishing of the field variables on the boundary Ω, the action is not related in a unique way to the Lagrangian density. We can in fact add to $\mathcal{L}$ a term of the form $\partial_\mu R^\mu$, with R^μ any sufficiently regular function of the fields and their derivatives, since by Gauss's theorem its integral over Ω vanishes.

The Hamiltonian density. With the Lagrangian density being a Lorentz scalar, the entire formulation we presented so far is explicitly Lorentz covariant. The Hamiltonian on the other hand is supposed to transform like the zero component of a four vector, therefore we must single out the time direction in x: $x = (t, \boldsymbol{x})$ and perform a Legendre transformation with respect to $\dot{\phi}(x) = \partial_0 \phi(x)$. Let us define

$$\pi_a(x) = \frac{\partial \mathcal{L}}{\partial \dot{\phi}_a(x)} \tag{3.18}$$

where we have assumed that the fields $\{\phi_a(x)\}_{a \in I}$ form a set of independent dynamical variables. We have seen already in Chapter 2 that this is not always true, for example when the index "a" denotes the components of a vector field. We will address this question in more detail in Chapter 7.

The Hamiltonian density is defined by the analogue of equation (3.2):

$$\mathcal{H} = \sum_{a \in I} \pi_a \dot{\phi}_a - \mathcal{L} \,, \tag{3.19}$$

and the Hamiltonian $H[\phi]$ by

$$H[\phi_a, \pi_a, t] = \int \mathrm{d}^3 x \, \mathcal{H} \left(\phi_a(x), \partial_\mu \phi_a(x), x \right) \tag{3.20}$$

from which Hamilton's equations of motion

$$\dot{\phi}_a(x) = \frac{\delta H}{\delta \pi_a(x)} \quad \text{and} \quad \dot{\pi}_a(x) = -\frac{\delta H}{\delta \phi_a(x)} \tag{3.21}$$

can be obtained.

Symmetry transformations. Following the formalism we developed for the case of classical mechanics with a finite number of degrees of freedom, we shall call "symmetry" a transformation that leaves invariant the equations of motion. Strictly speaking, this implies that only the action should be left invariant, but unless stated otherwise we will make here a stronger assumption, namely to assume that the transformations leave invariant the Lagrangian density. As we will see shortly, this assumption, apart from simplifying the formalism, will also have some important physical consequences. For practical and conceptual reasons, we will distinguish two types of transformations, *space-time transformations* and *internal symmetry transformations*.

• *The space-time transformations* are of the general form $x \to x' = f(x)$, so they are automorphisms of the Minkowski space $\mathbb{M}^4$. In this book we will consider only the ten-parameter group of Poincaré transformations which we will study briefly in

Chapter 5. They consist of the space-time translations and the Lorentz transformations. A translation is of the form $x'_\mu = x_\mu + \epsilon_\mu$ which, for ϵ infinitesimal, induces a transformation of the fields of the form $\phi_a(x) \to \phi'_a(x) = \phi(x-\epsilon) \simeq \phi_a(x) - \epsilon^\mu \partial_\mu \phi_a(x)$. A Lorentz transformation is of the form $x'^\mu = \Lambda^\mu{}_\nu x^\nu$ and we assume that the fields transform linearly. To these continuous transformations we should add the two discrete ones, the *space inversion*, or *parity* ($\mathcal{P}$): $t' = t$, $\boldsymbol{x}' = -\boldsymbol{x}$, and the *time reversal* ($\mathcal{T}$): $t' = -t$, $\boldsymbol{x}' = \boldsymbol{x}$.

• *Internal symmetry transformations.* These do not affect the space-time point x but transform the fields $\phi_a(x)$ among themselves. They play a very important role in particle physics and we will see many examples in this book. We will assume that all internal transformations form a compact Lie group[6] G under which the fields transform linearly. The action of an infinitesimal such transformation with parameters $\delta\theta_n$ on the fields $\phi_a(x)$ has the general form

$$\phi_a(x) \to \phi'_a(x) = \phi_a(x) + \delta\phi_a(x), \quad \delta\phi_a(x) = \mathrm{i}\,\delta\theta_n\,(T_n)_{ab}\,\phi_b(x) \tag{3.22}$$

The number N of parameters θ_n depends on the group G, while the form of the matrices T_n depends on the group and the transformation properties of the fields ϕ_a.

• *Global and local transformations.* Most transformations we are considering, space-time or internal symmetry ones, depend continuously on a set of parameters, such as ϵ^μ and θ_n in the above examples. We will call a transformation *global* if the corresponding parameters are constants, independent of the space-time point x. In the opposite case, if the parameters are arbitrary functions of x, we will call the transformation *local*. Local transformations are often also called *gauge transformations*. In Chapter 13 we shall study the physical consequences of local symmetries in more detail, but here let us just point out an obvious difference between the two cases. Let us consider the example of the internal transformation of equation (3.22). If the transformation is global, the derivative of the field $\phi_a(x)$ transforms the same way as the field itself. If, however, the parameters become functions $\delta\theta_n(x)$, the derivative picks up a second term proportional to $\partial_\mu\delta\theta_n(x)$. This will make the transformed theory look more complicated, but, as we will see, it will eventually result in a mathematically richer structure with amazing physical consequences.

Noether's theorem in classical field theory. We have seen in section 3.2 that to every continuous transformation that leaves the Lagrangian invariant there corresponds a conserved quantity. We will now prove that to every continuous change of the fields leaving invariant the action $S[\phi]$ there corresponds a conserved current.

We assume that the infinitesimal change of the fields, $\phi_a(x) \to \phi_a(x) + \delta\phi_a(x)$, leaves the action $S[\phi]$ invariant. The variation of the Lagrangian due to this change is still given by (3.16), but now the change $\delta\phi$ is not arbitrary. By assumption, it leaves the action invariant, which means that the change of the Lagrangian density has to be the 4-divergence of some quantity, i.e.

$$\delta\mathcal{L} = \partial_\mu R^\mu \tag{3.23}$$

---

[6]Some notions of group theory will be necessary in analysing the symmetry properties of a system and we will present a brief review in Chapter 5.

and this is true for arbitrary configurations of the fields ϕ_a. This is often called an "off-shell equality", because the fields do not need to satisfy the Euler–Lagrange equations.

Now suppose that the fields ϕ_a satisfy the Euler–Lagrange equations of motion. In this case, comparison of (3.23) and (3.16) leads to the relation

$$\delta\mathcal{L} = \partial_\mu R^\mu = \partial_\mu \left(\frac{\partial\mathcal{L}}{\partial(\partial_\mu\phi_a)} \delta\phi_a \right) \tag{3.24}$$

from which it follows that

$$J^\mu = \frac{\partial\mathcal{L}}{\partial(\partial_\mu\phi_a)} \delta\phi_a - R^\mu \tag{3.25}$$

is conserved (it is an "on-shell property"):

$$\partial_\mu J^\mu(x) = 0 \,. \tag{3.26}$$

The Noether current J^μ is conserved and this conservation is a consequence of the symmetry of the theory. The associated "charge" $Q(V)$ defined by the spatial integral of $J^0(x)$ over a volume V bounded by the surface $S(V)$ satisfies the relation

$$\frac{\mathrm{d}Q(V)}{\mathrm{d}t} = \int_V \mathrm{d}^3x \frac{\partial J^0}{\partial t} = -\int_V \mathrm{d}^3x \nabla \cdot \boldsymbol{J} = -\int_{S(V)} \mathrm{d}\boldsymbol{S} \cdot \boldsymbol{J} \tag{3.27}$$

expressing the fact that the amount of charge $Q(V)$ inside any volume changes per unit time by exactly the amount that flows through its boundary. Under the assumption that nothing flows to infinity, the total charge Q in the entire space is conserved, since it satisfies

$$\frac{\mathrm{d}Q}{\mathrm{d}t} = 0 \,. \tag{3.28}$$

There is a certain freedom in the definition of the Noether current because we can add to it and/or multiply it by any constant without changing the fact that its 4-divergence vanishes.

Examples of Noether currents. We shall give examples of space-time and internal symmetries.

• *Space-time symmetries: The energy-momentum tensor.* In classical mechanics, invariance under spatial translations gives rise to the conservation of the linear momentum while invariance under time translation gives rise to the conservation of energy. We will now derive the analogue of this in field theories. Specifically, we will show directly that the fact that the Lagrangian does not explicitly depend on x makes it possible to define, using the Euler–Lagrange equations, a conserved tensor. We choose Ω to be the entire four-dimensional space-time (a domain of integration which is invariant under translations) and assume that the fields decrease fast enough at infinity to ignore all surface terms.[7]

---

[7]The invariance under translations is the only case we will consider in this book in which the action is invariant but not the Lagrangian density. Indeed, $\mathcal{L}(\phi(x), \partial_\mu\phi(x))$ cannot be invariant under $x_\mu \to x_\mu + \epsilon_\mu$, unless it is a constant. However, under the assumption that the fields vanish sufficiently fast at infinity, its integral over the entire space-time, i.e. the action, is invariant.

Let us write in two different ways the total derivative of $\mathcal{L}$ with respect to x^μ. Since $\mathcal{L}$ depends on x^μ only through the dependence of the fields and their derivatives, we have

$$
\begin{aligned}
\frac{\mathrm{d}\mathcal{L}}{\mathrm{d}x^\mu} &= \frac{\partial\mathcal{L}}{\partial\phi_a(x)}\partial_\mu\phi_a(x) + \frac{\partial\mathcal{L}}{\partial[\partial_\nu\phi_a(x)]}\partial_\mu\partial_\nu\phi_a(x) \\
&= \partial_\nu\left(\frac{\partial\mathcal{L}}{\partial[\partial_\nu\phi_a(x)]}\right)\partial_\mu\phi_a(x) + \frac{\partial\mathcal{L}}{\partial[\partial_\nu\phi_a(x)]}\partial_\mu\partial_\nu\phi_a(x) \\
&= \partial_\nu\left(\frac{\partial\mathcal{L}}{\partial[\partial_\nu\phi_a(x)]}\partial_\mu\phi_a(x)\right)
\end{aligned}
$$

But, for any field configuration $\phi(x)$ the Lagrangian density is a function of x and its derivative with respect to x^μ is then just

$$
\frac{\mathrm{d}\mathcal{L}}{\mathrm{d}x^\mu} = \partial_\mu\mathcal{L} = \delta_\mu^\nu\,\partial_\nu\mathcal{L}
$$

Equating the above two expressions for $\mathrm{d}\mathcal{L}/\mathrm{d}x^\mu$ we find

$$
\partial_\nu\widetilde{T}_\mu^\nu = 0 \tag{3.29}
$$

where

$$
\widetilde{T}_\mu^\nu(x) = \frac{\partial\mathcal{L}}{\partial[\partial_\nu\phi_a(x)]}\partial_\mu\phi_a(x) - \mathcal{L}(x)\,\delta_\mu^\nu \tag{3.30}
$$

is called the *energy-momentum* tensor of the theory with Lagrangian density $\mathcal{L}$.

We deduce from it that the four-vector P^μ given by

$$
P^\mu = \int \mathrm{d}^3x\,\widetilde{T}^{0\mu}(t,\boldsymbol{x}) \tag{3.31}
$$

is time independent since

$$
\dot{P}^\mu = \int \mathrm{d}^3x\,\partial_t\,\widetilde{T}^{0\mu}(t,\boldsymbol{x}) = -\int \mathrm{d}^3x\,\partial_i\widetilde{T}^{i\mu}(t,\boldsymbol{x}) = 0 \tag{3.32}
$$

if the fields vanish fast enough at infinity, as it has been assumed.

The fact that P^μ is a four-vector can become manifest by changing the spatial integration in their definition from an integration over a surface of constant t to one over a space-like surface with d^3x being replaced by the associated covariant surface element $\mathrm{d}\sigma^\nu$ normal of the surface

$$
P^\mu = \int \mathrm{d}\sigma_\nu\widetilde{T}^{\nu\mu}(\boldsymbol{x},t) \tag{3.33}
$$

For reasons which will become clear later on, it is useful to replace this energy-momentum tensor by another conserved tensor $T^{\mu\nu}$ symmetric in its two indices, obtained from $\widetilde{T}^{\mu\nu}$ by adding to it the 4-divergence of a three-index tensor. Alternatively, it can be shown that $T^{\mu\nu}$ can be obtained by varying with respect to the space-time metric $g_{\mu\nu}$ the action obtained when the "matter" theory with the given $\mathcal{L}$ is coupled to gravity.

• *Internal symmetry.* Let us consider a global internal symmetry of the Lagrangian density with the fields transforming as in equation (3.22). From the definition of the Noether current (3.25), it follows

$$J^\mu = \frac{\partial \mathcal{L}(x)}{\partial(\partial_\mu \phi_a(x))} \delta\phi_a(x) = \mathrm{i}\, \delta\theta_n \frac{\partial \mathcal{L}(x)}{\partial(\partial_\mu \phi_a(x))} (T_n)_{ab}\, \phi_b(x) \qquad (3.34)$$

thus showing the existence of N conserved currents

$$J_n^\mu(x) = \mathrm{i}\, \frac{\partial \mathcal{L}(x)}{\partial(\partial^\mu \phi_a(x))} (T_n)_{ab}\, \phi_b(x)\,, \quad n = 1, \ldots, N \qquad (3.35)$$

where we made use of the previous remark, that we can add or multiply a current by any constant.

3.4 Problems

Problem 3.1 Consider a system of N particles with Lagrangian:

$$L = \sum_{a=1}^{N} \frac{1}{2} m_a \dot{\boldsymbol{q}}_a^2 - V(\boldsymbol{q}_1, \ldots, \boldsymbol{q}_N),\ \boldsymbol{q}_a \in \mathbb{R}^3,\ a = 1, \ldots, N$$

Using Noether's theorem prove that:

1. Invariance of L under translations implies the conservation of the total momentum, eq. (3.10).

2. Invariance of L under rotations implies the conservation of the total angular momentum, eq. (3.12).

Note that either one of these assumptions concerns in fact the potential V because the kinetic energy part, by construction, is invariant under both translations and rotations.

Problem 3.2 Find the equations of motion for the field $\phi(x)$ resulting from the following Lagrangian densities:

1. $\mathcal{L}_1 = \frac{1}{2} \partial^\mu \phi(x)\, \partial_\mu \phi(x) - \frac{1}{2} m^2 \phi^2(x) - \lambda \phi^3(x)$
2. $\mathcal{L}_2 = -\frac{1}{2} \phi(x) \Box \phi(x) - \frac{1}{2} m^2 \phi^2(x) - \lambda \phi^3(x)$
3. $\mathcal{L}_3 = \frac{1}{2} \partial^\mu \phi(x)\, \partial_\mu \phi(x) - \frac{1}{2} m^2 \phi^2(x) - \lambda \phi^3(x)$
$+ 2g\, \phi(x)\, \partial^\mu \phi(x)\, \partial_\mu \phi(x) + g\phi^2(x) \Box \phi(x)$

Comment on the results.

Problem 3.3 Let $\phi_a(x)$, $a = 1, 2, \cdots, N$ denote N real scalar fields. Find the group of internal transformations which leave invariant the Lagrangian density:

$$\mathcal{L} = \tfrac{1}{2} \sum_{a=1}^{N} \left[\partial^\mu \phi_a(x) \, \partial_\mu \phi_a(x) - m^2 \phi_a(x) \phi_a(x) \right] - \lambda \left[\sum_{a=1}^{N} \phi_a(x) \phi_a(x) \right]^2$$

where λ is a constant. Write the corresponding conserved currents.

Problem 3.4 *The conserved charges generate the corresponding symmetry transformations.* Consider a Lagrangian density $\mathcal{L}$ invariant under the global transformations (3.22) leading to the conserved currents (3.35). Show that the Poisson brackets of the corresponding conserved charges Q_n with the fields $\phi_a(x)$ generate the infinitesimal transformations (3.22).

4

Scattering in Classical and Quantum Physics

4.1 Introduction

Almost every act of observation involves a scattering experiment. In everyday life it is usually the scattering of visible light by the object under study and its resolution is limited by the light's wavelength. During the last century the quest for higher resolution forced us to abandon light as a probe and, following Rutherford's pioneering experiment, to use more and more energetic particle beams. Today all information we have about the structure of matter at the deepest accessed level comes from high energy scattering experiments. The probes are particles we can accelerate, which are essentially protons (or ions) and electrons. A particle accelerator is, in fact, a "microscope" whose spatial resolution is determined by its maximum energy. The CERN Large Hadron Collider (LHC), with proton beams up to 7 TeV, has a resolution reaching 10^{-19} m,[1] and is today – in 2021 – the most powerful microscope man has ever built. In this chapter we shall introduce the basic concepts necessary to describe and understand the results of scattering experiments in particle physics. Only the main ideas will be presented, with no detailed proofs. Some of the proofs are proposed as exercises at the end of the chapter.

4.2 The Scattering Cross Section

We start by considering the case of two colliding particles. The *laboratory frame* is defined to be the reference frame in which one of these particles, called *the target*, is at rest. The other is the projectile. By contrast, in the *centre of mass reference frame* nothing distinguishes the target from the beam.[2]

In the simplest case a projectile, idealised as a hard sphere of radius r in straight motion, will hit the target, a sphere of radius R, if the trajectory of the centre of the projectile intersects the disk of radius $r + R$ perpendicular to it and centred at the

---

[1]This is only a rough order-of-magnitude estimation of an "ideal resolution" δ obtained by setting $\delta \sim \lambda \sim h/E$. The actual resolution in a particular measurement will depend on the experimental conditions.

[2]The terminology has a purely historical origin. In the early experiments a beam of accelerated particles was hitting a target, which was fixed in the laboratory, hence the name of the reference frame. Today, however, most accelerators are colliders, in which two beams of particles are accelerated in opposite directions and are brought into a head-on collision. In these cases the "laboratory" frame is in fact the centre-of-mass one.

Elementary Particle Physics. John Iliopoulos and Theodore N. Tomaras, Oxford University Press.
© John Iliopoulos and Theodore N. Tomaras (2021). DOI: 10.1093/oso/9780192844200.003.0004

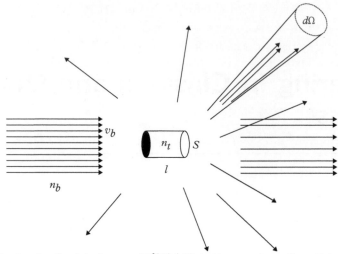

Fig. 4.1 Scattering in the lab frame. $\mathrm{d}\mathcal{N}(\Omega)/\mathrm{d}\Omega$ is the number of particles per unit time scattered inside $\mathrm{d}\Omega$ in the direction $\Omega(\theta, \phi)$.

centre of the target. The surface area of this disk, $\sigma_{\text{tot}} = \pi(r + R)^2$, is called *the total cross section* of this collision process.

In actual experiments the situation is more complex. First, we are not interested only in collisions of hard spheres. Even in classical physics we may want to compute the results of scattering among particles interacting through a potential, for example two electrically charged particles. Second, we do not usually consider the collision of just one particle against another. We send, instead, a beam of particles against a target containing many particles and place a detector in the direction $\Omega(\theta, \phi)$ with an acceptance of $\mathrm{d}\Omega$, which counts the number of particles $\mathrm{d}\mathcal{N}(\Omega)/\mathrm{d}\Omega$ going through it per unit solid angle and per unit time. We want to extract out of such a measurement a quantity, like the cross section we introduced previously, which refers to the collision of one particle in the beam and one particle in the target.

Let us consider a unidirectional monoenergetic beam of particles scattered off a target containing n_t centres of collision per unit volume. Suppose, for concreteness, that this target is a cylinder of length l and base area S perpendicular to the direction of the beam (see Figure 4.1). If the target is thin enough, the density n_t not too high and the number of incoming particles per unit time large enough but not too large,[3] then this experiment yields information about all processes involving one beam particle and one particle in the target. The reason is that the number $\mathrm{d}\mathcal{N}(\Omega)/\mathrm{d}\Omega$ measured in our detector is proportional to the number $n_t S l$ of scattering centres in the target and to the relative flux Φ_b of the incoming particles in the beam, i.e. the number of particles per unit transverse surface area, which reach the target per unit time. Thus, we write

---

[3] All these restrictions amount to assuming that the probability of multiple scattering, i.e. the same incident particle hitting several particles in the target, is small and can be neglected.

$$\frac{\mathrm{d}\mathcal{N}(\Omega)}{\mathrm{d}\Omega} = n_t Sl\Phi_b \frac{\mathrm{d}\sigma(\Omega)}{\mathrm{d}\Omega} \tag{4.1}$$

We shall call the proportionality factor $\mathrm{d}\sigma(\Omega)/\mathrm{d}\Omega$ *differential cross section* of the scattering process involving one beam particle and one target particle. If the density of particles in the beam is n_b and they all move with speed v_b with respect to the target, the incident flux is given by $\Phi_b = n_b v_b$. It is clear that by moving the detector around we can determine the function $\mathcal{N}(\Omega)$ and, assuming it is integrable, we can integrate over all angles and obtain *the integrated cross section σ*. Note that the differential cross section in the forward direction $\theta = 0$, is not directly measurable because a detector placed in the direction of the beam will count all particles in this direction, irrespective of whether or not they had a collision with a particle in the target. It follows that the forward cross section can be determined only by continuity.

From (4.1) we conclude that the total number $\mathrm{d}N$ of collisions that take place inside the volume element $\mathrm{d}V$ of the target in the time interval $\mathrm{d}t$ is given by

$$\mathrm{d}N = \sigma\, v_b\, n_b\, n_t\, \mathrm{d}V\mathrm{d}t \tag{4.2}$$

The differential cross section $\mathrm{d}\sigma/\mathrm{d}\Omega$ and the integrated cross section σ are defined in the lab frame, i.e. with the target at rest. The same number $\mathrm{d}N$ will be measured by any other observer, with respect to whom the colliding beams will have e.g. velocities $\boldsymbol{v}_A$ and $\boldsymbol{v}_B$, and particle densities n_A and n_B, respectively. In Problem 4.3 we invite the reader to prove that, in terms of quantities measured in that frame, $\mathrm{d}N$ is given by

$$\mathrm{d}N = \sigma\sqrt{(\boldsymbol{v}_A - \boldsymbol{v}_B)^2 - (\boldsymbol{v}_A \wedge \boldsymbol{v}_B)^2}\, n_A\, n_B\, \mathrm{d}V\mathrm{d}t \tag{4.3}$$

For collinear $\boldsymbol{v}_A$ and $\boldsymbol{v}_B$, e.g. in the centre of mass frame, $\mathrm{d}N = \sigma|\boldsymbol{v}_A - \boldsymbol{v}_B|\, n_A\, n_B\, \mathrm{d}V\mathrm{d}t$.

So far we have considered the collision of two classical particles but we can generalise the picture. Let us call A the particle in the beam and B the one in the target. The initial state, i.e. the state of the system before the collision, is characterised by the momenta p_A and p_B of the two particles (and possibly other quantities like spin orientations, which we will not discuss in this chapter). Assuming relativistic kinematics, they satisfy the on-shell condition, namely $p_A^2 = m_A^2$ and $p_B^2 = m_B^2$. We shall call *elastic scattering* the process described by $A(p_A) + B(p_B) \to A(p'_A) + B(p'_B)$, where the final momenta p'_A and p'_B are also on-shell and subject to the condition imposed by energy and momentum conservation: $p_A + p_B = p'_A + p'_B$. We know that in quantum physics elastic scattering is only part of the total process because, as we have seen already in Chapter 2, particles may be created or destroyed as a result of the interaction. Therefore the general process will be of the form $A(p_A) + B(p_B) \to C(q_1) + D(q_2) + \ldots$, where the particles C, D, etc. may or may not be one of the initial particles A or B. To give an example, at the LHC the average number of final particles in a proton–proton collision is on the order of 200. We often call the process, yielding a particular set of particles in the final state, *a channel* and we can define *a partial cross section σ_i* or *partial differential cross section* $\mathrm{d}\sigma_i$ corresponding to the channel i. The elastic cross section σ_{el} corresponds to the elastic scattering channel, which is one of the channels i. All the others correspond to *inelastic* processes. The sum over all channels gives the *total cross section* σ_{tot}.

Let us consider a particular channel consisting of r final particles with momenta $q_1, q_2, \ldots, q_r$. They all satisfy the mass shell conditions $q_a^2 = m_a^2$, $a = 1, 2, \ldots, r$, with m_a the mass of the a-th particle. The partial cross section σ_i will be given by an expression of the form:

$$\sigma_i = \int d\Omega_{m_1} \cdots \int d\Omega_{m_r} (2\pi)^4 \delta^4(p_A + p_B - q_1 - \cdots - q_r) \frac{d\sigma_i}{d\Omega_{m_1} \cdots} \qquad (4.4)$$

where $d\Omega_{m_a}$ is the integration measure on the mass hyperboloid of the a-th particle given by equation (2.19) and the four-dimensional delta function expresses the conservation of energy and momentum. The $d\sigma_i$ is the differential cross section which is a positive definite probability density depending on the initial and final momenta. It is defined as the number of the corresponding events per unit time and per unit target volume divided by the incident flux and by the initial densities.[4] It is useful to extract all these kinematic factors: If the incoming beams contain one particle per unit volume, the division factor reduces to the incoming flux whose value is

$$v = |\boldsymbol{v}_A - \boldsymbol{v}_B| \qquad (4.5)$$

where $\boldsymbol{v}_A$ and $\boldsymbol{v}_B$ are the velocities of the two incoming particles, assumed to be collinear, since this is the most interesting case in actual particle physics experiments. In general, the density of initial states depends on the way the beams are prepared, but here, although the discussion is classical, we shall adopt the normalisation given by $\varrho_{A,B} = 2\omega_{A,B} = 2\sqrt{\boldsymbol{p}_{A,B}^2 + m_{A,B}^2}$, as we found in equation (2.27). So, the overall normalisation factor is $(v2\omega_A 2\omega_B)^{-1}$. Let us compute the expression $\varrho_A \varrho_B v$ for collinear initial momenta $\boldsymbol{p}_A$ and $\boldsymbol{p}_B$ in the centre-of-mass frame:

$$(\varrho_A \varrho_B v)^2 = 16\omega_A^2 \omega_B^2 \left| \frac{\boldsymbol{p}_A}{\omega_A} - \frac{\boldsymbol{p}_B}{\omega_B} \right|^2 = 16|\boldsymbol{p}_A|^2 (\omega_A + \omega_B)^2 = 16\left((p_A \cdot p_B)^2 - m_A^2 m_B^2 \right) \qquad (4.6)$$

Therefore, the cross section can be written as:

$$\sigma_i = \frac{1}{4\sqrt{(p_A \cdot p_B)^2 - m_A^2 m_B^2}} \int d\Omega_{m_1} \cdots \int d\Omega_{m_r} (2\pi)^4 \delta^4(p_A + p_B - \Sigma q) |\mathcal{M}|^2 \quad (4.7)$$

where $|\mathcal{M}|^2$ is a positive definite quantity that encodes the dynamics of the particular process. Computing this function for any given scattering experiment is one of the main goals of theoretical particle physics and in this book we shall develop the necessary tools to do it. The $3r$-dimensional integral is called *the phase space* of the process and gives the cross section under the simplified assumption that $|\mathcal{M}|$ is a constant. We see that the cross section increases with the total volume of the phase space, which is

---

[4]The expression (4.4) is only formal. It assumes that we are able to identify all particles in the final state and define precisely the channel i. But in actual experiments there is no *universal detector* capable of identifying all particles in all kinematical regions. Especially at very high energies where the number of final particles is often very large, what we measure is a so-called *inclusive cross section* i.e. one summed over a large number of channels. We will ignore this problem for the moment and we shall give a more precise definition of inclusive processes in Chapter 21.

limited by the energy-momentum conservation. In Problem 12.10 we ask the reader to compute such integrals.

Before closing this section we want to justify the use of the term "cross section". Let us consider the total cross section σ_{tot} measured in the lab frame, although the argument applies also to any partial cross section. According to the definition (4.1), the total number of scattering events per unit time is given by

$$\mathcal{N}_{\text{tot}} = v_b n_b n_t S l \, \sigma_{\text{tot}} \tag{4.8}$$

On the other hand, $\mathcal{N}_{\text{tot}}$ should also be equal to the number of beam particles which hit the target per unit time $n_b v_b$, times the probability for each one of them to be scattered. Going back to the classical picture of hard spheres we can write this probability as:

$$\mathcal{P} = n_t S l \, \pi (r + R)^2 \tag{4.9}$$

and comparing equations (4.8) and (4.9), we conclude that, as advertised, σ_{tot} can be interpreted as the *effective cross sectional area* associated with each target particle as viewed by the particles of the beam.

A convenient unit for cross sections in particle physics is the *barn* (b), with $1\,\text{b} \equiv 10^{-28}\,\text{m}^2$. The magnitude of the cross section measures the probability of two particles to be scattered and can be used to characterise the strength of their interaction. Some typical examples: The proton–proton total cross section at a centre-of-mass energy of order 10 GeV is $\sigma_{\text{tot}}^{pp} \approx 40$ mb. The electron–positron total cross section at the same energy is $\sigma_{\text{tot}}^{e^+e^-} \approx 40 \times 10^{-6}$ mb, i.e. around six orders of magnitude smaller. The neutrino–nucleon cross section for a neutrino beam of 10 GeV is $\sigma^{\nu p} \approx 10^{-11}$ mb. As we will see later, these three reactions are due, respectively, to strong, electromagnetic and weak interactions. So, at least qualitatively, the experimentally determined values of cross sections give a measure of the strength of the underlying interaction. We will come back to this point presently.

A useful quantity in the design of a scattering experiment is *the mean free path* of a particle in a medium. It is defined as the average distance the particle moves in the medium between two successive collisions. From the definition of the cross section, the mean free path l is given by: (see Problem 4.2)

$$l = (\sigma_{\text{tot}} n)^{-1} \tag{4.10}$$

where n is the density of scattering centres in the medium and σ_{tot} the total cross section of the incident particle with one scattering centre. An order of magnitude of n is given by the Avogadro number which, for water, equals 6×10^{23} molecules/mol. This gives, for a 10 GeV proton, a mean free path in water on the order of 40 cm.[5] Note also that, since the cross section increases with the energy of the incident particle, the mean free path decreases.

---

[5]To a good approximation a 10 GeV incident proton sees each water molecule as a collection of 18 nucleons.

4.3   Collisions in Non-Relativistic Quantum Mechanics

4.3.1   The range and the strength of the interactions

Both in classical, as well as in non-relativistic quantum mechanics, the interaction between two particles can be modelled by an interaction potential U. For simplicity we shall assume that U depends only on the distance r between the two particles, even though the following discussion can easily be generalised to arbitrary asymmetric and/or spin-dependent potentials.

We know that the electromagnetic interactions of a point charge q and the gravitational interactions of a point mass m can be approximated by the corresponding Coulomb and Newton potential

$$U_C = \frac{q}{r} \ , \qquad U_G = -\frac{G_N m}{r} \tag{4.11}$$

These expressions allow us to give a first quantitative definition of the concepts of *range* and *strength* of the corresponding interactions.

Let us start with the strength. In the previous section we noted that the measured values of the cross section give a good estimate of the strength of the corresponding interactions. In the framework of the non-relativistic potential theory we can make this statement more explicit. Let us look at the expressions (4.11). They are identical up to the constant which multiplies the $1/r$ term. In our system of units the coefficient of the Coulomb term is dimensionless and, for e, the electron charge, it is given by the fine structure constant $\alpha = e^2/4\pi \simeq 1/137$. We called it the *coupling constant* in Chapter 2. It is considerably smaller than one, so we say that the electromagnetic interactions are relatively weak. The Newton coupling constant G_N, on the other hand, has dimensions of inverse mass squared $G_N \simeq 6.7 \times 10^{-39}$ $[\text{GeV}]^{-2}$. We see immediately that, if we take two protons and place them at a distance r apart, the gravitational potential energy is 36 orders of magnitude smaller than their Coulomb electrostatic energy. It is in this sense that we say that the gravitational interactions are extremely weak. In the following sections we will compute the corresponding cross sections for various potentials and show that the two measures of the strength of an interaction, namely the cross section and the static energy, indeed give identical results. Note also that, since the two constants have different dimensions, they are not directly comparable. If the masses relevant for our experiment are on the order of 10^{19} GeV, the quantity $G_N m_1 m_2$ is of order one and their gravitational interaction becomes strong. But masses of that scale are beyond the reach of any conceivable particle physics experiment and, as a result, we shall not have much occasion to talk about the gravitational interactions in this book.[6] The gravitational force becomes dominant in fields such as planetary physics, or astrophysics and cosmology which deal with very massive bodies.

---

[6]In the framework of Einstein's gravity theory, an estimate of the interaction energy of two particles with energies E_1 and E_2 is obtained by the formula $V_G \sim G_N E_1 E_2/r$. The most energetic particles observed until now on Earth are in the cosmic rays, and their energies reach 10^{11} GeV. Even if we consider a head-on collision of two such charged particles, their gravitational energy is still 14 orders of magnitude smaller than their electrostatic one.

Let us come now to the range. In both the electromagnetic and gravitational interactions the resulting forces between the two particles fall off as $1/r^2$ at large distances. We shall call them *long range*, in fact infinite range interactions, because there is no characteristic distance beyond which the forces can be neglected. In order to get a better understanding, we recall that the Coulomb potential (4.11) is given by $A_0(r)$ which is the solution of Maxwell's equations with a source created by a point charge q sitting at the origin, see equation (2.12). If, instead of the Coulomb gauge condition we used in Chapter 2, we use the Lorenz condition $\partial_\mu A^\mu = 0$, equation (2.11) becomes:

$$\Box A_0(x) = j_0(x) \tag{4.12}$$

with $\Box$ the d'Alembertian defined by $\Box = \partial^2/\partial t^2 - \Delta = \eta^{\mu\nu}\partial_\mu\partial_\nu$.

We can repeat this exercise starting from a more general wave equation. Consider a classical field $\Phi(x)$, which satisfies a linear wave equation. In order to simplify the discussion, we assume that it is a real scalar field with no vector, or tensor, indices. The most general linear, second order, Lorentz covariant wave equation is

$$(\Box + m^2)\Phi(x) = j(x)\,, \tag{4.13}$$

where $j(x)$ is an external Lorentz scalar source of the field Φ and m is a constant with the dimensions of inverse length. In our system of units it has the dimensions of mass. Note that (4.13) generalises Maxwell's equation (4.12) to non-zero values of m.

In order to compute the Φ-analogue of the Coulomb potential we obtained in equation (2.12), we solve the wave equation (4.13) with a static point source at the origin, i.e. for $j(x) = g\,\delta^3(\boldsymbol{x})$ with g a constant. To that effect, we Fourier transform (4.13) to obtain

$$\widetilde{\Phi}(\boldsymbol{k}) = \frac{g}{\boldsymbol{k}^2 + m^2} \tag{4.14}$$

The resulting potential, obtained by the inverse Fourier transform of $\widetilde{\Phi}$, is

$$U_Y(\boldsymbol{x}) = -g\,\frac{\mathrm{e}^{-mr}}{4\pi r}\,. \tag{4.15}$$

This is the well-known Yukawa potential. It falls off exponentially at distances $r \gg 1/m$. Interactions with this property are called *short range* interactions. At the limit $m \to 0$ we recover the long range $1/r$ Coulomb-type potential.

We just established that in classical physics the behaviour of the potential at large distances is controlled by the parameter m in its wave equation. This is just a mathematical result. If we want to give to it a physical meaning, we recall that the parameter m also appears in the dispersion relation $k_0^2 = \boldsymbol{k}^2 + m^2$ of the plane wave solutions of the free wave equation, which upon quantisation, along the steps followed for the electromagnetic field, relates the energy and momentum of the quanta of Φ. Thus, m is the mass of the Φ-particles. In a perturbation expansion similar to the one we developed for the electromagnetic interactions in Chapter 2, the interactions due to the exchange of Φ-particles, reduce in the non-relativistic limit to the Yukawa force (4.15) with range on the order of $\mathcal{O}(1/m)$. This analysis gives us a first justification of the connection between the range of an interaction and the mass of the quanta of the corresponding force carrier field.

A final remark: the Yukawa potential (4.15) follows from the wave equation (4.13). Since this is the most general second order linear, relativistic wave equation, we conclude that this form of a potential is quite general. We will see that the large distance behaviour, the exponential fall - off for massive theories and the $1/r$ slow decrease for massless ones, is always correct to any order in a perturbation expansion, as long as the classical approximation is reliable. However, we shall also encounter cases in which this semi-classical picture breaks down and the effective potential takes a completely different form.

We end with a historical note: The above reasoning, combined with the fact that the range of the nuclear forces was measured to be of order $\mathcal{O}(10^{-13}\mathrm{cm})$, prompted H. Yukawa in 1935 to formulate the first dynamical model for the interactions among nucleons. We will come back to this point in Chapter 6.

4.3.2   Potential scattering

This section does not pretend to be a full presentation of the theory of scattering in non-relativistic quantum mechanics. It is rather a review of basic concepts and a collection of well-known results, mainly about elastic scattering processes, often presented without proof, which will turn out to be useful in particle physics.

Consider the elastic scattering of two particles A and B. Denote their masses and positions m_A, m_B and $\boldsymbol{x}_A$, $\boldsymbol{x}_B$, respectively, and let $V(\boldsymbol{x}_A - \boldsymbol{x}_B)$ be their interaction potential energy. In terms of the centre of mass and the relative position variables the problem factorises into two parts, namely the free motion of the centre of mass and the motion of a "particle" of mass $m = m_A m_B/(m_A+m_B)$ with position vector $\boldsymbol{r}$ and potential energy $V(\boldsymbol{r})$. Thus, the elastic scattering of A and B, reduces to the study of potential scattering in the centre of mass frame of A and B, based on the Schrödinger equation

$$i\frac{\partial}{\partial t}\Psi(t, \boldsymbol{r}) = H\Psi(t, \boldsymbol{r}) = \left(-\frac{1}{2m}\triangle + V(\boldsymbol{r})\right)\Psi(t, \boldsymbol{r}) \tag{4.16}$$

The wave function of the incoming particle with kinetic energy $E = \boldsymbol{k}^2/2m$ is of the form $\Psi(t, \boldsymbol{r}) = \psi(\boldsymbol{r})\,\mathrm{e}^{-\mathrm{i}Et}$ with $\psi(\boldsymbol{r})$ obeying the equation

$$\left(-\frac{1}{2m}\triangle + V(\boldsymbol{r})\right)\psi(\boldsymbol{r}) = E\psi(\boldsymbol{r}) \tag{4.17}$$

Let us assume that the potential energy $V(\boldsymbol{r})$ decreases sufficiently rapidly at infinity, i.e. that $rV(\boldsymbol{r}) \to 0$ as $|\boldsymbol{r}| = r \to \infty$.[7] Then, for r large enough the potential term can be dropped and (4.17) reduces to the free Schrödinger equation.

We have chosen to describe the scattering experiment having in mind the picture shown in Figure 4.1, which leads to the appropriate boundary condition for the wave function $\psi(\boldsymbol{r})$

---

[7]Even though this condition excludes the Coulomb potential, which requires special care, it allows other most important cases for particle physics such as the Yukawa potential (4.15).

$$\psi(\boldsymbol{r}) \xrightarrow[r \to \infty]{} A \left(e^{i\boldsymbol{k} \cdot \boldsymbol{r}} + f(\theta, \phi) \frac{e^{ik\,r}}{r} \right) \tag{4.18}$$

with $k = |\boldsymbol{k}|$ and $\Omega = (\theta, \phi)$ the polar angles of the vector $\boldsymbol{r}$ with the polar axis $\theta = 0$ taken in the direction of the initial momentum $\boldsymbol{k}$. A is a normalisation constant. The asymptotic form of $\psi(\boldsymbol{r})$ describing the scattering state has two terms: a plane wave solution of the free Schrödinger equation corresponding to the incoming and undeflected part of the beam and an outgoing spherical wave, which is the leading asymptotic term of the solution of the free Schrödinger equation in spherical coordinates and represents the asymptotic form of the scattered part of the beam. It follows that the *differential cross section* $d\sigma(\Omega)$ which, by definition, is the number of particles emitted in the solid angle $d\Omega = \sin\theta d\theta d\phi$ in the direction (θ, ϕ) per unit time and per unit incident flux, is equal to the ratio of the flux of scattered particles crossing the surface element $r^2\, d\Omega$ for $r \to \infty$, divided by the incident flux. The two relevant fluxes are computed using the expression

$$\boldsymbol{j}(\boldsymbol{r}) = \frac{1}{2mi}\left[\psi^*(\boldsymbol{r})\boldsymbol{\nabla}\psi(\boldsymbol{r}) - (\boldsymbol{\nabla}\psi^*(\boldsymbol{r}))\psi(\boldsymbol{r})\right] \tag{4.19}$$

for the probability current density described by the wave function $\psi(\boldsymbol{r})$. The incident flux, obtained from the $A\exp(i\boldsymbol{k} \cdot \boldsymbol{r})$ plane wave piece of the asymptotic form of the wave function, is

$$\Phi_{\text{in}} = \boldsymbol{j}_{\text{in}} \cdot \hat{\boldsymbol{k}} = \frac{1}{m}AA^*k = |A|^2\, v \tag{4.20}$$

where v is the speed of the beam particles. Similarly, the flux of the scattered part of the beam is given by

$$\boldsymbol{j} \cdot \hat{\boldsymbol{r}} = \text{Re}\left[AA^* \frac{1}{mi}f^*(\theta, \phi)\frac{e^{-ikr}}{r}\frac{\partial}{\partial r}\left(f(\theta, \phi)\frac{e^{ikr}}{r} \right) \right] = \frac{|f(\theta, \phi)|^2}{r^2}|A|^2\, v \tag{4.21}$$

Consequently, for $\theta \neq 0$, the differential cross section, is given by

$$d\sigma(\theta, \phi) = \lim_{r \to \infty} \frac{\boldsymbol{j} \cdot \hat{\boldsymbol{r}}}{\boldsymbol{j}_{\text{in}} \cdot \hat{\boldsymbol{k}}} r^2\, d\Omega = |f(\theta, \phi)|^2\, d\Omega \tag{4.22}$$

or, equivalently[8]

$$\frac{d\sigma(\theta, \phi)}{d\Omega} = |f(\theta, \phi)|^2 \tag{4.23}$$

The function $f(\theta, \phi)$ is called *the scattering amplitude*. It is computed by solving the Schrödinger equation (4.17) subject to the asymptotic condition (4.18) and, according to (4.23), it contains all the information we need in order to predict the results of a

---

[8]This result can be further justified in the context of a more rigorous formulation of the scattering process, with the initial state taken to be a wave packet, that is to say a superposition of plane waves centred around $\boldsymbol{k}$. In this case, the asymptotic state is a superposition of asymptotic states of the form (4.18).

scattering experiment.[9] Let us next apply this programme to the case of a spherically symmetric potential.

If the potential energy V is spherically symmetric, it is convenient to work in spherical polar coordinates and look for solutions of the Schrödinger equation (4.17) of the form

$$\psi(\boldsymbol{r}) = R_l(r)Y_{lm}(\theta, \phi) \tag{4.24}$$

It is straightforward to show that the radial part $R_l(r)$ of the wave function is independent of the magnetic quantum number m and satisfies the equation

$$\frac{1}{r^2}\frac{\mathrm{d}}{\mathrm{d}r}\left(r^2\frac{\mathrm{d}R_l}{\mathrm{d}r}\right) - \frac{l(l+1)}{r^2}R_l + 2m\left(E - V(r)\right)R_l = 0 \tag{4.25}$$

which, with the substitution $Q_l(r) = rR_l(r)$, becomes

$$\frac{d^2Q_l}{\mathrm{d}r^2} - \frac{l(l+1)}{r^2}Q_l + 2m\left(E - V(r)\right)Q_l = 0 \tag{4.26}$$

At large distances the centrifugal and the potential terms drop out and the asymptotic form of R_l is conventionally written in the form

$$R_l(r) \underset{r\to\infty}{\sim} \frac{\sin\left(kr - \frac{1}{2}l\pi + \delta_l\right)}{r} \tag{4.27}$$

The so-called *phase shifts* δ_l are k-dependent constants and in principle can be computed given the potential.[10] It is instructive to express the scattering amplitude in terms of the phase shifts. Since the scattering problem has an axial symmetry around the $\boldsymbol{k}$-axis, we can integrate over the angle ϕ and the scattering amplitude depends only on θ. The asymptotic form (4.27) of the radial part of the solution of (4.17) contains both incoming and outgoing spherical waves. In the scattering process of interest here, the asymptotic condition (4.18) implies that only an outgoing wave should exist in the difference $\psi(\boldsymbol{r}) - A\exp(\mathrm{i}\boldsymbol{k}\cdot\boldsymbol{r})$. We can expand $f(\theta)$ in Legendre polynomials $P_l(\cos\theta)$

$$f(\theta) = \sum_{l=0}^{\infty}(2l+1)f_l P_l(\cos\theta) \tag{4.28}$$

and show (see Problem 4.4) that this requirement leads to the following expression for the *partial wave amplitudes* f_l in terms of the phase shifts

$$f_l = \frac{1}{2\mathrm{i}k}(S_l - 1) = \frac{1}{k\cot\delta_l - \mathrm{i}k}, \quad \text{with } S_l = e^{2\mathrm{i}\delta_l} \tag{4.29}$$

---

[9]The so-called *inverse scattering problem*, namely the possibility of using the scattering data to determine the potential, is obviously very important, but we shall not address it in this book.

[10]The presence of the term $-\frac{1}{2}l\pi$ is a convention. Since the overall sign of the wave function is arbitrary, the phase shifts are determined modulo π. Here we have adopted the convention followed in L.D. Landau and E.M. Lifshitz *Quantum Mechanics*, in which the phase shifts satisfy $0 \le \delta_l < \pi$ and vanish for a free particle.

The total cross section is then

$$\sigma = 2\pi \int_0^\pi |f(\theta)|^2 \sin\theta d\theta = \frac{4\pi}{k^2} \sum_{l=0}^\infty (2l+1)\sin^2\delta_l \tag{4.30}$$

It is the sum of *partial cross sections* $\sigma_l = 4\pi(2l+1)|f_l|^2$, each one corresponding to scattering with a given orbital angular momentum.

• *The optical theorem.* Using (4.29) it is easy to verify the relation

$$\sigma = \frac{4\pi}{k}\mathrm{Im}f(0) \tag{4.31}$$

which is known from classical optics. It connects the total cross section with the imaginary part of the forward scattering amplitude. We shall derive it under more general assumptions in Chapter 12.

• *Scattering at low energies.* In scattering experiments we are particularly interested in the dependence of the scattering amplitude and of the phase shifts on the incident energy E. For a general potential this can only be obtained numerically. However, this dependence simplifies considerably in the special case of scattering at very low energies, $E \to 0$, or, equivalently, very small momenta $k \to 0$. In this case, $1/k$ becomes larger than the range of the potential, so it is natural to study the general radial equation in three different domains: (i) the region with $r > 1/k$, which is our previous asymptotic region in which we could drop the potential and the centrifugal term. (ii) $r \sim 1/k$ in which we can drop only the potential and (iii), the region $r < 1/k$, but still larger than the range of the potential. In this case we can drop both V and E. We solve the equation inside each region and join the solutions by imposing the continuity of the wave function and its first derivatives. The result, which is rather easy to establish, is that, for a generic potential, all phase shifts are small, which is not surprising since we expect them to vanish at zero energy, and their leading order behaviour for small k is

$$\delta_l(k) \sim k^{2l+1} \tag{4.32}$$

which in turn implies for the partial wave amplitudes

$$f_l \sim k^{2l} \tag{4.33}$$

We conclude that at very low energies the scattering is dominated by the s-wave, which yields an isotropic angular distribution. It is customary to write

$$f(\theta) \sim f_0 = \frac{\delta_0(k)}{k} \equiv -a \tag{4.34}$$

where we have denoted by $-a$ the value of the s-wave scattering amplitude at zero energy. This value is called *the scattering length*. The corresponding total cross section is then

$$\sigma \simeq 4\pi a^2$$

This result is obtained for all potentials, which satisfy the asymptotic condition $rV(r) \to 0$ at large distances. Under some more specific assumptions we can go one step further and consider the first corrections to the formula (4.34) for the expansion of $\delta_0(k)$ for small k. We shall only give the result

$$k \cot \delta_0(k) \simeq \frac{k}{\delta_0(k)} \simeq -\frac{1}{a} + \frac{1}{2} r_0 k^2 \tag{4.35}$$

The left-hand side is roughly the inverse of the amplitude, and this formula shows that the first correction for small but finite k is proportional to k^2. The coefficient r_0 is another constant, whose value depends on the potential, but we can show that it is positive. It is called *the effective range*.

• *Scattering in the Born approximation.* In particle physics we often use the *weak potential approximation*, i.e. the approximation in which the entire interaction potential energy can be treated as a perturbation. In such cases we write the Hamiltonian $H = H_0 + V$, with H_0 the Hamiltonian of a free particle, and the wave function $\Psi = \Psi_0 + \Psi_1 + \ldots$, where the index denotes the order in the perturbation expansion. To zeroth order the solution of the Schrödinger equation with the given positive energy $E = \boldsymbol{k}^2/2m$ is chosen to be the incoming plane wave $\Psi_0 = A \exp(i\boldsymbol{k} \cdot \boldsymbol{r})$. Then, its first order correction Ψ_1 obeys the equation

$$\frac{1}{2m}(\triangle + k^2)\Psi_1 = V\Psi_0 \tag{4.36}$$

We shall use the Green function method to solve it. Taking into account the asymptotic condition, we obtain for the Green function

$$G(\boldsymbol{r}, \boldsymbol{r}') = -\frac{1}{4\pi} \frac{e^{ik|\boldsymbol{r}-\boldsymbol{r}'|}}{|\boldsymbol{r} - \boldsymbol{r}'|} \tag{4.37}$$

and, consequently, for Ψ_1

$$\Psi_1(\boldsymbol{r}) = -A\frac{m}{2\pi} \int \frac{e^{ik|\boldsymbol{r}-\boldsymbol{r}'|}}{|\boldsymbol{r} - \boldsymbol{r}'|} V(\boldsymbol{r}')e^{i\boldsymbol{k}\cdot\boldsymbol{r}'} \mathrm{d}^3r' \tag{4.38}$$

The Green function (4.37) can be written at large r as

$$G(\boldsymbol{r}, \boldsymbol{r}') \underset{r\to\infty}{\to} -\frac{1}{4\pi} \frac{e^{ikr}}{r} e^{-i\boldsymbol{k}'\cdot\boldsymbol{r}'} \tag{4.39}$$

where $\boldsymbol{k}'$ is a vector in the direction of $\boldsymbol{r}$, in the direction of observation, and with modulus equal to that of $\boldsymbol{k}$, $|\boldsymbol{k}'| = |\boldsymbol{k}| = k$. Using this expression in (4.38) we obtain the asymptotic form of Ψ_1 as $r \to \infty$, whose comparison with the asymptotic form of the scattered wave in (4.18) leads to the first order scattering amplitude

$$f_1(\theta) = -\frac{m}{2\pi} \int e^{-i(\boldsymbol{k}'-\boldsymbol{k})\cdot\boldsymbol{r}'} V(\boldsymbol{r}')\mathrm{d}^3r' \tag{4.40}$$

where $\boldsymbol{k}$ is the momentum of the incident particle and $\boldsymbol{k}'$ that of the scattered one. The vector $\boldsymbol{q} = \boldsymbol{k}' - \boldsymbol{k}$, whose magnitude is $q = 2k\sin(\theta/2)$, is the change of the momentum of the scattered particle and is called *the momentum transfer*. We conclude that the first order scattering amplitude is given by the Fourier transform of the potential. This simple formula is called *the Born scattering amplitude* and the first order perturbation theory for scattering is known as *the Born approximation*. The condition for its validity is $|\Psi_1(\boldsymbol{r})|/|\Psi_0(\boldsymbol{r})| \ll 1$. With a the range of the potential and $|V|$ its typical value inside its domain, this condition becomes either $|V| \ll 1/ma^2$ for $ka \ll 1$ (low energy scattering) or $|V| \ll v/a = ka/ma^2$ for $ka > 1$ (high energy scattering).

We shall use the Born approximation quite often in this book. In Problem 4.5 the reader is asked to compute the Born amplitude for $V_Y = g\,U_Y(\boldsymbol{r})$ (4.15). The result is

$$f_B(\theta) = \frac{mg^2}{2\pi} \frac{1}{M^2 + q^2} \tag{4.41}$$

where M is the mass parameter in the Yukawa potential.

• *An infrared divergence in Coulomb scattering.* In discussing the Yukawa potential in the previous section we noted that it gives the Coulomb potential at the limit $M \to 0$. We may be tempted to interpret the corresponding result obtained from (4.41) for $M = 0$ as the Born approximation of the Coulomb amplitude. We expect, however, that something must be wrong because the Coulomb potential does not satisfy the fast decrease condition $rV(r) \to 0$ for $r \to 0$, which we used in all previous derivations. Indeed, we see that the resulting expression for $f_B(\theta)$ has a small angle singularity of the form $\sim \theta^{-2}$, in other words the Coulomb scattering amplitude diverges in the forward direction. This singularity is not integrable, so the total cross section diverges as well. We conclude that, although formally the Coulomb potential can be viewed as the small mass limit of the Yukawa potential, when we compute quantities such as the total cross section, the limit does not exist. We refer to this singularity as *an infrared divergence* and we shall discuss it again in Chapter 17. Here we want only to point out that it has a rather simple physical origin;[11] the fact that the potential does not vanish fast enough at large distances means that there is a non-zero probability for a particle to be scattered by the potential, no matter how far away it passes from the scattering centre. It is the sum of all these small amplitudes, which add up to an infinite scattering probability.[12]

• *Resonance scattering.* Next we would like to discuss the phenomenon of *resonant scattering*. It is the strong enhancement of the scattering cross section, which appears in low energy scattering ($kr_0 \ll 1$) off an attractive potential $V(r)$ of range r_0 and characteristic depth V_0 much greater than the energy $E = k^2/2m$, when $V(r)$ has a discrete or quasi-discrete eigenstate with energy near zero. In this case we say that the scattering energy E is almost *in resonance* with that state. We shall describe the phenomenon in the context of the square well potential $V(r) = -V_0, r < r_0$ and $V(r) = 0, r > r_0$.

(i) We start with *s-wave scattering* ($l = 0$). The radial equation for $Q_0(r)$ is $Q_0'' + 2m(E - V(r))Q_0 = 0$. Its solution for $r < r_0$, which vanishes at the origin, is $Q_0^< =$

---

[11]The correct analysis of scattering from a Coulomb potential in non-relativistic quantum mechanics is more complicated and we will not present it here. It can be found, for example, in the book *quantum mechanics* by L.D. Landau and E.M. Lifshitz.

[12]Remember that in classical mechanics the cross section is always infinite if the potential is not strictly equal to zero outside a certain radius. In quantum mechanics this finite support condition is replaced by the condition of a decrease faster than $1/r$.

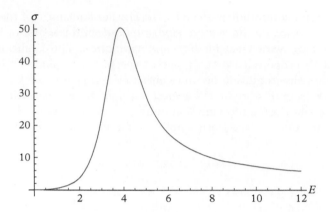

Fig. 4.2 The cross section σ in units of $12\pi r_0^2$ versus kinetic energy in units of 200 MeV for the $l = 1$ scattering of a mass $m = 100$ MeV, off the spherical potential well with radius $r_0 = 1$ fermi and depth $V_0 = 1970$ MeV, about 4 MeV lower than the minimum necessary for the potential to have one $l = 1$ bound state. Note the sharp maximum in the cross section. The resonance energy is about 4 MeV and its width about 2 MeV, corresponding to a mean lifetime of 0.33×10^{-21} s.

$A \sin(\kappa r)$ with $\kappa \simeq \sqrt{2mV_0}$ and in the exterior region $r > r_0$ it is $Q_0^> = B \sin(kr + \delta_0)$. Continuity of $Q'(r)/Q(r)$ at $r = r_0$ gives (for $\kappa \gg k$)

$$k \cot(kr_0 + \delta_0) = \kappa \cot \kappa r_0 \tag{4.42}$$

We solve this for $\delta_0(k)$. The answer depends crucially on the value of κr_0. If κr_0 is not close to a zero of $\cot(\kappa r_0)$, i.e. for $\kappa r_0 \neq (2n+1)\pi/2$, then $\delta_0 \ll 1$, (4.42) simplifies to $k/(kr_0 + \delta_0) \simeq \kappa \cot \kappa r_0$, its solution is $\delta_0 \simeq (k/\kappa)(\tan \kappa r_0 - \kappa r_0)$ and the corresponding partial cross section is

$$\sigma_0 = 4\pi |f_0|^2 = 4\pi \frac{1}{k^2} \frac{1}{|\cot \delta_0 - \mathrm{i}|^2} \simeq 4\pi r_0^2 \left(\frac{\tan \kappa r_0}{\kappa r_0} - 1 \right)^2 \tag{4.43}$$

independent of E and on the order of $\mathcal{O}(r_0^2)$.

On the other hand, for κr_0 near a zero of $\cot(\kappa r_0)$, e.g. near $\pi/2$, we parameterise $\kappa r_0 = \pi/2 + (2/\pi)qr_0$ with $qr_0 \ll 1$, and repeat the analysis of the previous paragraph. We solve for δ_0 and compute the cross section. We find

$$\sigma_0^R = \frac{4\pi}{k^2 + q^2} \tag{4.44}$$

which depends on E and, furthermore, for $qr_0, kr_0 \ll 1$ is much greater than $\mathcal{O}(r_0^2)$. The same is true for κr_0 near $(2n+1)\pi/2$, for any value of n.

What is special about the values $\kappa r_0 = (2n+1)\pi/2$? These are the values of the parameters of the potential $V(r)$ for which a bound s-state exists with zero binding energy. So, for κr_0 near $(2n+1)\pi/2$, the low energy scattering is almost in resonance with an $l = 0$ bound state and this leads to strong enhancement of the corresponding cross section. The crucial relevant property of $V(r)$ for this behaviour is the presence of the bound s-state with energy near zero, and the conclusion generalises to any potential with this property.

(ii) Similar features characterise resonance scattering with $l > 0$. The basic observation follows from (4.29), (4.32) and (4.35), according to which for low energy scattering we have

$$f_l(k) \simeq \frac{1}{\beta E^{-l}(\epsilon - E) - \mathrm{i}k} \tag{4.45}$$

where we kept the first two terms in the expansion of $\cot \delta_l$ about energy equal to ϵ. As already mentioned for $l = 0$, it can be shown that $\beta > 0$. For energies near the pole of f_l, the latter takes the form

$$f_l \simeq -\frac{\Gamma}{2k} \frac{1}{E - \epsilon + \mathrm{i}\Gamma/2}, \quad \Gamma = \frac{2\sqrt{2m}}{\beta} \epsilon^{l+1/2} \tag{4.46}$$

The corresponding partial cross section is

$$\sigma_l(E) \simeq \pi(2l + 1) \frac{\Gamma^2}{2m\epsilon} \frac{1}{(E - \epsilon)^2 + \Gamma^2/4} \tag{4.47}$$

which for $kr_0 \ll 1$ has a sharp maximum at $E = \epsilon = k^2/2m > 0$. This is another instance of resonance scattering. It can be interpreted as taking place in resonance with a metastable state with energy ϵ and angular momentum l in the effective potential $V_{\mathrm{eff}}(r) = V(r) + l(l+1)/r^2$.

The above discussion and the parametrisation

$$f_l \simeq -\frac{\Gamma_R}{2k} \frac{1}{E - E_R + \mathrm{i}\Gamma_R/2} \tag{4.48}$$

near a resonance is general, valid for any resonance in any potential that exhibits this phenomenon. $E = E_R$ is the *resonance energy* and Γ_R is the *resonance width*.

The physics of resonant scattering becomes more transparent if we go back to the original picture at the beginning of this section in which two particles A and B scatter one against the other. The potential model we studied is the non-relativistic approximation of this process. We see that, for particular values E_R of the total energy, the scattering cross section exhibits sharp peaks like the one shown in Figure 4.2. We interpret them as the production of short-lived "particles", called precisely *resonances*, with mass given in terms of E_R[13] and lifetime given by the inverse of Γ_R. Many unstable particles which we shall study in this book have been discovered through this effect.

Formula (4.48) is often called *the Breit–Wigner formula*. We can compare it with equation (2.59) which gives the decay probability of an excited state of the atom. The use of the same symbol Γ is not accidental: in both cases it is proportional to the decay probability of the metastable state.

---

[13]Because of interference with non-resonant scattering, the maximum of the peak may be displaced from the value of E_R, but the calculation of this effect depends on the details of the potential.

4.4 Problems

Problem 4.1 *The resolution of a "microscope".* We want to use a beam of particles with momentum p as a microscope to detect two very thin slits on a thin surface and measure their distance d. For that we hit the surface with the beam and watch the interference pattern on a screen behind it in our laboratory.

1. Think of the beam as a wave with de Broglie wavelength $\lambda = h/p$, with h the Planck's constant and show that the positions of the maxima of intensity on the screen are at angles θ_n with $\sin \theta_n = n\lambda/d$.

2. Argue that the minimum slit distance for which our apparatus is most efficient is $\delta \sim h/p$. This is the *resolution* of this primitive microscope. For an ultra-relativistic beam in particular, $p \simeq E$ and this becomes $\delta \sim h/E$.

Problem 4.2 1. Show that the probability that a beam particle will travel inside the target a distance in the interval $(x, x + \mathrm{d}x)$ before it scatters is $\mathrm{d}P(x) = n\sigma \exp(-n\sigma x)\mathrm{d}x$.

2. Use this probability distribution function to verify that the average distance travelled by the beam particles inside the target is given by (4.10).

Problem 4.3 Derive formula (4.3).

Problem 4.4 Derive formula (4.29).

Problem 4.5 Consider the scattering of a particle in a Yukawa potential. Assume a potential energy $V_Y = gU_Y$ with U_Y given by (4.15). Compute the Born amplitude and derive formula (4.41).

Problem 4.6 Solve (4.42) for δ_0 and show that the corresponding partial cross section $\sigma_0 = (4\pi/k^2)\sin^2 \delta_0$ is

$$\sigma_0 = 4\pi r_0^2 \, \frac{\sin^2 kr_0}{(kr_0)^2} \, \frac{(\cot kr_0 - (\kappa/k)\cot \kappa r_0)^2}{1 + (\kappa^2/k^2)\cot^2 \kappa r_0} \tag{4.49}$$

Problem 4.7 A wide uniform monoenergetic π^- beam with flux $\Phi_b = 10^7 \, \mathrm{m}^{-2} \sec^{-1}$ hits a liquid hydrogen target of volume $V = 10^{-4} \, \mathrm{m}^3$ and density $n_t = 71 \, \mathrm{kg}\,\mathrm{m}^{-3}$. We want to study the reaction $\pi^- p \to K^0 \Lambda$, whose cross section at the given energy is 0.4 mb. What is the rate of production of Λs ?

5
Elements of Group Theory

5.1 Introduction

The traditional applications of group theory to particle physics are based on the interconnection between symmetry and conservation laws.[1] In Chapter 3 we derived Noether's theorem which states that to every continuous symmetry G of a theory described by a Lagrangian L, corresponds a conserved quantity Q, the generator of the symmetry G. In turn, this often leads to a classification of the physical states in multiplets determined by the symmetry. A well-known example is provided by a system invariant under rotations in non-relativistic quantum mechanics. Noether's theorem implies that the three components of angular momentum commute with the Hamiltonian. The physical states are denoted by $|\nu, j, m\rangle$, with $j(j+1)$ the eigenvalue of $\boldsymbol{J}^2$, m that of J_z and ν collectively all other quantum numbers characterising the state. The symmetry tells us that, if $H|\nu, j, m\rangle = E|\nu, j, m\rangle$, $E = E(\nu, j)$ depends on ν and j but not on m. This is a consequence of the fact that m changes under rotations, but the energy does not. In this chapter we shall summarise the concepts and results of group theory which are absolutely essential in the physics of elementary particles.

5.2 Groups and Representations

Definition: **Group** is a set G of elements together with a composition law $G \times G \to G$. It is usually denoted by a dot "·" and conventionally called *multiplication*. It acts on ordered pairs of G and has the following properties:

(a) For every ordered pair (f, g) of elements of G, their product $f \cdot g$ also belongs to G.

(b) The associativity relation $f \cdot (g \cdot h) = (f \cdot g) \cdot h$ is valid for any three elements f, g and h of G.

(c) G has an element e (usually called *identity*) such that $f \cdot e = e \cdot f = f$ for all elements f of G.

(d) For every element $f \in G$, its *inverse* f^{-1}, defined by $f \cdot f^{-1} = f^{-1} \cdot f = e$, also belongs to G.

When the group operation is commutative, i.e. it satisfies $f \cdot g = g \cdot f$ for all f, g in G, the group is called *Abelian* in honour of the Norwegian mathematician N.H. Abel. Otherwise, it is *non-Abelian*.

---

[1] We will see later in this book that the role of symmetry has become more fundamental in the modern theories of elementary particles.

Elementary Particle Physics. John Iliopoulos and Theodore N. Tomaras, Oxford University Press.
© John Iliopoulos and Theodore N. Tomaras (2021). DOI: 10.1093/oso/9780192844200.003.0005

Definition: When G has a finite number of elements, the corresponding group is *finite*. Otherwise, it is an *infinite group*. The number of elements of a finite group G is called the *order* of G. A finite group can be defined by simply giving its *multiplication table*, which contains the result of the composition of all ordered pairs of its elements.

Examples of finite groups
- *The group Z_2.* The simplest non-trivial group will have just two elements. One must be the identity e, so it will be $G = \{e, a\}$. The only entry in the multiplication table which is not obvious is the $a \cdot a$. To satisfy the group property it must be either e or a. In the second case, a is also a unit element, so the group is a trivial combination of two units. The first case yields a table of the form:

G	e	a
e	e	a
a	a	e

We call this group Z_2. We can visualise it in several ways. For example, we can identify it with the set $G = \{1, -1\}$ with the usual multiplication of real numbers. Alternatively, we have seen in physics the groups of space, or time, inversions: $\mathcal{P} = \{\boldsymbol{x} \to \boldsymbol{x}, \boldsymbol{x} \to -\boldsymbol{x}\}$, or $\mathcal{T} = \{t \to t, t \to -t\}$. Similarly, we can identify it with S_2, the group of permutations of two elements.
- *The group Z_n, n integer ≥ 2,* the group whose elements can be identified with the nth roots of unity, with the natural multiplication rule.
- *The group S_n,* the group of permutations of n objects. Its order is $n!$.

Definition: **Homomorphism** is a mapping f from a group G into another G', which preserves the group structure, i.e. $f : G \to G'$, $\forall g_1, g_2 \in G$ $f(g_1 g_2) = f(g_1)f(g_2)$. The product $g_1 g_2$ on the left is given by the composition law of G, while $f(g_1)f(g_2)$ on the right is meant with respect to the operation in G'.

From the definition, we conclude that the identity element e of G is mapped to the identity e' of G', i.e. $f(e) = e'$. But, there may be more elements of G whose image is e'. The subset K of G which is mapped to the identity e' of G' is called the *kernel* of the homomorphism.

If, in addition, the homomorphism f is 1 to 1, then it is an *isomorphism*. G and G' are then *isomorphic*, which is denoted by $G \cong G'$. An isomorphism on G itself is called *automorphism*.

An immediate consequence to the above definitions is that if the kernel of a homomorphism f contains only the identity element e of G, then f is an isomorphism.

Definition: **Subgroup** H of G is a subset H of G which itself is a group with the composition law of G.

Given a subgroup H of G, we define the *left (right) coset* gH (Hg) of an element $g \in G$ to be the set of elements of G of the form gh (hg) with all $h \in H$, i.e.

$$gH \equiv \{gh \,|\, \forall h \in H\} \ \text{ and } \ Hg \equiv \{hg \,|\, \forall h \in H\} \tag{5.1}$$

A subgroup H of G, whose left cosets coincide with the right ones is called *normal or invariant subgroup*.

Every group G has two normal subgroups: (a) G itself and (b) its subgroup $\{e\}$. A group that does not have any other normal subgroup is called a *simple group*. A group that has normal subgroups, none of which is Abelian, is called *semisimple*.

Using the notion of left or right cosets we can define an *equivalence relation* among the elements of G (i.e. a relation that is reflexive, symmetric and transitive) as follows: We define the elements g, g' to be equivalent $(g \sim g')$ if one is in the left coset of the other, i.e. if $g' \in gH$. Like any equivalence relation among the elements of a set, this too partitions G into distinct equivalence classes of the form $g_1 H, g_2 H, \ldots$. The set of cosets is denoted by G/H.

In the special case in which H is a normal subgroup of G, the set G/H is a group, with operation mapping the cosets $g_1 H$ and $g_2 H$ to the coset $g_1 g_2 H$. The group G/H is called the *quotient group*.

A *simple example* demonstrating the above concepts is the following. Consider $G = Z_6$. Its elements can be identified with the sixth roots of unity $e^{2i\pi n/6}$. The set $H = Z_2 = \{1, -1\}$ is a normal subgroup of G, since the set of left cosets of G being $G/H = \{H, e^{i\pi/3}H, e^{2i\pi/3}H\}$ is identical to its right cosets. Finally, the set G/H is a group with operation the one defined above. The product of any two elements of G/H also belongs to G/H, the neutral element is H and the inverse of $e^{i\pi/3}H$ is $e^{2i\pi/3}H$.

The above example brings us naturally to the following definition: A group G is defined to be the *direct product* $G = G_1 \times G_2$ of its subgroups G_1 and G_2 iff (i) all elements of G_1 commute with those of G_2 and (ii) any element $g \in G$ is the product $g = g_1 g_2$, with $g_1 \in G_1$ and $g_2 \in G_2$.

It is straightforward to show that consequences of this definition are (a) that both G_1 and G_2 are then normal subgroups of G, and (b) that the quotient groups G/G_1 and G/G_2 are isomorphic to G_2 and G_1, respectively. We write

$$G/G_1 \cong G_2 \quad \text{and} \quad G/G_2 \cong G_1 \tag{5.2}$$

In the previous example notice that the subgroups Z_2 and Z_3 of Z_6 satisfy the conditions of the above definition, so that $Z_6 = Z_2 \times Z_3$. The reader can be easily convinced that Z_2 and Z_3 are normal subgroups of Z_6 and that indeed $Z_6/Z_3 \cong Z_2$, and $Z_6/Z_2 \cong Z_3$.

From the above definitions it follows that the kernel K of an homomorphism f from G to G' is a normal subgroup of G and, consequently, $G' \cong G/K$.

Definition: A d-dimensional **Representation** of a group G is a mapping $\hat{D}$ of G into a set of operators acting on a d-dimensional linear vector space $\mathcal{V}$ which preserves the group multiplication law. That is, for all pairs of elements of G, $\hat{D}$ satisfies

$$\hat{D}(f)\hat{D}(g) = \hat{D}(f \cdot g) \tag{5.3}$$

with the product on the left being the "synthesis" of the two operations.

Given an orthonormal basis $\{|m\rangle, m = 1, 2, \ldots\}$ in $\mathcal{V}$, the operator $\hat{D}(g)$ is represented by a matrix $D(g)$, whose elements are

$$D(g)_{mn} = \langle m| \hat{D}(g) |n\rangle$$

and (5.3) translates into the relation

$$D(f)D(g) = D(f \cdot g) \tag{5.4}$$

among matrices, with the usual matrix multiplication on the left, and defines a *matrix representation* of G. If the matrices D are unitary, we say that we have a *unitary representation* of G.

To avoid unnecessary proliferation of symbols, in what follows we shall often use the same symbol $D(g)$, for both the "operator" and the "matrix" representing g in a given basis. Which is which will be clear from the context. Most of the time, though, we shall be dealing with matrices.

The trivial representation. The one-dimensional representation $D(g) = 1$ for all group elements g, is a trivial representation of any group. It trivially satisfies the definition. Similarly, the representation with $D(g) = \mathbf{1}_n, \forall g \in G$, with $\mathbf{1}_n$ the $n \times n$ unit matrix, is yet another trivial representation of any group.

Equivalent representations. Given a matrix representation D of a group G, the matrices

$$D'(g) = S\,D(g)\,S^{-1}\,, \quad \forall g \in G \tag{5.5}$$

for any invertible matrix S, also satisfy (5.4) and consequently they too define a representation of G. The matrix representations D and D', which are related by a so-called *similarity transformation* S are called *equivalent*. It is natural to consider them "equivalent" since, as we will show, they are two matrix realisations of the same "operator" representation $\hat{D}$ of G, with respect to two different bases of $\mathcal{V}$.

Indeed, consider two orthonormal bases $\{|m\rangle\}$ and $\{|m\rangle'\}$ of $\mathcal{V}$, with $m = 1, 2, \ldots, N$, related by the linear transformation $|m\rangle' = S^{-1}|m\rangle$. The matrices representing the same operator $\hat{D}(g)$ in the two bases are related for all $g \in G$ by

$$D'(g)_{mn} \equiv {}'\langle m|\,\hat{D}(g)\,|n\rangle' = \langle m|\,S\hat{D}(g)S^{-1}\,|n\rangle = (SD(g)S^{-1})_{mn} \tag{5.6}$$

which, as promised, is identical to (5.5).

Invariant subspace. A subspace $\widetilde{\mathcal{V}}$ of the representation space $\mathcal{V}$ of a group is *invariant* under the action of the group, iff the result of the action of $D(g)$ on $\widetilde{\mathcal{V}}$ also belongs to $\widetilde{\mathcal{V}}$ for all $g \in G$. Thus, if $\mathcal{P}$ is the projector on a subspace, the condition that the latter is invariant under G is:

$$\mathcal{P}D(g)\mathcal{P} = D(g)\mathcal{P}\,, \quad \forall g \in G \tag{5.7}$$

Reducible and irreducible representations. A representation D of G is *reducible* if the corresponding representation space has a subspace invariant under G, or equivalently, if the representation matrices of all group elements can be written in the form

$$D(g) = \begin{pmatrix} D_1(g) & X(g) \\ 0 & D_2(g) \end{pmatrix} \tag{5.8}$$

Otherwise, D is called *irreducible*. The dimension of $D(g)$ is the sum of the dimensions of $D_1(g)$ and $D_2(g)$, which, it is easy to prove, are also representations of the group. A representation $D(g)$, which is equivalent to one of the block diagonal form

$$\begin{pmatrix} D_1(g) & 0 & \dots & 0 \\ 0 & D_2(g) & \dots & 0 \\ \dots & \dots & \dots & \dots \\ 0 & 0 & \dots & D_n(g) \end{pmatrix} \quad (5.9)$$

with $D_i(g)$ an N_i-dimensional irreducible representation of the group for all $i = 1, 2, \dots, n$, is called *completely reducible*. Clearly, the vector space $\mathcal{V}$ of the representation D is the direct sum of the invariant subspaces $\mathcal{V}_i$, the representation spaces of $D_1, D_2, \dots D_n$ with dimensions N_i, respectively. We write: $\mathcal{V} = \mathcal{V}_1 \oplus \mathcal{V}_2 \oplus \dots \oplus \mathcal{V}_n$ and, correspondingly,

$$D(g) = D_1(g) \oplus D_2(g) \oplus \dots \oplus D_n(g) \quad (5.10)$$

This way we decompose a reducible representation into its irreducible parts.

The reverse process is also possible. We can combine irreducible representations into bigger reducible ones.

Tensor product of representations. Consider two irreducible representations D_1 and D_2 of G, with dimensions N_1 and N_2, respectively. Choose orthonormal bases $\{|a\rangle$, $a = 1, 2, \dots, N_1\}$ and $\{|k\rangle, k = 1, 2, \dots, N_2\}$ in the corresponding representation spaces, form the tensor product space of dimension $N_1 \times N_2$ with basis vectors $|a, k\rangle \equiv |a\rangle \otimes |k\rangle$, and define for every group element g the $N_1 \times N_2$-dimensional matrix with elements

$$\langle a, k| \, D(g) \, |b, l\rangle \equiv \langle a| \, D_1(g) \, |b\rangle \, \langle k| \, D_2(g) \, |l\rangle \quad (5.11)$$

That is, the matrix $D(g)$ is the tensor product of the matrices $D_1(g)$ and $D_2(g)$. Readers can easily convince themselves that the matrices D, defined this way, satisfy the group property

$$\langle a, k| \, D(g_1 g_2) \, |b, l\rangle = \langle a, k| \, D(g_1) D(g_2) \, |b, l\rangle \, , \forall g_1, g_2 \in G \quad (5.12)$$

and conclude that they form a representation of G. We call this *tensor product of the representations* D_1 and D_2 and we write: $D \equiv D_1 \otimes D_2$. In general, $D_1 \otimes D_2$ is reducible.

Reminder: The tensor product of two matrices $A(m \times m)$ and $B(n \times n)$, with matrix elements $A_{\alpha\beta}$ and B_{ij}, respectively, is the $mn \times mn$ matrix whose $(\gamma k, \delta l)$ matrix element is $A_{\gamma\delta} B_{kl}$. Using the above definition, readers can easily convince themselves that the product of two tensor product matrices is given by

$$[(A \otimes B)(C \otimes D)]_{\alpha i, \beta j} = (A \otimes B)_{\alpha i, \gamma k} (C \otimes D)_{\gamma k, \beta j} = A_{\alpha\gamma} B_{ik} C_{\gamma\beta} D_{kj} = (AC)_{\alpha\beta} (BD)_{ij}$$

that is, in matrix notation

$$(A \otimes B)(C \otimes D) = (AC) \otimes (BD).$$

The pair of labels (αk) with $1 \leq \alpha \leq m$ and $1 \leq k \leq n$ is in 1-to-1 correspondence with the label $K = (\alpha - 1)n + k$, which takes values $1 \leq K \leq mn$. The $(\gamma k, \delta l)$ element $A_{\gamma\delta} B_{kl}$ of $A \otimes B$ can equivalently be described as the element $(A \otimes B)_{IJ}$ with $I = (\gamma - 1)n + k$ and $J = (\delta - 1)n + l$.

Example. The above definitions and concepts are general and apply to finite as well as infinite groups. Later we shall deal with examples of irreducible, reducible and completely reducible representations of infinite groups. However, it is instructive to give an example of these notions in a finite group.

Consider the group Z_3. It has three elements $\{e, a, b\}$ and multiplication table

	e	a	b
e	e	a	b
a	a	b	e
b	b	e	a

- A one-dimensional non-trivial irreducible unitary representation of Z_3 is

$$D_1(e) = 1, \quad D_1(a) = e^{2\pi i/3}, \quad D_1(b) = e^{4\pi i/3} \tag{5.13}$$

It is through such representations that we defined Z_n previously.

- Another representation of Z_3 is the so-called **regular representation**. Every finite group has a corresponding regular representation. Its dimension is equal to the *order of the group*, defined above as the number of its elements. The construction of the regular representation is described here for the special case of Z_3. Its generalisation to any finite group is straightforward.

Consider an orthonormal basis of the representation space labeled by the elements of the group, i.e. define

$$|e_1\rangle \equiv |e\rangle = \begin{pmatrix} 1 \\ 0 \\ 0 \end{pmatrix}, \quad |e_2\rangle \equiv |a\rangle = \begin{pmatrix} 0 \\ 1 \\ 0 \end{pmatrix}, \quad |e_3\rangle \equiv |b\rangle = \begin{pmatrix} 0 \\ 0 \\ 1 \end{pmatrix} \tag{5.14}$$

and define the 3×3 matrices $D_R(g)$ so that they satisfy

$$D_R(g_1)|g_2\rangle = |g_1 g_2\rangle \tag{5.15}$$

for $g_1, g_2 \in \{e, a, b\}$ and $g_1 g_2$ obtained from the multiplication table of the group. The construction is straightforward and the answer is

$$D_R(e) = \begin{pmatrix} 1 & 0 & 0 \\ 0 & 1 & 0 \\ 0 & 0 & 1 \end{pmatrix}, \quad D_R(a) = \begin{pmatrix} 0 & 0 & 1 \\ 1 & 0 & 0 \\ 0 & 1 & 0 \end{pmatrix}, \quad D_R(b) = \begin{pmatrix} 0 & 1 & 0 \\ 0 & 0 & 1 \\ 1 & 0 & 0 \end{pmatrix} \tag{5.16}$$

- The fact that these matrices form a representation follows from their definition and the group properties of G, since

$$D_R(g_1)D_R(g_2)|g\rangle = |g_1(g_2 g_3)\rangle = |(g_1 g_2)g\rangle = D_R(g_1 g_2)|g\rangle, \quad \forall g_1, g_2, g \in G$$

- Furthermore, it is easy to verify that D_R is completely reducible, since it is equivalent to the representation

$$D_3(e) = \begin{pmatrix} 1 & 0 & 0 \\ 0 & 1 & 0 \\ 0 & 0 & 1 \end{pmatrix}, \quad D_3(a) = \begin{pmatrix} 1 & 0 & 0 \\ 0 & \omega & 0 \\ 0 & 0 & \omega^2 \end{pmatrix}, \quad D_3(b) = \begin{pmatrix} 1 & 0 & 0 \\ 0 & \omega^2 & 0 \\ 0 & 0 & \omega \end{pmatrix} \tag{5.17}$$

with $\omega = \exp(2\pi i/3)$. D_3 is the direct sum of three one-dimensional representations, with the trivial being one of them, and is obtained from D_R by the similarity transformation

$$S = \frac{1}{3} \begin{pmatrix} 1 & 1 & 1 \\ 1 & \omega^2 & \omega \\ 1 & \omega & \omega^2 \end{pmatrix} \tag{5.18}$$

5.3   Lie Groups

The examples we studied in the previous section were all examples of finite groups which can be explicitly defined through their multiplication table. For infinite groups we should give the abstract composition law, but through their representations, we can realise them as groups of transformations acting on a given vector space. In physics we have encountered several examples, such as rotations, translations and Lorentz transformations.

Were we to give our imagination free rein, we could define a large variety of groups because the definition we gave is very general. It was the great merit of the Norwegian mathematician M.S. Lie to choose a particular class among infinite groups,[2] which is large enough to cover most of the interesting cases and precise enough to allow for a detailed study. It is based on the property of continuity. The precise formulation of this property in a group manifold follows the general methods of elementary topology, but we will never need it in this book. We can use instead the theory of representations, which map the group into a set of matrices. The elements of an infinite group are labelled by a set of parameters, which we can view as angles for rotations, or vectors for translations, and so on. Let us assume we have N such parameters, which we call collectively $\boldsymbol{\theta}$. They take values in some region, not necessarily bounded, of $\mathbb{R}^N$ which we choose to contain the origin $\boldsymbol{\theta} = 0$. The group element corresponding to the parameter value $\boldsymbol{\theta}$ is $g(\boldsymbol{\theta})$ and, by convention, we choose the identity to be $e = g(0)$. In a given representation the matrix elements will be functions of the same parameters.

Definition: A **Lie group** is a group for which the matrices in any non-trivial representation are continuous and differentiable functions of the parameters which label the group elements. We can prove that this property does not depend on the chosen representation and we can lift it to the group manifold.

Following this property of continuity we can introduce other topological notions. We will call a group G *connected* if, given any two elements g_1 and g_2 and a real variable $t_1 \le t \le t_2$, there exists a continuous curve $g(t)$ in the group joining these two elements, i.e. such that $g(t_1) = g_1$ and $g(t_2) = g_2$. A closed curve is one for which $g_1 = g_2$. A group will be called *simply connected* if, given any two elements of the group g_1 and g_2 and any two continuous curves $g(t)$ and $\tilde{g}(t), t_1 \le t \le t_2$, joining these elements, there exists a continuous function $g(t, s)$, $s_1 \le s \le s_2$ on the group such that $g(t, s_1) = g(t)$ and $g(t, s_2) = \tilde{g}(t)$; in other words, the two curves can be continuously

---

[2]The study and the complete classification of finite groups turned out to be a very interesting field of modern mathematics and it was completed only during the second half of the 20th century. We will not have the occasion to use it in this book.

deformed one to the other. It follows that in a simply connected group every closed curve is continuously reducible to the identity.

Another useful notion is that of a *limiting point* in the group manifold. A limiting point is the limit $\bar{\boldsymbol{\theta}}$ of a convergent sequence of elements $\boldsymbol{\theta}^{(i)}$, $i = 1, 2, \ldots$, of the group manifold. A limiting point may or may not belong to the group manifold. To decide, we may follow the corresponding sequence of matrices $g(\boldsymbol{\theta}^{(i)})$ and check whether the matrix $g(\bar{\boldsymbol{\theta}})$ exists or not.

Definition: A Lie group is called **compact** if it contains all its limiting points.

Examples. We give here a few examples of Lie groups that we have seen already as symmetry transformations in physics.

• $\mathbb{R}$: The set of real numbers is an Abelian Lie group with composition law the operation of addition. It is obviously non-compact.

• $U(1)$: The phase transformations $\Psi(x) \to e^{i\theta} \Psi(x)$ of the wave function in quantum mechanics. It is an Abelian group whose elements are labelled by the real parameter θ which varies from 0 to 2π. They are unitary one-parameter transformations and we denote the abstract group as $U(1)$. It is obviously compact.

• $SO(2)$: The rotations in a plane. In Cartesian coordinates, a vector $\boldsymbol{x}$ is rotated by an orthogonal matrix with unit determinant of the general form:

$$R(\theta) = \begin{pmatrix} \cos\theta & \sin\theta \\ -\sin\theta & \cos\theta \end{pmatrix} \tag{5.19}$$

where θ varies also from 0 to 2π. It is an Abelian compact Lie group. We call it $SO(2)$, meaning the group is isomorphic to that of the 2×2 orthogonal matrices with determinant one. The prefix S (for "special") is a general notation meaning that we consider only matrices with determinant equal to $+1$.

• $SO(3)$: The group of rotations in our familiar three-dimensional Euclidean space. It is isomorphic to the group of 3×3 orthogonal matrices with determinant equal to $+1$. Its elements depend on the three Euler angles. It is also a compact Lie group but, unlike to the previous ones, it is non-Abelian. It is easy to verify that the result of two successive rotations around e.g. two orthogonal axes depends on the order we apply them.

• P_4: The group of translations in four-dimensional Minkowski space-time: $x_\mu \to x_\mu + a_\mu$, with a_μ an arbitrary constant four-vector. It is an Abelian Lie group (the translations along different axes commute) but it is non-compact because we can consider sequences of ever increasing translations. It is essentially the product of the group $\mathbb{R}$ of real numbers with itself four times.

• The Lorentz group. It is the group of transformations which have a fixed point (the origin $x^\mu = 0$) and leave invariant the Minkowski line element $ds^2 = c^2 dt^2 - dx^2 - dy^2 - dz^2$. The transformations depend on six parameters, the three Euler angles for rotations, and the three parameters v_x/c, v_y/c and v_z/c for the three Lorentz boosts which vary from 0 to 1. c is the speed of light in the vacuum. The Lorentz group is a non-Abelian and non-compact Lie group. The limiting point of any sequence of boosts with $\boldsymbol{v}^{(i)}/c \to \boldsymbol{n}$, $i = 1, 2, \ldots$, with $\boldsymbol{n}^2 = 1$, does not belong to the group. As we have seen in special relativity, the matrices representing the Lorentz transformation contain the factor $(1 - v^2/c^2)^{-1/2}$ which diverges in that limit.

5.4 Lie Algebras

Lie's initial motivation was to study geometrical problems. It was during a visit in Paris around 1870, in contact with the French mathematician C. Jordan, that he discovered the importance of group theory in geometry. Between 1888 and 1893, he published a three-volume monumental work under the title *The theory of group transformations.* His reasoning follows a strong geometrical intuition, not always appreciated by his fellow mathematicians. Today it is essentially of historical interest, but our approach, which follows the physicist's motivations, respects Lie's geometrical approach. Let us look again at the first three examples of Lie groups we gave in section 5.3: $\mathbb{R}$, $U(1)$ and $SO(2)$. They are all one-parameter groups and the composition law is $(\theta_1, \theta_2) \to \theta = \theta_1 + \theta_2$ for all three of them. Lie understood that there exists a unified way to study apparently very different groups and, by doing so, he opened a new chapter of modern mathematics. The key remark is that the property of continuity makes it possible to restrict the study of the group to the vicinity of the identity element. Lie showed that this way we can extract several of the main features of the entire group. Let us develop this point in more detail.

Corresponding to the condition $g(\boldsymbol{\theta} = 0) = e$ set earlier, the linear operator which represents e in any representation $\hat{D}$ is the identity operator $\hat{D}(\boldsymbol{\theta} = 0) = I$, while in the vicinity of the identity element, i.e. for infinitesimal values $\delta\theta_a$ of its parameters, we may expand the corresponding representation matrix $D(\delta\boldsymbol{\theta})$ as:

$$D(\delta\boldsymbol{\theta}) = \mathbf{1} + \mathrm{i}\,\delta\theta_a\, T_a + \ldots \tag{5.20}$$

with $\mathbf{1}$ the unit matrix with dimensionality equal to the dimension of the representation. For reasons, that will become clear shortly, the matrices

$$T_a \equiv -\mathrm{i}\frac{\partial}{\partial\theta_a} D(\theta)\Big|_{\theta=0}, \quad a = 1, 2, \ldots, N \tag{5.21}$$

are called **generators** of the group in the given representation.[3] The generators will play a central role in our analysis, so we will briefly describe some of their properties.

• From the definition (5.20) we conclude that the generators obtained from unitary representations, that is representations satisfying $DD^\dagger = D^\dagger D = \mathbf{1}$, are Hermitian, i.e. they satisfy $T_a^\dagger = T_a$

• Consider a generic linear combination $c_1 T_1 + c_2 T_2$ of the generators corresponding to the parameters θ_1 and θ_2, respectively. The matrix $D(\delta\theta) = \mathbf{1} + \delta\theta(c_1 T_1 + c_2 T_2) + \mathcal{O}(\delta\theta^2)$ is the representation of a group element infinitesimally close to the identity, and according to (5.21) $c_1 T_1 + c_2 T_2$ is the generator corresponding to the parameter θ. We conclude that linear combinations of generators are also generators, only corresponding to a different parametrisation of its elements and, consequently, *the generators form a real linear space.*

---

[3]Equivalently, we could stick to the operator representation $\hat{D}$, expand the operator $\hat{D}(\delta\boldsymbol{\theta})$ around zero and define the generators as operators $\hat{T}_a$ from the linear term in the expansion $\hat{D}(\delta\boldsymbol{\theta}) = I + \mathrm{i}\,\delta\theta_a\,\hat{T}_a + \ldots$. The matrix representation of the generators could then be introduced at a later stage. Even though here we choose to work with matrices, we should keep this in mind, because we shall need this language later.

• *The Lie algebra:* We obtained the generators by looking at group elements infinitesimally close to the identity. In the opposite direction, the representation of a generic finite group element corresponding to the values $\{\theta_a\}$ in the parameter space can be obtained by the multiplication of an infinite number of infinitesimal steps in the given direction according to the formula

$$D(\boldsymbol{\theta}) = \lim_{k \to \infty} \left(1 + i\frac{\theta_a}{k} T_a \right)^k = e^{i\,\theta_a T_a} \equiv e^{i\,\boldsymbol{\theta}\cdot\mathbf{T}} \tag{5.22}$$

Thus, the group elements in the so-called *exponential parametrisation* are written in terms of the generators by exponentiation.

The representation of the group elements in a given direction $\{\theta_a\}$ is

$$D(\lambda) = e^{i\lambda\,\boldsymbol{\theta}\cdot\mathbf{T}} \tag{5.23}$$

with λ a real parameter. The representation of the synthesis of two such elements is easy. It is given simply by

$$D(\lambda_1)D(\lambda_2) = e^{i\lambda_1\boldsymbol{\theta}\cdot\mathbf{T}}e^{i\lambda_2\boldsymbol{\theta}\cdot\mathbf{T}} = D(\lambda_1 + \lambda_2) \tag{5.24}$$

But how about the synthesis of two group elements with generic values of the parameters $\{\theta_a\}$ and $\{\phi_a\}$, respectively? By definition of the group and the group representation there exists a set of N real values α_a of the parameters, such that

$$D(\boldsymbol{\theta})D(\boldsymbol{\phi}) = e^{i\theta_a T_a}e^{i\phi_b T_b} = e^{i\alpha_c T_c} = D(\boldsymbol{\alpha}) \tag{5.25}$$

However, since in general the generators $\boldsymbol{\theta}\cdot\mathbf{T}$ and $\boldsymbol{\phi}\cdot\mathbf{T}$ do not commute, the parameters α_a *are not* given simply as $\alpha_a = \theta_a + \phi_a$. Instead, a central result of Lie group theory is the following:

Theorem: The necessary and sufficient condition for the group property (5.25) to be satisfied is that the generators obey the commutator algebra, called *Lie algebra*

$$[T_a, T_b] = if_{abc} T_c\,, \qquad a, b, c = 1, 2, \dots, N \tag{5.26}$$

for some set of complex (in general) numbers $f_{abc} = -f_{bac}$, the same for all representations. Furthermore, as we can verify using the Baker–Campbell–Hausdorff formula, the parameters α_a of the product group element are then given for small θ_a and ϕ_a, with arbitrary accuracy as a power series in θ_a and ϕ_a, whose first terms are:

$$\alpha_a = \theta_a + \phi_a - \frac{1}{2}f_{bca}\theta_b\phi_c + \dots \tag{5.27}$$

The quantities f_{abc} are called *structure constants* of the group. They depend only on the Lie group at hand and can be computed explicitly given any non-trivial representation. By taking the Hermitian conjugate of (5.26) we conclude that if there exists a unitary representation of (5.26) the structure constants are all real. For $\phi_a = \lambda\,\theta_a$ (5.27) reduces to (5.24). The number N of generators is called the *dimension* of the Lie algebra.

There is a trivial case in which all structure constants vanish. We call the corresponding Lie algebra *Abelian*. In particular, a Lie algebra with just one generator is Abelian.

• *The Jacobi identity:* We obtained the commutation relation (5.26) as a condition for the set of operators T_a to be the generators of a Lie group. We can reverse the order of our reasoning: Let T_a, $a = 1, 2, \cdots, N$ be a set of N operators acting linearly on a given vector space $\mathcal{V}$. We say that they generate a Lie algebra $\mathfrak{g}^4$ if we can find a set of numbers f_{abc} such that the relation (5.26) is satisfied. In other words, the set $\mathfrak{g}$ is endowed with a mapping $\mathfrak{g} \times \mathfrak{g} \to \mathfrak{g}$ given by the relation (5.26). Obviously, these numbers cannot be arbitrary. Using the fact that a commutator is anti symmetric in the two operators, we see that any three among the operators T_a must satisfy the Jacobi identity

$$[T_a, [T_b, T_c]] + [T_b, [T_c, T_a]] + [T_c, [T_a, T_b]] = 0 \tag{5.28}$$

that translates into an identity, which the numbers f_{abc} must satisfy, namely

$$f_{bcd}f_{ade} + f_{cad}f_{bde} + f_{abd}f_{cde} = 0 \tag{5.29}$$

From these relations we can go back and reconstruct the group property of the operators obtained by exponentiation as in (5.22).

• *Representations of the Lie algebra:* By definition, the algebra captures the properties of the group near the identity element. Thus, by restricting a representation of the group G to group elements near the identity we obtain a representation of the corresponding algebra $\mathfrak{g}$.

However, the converse, in general, is not true. Not all representations of the algebra $\mathfrak{g}$ lead via exponentiation to representations of the group G, because of constraints imposed by the global properties of the group. Equivalently, the same algebra can be shared by more than one group, which differ in the global properties of their parameter space, in exactly the same way that the infinite straight line and the circle are different spaces, which nevertheless are identical locally. Two groups G_1 and G_2 which have the same Lie algebra are called *locally isomorphic*. They are denoted by $G_1 \approx G_2$. A very well known example is the case of the groups $SO(3)$ and $SU(2)$, which will be analysed in a later section. We will show that they share the same Lie algebra, but not all representations of $SU(2)$ are representations of $SO(3)$. We remind the reader that, when studying the spin states in non-relativistic quantum mechanics, we found that half-integer spin states were absent in the pure rotation group which is $SO(3)$.

As a trivial but still instructive demonstration of the above statements consider the case of (a) the compact group $G_1 = U(1)$, i.e. the set of 1×1 unitary matrices $e^{ix}, x \in [0, 2\pi]$, with operation the standard matrix multiplication, and (b) the group $G_2 = \mathbb{R}$ of real numbers $x \in (-\infty, +\infty)$ with the usual addition of arithmetic. They are both Abelian, with generator Q satisfying the algebra $[Q, Q] = 0$. Any constant matrix Q forms a representation of the algebra, which via exponentiation leads to the representation $q : x \to e^{iQx}$ of $\mathbb{R}$, since with $q(x) = e^{iQx}$ and $q(y) = e^{iQy}$ we obtain $q(x+y) = e^{iQ(x+y)} = e^{iQx}e^{iQy} = q(x)q(y)$. But only those matrices Q, that satisfy the condition $e^{2\pi iQ} = \mathbf{1}$, form representations of the group $U(1)$.

---

[4]We will denote the algebra of a group G by the corresponding lowercase symbol $\mathfrak{g}$.

We conclude that it will be instructive to study the representations of a Lie algebra without reference to a particular group. Among the representations of a Lie algebra two characteristic ones are:

• *The trivial representation.* An obvious representation of any Lie algebra (5.26) is the one with all generators equal to the number zero ($T_a = 0$). Upon exponentiation it gives the trivial representation of the group, namely all group elements are represented by the 1×1 matrix $D(\boldsymbol{\theta}) = 1$. This generalises trivially to generators equal to zero square matrices of any dimensionality.

• *The adjoint representation.* Define the set of matrices $\{t_a, a = 1, 2, ..., N\}$ whose matrix elements are

$$(t_a)_{bc} \equiv -\mathrm{i}f_{abc} \tag{5.30}$$

From the Jacobi identity it follows that they form a representation of the algebra. Its dimension is equal to the dimension N of the algebra and it is called *the adjoint representation.*

Two concepts related to the representations of an algebra are:

• *The conjugate representation.* Given a representation $\{T_a, a = 1, 2, \ldots, N\}$ of the algebra (5.26) the matrices $\{-T_a^*\}$ also satisfy the algebra. They form what is called the *conjugate* of the representation T_a. If T_a generate via exponentiation (5.22) the matrix representation $D(g)$ of G, the $-T_a^*$ generate the complex conjugate representation $D(g)^*$ of G.

• *Equivalent representations.* From their definition, it is apparent that the generators of two equivalent representations of the group, related by a similarity transformation S, are also related by the same similarity transformation

$$T_a' = S\, T_a\, S^{-1} \tag{5.31}$$

and that both sets T_a and T_a' satisfy the Lie algebra (5.26).

Choice of the structure constants: Although the structure constants are characteristic of the algebra, they are not unique. As we argued above, the algebra forms a linear space, in the sense that any linear combination $T_a' = L_{ab}T_b$ of generators is a generator too. However, their commutation relations become $[T_a', T_b'] = L_{ac}\, L_{bd}\,\mathrm{i}\, f_{cde}\, L_{ef}^{-1}\, T_f'$, that is they are of the same form, as for the T_a, but with the new structure constants $f_{abc}' = L_{ag}\, L_{bd}\, L_{ec}^{-1} f_{gde}$, which in turn, imply a corresponding change of the adjoint representation

$$t_a' = L_{ab}\, L\, t_b\, L^{-1} \tag{5.32}$$

We can promote the space of generators from a linear space to a vector space by defining an *inner product* in it, that is a mapping of pairs of algebra elements to, in general, the field of complex numbers: $\mathfrak{g} \times \mathfrak{g} \to \mathbb{C}$. A convenient choice is given by the trace of the product of the two elements in the adjoint representation. With appropriate choice of the matrix L in (5.32) we can bring the trace of the product of any two basis generators $\{t_a\}, a = 1, 2, \ldots, N$ to the form

$$\mathrm{Tr}(t_a t_b) = k_a \delta_{ab} \tag{5.33}$$

An algebra with $k_a > 0, \forall a$ is called *compact*. For such an algebra one can perform appropriate rescalings of the generators to make all k_as equal to a convenient value κ, so that

$$\mathrm{Tr}(t_a t_b) = \kappa\, \delta_{ab} \tag{5.34}$$

It is then straightforward to show that with this normalisation the structure constants are fully antisymmetric,

$$f_{abc} = -f_{bac} = -f_{acb} \tag{5.35}$$

An equivalent way to define the adjoint representation, reminiscent of the regular representation of finite groups, starts with the observation that its dimension is equal to the dimension of the Lie algebra. In other words, the states in its representation space can be labelled by the generators themselves. The basis vectors are then $\{|t_a\rangle, a = 1, 2, \ldots, N\}$, the action of a generator on any basis vector is given by

$$t_a\,|t_b\rangle = |[t_a, t_b]\rangle = \mathrm{i} f_{abc}\,|t_c\rangle \tag{5.36}$$

and the inner product of any two states in the adjoint representation space inherited from (5.34) is

$$\langle T_a | T_b \rangle = \frac{1}{\kappa}\mathrm{Tr}(T_a T_b) \tag{5.37}$$

With these definitions we obtain for the basis vectors in particular

$$(t_a)_{bc} = -\mathrm{i} f_{abc}\,, \quad \langle t_a | t_b \rangle = \frac{1}{\kappa}\mathrm{Tr}(t_a t_b) = \delta_{ab}$$

as required for consistency with our conventions. The vector space structure thus defined on the representation space of the adjoint representation gives the action of any generator $T = x_a t_a$ on any vector $|T' = y_b t_b\rangle$

$$T\,|T'\rangle = |[T, T']\rangle = x_a y_b\,|[t_a, t_b]\rangle = \mathrm{i} f_{abc} x_a y_b\,|t_c\rangle \tag{5.38}$$

The Cartan classification: As we mentioned above, in his work on transformation groups, Lie had identified the crucial role of the infinitesimal transformations in the vicinity of the identity. He called the resulting algebraic structure an *infinitesimal group*. Today we call it more appropriately *Lie algebra*, following a terminology introduced by Hermann Weyl around 1930. It is defined through the commutation relations (5.26). This shift of emphasis from the group to the algebra is mainly due to the work of a young French mathematician E. Cartan. In 1894, he published his first results on the complete classification of all finite-dimensional Lie algebras, one of the most remarkable mathematical results of the end of the 19th century.[5] He was 25 years old.

Cartan addressed the question of what are the possible choices of the structure constants, sets of numbers satisfying the Jacobi identity. Incredible as it may sound, he succeeded in finding the complete answer to this question; namely, he found the general solution, modulo rescalings and reparametrisations for any finite N. The original proof was rather lengthy, but in principle straightforward. We will only state the results:*All possible finite-dimensional simple Lie algebras over the complex numbers can be classified in four infinite series plus five exceptional cases.* An important concept

---

[5] A first, although not rigorous, step in this direction was made between 1888 and 1890 by W. Killing. In today's textbooks we mainly follow the method introduced by the 22-year-old E. Dynkin in 1947.

which will be useful to us in the examples we will study is the *rank* of an algebra. It is defined as the maximum number of commuting generators. In the following Cartan classification n is the rank of the algebra:

1. A_n, $n \geq 1$. In our language they are related to the groups $SU(n + 1)$, i.e. the groups which are isomorphic to the groups of $(n + 1) \times (n + 1)$ unitary matrices with determinant equal to +1. Their dimension $d = n(n + 2)$.

2. B_n, $n \geq 2$. Similarly, they correspond to odd-dimensional orthogonal groups $SO(2n + 1)$. They have $d = n(2n + 1)$.

3. D_n, $n \geq 4$, corresponding to even-dimensional orthogonal groups $SO(2n)$. $d = n(2n - 1)$.

4. C_n, $n \geq 3$. They correspond to the so-called *symplectic groups* $SP(2n)$, which we will not use in this book: $d = n(2n + 1)$.

5. In addition to these four infinite series we have five *exceptional algebras* denoted, for historical reasons, by: G_2 ($d = 14$), F_4 ($d = 52$), E_6 ($d = 78$), E_7 ($d = 133$), and E_8 ($d = 248$), with the index of the symbol being the rank of the corresponding algebra. Although they have been used in particle physics in attempts to unify the fundamental forces beyond the Standard Model, we will not describe them here.

5.5 Examples of Lie Groups and their Algebras

In this section we want to introduce some compact Lie groups which are often used in particle physics. They also serve as illustrations of the general concepts we presented earlier.

5.5.1 The group $U(1)$

By definition, the group $U(1)$ is the set of 1×1 unitary matrices. Thus, its elements are the complex phases $e^{i\theta}, 0 \leq \theta < 2\pi$, in 1-to-1 correspondence with the points of a circle S^1. The group is Abelian, since the product of any two of its elements $e^{i\theta_1} e^{i\theta_2} = e^{i(\theta_1 + \theta_2)} = e^{i\theta_2} e^{i\theta_1}$ is commutative.

The set of phases constitutes the only irreducible representation of $U(1)$. It is the trivial example of a group which is simple but not semi-simple, since it has an Abelian invariant subgroup, namely itself. Its Lie algebra having only one generator does not appear in the Cartan classification. As we have pointed out in section 5.4, $U(1)$ is locally isomorphic to $SO(2)$ and $\mathbb{R}$.

5.5.2 The group $SU(2)$

The group $SU(2)$ is familiar from the study of angular momentum in quantum mechanics. Here we review its properties and associated techniques, because they are important and useful for the analysis of all Lie groups.

By definition, $SU(2)$ is the group that is isomorphic to that of 2×2 unitary matrices with determinant equal to +1. The general form of such a matrix is

$$u = \begin{pmatrix} a & b \\ -b^* & a^* \end{pmatrix}, \quad \text{with } |a|^2 + |b|^2 = 1 \tag{5.39}$$

As usual, these matrices can be thought of as representing linear transformations of the two-dimensional complex space $\mathcal{C}^2$. We can also write the general 2×2 matrix in the basis of the unit matrix I and the three Pauli matrices $\boldsymbol{\sigma}$, and we find the equivalent parametrisation

$$u = \alpha_0 \, \mathbf{1} + \mathrm{i}\boldsymbol{\alpha} \cdot \boldsymbol{\sigma} \, , \quad \sum_{i=0}^{3} \alpha_i^2 = 1 \tag{5.40}$$

where α_i ($i = 0, 1, 2, 3$) are four real numbers constrained by the relation (5.40.)[6] We have used the notation $\boldsymbol{\alpha} \cdot \boldsymbol{\sigma}$ which is reminiscent of a three-dimensional scalar product although, at this stage, there is no justification for it, and this expression is just a shorthand for $\sum_{i=1}^{3} \alpha_i \sigma_i$. We will come back to this point later in this section. Alternatively, we can write:

$$u = \mathbf{1} \cos \gamma + \mathrm{i}\hat{\boldsymbol{\beta}} \cdot \boldsymbol{\sigma} \sin \gamma = \exp\!\left(\mathrm{i}\gamma \hat{\boldsymbol{\beta}} \cdot \boldsymbol{\sigma}\right), \quad \hat{\boldsymbol{\beta}}^2 = 1, \quad \gamma \in [0, 2\pi) \tag{5.41}$$

The components of the unit vector $\hat{\boldsymbol{\beta}}$ can be written in the form $\hat{\beta}_1 = \sin\theta\cos\phi$, $\hat{\beta}_2 = \sin\theta\sin\phi$ and $\hat{\beta}_3 = \cos\theta$ in the standard parametrisation (θ, ϕ) of the unit sphere S^2.

All these parametrisations show that the general $SU(2)$ matrix depends on three real parameters, which can be chosen to be three angles: $\theta \in [0, \pi]$, $\phi \in [0, 2\pi)$ and $\gamma \in [0, 2\pi)$. They also show that $SU(2)$ is a Lie group, it is compact and it is non-Abelian.

5.5.2.1 Representations of $SU(2)$. • *The singlet representation.* Every group has the trivial one-dimensional representation in which every group element u is represented by the number 1. It is also called the *singlet* representation and is denoted by **1**, its dimensionality in bold.
• *The spinor representation.* By definition $SU(2)$ has a two-dimensional representation with the matrices given by equations (5.40) or (5.41). It is called the *doublet* representation or, simply, **2**. For reasons which will become clear shortly, it is also called the *fundamental* representation. The matrices act in a two-dimensional space whose vectors are called *spinors*. Two quantities $\psi^\alpha, \alpha = 1, 2$ are defined to be the components of a *spinor* of $SU(2)$, if they transform according to

$$\psi^\alpha \to \psi'^\alpha = u^\alpha_\beta \, \psi^\beta \quad (\text{or in matrix notation } \psi' = u\,\psi) \tag{5.42}$$

for all $u \in SU(2)$. In order for the spinors to form a vector space we must introduce an $SU(2)$ invariant scalar product, which amounts to introducing an appropriate metric ζ for spinors. Using the property $\det u = 1$ we can easily verify that such a metric can be written as

$$\zeta_{\alpha\beta} = \begin{pmatrix} 0 & 1 \\ -1 & 0 \end{pmatrix} = \mathrm{i}(\sigma_2)_{\alpha\beta} = \epsilon_{\alpha\beta} \tag{5.43}$$

where $\epsilon_{\alpha\beta}$ is the two-index completely antisymmetric tensor. The scalar product of two arbitrary spinors ψ and χ is then defined by the antisymmetric quadratic form

$$< \chi, \psi > = \chi^T \epsilon \psi = \chi^\alpha \epsilon_{\alpha\beta} \psi^\beta = \chi^1 \psi^2 - \chi^2 \psi^1 \tag{5.44}$$

---

[6]Relation (5.40) implies that the elements of the group $SU(2)$ are in 1-to-1 correspondence with the points of a three-dimensional unit sphere S^3.

We can observe that the scalar product of any spinor with itself is zero

$$< \psi, \psi > = \psi^1 \psi^2 - \psi^2 \psi^1 = 0 \tag{5.45}$$

which means that we cannot use the invariant scalar product $<,>$ to define a norm for spinors.

• *The conjugate representation.* As we explained in a previous section, it is generally true and easy to verify that, if a set of matrices U forms a representation of a Lie group with operation the standard matrix multiplication, so do their complex conjugates. The latter is called the *complex conjugate representation* of the first.[7]

In particular, the matrices u^*, the complex conjugates of those defined in (5.39), form also a two-dimensional representation of $SU(2)$, called the conjugate representation of **2** and denoted by $\bar{\mathbf{2}}$.

Given the ψ^α which transform by the representation **2**, their complex conjugates $\psi^{\alpha*}$ transform according to the representation $\bar{\mathbf{2}}$: $\psi^{*\prime} = u^* \psi^*$.

It is easy to see that the representation **2** and its complex conjugate are equivalent: $\bar{\mathbf{2}} \cong \mathbf{2}$. Indeed, as can be checked explicitly using (5.41), the matrices u and u^* are related by the similarity transformation

$$\epsilon \, u \, \epsilon^{-1} = u^* \tag{5.46}$$

with $\epsilon = i\sigma_2$ given in equation (5.43). In fact, (5.46) defines an inner automorphism of $SU(2)$, since ϵ is itself a member of $SU(2)$. We call *real* any representation which is equivalent to its conjugate.

A final remark: since the complex conjugates $\psi^{\alpha*}$ of the components of a spinor transform according to the representation $\bar{\mathbf{2}}$, it follows that the sum of the squares of the moduli of any spinor is an invariant. Therefore, we can use it to define a norm. In matrix notation: $\psi^{*\prime T} \psi' = \psi^\dagger u^\dagger u \psi = \psi^\dagger \psi$.

• *Covariant and contravariant components.* The introduction of the metric ϵ in the space of spinors makes it possible to apply the usual rules of raising and lowering indices. Thus the components ψ^α of a spinor are called "contravariant" and the components $\psi_\alpha = \epsilon_{\alpha\beta}\psi^\beta$ are called "covariant". We can rewrite the scalar product of two spinors in the form

$$< \chi, \psi > = -\chi_\alpha \psi^\alpha = \chi^\alpha \psi_\alpha \tag{5.47}$$

Using the definition of the covariant spinor and (5.46), we find that

$$\chi'_\alpha = \epsilon_{\alpha\beta}\chi'^\beta = -\epsilon_{\alpha\beta}u^\beta_\gamma \epsilon^{\gamma\delta}\chi_\delta = \chi_\delta (u^\dagger)^\delta_\alpha$$

which implies that χ_α transforms under $SU(2)$ in the same way as $\psi^\dagger_\alpha$.

• *Tensor calculus for spinors.* Combining spinors, we can obtain higher tensors of $SU(2)$. For example, consider two spinors ψ, χ, and combine their components to form the four quantities $\psi^\alpha \chi^\beta$. These can be thought of as the components of a vector in the four-dimensional tensor product space of $\mathcal{C}^2$ with itself.

---

[7]If T_a are the generators of the original representation, the generators of its conjugate are $-T_a^*$. We showed in section 5.4 that they satisfy the same Lie algebra as the T_a.

How do these quantities transform under $SU(2)$? This is obtained using the way ψ and χ transform, namely

$$(\psi^\alpha \chi^\beta)' = u^\alpha_\gamma u^\beta_\delta (\psi^\gamma \chi^\delta) \tag{5.48}$$

Also, if u is followed by another $SU(2)$ transformation v, the end result is

$$(\psi^\alpha \chi^\beta)'' = v^\alpha_\gamma v^\beta_\delta (\psi^\gamma \chi^\delta)' = v^\alpha_\gamma v^\beta_\delta u^\gamma_\epsilon u^\delta_\zeta (\psi^\epsilon \chi^\zeta) = (vu)^\alpha_\epsilon (vu)^\beta_\zeta (\psi^\epsilon \chi^\zeta) \tag{5.49}$$

which demonstrates that (5.48) defines a new representation of $SU(2)$. In general, we say that the four quantities $T^{\alpha\beta}$ belong to the *two-index tensor* of $SU(2)$, if they transform according to

$$(T^{\alpha\beta})' = u^\alpha_\gamma u^\beta_\delta T^{\gamma\delta} \tag{5.50}$$

This is not a new result. We have shown the general rule, that the tensor product of the matrices corresponding to two representations defines a new representation. However, the latter is, in general, reducible. Let us look at the one we obtained as $\mathbf{2} \otimes \mathbf{2}$: we can split $T^{\alpha\beta}$ into its symmetric and antisymmetric parts

$$T^{\alpha\beta} = \frac{1}{2}(T^{\alpha\beta} + T^{\beta\alpha}) + \frac{1}{2}(T^{\alpha\beta} - T^{\beta\alpha}) \equiv T^{(\alpha\beta)} + T^{[\alpha\beta]} \tag{5.51}$$

Using (5.50) you may easily verify that

$$T^{(\alpha\beta)'} = u^\alpha_\gamma u^\beta_\delta T^{(\gamma\delta)} \text{ and } T^{[\alpha\beta]'} = u^\alpha_\gamma u^\beta_\delta T^{[\gamma\delta]} \tag{5.52}$$

i.e. the symmetric and antisymmetric parts of T transform independently under the group. Note that the quantity $T^{[\alpha\beta]}$, being antisymmetric in its indices, has only one component $T^{[12]}$ which is invariant under $SU(2)$, since according to (5.52), it transforms to

$$T^{[12]'} = (u^1_1 u^2_2 - u^1_2 u^2_1)T^{[12]} = (\det u)T^{[12]} = T^{[12]} \tag{5.53}$$

The three components $T^{(\alpha\beta)}$ of a *two-index symmetric tensor* also transform irreducibly. They all mix among themselves by the generic $SU(2)$ transformation. Thus, they carry an irreducible three-dimensional representation of $SU(2)$, denoted by **3**.

What has happened is that the four-dimensional representation space of the tensor product of two spinor representations split into the direct sum of a three- and a one-dimensional subspace, each of which carries an irreducible representation of $SU(2)$. We denote this reduction by [8]

$$\mathbf{2} \otimes \mathbf{2} = \mathbf{3} \oplus \mathbf{1} \tag{5.54}$$

Combining an arbitrary number of **2**s and **2̄**s, we obtain a tensor of the general form $T^{\alpha_1 \alpha_2 ... \alpha_n}_{\beta_1 \beta_2 ... \beta_m}$.

• *The complete list of $SU(2)$ irreducible representations.* We are now in a position to generalise the construction which gave us the **3** representation to tensors with any

---

[8]Another notation for the representations of $SU(2)$ comes from quantum mechanics and uses their "spin" value. For example, instead of **1,2,3** we write **0,1/2** and **1**, respectively. Correspondingly, (5.54) is written as $\mathbf{1/2} \otimes \mathbf{1/2} = \mathbf{1} \oplus \mathbf{0}$ and is described as: "Combining two spin-$\frac{1}{2}$ we obtain either spin 1 or spin 0".

number of indices. We first note that we can raise all lower indices by acting with the appropriate number of $\epsilon^{\alpha_i \beta_j}$. Thus the representation of the original tensor with n upper and m lower indices is equivalent to one with $n+m$ upper indices. Thus we can restrict our discussion to tensors with only upper indices. We proceed inductively. We have shown that for a two-index tensor the process of symmetrisation and contraction of indices gives the reduction shown in equation (5.54). We assume that this is true for a tensor of $n-1$ indices and we consider a tensor with n indices. Contracting two upper indices α_k and α_l with the invariant tensor $\epsilon_{\alpha_k \alpha_l}$ produces a new tensor with two indices less and has been counted in the previous step of our iterative construction. The end result is that the independent irreducible representations of $SU(2)$ correspond to symmetric tensors with any number of spinor indices because they are the only ones which cannot be contracted any further. Each index can take two values, 1 and 2, so the dimension of such a representation with n indices is equal to the ways we can write a sequence of a total number of n 1s and 2s:

$$D(n) = n + 1, \quad n = 0, 1, 2, \ldots \tag{5.55}$$

or, following the quantum mechanical notation with $n \equiv 2j, j = 0, 1/2, 1, \ldots$, we obtain the well-known result for the dimension of the "spin-j" representation

$$D(j) = 2j + 1, \quad j = 0, 1/2, 1, \ldots \tag{5.56}$$

We can show that this way we obtain the complete list of the $SU(2)$ irreducible representations, and, furthermore, since we know that **2** is real, we conclude that all these representations are real.

5.5.2.2   The Clebsch–Gordan decomposition.   As demonstrated with the example of the product of two doublets, the tensor product of two irreducible representations in general is reducible and can be reduced to the direct sum of irreducible representations, written as

$$D \otimes D' = \oplus \sum_i n_i R_i \tag{5.57}$$

with integers n_i. This relation is called *Clebsch–Gordan* decomposition of the tensor product $D \otimes D'$.

Using the tensor language it is easy to perform this decomposition. Let n and n' be the orders of the tensors belonging to the irreducible representations D and D', respectively. The general element of their product has the structure $T^{\alpha_1 \cdots \alpha_n \alpha_{n+1} \cdots \alpha_{n+n'}}$, which is completely symmetric in the first n and in the last n' indices. Let us choose any pair of indices, one from the first n and the other from the last n', for instance α_1 and α_{n+1}. The generic component of T can be written as the sum of one symmetric under these two indices and one antisymmetric. The symmetric is now fully symmetric over the $n+n'$ indices and belongs to the irreducible representation with $n+n'$ indices, while the antisymmetric can be written as the product of $\epsilon^{\alpha_1 \alpha_{n+1}}$ times a new tensor with $n + n' - 2$ indices, still symmetric over the $n-1$ and over the $n'-1$ remaining indices. We continue this process of symmetrisation and antisymmetrisation and reduction of the indices in pairs a $\min(n, n')$ number of times, i.e. until the smaller one of the two sets of indices is exhausted. At each step we obtain a fully symmetric tensor plus a new one with two fewer indices. The final element of the series is a tensor with $|n - n'|$ indices.

Thus we demonstrated the well-known result for $SU(2)$, namely

$$D(n) \otimes D(n') \cong D(n+n') \oplus D(n+n'-2) \oplus \ldots \oplus D(|n-n'|) = \oplus \sum_{k=0}^{min(n,n')} D(n+n'-2k)$$

$$(5.58)$$

or, in the standard quantum mechanical notation

$$j_1 \otimes j_2 \cong (j_1 + j_2) \oplus (j_1 + j_1 - 1) \oplus \ldots \oplus |j_1 - j_2| \cong \oplus \sum_{j=|j_1-j_2|}^{j_1+j_2} j \qquad (5.59)$$

Example:

Let us demonstrate this method with the tensor product $\mathbf{3} \otimes \mathbf{2}$. Consider the product

$$\psi^{(\alpha\beta)}\chi^\gamma = \frac{1}{3}(\psi^{(\alpha\beta)}\chi^\gamma + \psi^{(\alpha\gamma)}\chi^\beta + \psi^{(\beta\gamma)}\chi^\alpha)$$

$$+ \frac{1}{3}(\psi^{(\alpha\beta)}\chi^\gamma - \psi^{(\alpha\gamma)}\chi^\beta) + \frac{1}{3}(\psi^{(\alpha\beta)}\chi^\gamma - \psi^{(\beta\gamma)}\chi^\alpha)$$

$$= \frac{1}{3}(\psi^{(\alpha\beta)}\chi^\gamma + \psi^{(\alpha\gamma)}\chi^\beta + \psi^{(\beta\gamma)}\chi^\alpha) + \frac{1}{3}\left(\epsilon^{\beta\gamma}\delta^\alpha_\rho + \epsilon^{\alpha\gamma}\delta^\beta_\rho\right)(\psi^{\rho\sigma}\chi_\sigma)$$

$$\equiv S^{(\alpha\beta\gamma)} + \frac{1}{3}\left(\epsilon^{\beta\gamma}\delta^\alpha_\rho + \epsilon^{\alpha\gamma}\delta^\beta_\rho\right)\Psi^\rho \qquad (5.60)$$

with S a three index symmetric tensor transforming as a **4** and Ψ a doublet. As anticipated, we conclude: $\mathbf{3} \otimes \mathbf{2} \cong \mathbf{4} \oplus \mathbf{2}$.

5.5.2.3 Young tableaux for SU(2). There is a pictorial way to represent the Clebsch–Gordan decomposition described above and which generalises in a straightforward way to higher unitary groups. Let us associate to the n-index tensor $\psi^{(\alpha_1\alpha_2\cdots\alpha_n)}$ the array of n boxes, arranged in rows and columns, under the following conventions:

1. Boxes belonging to the same row correspond to symmetrised indices.

$$(5.61)$$

2. Boxes belonging to the same column correspond to antisymmetrised indices. In particular, for $SU(2)$, two boxes arranged one on top of the other

$$(5.62)$$

correspond to the two-index antisymmetric tensor, which as we have explained already, is a singlet of $SU(2)$. For $SU(2)$ more than two boxes in a column are zero, since there is no fully antisymmetric tensor with more than two indices, ranging from 1 to 2.

The defining representation is associated with the one-box "tableau" ((5.63), left), while a specific component ψ^α of the corresponding one-index tensor is described by (5.63), right.

$$\Box \qquad \boxed{\alpha} \qquad\qquad (5.63)$$

Translating to pictures the tensor product of two spinor representations is represented by

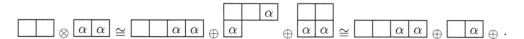

$$\Box \otimes \boxed{\alpha} \cong \boxed{\;\;\boxed{\alpha}} \oplus \boxed{\alpha} \qquad\qquad (5.64)$$

That is, we place the labelled box of the second tableau either to the right of the first to form the symmetric combination or below to form the antisymmetric one. For $SU(2)$ the latter is the singlet which we denote by a simple "dot". Thus we write

$$\Box \otimes \Box \cong \boxed{\;\;} \oplus \cdot . \qquad\qquad (5.65)$$

This is pictorially the decomposition $\mathbf{2} \otimes \mathbf{2} \cong \mathbf{3} \oplus \mathbf{1}$.

Making use of such pictures, we can describe the tensor product of two arbitrary irreducible representations of $SU(2)$ in exactly the way described above. The rules of the game are: (i) Denote the entries of the second Young tableau by the same letter α. (ii) Stick one-by-one the boxes of the second tableau on the right or below the ones of the first, in such a way that two boxes indexed by the same letter never appear in the same column. (iii) In the resulting tableaux ignore the columns with two vertical boxes. This corresponds to contracting with the ϵ symbol.

Let us work out an explicit example. Consider the tensor product of two *two*−index tensors.

$$\boxed{\;\;\;} \otimes \boxed{\alpha|\alpha} \cong \boxed{\;\;|\alpha|\alpha} \oplus \boxed{\alpha\;|\;\;\alpha} \oplus \boxed{\alpha|\alpha} \cong \boxed{\;\;|\alpha|\alpha} \oplus \boxed{\;\;|\alpha} \oplus \cdot .$$

This is the decomposition $\mathbf{3} \otimes \mathbf{3} \cong \mathbf{5} \oplus \mathbf{3} \oplus \mathbf{1}$ in the language of Young tableaux.

5.5.2.4 The Lie algebra of $SU(2)$. According to the general theory of section 5.4, the representations of a Lie group are obtained via exponentiation from the representations of the corresponding Lie algebra. So it is interesting to study the latter without referring to the Lie group we started from. We will do this exercise here for the case of the $SU(2)$ Lie algebra $\mathfrak{su}(2)$.

We know that for a Lie group G, the dimension of the corresponding Lie algebra equals the number of independent parameters needed to label the group elements. We saw that for $SU(2)$ this number is 3, so $\mathfrak{su}(2)$ will have three generators. It is the algebra A_1 in the Cartan classification scheme, the only one with three generators. To compute them and determine the corresponding Lie algebra we use the definition (5.21). We start from the general expression for an $SU(2)$ group element (5.41), which we choose to write in the more conventional parametrisation[9]

---

[9]It amounts to a specific choice of the normalisation of the generators.

$$u = \exp\left(i\boldsymbol{\theta} \cdot \frac{\boldsymbol{\sigma}}{2}\right), \quad \boldsymbol{\theta} = 2\gamma\hat{\boldsymbol{\beta}}, \quad 0 \le |\boldsymbol{\theta}| < 4\pi \qquad (5.66)$$

from which we see that the generators are $J_a = \sigma_a/2$ and satisfy the commutation relations:

$$[J_a, J_b] = i\epsilon_{abc}J_c, \quad \mathrm{Tr}J_aJ_b = \frac{1}{2}\delta_{ab} \qquad (5.67)$$

• *Representations of* $\mathfrak{su}(2)$. Now we can look for representations of this algebra ignoring the parent $SU(2)$ group. We have solved this problem in elementary quantum mechanics so we will only sketch the construction here. It should be stressed, however, that the analysis is pure mathematics. For convenience, we shall be using the standard notation for the angular momentum and the vectors of linear spaces used in quantum mechanics textbooks but, a priori, there is nothing quantum mechanical in the discussion. In particular, there is no $\hbar$ present in any of the formulae.

On the basis of our results on $SU(2)$ representations, we are after finite-dimensional sets of Hermitian and traceless matrices $J_a, a = 1, 2, 3$, satisfying the algebra (5.67). They can be thought of as operators acting on a finite-dimensional linear vector space. According to (5.67) the rank of $\mathfrak{su}(2)$ is one, since at most one of the three operators J_a can be diagonalised. We choose it to be J_3. We denote the corresponding orthonormal basis vectors as $\{|m\rangle\}$, with the real number "m" being an eigenvalue of J_3. The eigenvalues of J_3 are called *weights* of the representation under discussion. They satisfy

$$J_3 |m\rangle = m |m\rangle \qquad (5.68)$$

It is convenient to define the *raising* and *lowering* operators $J_\pm$, respectively, by

$$J_\pm \equiv J_1 \pm iJ_2, \quad J_\pm^\dagger = J_\mp \qquad (5.69)$$

in terms of which the algebra takes the form

$$[J_3, J_\pm] = \pm J_\pm \quad \text{and} \quad [J_+, J_-] = 2J_3 \qquad (5.70)$$

As a consequence, $J_\pm$ acting on eigenvectors of J_3 raises (lowers) their eigenvalues by one unit. Thus,

$$J_\pm |m\rangle = C(m) |m \pm 1\rangle \qquad (5.71)$$

For finite-dimensional representations m must be bounded from above and below, i.e. there is a *highest weight* $m \equiv j$ and a *lowest* one $m \equiv j'$ with $j' < j$. The corresponding eigenstates satisfy

$$J_+ |j\rangle = 0; \ J_- |j'\rangle = 0 \qquad (5.72)$$

Acting on $|j\rangle$ successively with J_- we construct one by one all the basis vectors of the representation space. For finite-dimensional representations this process has to terminate. Let us write for the non-vanishing matrix elements of $J_\pm$

$$\langle m+1| J_+ |m\rangle = \alpha_m \quad \text{and} \quad \langle m| J_- |m+1\rangle = \alpha_m^* \qquad (5.73)$$

and compute the $\langle m|-|m\rangle$ matrix element of both sides of the relation $J_+J_- - J_-J_+ = 2J_3$. The result is

$$|\alpha_{m-1}|^2 - |\alpha_m|^2 = 2m \tag{5.74}$$

whose general solution is

$$|\alpha_m|^2 = C - m(m+1) \tag{5.75}$$

with C a constant. Comparing equations (5.72) and (5.73) we obtain the constraints $\alpha_j = \alpha_{j'-1} = 0$. We use the first to fix C and we obtain

$$|\alpha_m|^2 = j(j+1) - m(m+1) \tag{5.76}$$

and the second gives $j' = -j$. But j and j' differ by the number of steps, which is an integer. We conclude that $2j$ has to be an integer. This implies that the dimension of the representation with highest weight j is $2j+1$ and in addition it is irreducible, since all its basis vectors are interconnected by the action of the generators of the algebra.

To summarise: The irreducible representations of the algebra $\mathfrak{su}(2)$ are labelled by an index j, which takes the values $j = 0, 1/2, 1, 3/2, 2, \ldots$. The orthonormal basis of the representation space for a given j consists of the $2j+1$ vectors $|m\rangle$, with $m = -j, -j+1, \ldots, j-1, j$. Thus, the dimensionality of the representation j is $2j+1$, and the non-vanishing elements of the corresponding representation matrices in the above basis are [10]

$$\langle m \pm 1 | J_\pm | m \rangle = \sqrt{j(j+1) - m(m \pm 1)}, \quad \langle m' | J_3 | m \rangle = m\, \delta_{mm'} \tag{5.77}$$

Finally, according to a general theorem due to Racah, in a given Lie algebra of rank k, there are exactly k independent operators constructed out of the generators of the algebra, which commute with all its generators and thus can also be diagonalised. One of those is always the sum of the squares of the generators, which for $\mathfrak{su}(2)$ is

$$J^2 \equiv J_a J_a = J_1^2 + J_2^2 + J_3^2 = \frac{1}{2}(J_+ J_- + J_- J_+) + J_3^2 \tag{5.78}$$

Its matrix elements in the representation j can be computed using (5.77) and are

$$\langle m' | J^2 | m \rangle = j(j+1)\, \delta_{mm'} \tag{5.79}$$

According to our definition, the numbers $\{-j, -j+1, \ldots, j-1, j\}$ of J_3 are the *weights* of the representation j. In $\mathfrak{su}(2)$, the weight diagrams of the representations are one-dimensional, since only one of its generators can be diagonalised. In general, the dimension of the weight diagrams of an algebra is equal to its rank.

Before closing this section let us give the explicit matrix representations of the generators J of the $\mathfrak{su}(2)$ algebra for the first few low lying representations. They are easily obtained using equations (5.77). By exponentiation of $i\theta_a J_a$ we obtain the corresponding representations of $SU(2)$.

- For the trivial representation $j = 0$ all generators vanish: $J_a^{(0)} = 0$.
- For the $j = \frac{1}{2}$ representation they are given by $\sigma_a/2$.
- For $j = 1$ we obtain the matrices

---

[10]By convention we take the phases of all the parameters α_m equal to zero. Note that any set of phases can be absorbed into the definition of the basis vectors $|m\rangle$.

$$J_1^{(1)} = \frac{1}{\sqrt{2}} \begin{pmatrix} 0 & 1 & 0 \\ 1 & 0 & 1 \\ 0 & 1 & 0 \end{pmatrix}, \quad J_2^{(1)} = \frac{1}{\sqrt{2}} \begin{pmatrix} 0 & -i & 0 \\ i & 0 & -i \\ 0 & i & 0 \end{pmatrix}$$

$$J_3^{(1)} = \begin{pmatrix} 1 & 0 & 0 \\ 0 & 0 & 0 \\ 0 & 0 & -1 \end{pmatrix}, \quad (J^{(1)})^2 = 2 \tag{5.80}$$

5.5.2.5 Clebsch–Gordan coefficients. In equations (5.58) or (5.59) we obtained the Clebsch–Gordan decomposition of the product of two irreducible representations labelled by j_1 and j_2. The corresponding vector space has dimension $d = (2j_1 + 1) \times (2j_2 + 1)$ and the basis vectors are $|j_1, m_1\rangle \otimes |j_2, m_2\rangle$. This space is split into the direct sum of irreducible invariant subspaces as shown in equation (5.59) with basis vectors $|J, M\rangle$. The two sets of basis vectors are related by a linear transformation whose elements are called *Clebsch–Gordan coefficients*. We want here to show how they can be computed. Let us write:

$$|j_1, m_1\rangle \otimes |j_2, m_2\rangle = \sum_{J=|j_1-j_2|}^{j_1+j_2} \sum_{M=-J}^{+J} C(j_1, m_1; j_2, m_2|J, M) |J, M\rangle \tag{5.81}$$

The coefficients $C(j_1, m_1; j_2, m_2|J, M)$ are written following the standard notation. Let us compute them explicitly for the simple case $j_1 = j_2 = 1/2$ using the method of highest weight. We start with the highest weight vector $|1/2, 1/2\rangle \otimes |1/2, 1/2\rangle$. It has $J_3 = +1$. Thus, it is the vector $|1, 1\rangle$ on the right-hand side of (5.81). With a standard choice of phases, we then write: $C(1/2, 1/2; 1/2, 1/2|1, 1) = 1$ and

$$|1/2, 1/2\rangle \otimes |1/2, 1/2\rangle = |1, 1\rangle$$

We then act on both sides with $J_- = (J_1)_- + (J_2)_-$. Using (5.77) we obtain for the action of J_- on the right: $J_- |1, 1\rangle = \sqrt{2} |1, 0\rangle$ and for that of $(J_1)_- + (J_2)_-$ on the left[11]

$$((j_1)_- + (j_2)_-)(|1/2, 1/2\rangle |1/2, 1/2\rangle) = |1/2, -1/2\rangle |1/2, 1/2\rangle + |1/2, 1/2\rangle |1/2, -1/2\rangle$$

so that

$$|1, 0\rangle = \frac{1}{\sqrt{2}} (|1/2, -1/2\rangle |1/2, 1/2\rangle + |1/2, 1/2\rangle |1/2, -1/2\rangle) \tag{5.82}$$

Acting once more on both sides with J_- we obtain

$$|1, -1\rangle = |1/2, -1/2\rangle |1/2, -1/2\rangle$$

while the orthogonal combination to (5.82) is

$$|0, 0\rangle = \frac{1}{\sqrt{2}} (|1/2, -1/2\rangle |1/2, 1/2\rangle - |1/2, 1/2\rangle |1/2, -1/2\rangle) \tag{5.83}$$

---

[11]To simplify the notation we omit the symbol $\otimes$ in various formulae if there is no danger of confusion.

It is straightforward to invert the above relations to express $|1/2, m_1\rangle\,|1/2, m_2\rangle$ in terms of the new basis $|J, M\rangle$. We obtain for the C–G coefficients

$$C(1/2, 1/2; 1/2, 1/2|1, 1) = 1, \quad C(1/2, -1/2; 1/2, -1/2|1, -1) = 1,$$

$$C(1/2, -1/2; 1/2, 1/2|1, 0) = 1/\sqrt{2}, \quad C(1/2, -1/2; 1/2, 1/2|0, 0) = 1/\sqrt{2},$$

$$C(1/2, 1/2; 1/2, -1/2|1, 0) = 1/\sqrt{2}, \quad C(1/2, 1/2; 1/2, -1/2|0, 0) = -1\sqrt{2}$$

This calculation can be repeated for any values of j_1 and j_2 and the result can be found in standard books of quantum mechanics or in math tables. The procedure can also be generalised to the tensor product of three or more representations.

5.5.2.6 Tensor operators. A set of $2j+1$ operators $\mathcal{O}_m^{(j)}, m = -j, \ldots, +j$ are defined to be the components of an irreducible tensor operator in the j representation of $SU(2)$, or with spin-j for short, if they satisfy the commutation relations

$$\left[J_a, \mathcal{O}_m^{(j)}\right] = \mathcal{O}_{m'}^{(j)} \left(J_a^{(j)}\right)_{m'm} \tag{5.84}$$

where $(J_a^{(j)})_{m'm}$ are the matrix elements of the generators J_a of $SU(2)$ in the j representation.

It is easy to show that this definition is equivalent to

$$U(\boldsymbol{\theta})\,\mathcal{O}_m^{(j)}\,U^\dagger(\boldsymbol{\theta}) = \mathcal{O}_{m'}^{(j)}\,(U^{(j)}(\boldsymbol{\theta}))_{m'm} \tag{5.85}$$

with $U(\boldsymbol{\theta})$ the unitary representation of the element $\boldsymbol{\theta}$ of $SU(2)$.

The Wigner–Eckart theorem. This is an important theorem, which makes possible the extraction of physical information on the basis of symmetry alone with no reference to the specific dynamics.

The theorem states that the matrix elements of a tensor operator $\mathcal{O}_M^{(J)}$ of spin-J between two $SU(2)$ eigenstates $|j_1, m_1\rangle$ and $|j_2, m_2\rangle$ is of the form

$$\langle j_2 m_2|\,\mathcal{O}_M^{(J)}\,|j_1 m_1\rangle = (-1)^{J+j_2-j_1}(2j_2 + 1)^{-1/2}\,\langle j_2||\mathcal{O}^{(J)}||j_1\rangle\,C(j_2 m_2; JM|j_1 m_1) \tag{5.86}$$

$\langle j_2||\mathcal{O}^{(J)}||j_1\rangle$ is the so-called *reduced matrix element*, which depends only on j_1, j_2, J and the tensor operator $\mathcal{O}^{(J)}$. All dependence on M, m_1 and m_2 is encoded in the C–G coefficients.

Consider the $(2J + 1)(2j_1 + 1)$ states $\mathcal{O}_M^{(J)}\,|j_1, m_1\rangle$. Under $SU(2)$ they transform in exactly the same way as the states $|J, M\rangle \otimes |j_1, m_1\rangle$ of the tensor product of the representations J and j_1. This follows from the identical action of the generators J_a on these two sets of states. For instance, the action of J_3 gives

$$J_3 \mathcal{O}_M^{(J)}\,|j_1, m_1\rangle = \left[J_3, \mathcal{O}_M^{(J)}\right]|j_1, m_1\rangle + \mathcal{O}_M^{(J)} J_3\,|j_1, m_1\rangle = (M + m_1)\mathcal{O}_M^{(J)}\,|j_1, m_1\rangle$$

and is identical to the $J_3\,|J, M\rangle \otimes |j_1, m_1\rangle = (M + m_1)\,|J, M\rangle \otimes |j_1, m_1\rangle$. Similarly, with $J_\pm$. The state $\mathcal{O}_J^{(J)}\,|j_1 j_1\rangle$ is the highest weight state, being annihilated by J_+. Acting on it successively with J_-, we construct the ladder of the $2(J + j_1) + 1$ states in the $J + j_1$

representation. Then, starting from the state orthogonal to the $|J+j_1, J+j_1-1\rangle$, which is the highest weight state in the $J + j_1 - 1$ representation and acting on it successively with J_-, we build the tower of the $2(J + j_1 - 1) + 1$ states of the representation $J + j_1 - 1$. And so on. We end up with

$$\widetilde{\mathcal{N}}_j \, |j, m = M + m_1\rangle = \sum_{M=-J}^{+J} \sum_{m_1=-j_1}^{+j_1} C(JM; j_1 m_1 | j, m = M + m_1) \, \mathcal{O}_M^{(J)} |j_1 m_1\rangle \quad (5.87)$$

with $\widetilde{\mathcal{N}}_j$ a normalisation constant, which depends on $\mathcal{O}^J$ and on j, J and j_1, but does not depend on M, m_1 and m_2.

Making use of the orthonormality of the Clebsch–Gordan coefficients, we can invert this relation to obtain

$$\mathcal{O}_M^{(J)} |j_1 m_1\rangle = \sum_{j=|J-j_1|}^{J+j_1} \mathcal{O}_j \, C(j_1 m_1; JM | jm) \, |jm\rangle \quad (5.88)$$

Finally, taking the inner product with $\langle j_2 m_2|$ and writing $\mathcal{N}_j = (-1)^{J+j_2-j_1}$ $(2j_2 + 1)^{-1/2} \langle j_2 || \mathcal{O}^{(J)} || j_1 \rangle$ in order to follow standard convention, we end up with (5.86).

We shall use this theorem and its generalisation to higher groups many times in the following chapters. In order to assess its power, notice that for given j_1, j_2, J and operator $\mathcal{O}$, on the left-hand side of (5.86) we have $(2j_1+1)(2j_2+1)(2J+1)$ quantities. Formula (5.86) tells us that all these quantities are given by the corresponding C–G coefficient, up to just one overall common coefficient, the reduced matrix element, which in physics applications depends on the dynamics and cannot be determined by symmetry alone.

In particular, we conclude that spherical symmetry implies that the matrix elements of any vector operator $\boldsymbol{V}$ between angular momentum eigenstates are proportional to the matrix elements between the same states of, e.g. the angular momentum operator $\boldsymbol{J}$, i.e.

$$\langle j_2 m_2 | V_M^{(1)} |j_1 m_1\rangle = v(j_1, j_2) \langle j_2 m_2 | J_M |j_1 m_1\rangle \quad (5.89)$$

where $v(j_1, j_2)$ depends on the specific operator $\boldsymbol{V}$, the angular momenta of the two states and all quantum numbers necessary to specify the two states involved, but not on m_1, m_2 and M.

5.5.3 The group $O(3)$

It is the group which is isomorphic to that of 3×3 orthogonal matrices R. It can be viewed as the group of rotations in $\mathbb{R}^3$, i.e. the group of transformations which leave invariant the three-dimensional Euclidean metric and one point of the space, the origin of the coordinate system, fixed. Thus, the matrices R satisfy

$$\delta_{ij} R_k^i R_l^j = \delta_{kl} \quad (5.90)$$

or, equivalently,

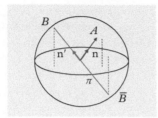 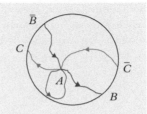

Fig. 5.1 Left: The parameter space of SO(3). Right: Closed loops in SO(3). The loop $(AB\bar{B}A)$ can be continuously deformed to the $(AC\bar{C}A)$. Viewed as one (combined) closed loop, the loop $(AB\bar{B}AC\bar{C}A)$ is contractible to a point.

$$R^T R = 1 = R\,R^T \tag{5.91}$$

with R^T the transpose of R. Given that $\det R^T = \det R$, we conclude that $\det R = \pm 1$. Any orthogonal matrix with $\det R = -1$ can be written as the product of the matrix $-\mathbf{1}$, which represents *space inversion*, times an orthogonal matrix with $\det R = +1$, which describes a so-called *proper rotation*. The latter form a subgroup of $O(3)$ denoted by $SO(3)$. Since there is no continuous path lying entirely in $O(3)$, which connects two elements with determinant $+1$ and -1 respectively, it follows that $O(3)$ is *disconnected*.

5.5.3.1 The general rotation matrix. A general rotation is described by a vector $\boldsymbol{\alpha} \equiv \boldsymbol{n}\,\alpha$. The unit vector $\boldsymbol{n}$ determines the axis of rotation and α the rotation angle in the direction fixed by convention by the right-hand rule. The matrix $R(\boldsymbol{\alpha})$ which represents this generic rotation is obtained by the succession of $N \to \infty$ infinitesimal rotations around $\boldsymbol{n}$, each by an angle $\delta\alpha = \alpha/N$. Under such an infinitesimal rotation the point $\boldsymbol{x}$ changes by

$$\delta\boldsymbol{x} = -\delta\alpha\,\boldsymbol{n} \times \boldsymbol{x} \tag{5.92}$$

and the infinitesimal rotation matrix is obtained from $x'^i = x^i + \delta x^i = R^i_j\,x^j$. Comparing with (5.92) we conclude that the matrix R is given by

$$R = 1 + \mathrm{i}\frac{\alpha}{N}\,J_i\,n^i \tag{5.93}$$

where the jk element of the matrix J_i is [12]

$$(J_i)_{jk} = -\mathrm{i}\,\epsilon_{ijk} \tag{5.94}$$

Thus, the matrix which represents the generic finite rotation $\boldsymbol{\alpha}$ is given by

$$R(\boldsymbol{\alpha} = \boldsymbol{n}\,\alpha) = \lim_{N\to\infty}\left(1 + \mathrm{i}\frac{\alpha}{N}\,J_i\,n^i\right)^N = \mathrm{e}^{\mathrm{i}\,\alpha\,\boldsymbol{n}\cdot\boldsymbol{J}} = \mathrm{e}^{\mathrm{i}\,\boldsymbol{\alpha}\cdot\boldsymbol{J}} \tag{5.95}$$

---

[12]The imaginary unit i is inserted here for later convenience.

5.5.3.2 ***SO(3)*** *is doubly connected.* The parameter space $(\mathbf{n}, \alpha)$ of $SO(3)$ rotations is a solid ball with radius π, as shown in Figure 5.1 (left). The centre $(\alpha = 0)$ corresponds to the identity element, while any point inside the ball defines a vector, which connects it to the origin. Its direction gives the axis of rotation and its magnitude the rotation angle. However, notice that the rotation $(\mathbf{n}, \pi)$ is identical to $(-\mathbf{n}, \pi)$. This means that any two antipodal points on the surface of the ball are identified. Thus, as shown in Figure 5.1 (right), there are two kinds of laces (closed loops) in $SO(3)$ beginning from and ending at any point A inside the ball: One which starts at A and returns to A without hitting the surface. Any such loop can be continuously deformed to the trivial loop, which never leaves A. The other starts at A, hits the surface at a point e.g. B and returns to A from the antipodal point $\bar{B}$ of B. This loop cannot be deformed to the trivial loop in a continuous manner. A loop which hits the surface successively at several points before reaching A can be continuously deformed to the one in which all these points are brought to coincide with B and $\bar{B}$ respectively. For example, in Figure 5.1 (right) we can bring C to $\bar{B}$ and $\bar{C}$ to B. Therefore we conclude that any closed loop which hits the surface and continues from its antipode an even (odd) number of times can be continuously deformed to any other of the same kind and, consequently, can (cannot) be deformed to the trivial loop. Thus, the closed loops form two distinct equivalence classes and, correspondingly, the group $SO(3)$ is doubly connected.

5.5.3.3 *Tensors of* ***SO(3)***. We have seen already in the course of geometry (or classical mechanics) the elements of tensor calculus in three-dimensional Euclidean space. Here we briefly review the relevant concepts.

• A quantity S is a *scalar* if it is invariant under rotations, i.e. iff under rotations $S' = S$. For example, the quantity $l^2 \equiv x_1^2 + x_2^2 + x_3^2$ is a scalar.

• Any three quantities $V^i = V_i, i = 1, 2, 3$ are said to be the components of a *vector* if under a rotation R they transform in the same way as x_i, i.e. iff $V'_i = R_{ij}V_j$. An example of a vector other than x_i is the momentum $p_i = mv_i$ of a particle.

• Generalising further, the 3^n quantities $T_{i_1 i_2 \dots i_n}$ are the components of an *n-index tensor*, if their transformation rule under rotations is

$$T'_{i_1 i_2 \dots i_n} = R_{i_1 j_1} R_{i_2 j_2} \dots R_{i_n j_n} T_{j_1 j_2 \dots j_n} . \tag{5.96}$$

Definition. Tensors of $SO(3)$ which change sign under the spatial inversion $-\mathbf{1} \in O(3)$ are given the name *pseudo-tensors*.

Definition. If the values of the components $T_{ij\dots}$ of a tensor (pseudo-tensor) do not change under proper rotations, the tensor is called an *invariant tensor (pseudo-tensor)*.

• *Invariant tensors of* $SO(3)$. (a) The orthogonality property (5.90) of the generic rotation matrix R shows that the quantities δ_{ij} are the components of an invariant two-index tensor.

Using the invariant tensor δ_{ij} and its inverse δ^{ij}, we can lower and raise the indices of any tensor. The fact that δ_{ij} and δ^{ij} are the unit matrices, means that there is no difference between the upper and the lower indices. So, in our discussion of $SO(3)$ tensors in this section, we use upper and lower indices interchangeably.

(b) Similarly, the 27 components ϵ_{ijk} of the fully antisymmetric symbol form an invariant three-index pseudo-tensor (see Problem 5.2).

5.5.3.4 Tensor decomposition. Irreducible tensors. Consider a two-index tensor T_{ij}. It has nine components. It can be decomposed into a symmetric and traceless part S, an antisymmetric part A and the trace T_{kk}, with five, three and one independent components, respectively, as follows

$$T_{ij} = \frac{1}{2}(T_{ij} + T_{ji}) - \frac{1}{3}\delta_{ij}T_{kk} + \frac{1}{2}(T_{ij} - T_{ji}) + \frac{1}{3}\delta_{ij}T_{kk}$$

$$\equiv S_{ij} + A_{ij} + \frac{1}{3}\delta_{ij}T_{kk} \tag{5.97}$$

The new objects S, A and T_{kk} have the following properties: (a) The trace T_{kk} of T is a scalar under $O(3)$. Indeed, $T'_{kk} = R_{kl}R_{km}T_{lm} = (R^T R)_{lm}T_{lm} = \delta_{lm}T_{lm} = T_{mm}$.

(b) The A_{ij} transform linearly among themselves under $O(3)$ and form a two-index antisymmetric tensor, which is equivalent to a pseudo-vector.

Indeed, defining $\mathcal{A}_i \equiv \epsilon_{ijk}A_{jk}$, and making use of the properties of the ϵ-symbol, we verify that $\mathcal{A}'_i = (\det R)R_{ik}\mathcal{A}_k$.

(c) The five independent quantities $S_{ij} \equiv (1/2)(T_{ij} + T_{ji}) - (1/3)\delta_{ij}T_{kk}$, form a symmetric traceless two-index tensor.[13]

Thus, under the action of the rotation group the five components of S, the three components of A and the one component T_{kk} transform independently. We say that the generic two-index tensor decomposes into a sum of three irreducible tensors of ranks 2, 1 and 0 respectively. We write

$$\mathbf{3} \otimes \mathbf{3} \cong \mathbf{5} \oplus \mathbf{3} \oplus \mathbf{1}$$

and, as will become clear soon, this is an example of the general reduction formula (5.58).

The same procedure applies to the decomposition of higher tensor quantities. The operations of symmetrisation, anti symmetrisation and tracing give seven-, nine-, etc. dimensional irreducible representations of the rotation group.

*5.5.3.5 The Lie algebra of **SO(3)**.* The group elements of $SO(3)$ depend on three independent variables, the three rotation angles and therefore the Lie algebra $\mathfrak{so}(3)$ has three generators. On the other hand we know from the Cartan classification that there exists only one independent three-dimensional Lie algebra, the one we studied in association with the group $SU(2)$. It follows that we have an isomorphism $\mathfrak{su}(2) \cong \mathfrak{so}(3)$. In Problem 5.3, we ask the reader to verify this by explicit calculation. This is another example of the statement that two different groups may have identical Lie algebras.

We have studied the representations of this algebra and have found that they are labelled by an index j which can take integer, or half-integer, values. Which of these representations will give, by exponentiation, representations of $SO(3)$? The group $SO(3)$ is characterised by the fact that a rotation by 2π is identical to no rotation. This property should be preserved by all its representations. This means that the rotation matrix in any representation should satisfy $\exp\{2i\pi\, \boldsymbol{n} \cdot \boldsymbol{J}\} = 1$. Applied to rotations around the z-axis, it implies that J_3 should have integer eigenvalues. Thus,

---

[13]If T is a pseudo-tensor, the corresponding S and T_{kk} are, respectively, pseudo-tensor and pseudo-scalar, while $\mathcal{A}$ is a vector.

of all representations of the algebra, only the ones corresponding to integer j yield by exponentiation representations of $SO(3)$. In order to obtain this result we have to use the global property of $SO(3)$ and identify a rotation of 2π with the identity.

*5.5.3.6 The relation of **SO(3)** and **SU(2)**.* As shown above, the groups $SO(3)$ and $SU(2)$ are locally isomorphic. Via exponentiation we obtain the general elements u and R of these groups, which are both parametrised by a direction $\boldsymbol{n}$ and an angle α. So the mapping $u(\boldsymbol{n}\alpha) \to R(\boldsymbol{n}\alpha)$ from $SU(2)$ to $SO(3)$, which preserves the group property, is a group homomorphism. This mapping is not one-to-one. The elements of $SU(2)$ with $\alpha = 0$ and $\alpha = 2\pi$ are both mapped to the identity element of $SO(3)$. Thus, the centre of the mapping is $\{I, -I\} \cong Z_2$ and we conclude that $SO(3)$ is the quotient group

$$SO(3) \cong SU(2)/Z_2 \tag{5.98}$$

We say that $SU(2)$, which, being isomorphic to S^3, is simply connected, is the *double covering group* of $SO(3)$. In physics, these considerations are translated into the fact that, if we take into account the global properties of $SO(3)$, we find only integer–j representations, which describe bosons, and we must look at the Lie algebra in order to find half-integer ones describing fermions. This is related to the well-known fact that fermion wave functions change sign under a 2π rotation. On those the identity is equivalent to a 4π rotation.

Let us give an example of the relation between $O(3)$ and $SU(2)$, which, historically, was the first discovery of the transformation laws of objects we call today "spinors". Let $\boldsymbol{x}$ be a vector $\boldsymbol{x} \in \mathbb{R}^3$. Under a rotation $\boldsymbol{\alpha} = \boldsymbol{n}\alpha$ it transforms to $x_i' = R(\boldsymbol{\alpha})_{ij}x_j$. We can map $\mathbb{R}^3$ into the space of 2×2 Hermitian and traceless matrices by writing

$$\boldsymbol{x} \to \tilde{x} = \sum_{i=1}^{3} x_i \sigma_i \tag{5.99}$$

The element $u(\boldsymbol{\alpha})$ of $SU(2)$ satisfies $u(\boldsymbol{\alpha})\sigma_i u^\dagger(\boldsymbol{\alpha}) = \sigma_k R(\boldsymbol{\alpha})_{ki}$, which implies that the action of rotations on $\tilde{x}$ is given by

$$\tilde{x}' = u(\boldsymbol{\alpha})\tilde{x}u^\dagger(\boldsymbol{\alpha}), \text{ with } R(\boldsymbol{\alpha})_{ji} = \frac{1}{2}\operatorname{Tr}\big(u(\boldsymbol{\alpha})\sigma_i u^\dagger(\boldsymbol{\alpha})\sigma_j\big) \tag{5.100}$$

and establishes the two-to-one relation of $SU(2)$ and $SO(3)$.

5.5.4 The group $SU(3)$

In this chapter we shall describe the main properties of the group $SU(3)$ and its representations. As will become clear in the following chapters, $SU(3)$ is related to a symmetry of fundamental importance in particle physics. Here we shall present its mathematical structure.

$SU(3)$ is the group which is isomorphic to that of unitary 3×3 matrices of determinant equal to $+1$. Thus, the general element u satisfies the conditions

$$u^\dagger u = \mathbf{1} = uu^\dagger \quad \text{and} \quad \det u = 1 \tag{5.101}$$

and, consequently, is characterised by eight real parameters.

*5.5.4.1 The representations of **SU(3)**.* To construct the irreducible representations of $SU(3)$ and the corresponding Clebsch–Gordan decomposition rules, we shall follow the tensor method used for $SU(2)$. Modulo a few special features of $SU(2)$ and the combinatoric complications as we move on to higher groups and representations, the method is applicable to all members of the unitary series and even beyond.

• *The trivial (singlet) representation **1**.* As usual, this is the representation in which all elements of the group are represented by the number 1.

• *The defining (**3**) representation.* Three complex quantities $\psi^i, i = 1, 2, 3$, are said to be the components of a *triplet* (**3**) of $SU(3)$, if they transform according to

$$\psi^{i\,'} = u^i_j\,\psi^i \quad (\psi' = u\psi) \; \forall u \in SU(3) \tag{5.102}$$

• *The conjugate **$\bar{3}$** of the defining representation.* Three complex quantities $\chi_i, i = 1, 2, 3$, are defined to be the components of an *anti-triplet* **$\bar{3}$** of $SU(3)$, if they transform according to

$$\chi_i{}' = u^{\dagger\,j}_i\,\chi_j \quad (\chi' = \chi u^\dagger) \forall u \in SU(3) \tag{5.103}$$

An immediate consequence of this definition is that, if ψ^i is a triplet, the complex conjugates ψ^{i*} form an anti-triplet, since then, $\psi^\dagger$, whose components are exactly the ψ^{i*}, indeed transform according to $\psi^{\dagger\,'} = \psi^\dagger u^\dagger$. Thus, the consistent notation of the components of $\psi^\dagger$ is $\psi^\dagger_i$ with one lower index.

Also, notice that given two triplets ψ and χ, the quantities $\psi^\dagger \chi = \psi^{i*} \chi^i$ and its Hermitian conjugate $\chi^\dagger \psi$ are invariant (singlets) under $SU(3)$.

• *The invariant tensor δ^i_j.* The unitarity of the elements $u \in SU(3)$ is expressed by the relation

$$u^i_k\,\delta^k_l\,(u^\dagger)^l_j = \delta^i_j$$

and is equivalent to the statement that the Kronecker δ is an invariant tensor of $SU(3)$ with one upper and one lower index.

• *The invariant fully antisymmetric Levy-Civita symbols.* The three-index fully antisymmetric symbols are defined as usual by

$$\epsilon_{ijk} = -\epsilon_{jik} = -\epsilon_{ikj}, \epsilon_{123} = +1 \;\; \text{and} \;\; \epsilon^{ijk} = -\epsilon^{jik} = -\epsilon^{ikj}, \epsilon^{123} = +1 \tag{5.104}$$

The conditions $\det u = 1 = \det u^\dagger$ for $u \in SU(3)$, which are

$$\epsilon^{ijk} u^l_i u^m_j u^n_k = \epsilon^{lmn} \quad \text{and} \quad \epsilon_{ijk} u^i_l u^j_m u^k_n = \epsilon_{lmn} \tag{5.105}$$

imply that ϵ^{ijk} is a three-index invariant tensor of $SU(3)$, with ϵ_{ijk} its conjugate.

• *$\bar{3}$ is not equivalent to 3.* Unlike the situation in $SU(2)$, in $SU(3)$ the defining representation is not equivalent to its conjugate. There is no similarity transformation between the generic $u \in SU(3)$ and its complex conjugate.

• *General tensors of $SU(3)$.* Taking products of **3**s and **$\bar{3}$**s we can now build higher-dimensional irreducible representations of $SU(3)$. The procedure is analogous to the one we followed in $SU(2)$, but as we may expect, the resulting structure is richer, due to the inequivalence of **3** and **$\bar{3}$**.

Let us start with two **3**s ψ^i and χ^j. The nine quantities $\psi^i\chi^j$ are the components of a two upper index tensor. However, the nine-dimensional representation carried by this tensor is reducible. Exactly like in $SU(2)$ we split $\psi^i\chi^j$ into its symmetric and antisymmetric parts, which transform irreducibly under $SU(3)$, writing

$$\psi^i\chi^j = \frac{1}{2}(\psi^i\chi^j + \psi^j\chi^i) + \frac{1}{2}(\psi^i\chi^j - \psi^j\chi^i)$$

The six components of the symmetric piece transform among themselves according to the irreducible representation **6** of $SU(3)$. But the antisymmetric piece has three components, which also transform among themselves. Using the invariant tensor ϵ^{ijk}, we can write

$$\frac{1}{2}(\psi^i\chi^j - \psi^j\chi^i) = \epsilon^{ijk}h_k$$

in terms of a new object h, which carries one lower index and, consequently, transforms as a $\bar{\mathbf{3}}$. We express the above reduction of the initial $\mathbf{3} \otimes \mathbf{3}$ representation to its irreducible pieces as

$$\mathbf{3} \otimes \mathbf{3} \cong \mathbf{6} \oplus \bar{\mathbf{3}}$$

Another simple example is provided by the tensor product of a **3** and a $\bar{\mathbf{3}}$. Indeed, consider a tensor with the transformation properties of $\mathbf{3} \otimes \bar{\mathbf{3}}$. It has the form T^i_j. Making use of the invariant tensor δ^i_j we can split its nine components to its traceless piece $\widetilde{T}^i_j$ and the trace T, according to

$$T^i_j = T^i_j - \frac{1}{3}\delta^i_j T^k_k + \frac{1}{3}\delta^i_j T^k_k \equiv \widetilde{T}^i_j + \frac{1}{3}\delta^i_j T$$

No further reduction is possible. The eight components of $\widetilde{T}$ and the singlet T transform independently of one another, forming an eight-dimensional irreducible representation of $SU(3)$ and a singlet, respectively. We describe this reduction writing

$$\mathbf{3} \otimes \bar{\mathbf{3}} \cong \mathbf{8} \oplus \mathbf{1} \tag{5.106}$$

Two useful lessons we learn from these two examples are: (a) that using the invariant tensor δ^i_j we can contract one upper and one lower index, thus reducing the number of both upper and lower indices of a tensor by one, and (b) using the ϵ^{ijk} (ϵ_{ijk}) we can replace two antisymmetric upper (lower) indices of a tensor by one lower (upper). Thus, for the discussion of the induced representations of $SU(3)$ on the space of tensors we can restrict ourselves to tensors fully symmetric in the number of upper and lower indices and traceless. The representation induced on the tensors with symmetric n upper indices and m lower indices is labelled by the ordered pair (n, m). In this notation the representations we encountered above are $\mathbf{3} \sim (1,0)$, $\bar{\mathbf{3}} \sim (0,1)$, $\mathbf{6} \sim (2,0)$ and $\mathbf{8} \sim (1,1)$. According to its definition complex conjugation interchanges upper and lower indices. Thus, we immediately conclude that[14]

---

[14]We adopt the convention of using a bar instead of a star to denote the complex conjugate of a representation.

$$\overline{(n,m)} = (m,n)$$

- *To summarise: Tensors and tensor operators.* The quantities $T^{(i_1 i_2 ... i_n)}_{(j_1 j_2 ... j_m)}$, with each index taking the values $1, 2, 3$, are the components of an (n,m) tensor of $SU(3)$ if and only if (a) they are symmetric in the upper and in the lower indices, (b) they are traceless with respect to one upper and one lower index and (c) under $SU(3)$ they transform according to

$$T'^{(i_1 i_2 ... i_n)}_{(j_1 j_2 ... j_m)} = u^{i_1}_{k_1} ... u^{i_n}_{k_n} T^{(k_1 ... k_n)}_{(l_1 ... l_m)} u^{\dagger l_1}_{j_1} ... u^{\dagger l_m}_{j_m}, \qquad \forall u \in SU(3) \qquad (5.107)$$

In physics applications we are interested in operators acting on the Hilbert space of quantum states. In the presence of symmetries, both the states and the operators should be classified in representations of the relevant symmetry groups.

For $SU(3)$ in particular, an (n,m) *tensor operator* is defined in the same way as the (n,m) tensor in the previous paragraph. Namely, we say that the operators $\hat{O}^{(i_1 i_2 ... i_n)}_{(j_1 j_2 ... j_m)}$ transform according to its (n,m) representation, if and only if they are symmetric and traceless, and transform according to

$$\hat{O}'^{(i_1 i_2 ... i_n)}_{(j_1 j_2 ... j_m)} = u^{i_1}_{k_1} ... u^{i_n}_{k_n} \hat{O}^{(k_1 ... k_n)}_{(l_1 ... l_m)} u^{\dagger l_1}_{j_1} ... u^{\dagger l_m}_{j_m}, \qquad \forall u \in SU(3) \qquad (5.108)$$

- *The dimension of the representation* (n,m). Consider the representation induced on the product space of a tensor $(n,0)$ with a $(0,m)$. It is reducible, because it is not traceless. It is written as the traceless (n,m) part, plus the trace over one pair of indices - because of the symmetry of the upper and lower indices any pair is equivalent to any other - which removes one upper and one lower index. Thus,

$$(n,0) \otimes (0,m) \cong (n,m) \oplus \big((n-1,0) \otimes (0,m-1) \big)$$

from which we conclude that

$$\dim(n,m) = \dim \big((n,0) \otimes (0,m) \big) - \dim \big((n-1,0) \otimes (0,m-1) \big)$$

The dimension of $(n,0)$ is equal to the number of ways we can place n identical objects in three boxes carrying the labels 1, 2 and 3, respectively. We can place all n objects in 1, or $n-1$ objects in 1 and one in 2 or 3, etc. Counting all these we obtain $1 + 2 + 3 + ... + (n+1) = (n+1)(n+2)/2$. Consequently, the dimension of the representation induced on the space of traceless tensors with n upper and m lower symmetric indices is

$$\dim(n,m) = \frac{1}{2}(n+1)(m+1)(n+m+2)$$

Let us now state the following important theorem:[15]

---

[15]For the proof, see, for example, Sidney Coleman "Aspects of Symmetry"; Cambridge University Press.

Theorem: The representations (n, m) for all positive or zero values of n and m form the complete set of inequivalent irreducible representations of $SU(3)$.

That is, any two representations (n, m) and (n', m'), with $n \neq n'$ and/or $m \neq m'$, are inequivalent and, furthermore, any reducible representation of $SU(3)$ can be decomposed in a direct sum of such representations.

5.5.4.2 Tensor product representation. Clebsch–Gordan decomposition. Consider the direct product $(n, m) \otimes (n', m')$ of two irreducible representations of $SU(3)$. It is a representation induced on the space of tensors with $n + n'$ upper indices, $m + m'$ lower ones, symmetric separately over the n, the n', the m and the m' of them, and vanishing when one of the $n(n')$ indices is contracted with one of the $m(m')$. There is no symmetrisation among the n and n' upper indices, or among the lower m and m' ones. It is convenient to describe its decomposition in two steps:

First, denote by $(n, n'; m, m')$ the representation induced on the set of the above tensors with all traces removed, including the ones involving one of the $n(n')$ indices with one of the $m'(m)$. Removing one by one these remaining traces, until we run out of indices to contract, we obtain the decomposition

$$(n, m) \otimes (n', m') \cong \sum_{i=0}^{\min(n, m')} \sum_{j=0}^{\min(n', m)} \oplus(n - i, n' - j; m - j, m' - i) \qquad (5.109)$$

Second, we need to decompose the generic $(n, n'; m, m')$ into (N, M)s. To this end, we have to complete the symmetrisations over all the $n + n'$ upper and all the $m + m'$ lower indices. To be explicit, consider the tensor

$$\epsilon_{i_1 i_{n+1} k} T^{i_1 \cdots i_n i_{n+1} \cdots i_{n+n'}}_{j_1 \cdots j_m j_{m+1} \cdots j_{m+m'}}$$

with two fewer upper and one more lower indices. Note that this tensor is symmetric over any pair of its lower indices. Indeed, contracting e.g. with an $\epsilon^{j_1 j_{n+1} l}$ we obtain

$$\epsilon^{j_1 j_{n+1} l} \epsilon_{i_1 i_{n+1} k} T^{i_1 \cdots i_n i_{n+1} \cdots i_{n+n'}}_{j_1 \cdots j_{m+m'}} = \left(\delta^{j_1}_{i_1} \delta^{j_{n+1}}_{i_{n+1}} \delta^k_l \pm \text{permutations} \right) T^{i_1 \cdots i_n i_{n+1} \cdots i_{n+n'}}_{j_1 \cdots j_{m+m'}} = 0$$

as a consequence of the tracelessness of the tensors in $(n, n'; m, m')$. So, the antisymmetrisation of two upper indices leads automatically to a tensor in the irreducible representation of the type (N, M) with the two antisymmetrised upper indices replaced by one lower one. Similarly, with the antisymmetrisation of two lower indices. We continue this procedure for both upper and lower indices until we run out of pairs to (anti-)symmetrise. We, thus, end up with the decomposition series

$$(n, n'; m, m') \cong (n + n', m + m') \oplus \sum_{i=1}^{\min n, n'} (n + n' - 2i, m + m' + i)$$

$$\oplus \sum_{j=1}^{\min(m, m')} (n + n' + j, m + m' - 2j) \qquad (5.110)$$

The combination of (5.109) and (5.110) gives the decomposition of the product of any two representations of $SU(3)$.

Examples.

1.
$$(1,0) \otimes (1,0) \cong (1,1;0,0) \cong (2,0) \oplus (0,1)$$

derived before in the form $\mathbf{3} \otimes \mathbf{3} \cong \mathbf{6} \oplus \bar{\mathbf{3}}$.

2.
$$(2,0) \otimes (1,0) \cong (2,1;0,0) \cong (3,0) \oplus (1,1)$$

or, using the alternative notation, $\mathbf{6} \otimes \mathbf{3} \cong \mathbf{10} \oplus \mathbf{8}$.

In the tensor language this is described as

$$\psi^{ij}\chi^k = \frac{1}{3}(\psi^{ij}\chi^k + \psi^{ik}\chi^j + \psi^{kj}\chi^i) + \frac{1}{3}(\psi^{ij}\chi^k - \psi^{ik}\chi^j) + \frac{1}{3}(\psi^{ij}\chi^k - \psi^{kj}\chi^i)$$

$$= \psi^{(ij}\chi^{k)} + \frac{1}{3}(\epsilon^{jkl}\delta^i_m + \epsilon^{ikl}\delta^j_m)\Psi^m_l \qquad (5.111)$$

with $\Psi^m_l \equiv \epsilon_{jkl}(\psi^{mj}\chi^k - \psi^{mk}\chi^j)$. The three-index fully symmetric piece is the $(3,0)$ and the Ψ^m_l, which is traceless, is the $(1,1)$.

5.5.4.3 Young tableaux for SU(3).   The pictorial method of the Young tableaux we introduced for $SU(2)$ applies also to $SU(3)$, as well as to all the higher unitary groups.[16] We will only give the rules using some specific examples. Our presentation will proceed in two steps. First we shall associate a Young tableau to each irreducible representation (m,n) of $SU(3)$ and explain how to compute its dimension, and second we shall give the rules about how to combine two such tableaux into a direct sum of irreducible terms.

Associate representations to tableaux and vice versa. Consider the representation (m,n) of $SU(3)$. As we know, it is induced on traceless tensors of the form

$$\psi^{(i_1 i_2 \ldots i_m)}_{(j_1 j_2 \ldots j_n)}$$

which, in addition, are symmetric in their upper and lower indices. Contracting with ϵ tensors we can replace each lower index by a pair of antisymmetric upper indices. Thus, equivalently, the representation is induced on tensors of the form

$$\psi^{(i_1 i_2 \ldots i_m)}_{(j_1 j_2 \ldots j_n)} \rightarrow \psi^{([k_1 l_1][k_2 l_2] \ldots [k_n l_n])(i_1 i_2 \ldots i_m)}$$

symmetric under the exchange of any two antisymmetric pairs.

Rule 1: To this representation we associate the Young tableau with n vertical columns with two boxes each and m horizontal boxes on their right. The mapping of tableaux to representations is clearly one-to-one.

For example, the tableau associated to the representation $(m,n) = (3,2)$, is

$$\psi^{i_1 i_2 i_3}_{j_1 j_2} \quad \rightarrow \quad \psi^{[k_1 l_1][k_2 l_2] i_1 i_2 i_3} \quad \rightarrow \quad$$ $$\qquad (5.112)$$

Comments:

---

[16]The application to the groups of the other series is possible but requires modifications.

• The m single boxes on the right correspond to the symmetric upper indices of the tensor, while the n vertical pairs of boxes correspond to the n antisymmetric pairs, which replaced the original n lower indices.

• In $SU(3)$ we cannot have more than three anti symmetrised indices. Thus there is no tableau having columns with more than three boxes. We shall encounter such tableaux below at intermediate steps of e.g. Clebsch–Gordan decomposition, but they will be discarded.

• The three-box column corresponds to the trivial one-dimensional representation of $SU(3)$. We write:

 $\cong$.

where "·" stands for the trivial representation, a tensor with no free $SU(3)$ index.

• One or more three-box columns in a tableau can just be discarded. For example,

$\cong$

• When we refer to a "legal tableau" below, we mean a tableau in which the number of boxes in the rows does not increase as we move down the rows and the number of boxes in the columns does not increase as we move from left to right.

Rule 2: *Dimensionality from the tableau*. Any such tableau is associated to a representation whose dimensionality is the ratio N/D of two integers computed as follows:

The numerator N: is the product of the integers shown in the tableau

$$N = \boxed{\begin{array}{ccccc} 3 & 4 & 5 & 6 & 7 \\ 2 & 3 \end{array}} \qquad = 3 \times 4 \times 5 \times 6 \times 7 \times 2 \times 3$$

and which are fixed, starting with the 3 (for $SU(3)$) on the top left, increase by 1 as we move to the right and decrease by 1 as we move downwards.

The denominator D: is the product of the integers shown

$$D = \boxed{\begin{array}{ccccc} 6 & 5 & 3 & 2 & 1 \\ 2 & 1 \end{array}} \qquad = 6 \times 5 \times 3 \times 2 \times 1 \times 2 \times 1$$

The number inside each box is the total number of boxes to its right and below, including the box in question. Thus, the dimensionality of the representation $(3, 2)$ of this example is

$$\frac{N}{D} = \frac{3 \times 4 \times 5 \times 6 \times 7 \times 2 \times 3}{6 \times 5 \times 3 \times 2 \times 1 \times 2 \times 1} = 42$$

in agreement, of course, with the formula $(m+1)(n+1)(m+n+2)/2 = 42$.

Decompose a tensor product. We will describe the Clesch–Gordan decomposition of the product of two representations using a specific example.

Consider the decomposition of the product $\mathbf{8} \otimes \mathbf{8}$.

Step 1: We label the boxes in each row of the second tableau with the same letter, different from row to row

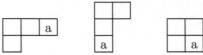

Step 2: Place in all possible ways the boxes of the second tableau, one by one on the first, starting with the boxes of the first row and going down the rows, making sure that at all steps the resulting tableau is legal, as defined previously.

(1) Thus, after we join in all possible ways the first "a" box onto the first tableau, we obtain

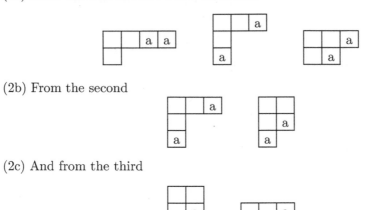

(2) We continue with the second "a" box, again in all possible ways. Since the tensor associated with a tableau is antisymmetric in the indices of the same column, *it is not allowed to put two "a"s in the same column*. Finally, *we count each arrangement of the resulting tableaux only once*.

Thus, the addition of the second "a" box on the first tableau leads successively to:

(2a) From the first tableau above we obtain

(2b) From the second

(2c) And from the third

(2a)+(2b)+(2c): Keeping the identical tableaux only once, the full list up to here is

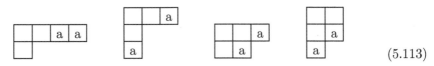

$$(5.113)$$

(3) Continue with the "b" box in the second row of the second original tableau. We shall place it in all possible positions, following the rules above, plus the additional one that *as we read a tableau down the rows, from right to left, at any moment the number of "a"s we meet should be larger than or equal to the number of "b"s*.

Let us apply all these rules when we add the "b" box on each tableau in (5.113). We obtain:

(3a) From the first

$$\cong \ (2,2) \cong \mathbf{27} \qquad\qquad \cong \ (3,0) \cong \mathbf{10}$$

(3b) From the second

$$\cong \ (1,1) \cong \mathbf{8}$$

(3c) From the third

$$\cong (1,1) \cong \mathbf{8} \qquad\qquad \cong \ (0,3) \cong \overline{\mathbf{10}}$$

(3d) Finally, from the fourth, the only new tableau we obtain is

$$\cong \ (0,0) \cong \mathbf{1}$$

We end up with

$$\mathbf{8} \otimes \mathbf{8} \cong \mathbf{27} \oplus \mathbf{10} \oplus \overline{\mathbf{10}} \oplus \mathbf{8} \oplus \mathbf{8} \oplus \mathbf{1} \tag{5.114}$$

5.5.4.4 *The Lie algebra* $\mathfrak{su}(3)$. According to our general discussion, the elements of $SU(3)$ can be obtained by exponentiation of the elements of the corresponding Lie algebra, the algebra of Hermitian and traceless 3×3 matrices. [17] A well-known basis in the linear space of these matrices are the eight Gell-Mann matrices $\lambda_a, a = 1, 2, \ldots, 8$. They are

$$\lambda_1 = \begin{pmatrix} 0 & 1 & 0 \\ 1 & 0 & 0 \\ 0 & 0 & 0 \end{pmatrix}, \quad \lambda_2 = \begin{pmatrix} 0 & -i & 0 \\ i & 0 & 0 \\ 0 & 0 & 0 \end{pmatrix}, \quad \lambda_4 = \begin{pmatrix} 0 & 0 & 1 \\ 0 & 0 & 0 \\ 1 & 0 & 0 \end{pmatrix},$$

$$\lambda_5 = \begin{pmatrix} 0 & 0 & -i \\ 0 & 0 & 0 \\ i & 0 & 0 \end{pmatrix}, \quad \lambda_6 = \begin{pmatrix} 0 & 0 & 0 \\ 0 & 0 & 1 \\ 0 & 1 & 0 \end{pmatrix}, \quad \lambda_7 = \begin{pmatrix} 0 & 0 & 0 \\ 0 & 0 & -i \\ 0 & i & 0 \end{pmatrix}, \tag{5.115}$$

$$\lambda_3 = \begin{pmatrix} 1 & 0 & 0 \\ 0 & -1 & 0 \\ 0 & 0 & 0 \end{pmatrix}, \quad \lambda_8 = \frac{1}{\sqrt{3}} \begin{pmatrix} 1 & 0 & 0 \\ 0 & 1 & 0 \\ 0 & 0 & -2 \end{pmatrix}$$

[17]With $u = \exp(iQ)$, the conditions $u^\dagger u = \mathbf{1}$ and $\det u = 1$ translate to $Q^\dagger = Q$ and $\mathrm{Tr}\,Q = 0$, respectively.

and the generators of the defining **3** representation of $SU(3)$, are

$$T^a = \frac{1}{2}\lambda_a, \quad a = 1, 2, \ldots, 8 \tag{5.116}$$

They are normalized to satisfy

$$\operatorname{Tr} T^a T^b = \frac{1}{2}\delta_{ab} \tag{5.117}$$

and the corresponding structure constants computed from

$$[T^a, T^b] = \mathrm{i} f^{abc} T^c \tag{5.118}$$

are fully antisymmetric, given explicitly by

$$f_{123} = 1, \ f_{147} = f_{246} = f_{257} = f_{345} = 1/2$$

$$f_{156} = f_{367} = -1/2, \ f_{458} = f_{678} = \sqrt{3}/2 \tag{5.119}$$

As seen explicitly in (5.115), $\mathfrak{su}(3)$ is a rank-two algebra. Let us call T_3 and T_8 the two generators which can be diagonalised simultaneously. The three basis vectors of the representation space are eigenvectors of T_3 and T_8. They are denoted as $|T_3, T_8\rangle$, labelled by the corresponding (T_3, T_8) eigenvalues and are shown here:

$$\begin{pmatrix} 1 \\ 0 \\ 0 \end{pmatrix} \to \left|1/2, \sqrt{3}/6\right\rangle, \ \begin{pmatrix} 0 \\ 1 \\ 0 \end{pmatrix} \to \left|-1/2, \sqrt{3}/6\right\rangle, \ \begin{pmatrix} 0 \\ 0 \\ 1 \end{pmatrix} \to \left|0, -\sqrt{3}/3\right\rangle \tag{5.120}$$

Their eigenvalues define three points on the $T_3 - T_8$ plane. They are the *weights* of the **3** representation of $SU(3)$ and are shown in the *weight diagram* Figure 5.2. [18]

By definition of the conjugate representation, the weights of the $\bar{\mathbf{3}}$ are opposite to the ones of the **3**. Its basis states are $|-1/2, -\sqrt{3}/6\rangle$, $|1/2, -\sqrt{3}/6\rangle$ and $|0, \sqrt{3}/3\rangle$.

5.5.4.5 The Cartan–Weyl basis, roots, simple roots, weight diagrams and all that. It is instructive to extend to $\mathfrak{su}(3)$ the method of raising and lowering operators we used to study the representations of $\mathfrak{su}(2)$. It will then become apparent how to extrapolate this method to all classical groups. It is essentially a change of basis in the space of the generators, which for $\mathfrak{su}(2)$ was $(J_1, J_2, J_3) \to (J_\pm, J_3)$.

We start with a change of the names of the commuting generators T_3 and T_8. The more or less standard names for them are $H_i, i = 1, 2$, with $H_1 \equiv T_3$ and $H_2 \equiv T_8$. They form a basis in the so-called *Cartan subalgebra* of $\mathfrak{su}(3)$

$$[H_i, H_j] = 0, \quad i, j = 1, 2 \tag{5.121}$$

The states in the representation space will be labelled by their simultaneous eigenvalues of H_i, called *weight vectors* of the representation. They will correspond to points in

---

[18]In general, the vectors with coordinates the eigenvalues of the Cartan generators in a given representation of a Lie algebra are called the *weights of the representation* and the corresponding graph analogous to Figure 5.2 its *weight diagram*.

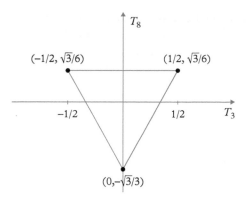

Fig. 5.2 The weight diagram of the representation **3**.

the (H_1, H_2) plane. The *weight diagram* of the **3** representation (5.118) is shown in Figure 5.2.

Then, in analogy with the relation $[J_3, J_\pm] = \pm J_\pm$, we ask for the so-called *step operators*, or *raising and lowering operators*, namely linear combinations $X = x_a T_a$ of the generators, which satisfy

$$[H_i, X] = \alpha_i\, X \qquad (5.122)$$

The eigenvalues $\boldsymbol{\alpha} = (\alpha_1, \alpha_2)$ of this system of equations are called *roots* of the algebra and according to the definition (5.38) of the adjoint representation, they are also the weight vectors of that representation.

Two solutions of (5.122) are obvious. They are $X = H_1$ and $X = H_2$. The corresponding roots are zero. Furthermore, a quick look at Figure 5.2 shows that the step generators, which, by definition, take us from one vertex of the triangle to the other, are expected to change the eigenvalues (H_1, H_2) of the states by

$$\boldsymbol{\alpha} = \pm(1, 0), \ \ \pm(1/2, \sqrt{3}/2) \ \ \text{and} \ \ \pm(1/2, -\sqrt{3}/2) \qquad (5.123)$$

Indeed, let's start with $H_1 = T_3$. The equation becomes $if_{3ab}x_a T_b = \kappa x_b T_b$. Multiplying both sides by T_c and taking the trace, we obtain the homogeneous linear system

$$(-if_{3ac} - \kappa \delta_{ac})\, x_c = 0$$

which has non-trivial solutions if and only if $\det(-if_{3ac} - \kappa \delta_{ac}) = 0$ or, using the structure constants (5.119),

$$\kappa^2(\kappa^2 - 1)\left(\kappa^2 - \frac{1}{4}\right)^2 = 0$$

The two zero eigenvalues correspond to the generators H_1, H_2, which commute with H_1. The generators corresponding to the eigenvalues $\kappa = \pm 1, \pm 1/2$ and $\kappa = \pm 1/2$ are proportional to $T_1 \pm iT_2$, $T_4 \pm iT_5$ and $T_6 \mp iT_7$, respectively.

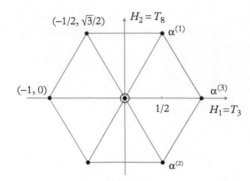

Fig. 5.3 The root system of $\mathfrak{su}(3)$. Their number is equal to the number of generators of the algebra. The two circles at the origin correspond to the two vanishing roots. The rest form a regular hexagon. By definition, it is the weight diagram of the adjoint representation **8**.

We repeat this procedure for $H_2 = T_8$. We are interested in solutions of $i f_{8ab} x_a T_b = \kappa x_b T_b$. The corresponding homogeneous linear system is $(-i f_{8ac} - \kappa \delta_{ac}) x_c = 0$, with eigenvalue equation

$$\det(-i f_{8ac} - \kappa \delta_{ac}) = 0$$

which, upon substitution of the structure constants, becomes

$$\kappa^4 \left(\kappa^2 - \frac{3}{4} \right)^2 = 0$$

The generators H_1, H_2 and $T_1 \pm i T_2$ correspond to the four zero eigenvalues, while the $T_4 \pm i T_5$ and $T_6 \mp i T_7$ correspond to the four eigenvalues $\pm \sqrt{3}/2$.

We now define the six step generators $E_{\boldsymbol{\alpha}}$ labelling them by the corresponding step vector in the (H_1, H_2) plane:

$$E_{\pm(1,0)} = \frac{1}{\sqrt{2}} \left(T_1 \pm i T_2 \right), \ E_{\pm(1/2, \sqrt{3}/2)} = \frac{1}{\sqrt{2}} (T_4 \pm i T_5), \ E_{\pm(-1/2, \sqrt{3}/2)} = \frac{1}{\sqrt{2}} (T_6 \pm i T_7) \tag{5.124}$$

They satisfy

$$[H_i, E_{\boldsymbol{\alpha}}] = \alpha_i E_{\boldsymbol{\alpha}} \tag{5.125}$$

which implies that the action of $E_{\boldsymbol{\alpha}}$ on a state changes its weight by $\boldsymbol{\alpha}$.

The remaining commutators of the algebra $\mathfrak{su}(3)$ are obtained by direct computation and are

$$[E_{\boldsymbol{\alpha}}, E_{-\boldsymbol{\alpha}}] = \alpha_i H_i, \quad [E_{\boldsymbol{\alpha}}, E_{\boldsymbol{\beta}}] = N_{\boldsymbol{\alpha\beta}} E_{\boldsymbol{\alpha+\beta}}, \ \boldsymbol{\alpha} + \boldsymbol{\beta} \neq 0 \tag{5.126}$$

The $N_{\boldsymbol{\alpha\beta}}$ are constants, obtained via direct evaluation of the commutators.

To summarise: The algebra $\mathfrak{su}(3)$ in the so-called *Cartan–Weyl basis* is

$$[H_i, H_j] = 0 \,, \ i, j = 1, 2,$$
$$[H_i, E_{\boldsymbol{\alpha}}] = \alpha_i E_{\boldsymbol{\alpha}},$$
$$[E_{\boldsymbol{\alpha}}, E_{-\boldsymbol{\alpha}}] = \alpha_i H_i, \quad [E_{\boldsymbol{\alpha}}, E_{\boldsymbol{\beta}}] = N_{\boldsymbol{\alpha\beta}} E_{\boldsymbol{\alpha+\beta}}, \ \boldsymbol{\alpha} + \boldsymbol{\beta} \neq 0 \tag{5.127}$$

with the six non-vanishing roots $\boldsymbol{\alpha}$ given by (5.123).[19] The eight roots of $\mathfrak{su}(3)$ are shown in Figure 5.3. In Problem 5.9 we ask the reader to find the weight diagrams for the representations **6** and **10**.

5.6 The Lorentz and Poincaré Groups

The *Poincaré transformations* consist of the spatial rotations, the boosts and the global translations of space-time. They are of the general form:

$$x'^{\mu}(x) = \Lambda^{\mu}{}_{\nu}\, x^{\nu} + a^{\mu}\,. \tag{5.128}$$

and leave the Minkowski metric $\eta_{\mu\nu}$ invariant. The invariance of the physical laws under these transformations has been tested with impressive accuracy. In this section we will study the group structure of Lorentz and Poincaré transformations and their applications in the physics of elementary particles.

The physical theories we will consider in this book are formulated in terms of a local Lagrangian density $\mathcal{L}$, which is a function of the dynamical fields $\Phi(x)$ and their first derivatives. Out of $\mathcal{L}$ we form the action $S[\Phi]$ as

$$S[\Phi] = \int \mathrm{d}^4 x \mathcal{L}(\Phi(x), \partial\Phi(x)) \tag{5.129}$$

We see that, by construction, the action, if it exists, is translationally invariant. Therefore, if we enforce the invariance of $\mathcal{L}$ under Lorentz transformations, the invariance of the action under the full Poincaré symmetry is guaranteed. This leads us to study first the representation theory of the Lorentz group. We will consider the fields $\Phi(x)$ belonging to irreducible representations and build out of them invariant Lagrangian densities. As a second step we shall study the representation theory of the full Poincaré group.

5.6.1 The Lorentz group

We have already seen that the Lorentz group preserves the Minkowski metric which, in our notation, is $\eta = \mathrm{diag}(1, -1, -1, -1)$, with one $+$ and three $-$ signs. It is often denoted by $O(1, 3)$. We have also seen that it is a non-compact Lie group. In order to build invariant Lagrangian densities we must know how the fields $\Phi(x)$ transform. We are only interested in fields with a finite number of components, so we will study the finite-dimensional representations of the Lorentz group. There is a general theorem according to which all non-trivial unitary representations of a non-compact group are infinite-dimensional. Therefore, contrary to what we did with the symmetry groups we have considered so far, we will study non-unitary representations of the Lorentz group.

A coordinate x^{μ} is transformed linearly as

$$x'^{\mu}(x) = \Lambda^{\mu}{}_{\nu}\, x^{\nu}. \tag{5.130}$$

[19]The relations (5.127) generalise to all classical groups and are the starting point for a complete classification of all classical algebras.

with the matrix Λ satisfying, by definition, the condition

$$\eta_{\mu\nu} \Lambda^{\mu}{}_{\alpha} \Lambda^{\nu}{}_{\beta} = \eta_{\alpha\beta} \tag{5.131}$$

From this relation we conclude that $(\det \Lambda)^2 = 1$. Thus, $\det \Lambda = \pm 1$. Furthermore, (5.131) for $\alpha = \beta = 0$ gives $(\Lambda^0{}_0)^2 = 1 + (\Lambda^i{}_0)^2$, from which it follows that either $\Lambda^0{}_0 \geq 1$ or $\Lambda^0{}_0 \leq -1$. Therefore, the Lorentz transformations split into four disconnected sectors:

(a) The subset of matrices Λ with

$$\det \Lambda = +1 \text{ and } \Lambda^0{}_0 \geq +1 \tag{5.132}$$

contains the identity matrix and forms a group. It is the subgroup $L_+^\uparrow$ of *proper orthochronous Lorentz transformations*. The remaining three sectors of Lorentz transformations are not subgroups. They do not contain the identity element. Instead, each of them contains a characteristic distinct representative, as follows:

(b) The sector with $\det \Lambda = -1$ and $\Lambda^0{}_0 \geq +1$ contains the *parity* transformation

$$\mathcal{P} = \begin{pmatrix} 1 & 0 & 0 & 0 \\ 0 & -1 & 0 & 0 \\ 0 & 0 & -1 & 0 \\ 0 & 0 & 0 & -1 \end{pmatrix} \tag{5.133}$$

(c) the sector with $\det \Lambda = -1$ and $\Lambda^0{}_0 \leq -1$ contains the *time reversal*

$$\mathcal{T} = \begin{pmatrix} -1 & 0 & 0 & 0 \\ 0 & 1 & 0 & 0 \\ 0 & 0 & 1 & 0 \\ 0 & 0 & 0 & 1 \end{pmatrix} \tag{5.134}$$

and, finally,

(d) the sector with $\det \Lambda = +1$ and $\Lambda^0{}_0 \leq -1$ contains the *space-time reversal*

$$\mathcal{PT} = \begin{pmatrix} -1 & 0 & 0 & 0 \\ 0 & -1 & 0 & 0 \\ 0 & 0 & -1 & 0 \\ 0 & 0 & 0 & -1 \end{pmatrix} \tag{5.135}$$

The generic element of each sector can be obtained as the product of a proper orthochronous Lorentz transformation times the corresponding representative $\mathcal{P}$, $\mathcal{T}$ or $\mathcal{PT}$ of that sector.

It should be pointed out that, in contrast to the proper Lorentz transformations, which are known to be a symmetry group of fundamental physics with impressive accuracy, $\mathcal{P}$, $\mathcal{T}$ and also $\mathcal{PT}$ are violated by the weak interactions. The laws of weak interactions are different viewed through a mirror, and/or with time running backwards.

5.6.1.1 Tensor calculus. The only noticeable difference between the tensor calculus in the Lorentz group and the corresponding one in any real orthogonal group is the non-Euclidean metric in $\mathbb{M}^4$ which distinguishes between upper and lower indices. So, a *contravariant four-vector* V^μ with $\mu = 0, 1, 2, 3$, is defined as a set of four quantities transforming as:

$$V'^\mu = \Lambda^\mu{}_\nu V^\nu \tag{5.136}$$

Similarly, a set of four quantities U_μ define a *covariant 4-vector* if they transform under a Lorentz transformation as

$$U'_\mu = U_\nu (\Lambda^{-1})^\nu{}_\mu \equiv \Lambda_\mu{}^\nu U_\nu \,, \quad \Lambda_\mu{}^\nu = \eta_{\mu\lambda} \Lambda^\lambda{}_\rho \eta^{\rho\nu} \tag{5.137}$$

The metric $\eta_{\mu\nu}$ and its inverse $\eta^{\mu\nu}$ can be used to lower and raise indices, respectively. These definitions extend to general contravariant or covariant or mixed tensors with arbitrary numbers of upper (contravariant) and/or lower (covariant) indices. The 4^{n+m} quantities $T^{\mu_1\mu_2\ldots\mu_n}{}_{\nu_1\nu_2\ldots\nu_m}$ are the components of a *mixed tensor* with n contravariant and m covariant indices, iff under Lorentz transformations they transform according to

$$T'^{\mu_1\mu_2\ldots\mu_n}{}_{\nu_1\nu_2\ldots\nu_m} = \Lambda^{\mu_1}{}_{\alpha_1} \Lambda^{\mu_2}{}_{\alpha_2} \ldots \Lambda^{\mu_n}{}_{\alpha_n} (\Lambda^{-1})^{\beta_1}{}_{\nu_1} (\Lambda^{-1})^{\beta_2}{}_{\nu_2} \ldots (\Lambda^{-1})^{\beta_m}{}_{\nu_m} T^{\alpha_1\alpha_2\ldots\alpha_n}{}_{\beta_1\beta_2\ldots\beta_m} \tag{5.138}$$

A quantity which is invariant under Lorentz transformations is called a *Lorentz scalar*. A tensor without any index or with all its indices contracted is a Lorentz scalar.

In analogy to tensors of the rotation group $O(3)$, Lorentz tensors decompose into irreducible ones transforming independently with the help of the operations of symmetrisation, anti symmetrisation and trace.

5.6.1.2 The Lie algebra of the Lorentz group. The proper Lorentz group contains the spatial rotations as a subgroup. Indeed, the matrices

$$\Lambda = \begin{pmatrix} 1 & \\ & R \end{pmatrix} \tag{5.139}$$

with the three-by-three matrices R being rotation matrices $R^T R = R R^T = 1$, satisfy (5.131). The rotation matrix in the "$x^1 - x^2$ plane" by angle $\omega_{12} = \theta$ is

$$R(\omega_{12} = \theta) = \begin{pmatrix} \cos\theta & \sin\theta & 0 \\ -\sin\theta & \cos\theta & 0 \\ 0 & 0 & 1 \end{pmatrix} \tag{5.140}$$

and the corresponding generator defined in general in (5.21) is

$$\mathcal{J}^{12} = \begin{pmatrix} 0 & 0 & 0 & 0 \\ 0 & 0 & i & 0 \\ 0 & -i & 0 & 0 \\ 0 & 0 & 0 & 0 \end{pmatrix} \tag{5.141}$$

Similarly, the generators of rotations in the 2-3 and 3-1 planes are

$$J^{23} = \begin{pmatrix} 0 & 0 & 0 & 0 \\ 0 & 0 & 0 & 0 \\ 0 & 0 & 0 & i \\ 0 & 0 & -i & 0 \end{pmatrix} \quad \text{and} \quad J^{13} = \begin{pmatrix} 0 & 0 & 0 & 0 \\ 0 & 0 & 0 & i \\ 0 & 0 & 0 & 0 \\ 0 & -i & 0 & 0 \end{pmatrix} \tag{5.142}$$

The Lorentz boost with velocity V in the x^1-direction is given by

$$t' = \gamma(V)(t - Vx), \ \ x' = \gamma(V)(x - Vt), \ \ y' = y, \ \ z' = z \tag{5.143}$$

with

$$\gamma(V) = \frac{1}{\sqrt{1 - V^2}} \tag{5.144}$$

and similarly for boosts in the other two directions. From (5.143) we obtain for the generators of boosts

$$J^{01} = -i \begin{pmatrix} 0 & 1 & 0 & 0 \\ 1 & 0 & 0 & 0 \\ 0 & 0 & 0 & 0 \\ 0 & 0 & 0 & 0 \end{pmatrix}, \ J^{02} = -i \begin{pmatrix} 0 & 0 & 1 & 0 \\ 0 & 0 & 0 & 0 \\ 1 & 0 & 0 & 0 \\ 0 & 0 & 0 & 0 \end{pmatrix}, \ J^{03} = -i \begin{pmatrix} 0 & 0 & 0 & 1 \\ 0 & 0 & 0 & 0 \\ 0 & 0 & 0 & 0 \\ 1 & 0 & 0 & 0 \end{pmatrix} \tag{5.145}$$

Calling $\omega_{\mu\nu} = -\omega_{\nu\mu}$ collectively the six independent parameters of the general Lorentz transformation, with $\theta^k = -(1/2)\epsilon^{ijk}\omega_{ij}$ the rotation angle around the axis x^k and ω_{i0} the boost parameter in the positive direction x^i, we write its infinitesimal form as

$$\Lambda(\delta\omega) = \mathbf{1} + \frac{i}{2}\delta\omega_{\mu\nu}J^{\mu\nu} \tag{5.146}$$

The group element with finite parameters $\omega_{\mu\nu}$ is obtained as the product of $N \to \infty$ infinitesimal ones with parameters $\delta\omega_{\mu\nu} = \omega_{\mu\nu}/N$, i.e.

$$\Lambda(\omega) = \lim_{N\to\infty} \left(1 + \frac{i}{2}\delta\omega_{\mu\nu}J^{\mu\nu}\right)^N = \exp\left(+\frac{i}{2}\omega_{\mu\nu}J^{\mu\nu}\right) \tag{5.147}$$

It is now straightforward to prove that the six generators $J^{\mu\nu} = -J^{\nu\mu}$ of Lorentz transformations satisfy the so-called *Lorentz algebra*

$$[J^{\mu\nu}, J^{\lambda\rho}] = i\left(\eta^{\mu\lambda}J^{\nu\rho} - \eta^{\mu\rho}J^{\nu\lambda} + \eta^{\nu\rho}J^{\mu\lambda} - \eta^{\nu\lambda}J^{\mu\rho}\right) \tag{5.148}$$

Now we can prove the following statement:

The complexified Lorentz algebra (5.148) $\mathfrak{so}(1,3)_C$ *is isomorphic to the algebra* $\mathfrak{su}(2)_C \oplus \mathfrak{su}(2)_C$. Practically, this means that one can define complex combinations of the Lorentz generators which satisfy the algebra of two commuting $\mathfrak{su}(2)$s.

Indeed, start by defining the rotation and boost generators by

$$J^i \equiv \frac{1}{2}\epsilon^{ijk}J^{jk} \quad \text{and} \quad K^i \equiv J^{0i} \tag{5.149}$$

in terms of which the Lorentz algebra (5.148) takes the form

$$[J^i, J^k] = i\epsilon^{ikl} J^l, \quad [J^i, K^j] = i\epsilon^{ijl} K^l, \quad [K^i, K^j] = -i\epsilon^{ijk} J^k \tag{5.150}$$

Define next the combinations

$$M^i \equiv \frac{1}{2}(J^i + iK^i) \quad \text{and} \quad N^i \equiv \frac{1}{2}(J^i - iK^i) \tag{5.151}$$

They satisfy the $\mathfrak{su}(2)\oplus\mathfrak{su}(2)$ algebra

$$[M^i, M^j] = i\epsilon^{ijk} M^k, \quad [N^i, N^j] = i\epsilon^{ijk} N^k, \quad [M^i, N^j] = 0 \tag{5.152}$$

It follows that the irreducible representations of the algebra will be labelled by two indices, j_1 and j_2, each taking integer, or half-integer values.

A few remarks are important here:

(i) As we have noted already, the Lorentz group is non-compact and this is reflected to the fact that the generators of the algebra $\mathfrak{so}(1,3)$ are not all Hermitian. Indeed, $\mathcal{J}^{0i}$ given by equation (5.145) are anti-Hermitian.

(ii) The relations (5.151) establish a relation between the Lie algebra $\mathfrak{su}(2)\oplus\mathfrak{su}(2)$ and the complexified Lorentz algebra $\mathfrak{so}(1,3)_C$. Therefore, when constructing a group element by exponentiation using the generators of $\mathfrak{su}(2)\oplus\mathfrak{su}(2)$, we should take into account the fact that a "rotation" describing a Lorentz boost is made with an imaginary angle. It follows that our result concerning the reality of the $SU(2)$ representations does not apply. Instead, we see that complex conjugation interchanges M and N. We conclude (see also Problem 5.10) that it relates $\overline{(j_1, j_2)} = (j_2, j_1)$. The only real irreducible representations are the ones with $j_1 = j_2$.

(iii) The generators for rotations J^i, like the components of angular momentum, are pseudo-vectors. Those of Lorentz boosts K^i are vectors. As a result, a parity operation also interchanges the representations $(j_1, j_2) \leftrightarrow (j_2, j_1)$. We show this explicitly in equation (5.168) below.

In the following we will study in some more detail the low-lying representations $(0,0)$ (scalar), $(\frac{1}{2},0)$ and $(0,\frac{1}{2})$ (spinors), and $(\frac{1}{2},\frac{1}{2})$ (vector).

• *The $(0,0)$ scalar representation.* It is often called trivial, but in fact it is the only representation which corresponds to a Lorentz invariant quantity. Therefore, when we want to build a Lorentz invariant action, we must make sure that the Lagrangian density belongs to this representation.

• *The two-dimensional spinor representations.* There are two inequivalent ones: the $(\frac{1}{2},0)$ and the $(0,\frac{1}{2})$.

1) We start with $(\frac{1}{2},0)$. It is a spinor with respect to the first $SU(2)$ factor and scalar with respect to the second. The generators take the form:

$$M^i = \frac{1}{2}\sigma_i = J^i = iK^i, \quad N^i = 0, \quad i = 1,2,3 \tag{5.153}$$

and the corresponding representation of the rotation $u_{(1/2,0)}(\hat{n}\theta)$ and the boost $v_{(1/2,0)}(\hat{m}\phi)$ [20] are the 2×2 matrices:

---

[20]For notational uniformity with the rotations, instead of denoting the boost parameter by $V_i = \omega_{i0}$, we write $\boldsymbol{V} = \hat{\boldsymbol{V}}|\boldsymbol{V}| \equiv \hat{\boldsymbol{m}} \tanh\phi$. The unit vector $\hat{\boldsymbol{m}}$ denotes the direction of the boost and $\tanh\phi$, with $\phi \in [0, \infty)$ the speed.

$$u_{(1/2,0)}(\hat{\boldsymbol{n}}\theta) = e^{i\theta\hat{n}_i J^i} = \cos\frac{\theta}{2} + i\hat{\boldsymbol{n}}\cdot\boldsymbol{\sigma}\sin\frac{\theta}{2} \tag{5.154}$$

$$= \begin{pmatrix} \cos(\theta/2) + i\hat{n}_3\sin(\theta/2) & (i\hat{n}_1 + \hat{n}_2)\sin(\theta/2) \\ (i\hat{n}_1 - \hat{n}_2)\sin(\theta/2) & \cos(\theta/2) - i\hat{n}_3\sin(\theta/2) \end{pmatrix}$$

and

$$v_{(1/2,0)}(\hat{\boldsymbol{m}}\phi) = e^{i\phi\hat{m}_i K^i} = \cosh\frac{\phi}{2} + \hat{\boldsymbol{m}}\cdot\boldsymbol{\tau}\sinh\frac{\phi}{2} \tag{5.155}$$

$$= \begin{pmatrix} \cosh(\phi/2) + \hat{m}_3\sinh(\phi/2) & (\hat{m}_1 - i\hat{m}_2)\sinh(\phi/2) \\ (\hat{m}_1 + i\hat{m}_2)\sinh(\phi/2) & \cosh(\phi/2) - \hat{m}_3\sinh(\phi/2) \end{pmatrix}$$

respectively.

In order to simplify the formulae of this section it is convenient to use the notation

$$u \equiv u_{(1/2,0)}(\hat{\boldsymbol{n}}\theta) \quad \text{and} \quad v \equiv v_{(1/2,0)}(\hat{\boldsymbol{m}}\phi) \tag{5.156}$$

for the $(\frac{1}{2},0)$ representation matrices of the generic rotation $\hat{\boldsymbol{n}}\theta$ and boost $\hat{\boldsymbol{m}}\phi$, respectively.

Definition. A set of two quantities $\xi^\alpha, \alpha = 1, 2$, are the contravariant components of a Lorentz spinor $(1/2, 0)$ if they transform under rotations (R) and boosts (B) according to

$$\xi \xrightarrow{R} u\xi \quad \text{and} \quad \xi \xrightarrow{B} v\xi \tag{5.157}$$

2) We now move to the second one, $(0,\frac{1}{2})$. It is a scalar with respect to the first $SU(2)$ factor and spinor with respect to the second. The generators take the form

$$M^i = 0, \quad N^i = \frac{1}{2}\sigma_i = J^i = -iK^i, \quad i = 1, 2, 3; \tag{5.158}$$

so that the corresponding representation of the rotation $u_{(0,1/2)}(\hat{\boldsymbol{n}}\theta)$ and the boost $v_{(0,1/2)}(\hat{\boldsymbol{m}}\phi)$ are

$$u_{(0,1/2)}(\hat{\boldsymbol{n}}\theta) = e^{i\theta\hat{n}_i J^i} = u_{(1/2,0)}(\hat{\boldsymbol{n}}\theta) = u \tag{5.159}$$

and

$$v_{(0,1/2)}(\hat{\boldsymbol{m}}\phi) = e^{i\phi\hat{m}_i K^i} = v_{(1/2,0)}(-\hat{\boldsymbol{m}}\phi) = v^{-1} \tag{5.160}$$

Definition. A set of two quantities $\eta^{\dot\alpha}, \dot\alpha = 1, 2$,[21] are the contravariant components of a Lorentz spinor $(0, 1/2)$ if they transform under rotations (R) and boosts (B) according to

$$\eta \xrightarrow{R} u\eta \quad \text{and} \quad \eta \xrightarrow{B} v^{-1}\eta \tag{5.161}$$

Raising and lowering indices. Inside each one of the two two-dimensional spinor spaces we can raise and lower indices with the metric ϵ given in equation (5.43)

$$\xi_\alpha \equiv \epsilon_{\alpha\beta}\xi^\beta, \quad \text{or equivalently} \quad \xi^\alpha = \epsilon^{\alpha\beta}\xi_\beta \tag{5.162}$$

$$\eta_{\dot\alpha} \equiv \epsilon_{\dot\alpha\dot\beta}\eta^{\dot\beta}, \quad \text{or equivalently} \quad \eta^{\dot\alpha} = \epsilon^{\dot\alpha\dot\beta}\eta_{\dot\beta} \tag{5.163}$$

ξ_α and $\eta_{\dot\alpha}$ are the covariant components of the spinors ξ and η.

---

[21]We follow the notation introduced by B.L. van der Waerden of undotted and dotted indices in order to distinguish quantities which transform according to the first, or the second $SU(2)$ factor, respectively. The metric $\epsilon_{\alpha\beta}$ satisfies the relation $\epsilon_{\alpha\beta}\epsilon^{\beta\gamma} = \delta_\alpha^\gamma$.

Complex conjugation. We have already pointed out that complex conjugation interchanges the two $SU(2)$ factors. Let us show this explicitly for the spinor representations. We obtain in matrix notation

$$\xi'^{\,*} = u^{*}\xi^{*} \quad \text{and} \quad \xi'^{\,*} = v^{*}\xi^{*} \tag{5.164}$$

But, as explained in the study of $SU(2)$, the matrices τ^{i} and $-(\tau^{i})^{*}$ are related by the similarity transformation $\epsilon\,\tau^{i}\,\epsilon^{-1} = -(\tau^{i})^{*}$. Thus, in the transformation formulae of ξ^{*}, one is free to replace $-\boldsymbol{\sigma}^{*} \to \boldsymbol{\sigma}$ and $\boldsymbol{\tau}^{*} \to -\boldsymbol{\tau}$, in which case they become identical to the transformation of the spinor η. We conclude that the complex conjugate of $(1/2,0)$ is the representation $(0,1/2)$. We denote this by

$$\overline{(1/2,0)} \cong (0,1/2) \tag{5.165}$$

Correspondingly, we write $(\xi^{\alpha})^{*} \sim \eta^{\dot{\alpha}}$ to express the fact that, as we just showed, the spinors $(\xi^{\alpha})^{*}$ and $\eta^{\dot{\alpha}}$ transform in the same way under Lorentz transformations.

Notice, in particular, that complex conjugation does not affect the way a Lorentz spinor transforms under rotations. The Lorentz spinor and its complex conjugate transform under rotations in equivalent ways, since the corresponding representation matrices are related by a similarity transformation.

Below we shall find it more convenient to work with the Hermitian conjugates of the spinors. Their transformation formulae under rotations and boosts are

$$\xi'^{\,\dagger} = \xi^{\dagger}u^{\dagger} = \xi^{\dagger}u^{-1}$$

and

$$\xi'^{\,\dagger} = \xi^{\dagger}v^{\dagger} = \xi^{\dagger}v \tag{5.166}$$

respectively.

The above argument generalizes to the complex conjugates of all representations of the Lorentz group. It is straightforward to show that

$$\overline{(j_1,j_2)} \cong (j_2,j_1) \tag{5.167}$$

The action of parity on spinors. The parity transformation $(\mathcal{P})$ has the following properties: (i) It commutes with spatial rotations, but it does not commute with boosts. Indeed, using (5.133) and the generators (5.145) in the defining representation of the Lorentz algebra, we obtain [22]

$$\mathcal{P}J^{i}\mathcal{P} = J^{i}\,, \quad \mathcal{P}K^{i}\mathcal{P} = -K^{i} \tag{5.168}$$

which imply that $\mathcal{P}\Lambda(\boldsymbol{V})\mathcal{P} = \Lambda^{-1}(\boldsymbol{V})$. This proves that parity also interchanges the two $SU(2)$ factors and relates $(j_1,j_2)\overset{\mathcal{P}}{\longrightarrow}(j_2,j_1)$. Thus, a priori for spinors we have

$$\xi \overset{\mathcal{P}}{\longrightarrow} \kappa\eta \quad \text{and} \quad \eta \overset{\mathcal{P}}{\longrightarrow} \kappa'\xi \tag{5.169}$$

with $|\kappa| = |\kappa'| = 1$. Double action of parity returns space back to itself. But, space returns "back to itself" either as a result of no rotation, or of a rotation by 360°.

---

[22]These relations are equivalently stated as: the angular momentum $\boldsymbol{J}$ is a pseudo-vector, while $\boldsymbol{K}$ is a vector.

In the first case, spinors return to themselves, while in the second they change sign. We conclude that $\mathcal{P}^2$ on spinors can consistently be taken equal to either $+1$ or -1.[23] Thus,

$$\mathcal{P}^2 = \pm 1 \quad \longrightarrow \quad \kappa\kappa' = \pm 1 \tag{5.170}$$

So there are two possible inequivalent choices, namely $\kappa = \kappa' = 1$ which implies

$$\mathcal{P}\xi = \eta \quad \text{and} \quad \mathcal{P}\eta = \xi, \quad \mathcal{P}^2 = +1 \tag{5.171}$$

or, $\kappa = \kappa' = i$ which implies

$$\mathcal{P}\xi = i\eta \quad \text{and} \quad \mathcal{P}\eta = i\xi, \quad \mathcal{P}^2 = -1 \tag{5.172}$$

We can consider the direct sum representation $(\frac{1}{2},0)\oplus(0,\frac{1}{2})$. It is real and invariant under parity. It is reducible under proper orthochronous Lorentz transformations, but it becomes irreducible if we include parity. It will be useful in building parity invariant spinor theories.

- *Higher representations. Tensor calculus.* It is straightforward to define tensors $T^{\alpha_1\cdots\alpha_n}_{\beta_1\cdots\beta_m}$ with undotted indices, n contravariant and m covariant ones, as well as the corresponding tensors with dotted indices $T^{\dot{\alpha}_1\cdots\dot{\alpha}_n}_{\dot{\beta}_1\cdots\dot{\beta}_m}$. The resulting tensor calculus follows exactly the principles we developed in section 5.5.2. By the operations of symmetrisation and anti symmetrisation we build quantities belonging to the irreducible representations $(j_1,0)$ and $(0,j_2)$.

We can also define mixed tensors with dotted and undotted indices and we build in this way elements of the representations (j_1,j_2). The $(n+1)(m+1)$ quantities $T^{\alpha_1\alpha_2\cdots\alpha_n\dot{\beta}_1\dot{\beta}_2\cdots\dot{\beta}_m}$ with the αs and the βs symmetrised, transform according to the $(n/2, m/2)$ representation of the Lorentz group. Of particular interest is the representation $(\frac{1}{2},\frac{1}{2})$ to which we turn next.

- *The vector representation.* The tensor product of a $(\frac{1}{2},0)$ and a $(0,\frac{1}{2})$ spinor gives

$$\left(\frac{1}{2},0\right) \otimes \left(0,\frac{1}{2}\right) = \left(\frac{1}{2},\frac{1}{2}\right)$$

Thus, in particular, the tensor product of a spinor with its complex conjugate gives the irreducible four-dimensional representation $(\frac{1}{2},\frac{1}{2})$. It is the only real four-dimensional irreducible representation of the algebra, so it is the representation in which four vectors like dx^μ belong. It will be useful in applications to make this correspondence explicit. The approach is the same as the one we used in establishing the relation between $O(3)$ and $SU(2)$ in section 5.5.3.

Let us denote by $\sigma^\mu = (\sigma^0, \sigma^i)$ the quadruplet of the matrices $\mathbf{1}$ and $\boldsymbol{\sigma}$.

Every 2×2 Hermitian matrix can be written as a linear combination with real coefficients of these four matrices. Therefore, to each four-vector V^μ, there corresponds a 2×2 matrix

$$\widetilde{V} = \sigma_\mu V^\mu = \begin{pmatrix} V^0 - V^3 & -V^1 + iV^2 \\ -V^1 - iV^2 & V^0 + V^3 \end{pmatrix} \tag{5.173}$$

---

[23] All physical observables are bilinear in a spinor and its conjugate, so no observable will be sensitive to the sign of $\mathcal{P}^2$.

$\widetilde{V}$ is Hermitian if and only if V^μ is real. It can be checked that

$$V^\mu = \frac{1}{2}\text{Tr}(\sigma_\mu \widetilde{V}) \qquad \text{and} \qquad \det \widetilde{V} = V^\mu V_\mu. \qquad (5.174)$$

Using the spinors ξ and Ξ of the type $(\frac{1}{2},0)$ and the spinors η and H both $(0,\frac{1}{2})$, and the matrices $\sigma^\mu \equiv (\mathbf{1},\sigma^i)$ and $\overline{\sigma}^\mu \equiv (\mathbf{1},-\sigma^i)$, we can construct the Lorentz vectors with contravariant components

$$\Xi^\dagger \sigma^\mu \xi \quad \text{and} \quad H^\dagger \overline{\sigma}^\mu \eta \qquad (5.175)$$

i.e. which under the generic Lorentz transformation $\Lambda^\mu{}_\nu$ satisfy

$$(\Xi^\dagger \sigma^\mu \xi)' = \Lambda^\mu{}_\nu \, \Xi^\dagger \sigma^\nu \xi \quad \text{and} \quad (H^\dagger \overline{\sigma}^\mu \eta)' = \Lambda^\mu{}_\nu \, H^\dagger \overline{\sigma}^\nu \eta \qquad (5.176)$$

in accordance with the composition rule $(\frac{1}{2},0) \otimes (0,\frac{1}{2}) \cong (\frac{1}{2},\frac{1}{2})$ in both cases.

As we see from these expressions, the matrices σ and $\overline{\sigma}$ have one dotted and one undotted index; when applied to a spinor ξ they yield a spinor which transforms like η. In particular, the derivative operator $p^\mu = i\partial^\mu$ can be written as a 2×2 matrix $\widetilde{p}$ which acts on spinor fields as

$$\widetilde{p}^{\alpha\dot{\beta}} \eta_{\dot{\beta}} \sim \xi^\alpha, \;\; \widetilde{p}_{\alpha\dot{\beta}} \xi^\alpha \sim \eta_{\dot{\beta}} \qquad (5.177)$$

5.6.1.3 The relation between the Lorentz group and $SL(2,\mathbb{C})$. In section 5.5.3 we established a relation between the locally isomorphic groups $O(3)$ and $SU(2)$. As we just saw, there is a similar relation between the Lorentz group and $SL(2,\mathbb{C})$, the group which is isomorphic to that of 2×2 complex matrices with determinant equal to 1. The elements of $SL(2,\mathbb{C})$ depend also on six arbitrary parameters, and the two groups have the same Lie algebra. We can lift to the groups the relations we can deduce based on their common Lie algebra.

There exists a group homomorphism of $SL(2,\mathbb{C})$ on the restricted Lorentz group, the kernel of which is $\mathbb{Z}_2$. The mapping is given by

$$A \in SL(2,\mathbb{C}) \mapsto \Lambda(A) \qquad (5.178)$$

with

$$\Lambda(A)^\mu{}_\nu = \frac{1}{2}\text{Tr}(\sigma_\mu A \sigma_\nu A^\dagger) \qquad (5.179)$$

If A is Hermitian, then $\Lambda(A)$ is symmetric. To the 4-vector x^μ we associate the 2×2 matrix $\widetilde{x} = \sum_0^3 x^\mu \sigma_\mu$. The action (5.130) of the Lorentz transformation $\Lambda(A)$ on x^μ translates to the action

$$\widetilde{x}' = A \widetilde{x} A^\dagger \qquad (5.180)$$

of the corresponding element A of $SL(2,\mathbb{C})$ on $\widetilde{x}$. The group $SL(2,\mathbb{C})$ is the covering group of the Lorentz group. It is a simply connected group.

We can easily show that $\Lambda(A)$ is a rotation if and only if A is unitary, and a boost if and only if A is Hermitian.

The properties we have derived previously concerning the action of a Lorentz transformation on spinors, equations (5.157) and (5.161), can be interpreted as automorphisms of the group $SL(2,\mathbb{C})$.

5.6.2   The Poincaré group

In the previous section we studied the finite dimensional representations of the Lorentz group. We intend to assign to them the fields $\Phi(x)$ which will be the dynamical variables of our theories. As we have explained already, if we build a Lorentz scalar Lagrangian density $\mathcal{L}(\Phi, \partial\Phi)$, the resulting action given in equation (5.129) will be automatically Poincaré invariant. In principle, the knowledge of the Lagrangian and the corresponding Hamiltonian of a system should be sufficient to determine also the quantum states of the system. However, as we have already seen in the specific example we studied in Chapter 2 and we will see in more detail in the following chapters, diagonalising the complete Hamiltonian for an interacting theory is beyond our capabilities. We have to postulate the existence and the physical properties of the Hilbert space of our theory and, in practice, we will assume that it has the form of the Fock space, like the one we studied in Chapter 2. The building block is the space of one-particle states and we will assume that they form an infinite-dimensional unitary representation of the Poincaré group. It is the subject of this section.

The Lie algebra $\mathfrak{P}$ of the Poincaré group has ten generators, the six generators $\mathcal{J}^{\mu\nu}$ of the Lorentz group and the four generators P^μ of translations. This algebra closes with the following commutation relations

$$\left[\mathcal{J}^{\mu\nu}, \mathcal{J}^{\lambda\rho}\right] = i\left(\eta^{\mu\lambda}\mathcal{J}^{\nu\rho} - \eta^{\mu\rho}\mathcal{J}^{\nu\lambda} + \eta^{\nu\rho}\mathcal{J}^{\mu\lambda} - \eta^{\nu\lambda}\mathcal{J}^{\mu\rho}\right)$$

$$\left[\mathcal{J}^{\mu\nu}, P^\lambda\right] = i\left(\eta^{\mu\lambda}P^\nu - \eta^{\nu\lambda}P^\mu\right) \tag{5.181}$$

$$[P_\mu, P_\nu] = 0$$

The first gives the Lie algebra of the Lorentz group, the second shows that the operators P_μ transform as the components of a four-vector under Lorentz transformations, and the third one, which completes the Poincaré algebra, says that the four translations commute among themselves. We are looking for unitary operators $U(a, A)$ where a_μ are the four parameters of translations and A the $SL(2, \mathbb{C})$ matrix of Lorentz transformations.

5.6.2.1   The irreducible unitary representations of the Poincaré group.   Let us first consider the subgroup of space-time translations. Since it is an Abelian group, its finite-dimensional irreducible representations are of dimension 1. Since they are unitary, they are phase factors of the form: $U(a, \mathbf{1}) = e^{ia \cdot p}$ where p is a real four-vector with components p_μ and $a \cdot p = a^\mu p_\mu$. It follows that the states in the Hilbert space which correspond to irreducible unitary representations of the Poincaré group will be characterised by the four-momentum p_μ. We still have to take into account the Lorentz transformations, but we see that, for every p_μ, there exists a subgroup of transformations which will play a special role: they are the ones which leave the particular p_μ invariant. They form the *little group* of p_μ. We must distinguish four cases:
  1. p^μ is *time-like*. The corresponding states have $p^2 = m^2 > 0$.
  2. p^μ is *light-like*. The corresponding states have $p^2 = 0$.
  3. p^μ is *space-like*. The corresponding states have $p^2 = m^2 < 0$.
  4. $p^\mu = 0$. The null vector.

The third case corresponds to states with imaginary values of the mass and, although it has been fully studied mathematically, it does not seem to describe any physically interesting states. It is the same with the case 4 of the null vector but with one notable exception: the trivial case of the one-dimensional representation $U(a, A) = 1$, the only finite-dimensional irreducible representation. It will be used for the vacuum state.

We are left with cases 1 and 2. The analysis simplifies if we take into account that the algebra $\mathcal{P}$ given by the commutation relations (5.181) admits two *Casimir* operators, namely two operators which commute with all generators of the algebra. Therefore, as we did with the operator J^2 of the rotation group, we can label the representations by the eigenvalues of these operators.

We can immediately find the first one: it is $P^2 = P_\mu P^\mu$. It obviously commutes with the four translations and, being a Lorentz scalar, it commutes also with the Lorentz transformations. To find the second Casimir operator we first introduce the Pauli–Lubanski four-pseudovector $W^\mu = \frac{1}{2}\epsilon^{\mu\nu\varrho\sigma}P_\nu \mathcal{J}_{\varrho\sigma}$. We can prove by explicit computation that it commutes with P^λ. Therefore its square $W^\mu W_\mu$ commutes with all generators of the algebra. It follows that the representations we are looking for are labelled by the eigenvalues of these operators.

1. For $p^2 = m^2 > 0$ we can choose a frame in which $p_\mu = (m,0,0,0)$. In this case the little group is the group $SU(2)$ of space rotations. Therefore in this frame the states will be labelled as $|m, s\rangle$ with m positive and s integer or half-integer taking $2s + 1$ values. We say in physics that the states of one massive particle are characterised by the values of its mass and spin. The action of the Poincaré transformations on $|m, s\rangle$ leads to an infinite-dimensional representation described by the spin s and momenta on the mass hyperboloid $p^2 = m^2$.

2. For $p^2 = 0$ we can choose the frame $p^\mu = (E,0,0,E)$. The little group is the *Euclidean group* E_2 of motions in the plane, i.e. rotations and translations in the $x - y$ plane. E_2 has an invariant Abelian subgroup, the translation subgroup. We can repeat the analysis as for the previous case. We first seek the representations of the translation subgroup. They are of dimension 1 and characterised by a two-vector $q = (q_1, q_2)$, with either $q^2 > 0$ or $q^2 = 0$. The first case is excluded since the representation is characterised by a continuous number q^2 which has to be an invariant associated to a particle; it could be a spin value but no continuous spin has been observed. The second case corresponds to $q_1 = q_2 = 0$. The little group reduces to the rotations around the z-axis. It is Abelian, its representations are of dimension 1 and, being unitary, they are of the form $e^{ih\phi}$. Since we must have the identity for $\phi = 4\pi$, we find $h = 0, \pm 1/2, \pm 1, \ldots$. The parameter h is called the *helicity* of the massless particle. It results from the invariance under rotations around the axis defined by the momentum $\boldsymbol{p}$, so it can be interpreted as the projection of spin in the direction of motion. Notice however that, contrary to what happens for massive particles where the spin projection takes $2s + 1$ values, helicity takes only one value h. Parity leaves the spin invariant and changes $\boldsymbol{p} \to -\boldsymbol{p}$. Thus, it changes $h \to -h$. So, irreducible representations of the full Poincaré group including parity contain states with both values $\pm h$.

We summarise: The one-particle states form infinite-dimensional unitary irreducible representations of the Poincaré group. Each state belonging to such a representation is characterised by a four-momentum p^μ on a hyperboloid with $p^2 \geq 0$ and an extra index, which for $p^2 = m^2 > 0$ is the spin $s = 0, \frac{1}{2}, 1, \ldots$ and takes $2s + 1$ values, while for $p^2 = 0$ it is the helicity h and takes one of the values 0, $\pm\frac{1}{2}$, ± 1, $\ldots$, or both $\pm h$ if parity is to be preserved. We obtain the well-known result that a massless particle like the photon has two helicity states.

5.7 The Space of Physical States

5.7.1 Introduction

In the previous sections we often referred to the space of physical states in which the operators corresponding to various symmetry transformations act. In this section we want to give a more precise definition of this space, which will be used throughout this book.

Classical physics is formulated in terms of the dynamical variables $q_a(t)$ and $p_a(t)$, $a = 1, \ldots, N$, where N is the number of degrees of freedom of the system. For example, if the system consists of point particles, the qs and the ps could be the positions and momenta of the particles. The physical states of such a system are just the possible values of these variables at a given time, so they do not require any special care in order to be specified. The situation does not change radically when we take the large N limit and consider the index i taking values in a continuum, for example, the points in the three-dimensional space. The theory becomes a classical field theory, such as the electromagnetic theory or the theory of general relativity. The physical states are still determined by the values of the fields and their first time derivatives at each point in space, which, for the electromagnetic field, correspond to a given configuration of the electric and magnetic fields.

Quantum mechanics profoundly changes this simple picture. Positions and momenta cannot be specified simultaneously, so the classical definition of a physical state is not directly applicable. In quantum mechanics we postulate instead that the states of a non-relativistic particle are determined by the possible wave functions, i.e. vectors in a Hilbert space, chosen to be that of square integrable functions. Special relativity brings a further, very important, complication: Already in classical physics, it establishes a connection between energy and mass with the famous relation $E = mc^2$. But it is in quantum physics that this relation reveals all its significance because it implies the possibility of creation, or annihilation, of an arbitrary number of particles. One expects therefore that establishing a relativistic quantum formalism for the states of a physical system will not be a simple exercise. In fact, such a formalism has to include in an essential way the non-conservation of the number of particles and, as we shall see in this book, the fact that to each particle is associated an antiparticle. In the old books this is often called *the formalism of second quantisation*, although it is, in fact, the theory of quantum fields. We described an example of such a formalism in Chapter 2 for the quantum states of the free electromagnetic field.

5.7.2 Particle states

From the remarks above, we see that a necessary step for understanding quantum mechanical interactions among relativistic particles is a good knowledge of the states consisting of an arbitrary number of particles. In the following chapters we shall show that this is in fact the space of physical states appropriate to a quantum field theory. However, studying the most general n-particle state, with arbitrary n, will turn out to be an impossible task for any physically interesting theory. Fortunately, for most applications we are going to consider in this book, it will be enough to restrict ourselves to a particular class of n-particle states, the states of non-interacting particles. They can be viewed as the states of n particles lying far away from each other and having only short range interactions. If the distances between them are much larger than the interaction range, the particles can be considered as free. The interactions among them will be treated separately.[24] Since the Poincaré group is the invariance group of any physical system that we are going to consider, a good description of these states will be provided by the irreducible representations of this group which we studied in section 5.6.2. As we mentioned there, our space of states will have the form of a Fock space, similar to the one we constructed in Chapter 2, but without assuming the existence of an underlying quantum field theory with creation and annihilation operators. The building blocks of such a space are the following:

5.7.2.1 The vacuum state $|0\rangle$. It belongs to the trivial representation of the Poincaré group with $p^\mu = 0$. Unless specified otherwise, the vacuum state is supposed to be unique and invariant under Poincaré transformations

$$U(a, A)|0\rangle = |0\rangle \tag{5.182}$$

5.7.2.2 The one-particle states. Following the discussion of section 5.6.2, we distinguish two cases:

• *The states of a massive particle with $p^2 = m^2$.*

These states are characterised by a three-vector p and a pair of indices s and s_z. s is integer or half-integer and s_z takes the $2s+1$ values $-s, -s+1, \ldots, s$. According to the general principles of quantum mechanics, we assume that these states correspond to vectors in a Hilbert space of square integrable functions. For example, for a spin-zero particle, a one-particle state Ψ is given by a complex-valued function $\Psi(p)$ in a Hilbert space $\mathcal{H}^{(1)}$ equipped with the norm

$$\|\Psi\|^2 = (\Psi, \Psi) = \int |\Psi(p)|^2 d\Omega_m(p) \tag{5.183}$$

and which transforms[25] under the Poincaré group as

$$(U(a, A)\Psi)(p) = e^{i\Lambda p \cdot a} \Psi(\Lambda(A)^{-1}p) \tag{5.185}$$

---

[24] According to our discussion in Chapter 4, this requirement of short range interactions implies the absence of zero mass particles and this seems to exclude physical theories such as electromagnetism. We will ignore this problem for the moment and we will come back to it in Chapters 12 and 17.

[25] We can use the bra and ket formalism by introducing a continuous basis $|p\rangle$, $p^2 = m^2$, such that $\langle q|p\rangle = \delta(q - p)$ with $\Psi(p) = \langle p|\Psi\rangle$, the transformation law of $|p\rangle$ being

$$U(a, A)|p\rangle = e^{i\Lambda p \cdot a}|\Lambda p\rangle \tag{5.184}$$

This formula generalises to states describing a particle with mass m and any spin s. If we have N different species of particles, each state will carry an additional index n, $n = 1, \ldots, N$.

Several remarks are necessary here:

(i) In the previous formulae we assumed the state Ψ to describe a particle with mass m and momentum $\boldsymbol{p}$. We know that such states are represented by plane waves and are not normalisable. For a correct description we must use wave packets. The extension of equations (5.183) to (5.185) to the transformation laws of wave packets is straightforward but, unless it is absolutely necessary, we will use the plane wave expressions, for notational simplicity.

(ii) In order to describe the one-particle states we used the irreducible representations of the Poincaré group corresponding to real and positive values of m and integer, or half-integer, values of s. Group theory does not give us any hint to choose the particular values of m and s which describe what we call "elementary" particles. This is a purely empirical question. It is an experimental fact that all elementary particles known so far have spin s satisfying $s \leq 1$. We believe that, if we include gravitation, this inequality will be extended to $s \leq 2$.

(iii) We remind ourselves that a physical state is represented by a ray in the Hilbert space, i.e. a vector $\Psi \in \mathcal{H}$ defined up to a multiplicative constant. The normalisation condition tells us that we need only to consider unit-norm vectors, i.e. $\Psi \in \mathcal{H}$ such that $(\Psi, \Psi) = 1$. Furthermore, experiment shows that not all these vectors of $\mathcal{H}$ correspond to physically realisable states. The linear superposition of two physical states, although it is a vector in the Hilbert space, does not always correspond to a physical state. For example, a state with one proton and a state with one electron are both physical states. The first has electric charge $+1$ and the second -1. A linear superposition of the two will not have a definite value of the electric charge, so it will not be an eigenstate of the charge operator. It is an empirical fact that there are operators such that all physical states are eigenstates of them with well-defined eigenvalues. We say that such an operator defines *a superselection rule*. These operators split the Hilbert space into sub spaces corresponding to definite eigenvalues. We will restrict our discussion to one of these subspaces, for which all the states are always physically realisable. In the example above, the electric charge operator Q defines a superselection rule: every physical state $|\Phi\rangle$ must be an eigenstate of Q in other words, it must have a definite value of the electric charge. In the following chapters we will encounter examples of other operators with the same property.

• *The states of a massless particle.*

We saw that the physically interesting representations corresponding to $p^2 = 0$ are characterised by the value of helicity which takes integer or half-integer values. For zero helicity the representation is one-dimensional and for all other values it is two-dimensional, if we include the operation of parity. The states of a massless particle present some complications which we will discuss presently.

5.7.2.3 The multi-particle states. Let us consider a system made of two non-interacting scalar particles. Their state Ψ is given by a complex valued function of two variables $\Psi(p_1, p_2)$, such that $p_1^2 = m_1^2$, $p_2^2 = m_2^2$, $p_1^0 > 0$ and $p_2^0 > 0$. This function transforms according to

$$(U(a, \Lambda)\Psi)(p_1, p_2) = e^{i\Lambda(p_1+p_2)\cdot a}\Psi(\Lambda^{-1}p_1, \Lambda^{-1}p_2) \tag{5.186}$$

and is an element of the Hilbert space $\mathcal{H}^{(2)}$ with the scalar product

$$(\Phi, \Psi) = \int\int d\Omega_{m_1}(p_1)d\Omega_{m_2}(p_2)\Phi^*(p_1, p_2)\Psi(p_1, p_2) \tag{5.187}$$

This shows that the transformation law of the state is the same as for $[m_1, 0] \otimes [m_2, 0]$. If the two particles are identical, $m_1 = m_2 = m$, Ψ is symmetrical with respect to the exchange of the variables and transforms according to the $([m, 0] \otimes [m, 0])_S$ representation. The scalar product defined above is Poincaré invariant.

The states of two non-interacting spinless particles are described by the mass M of the system they form, given by $M^2 = (p_1 + p_2)^2$, a number which varies from $m_1 + m_2$ to infinity, and by the angular momentum of the system, characterised by an integer L. In terms of our previous analysis of elementary systems, for given values of M and L, the two-particle state belongs to the (M, L) representation of the Poincaré group. Clearly, there exists a continuum of possible two-particle states.

Formally we can repeat the analysis for two massless particles. The complication we alluded to previously comes from the fact that in this case M^2 varies from zero to infinity; in other words, there is no mass gap between the single particle and the multiparticle states. This is known as *the infrared problem* and has a simple physical origin: Let us take a one-photon state. Following what we said previously, it has a momentum p with $p^2 = 0$. However, since every measurement has a finite resolution Δp, there is no conceivable experiment which can distinguish this single particle state from an infinity of others which will contain in addition an arbitrary number of photons, provided their total energy is smaller than Δp. In Problem 2.3 we introduced the formalism of *coherent states* each one of which contains an indefinite number of photons. We shall come back to this problem when we develop better tools to tackle it but, for the moment, we adopt a simple-minded recipe: we assume that all particles, including photons, are massive. At the end of a calculation we may have to consider the limit $m_\gamma \to 0$.

One can obviously do a similar analysis for a system of more than two[26] particles with or without spin (in the case of identical particles, a rule of symmetrisation or antisymmetrisation, depending on whether their spin is integer or half-odd integer, is understood).

5.7.3 The Fock space

The Fock space describes the states of an arbitrary number of non-interacting particles. To simplify the presentation we will first assume that we have only one kind of particle: a scalar particle of mass m.

The Fock space is built as a direct sum of spaces, each of which corresponds to a fixed number of particles

$$\mathcal{F} = \oplus_n \mathcal{H}^{(n)} \tag{5.188}$$

---

[26]If there are more than two particles, there exist not only one but an infinite number of states for given mass and angular momentum.

where $\mathcal{H}^{(n)}$ is the Hilbert space of n-particle states. The elements of $\mathcal{H}^{(n)}$ are the symmetric functions $\Phi^{(n)}(p_1, \cdots, p_n)$ and the scalar product, invariant under $([m,0])_S^{\otimes n}$, is given by

$$(\Phi^{(n)}, \Psi^{(n)}) = \int \cdots \int d\Omega_m(p_1) \cdots d\Omega_m(p_n) \Phi^{(n)*}(p_1, \cdots, p_n) \Psi^{(n)}(p_1, \cdots, p_n)$$
(5.189)

The space $\mathcal{H}^{(0)}$ is the one-dimensional space of complex numbers.

An element of $\mathcal{F}$ is a multiplet of the form

$$\Phi = (\Phi^{(0)}, \Phi^{(1)}, \Phi^{(2)}, \cdots)$$
(5.190)

and the scalar product is given by

$$(\Phi, \Psi) = \sum_{n=0}^{\infty} (\Phi^{(n)}, \Psi^{(n)})$$
(5.191)

thus $\Psi \in \mathcal{H}$ if $(\Psi, \Psi) = \|\Psi\|^2 < \infty$. Fock spaces of more than one species of particles can be built in a similar way.

In the case of particles with mass m and spin s, the elements of $\mathcal{H}^{(n)}$ are the functions $\Psi_{\alpha_1 \ldots \alpha_n}^{(n)}(p_1, \ldots, p_n)$ which are, according to the value of the spin (i.e. integer or half-odd integer), symmetric or antisymmetric under the exchanges of any pairs (p_i, α_i) and (p_j, α_j). The spin indices α_i take the $2s + 1$ values s, $s - 1$, $\ldots$, $-s$. Alternatively, we can view each index α_i as a collection of $2s$ spinorial, dotted and/or undotted, indices.

5.7.4 Action of internal symmetry transformations on $\mathcal{F}$

In Chapter 3 we introduced the concept of internal symmetry transformations, namely transformations which do not affect the space-time point x but transform linearly the dynamical variables, such as the fields $\phi_a(x)$, among themselves. We want to study here the realisations of such symmetry transformations in terms of operators acting on $\mathcal{F}$.

We start with the group Z_2. As we shall see in Chapter 8, relativistic quantum mechanics predicts, and experiment confirms, that to every particle A there corresponds an *anti-particle*, denoted by $\overline{A}$, which has the same mass, spin and lifetime as the particle, but carries the opposite of all charges, such as the electric charge. This leads to the introduction of a linear unitary operator $\mathcal{C}$ which maps the one-particle state $|A\rangle$ to the corresponding anti-particle state $|\overline{A}\rangle$. This mapping is unitary and involutory, i.e. $\mathcal{C}^2 = I$, so $\mathcal{C}|A\rangle = \eta_A |\overline{A}\rangle$ with $\eta_A = \pm 1$. By assumption, $\mathcal{C}|0\rangle = |0\rangle$. We can deduce the action of $\mathcal{C}$ on multiparticle states, see Problem 8.1. The operator $\mathcal{C}$ is called *charge conjugation* operator and its action on $\mathcal{F}$ implies corresponding transformation properties of operators $\mathcal{O}(x)$ acting on $\mathcal{F}$. We shall see them in Chapter 11.

Let us now turn to internal symmetry transformations forming a Lie group G and we call Q the operator acting on $\mathcal{F}$ representing one of its generators. In this chapter we assume that the vacuum state is invariant under any transformation in G, i.e. $Q|0\rangle = 0$.[27]

---

[27]In Chapter 14 we shall see what happens if this condition is not satisfied.

If G is not semi-simple, $G \approx U(1) \times G'$ and Q is the generator of $U(1)$, it can be used to define a superselection rule i.e. all physically realisable states of $\mathcal{F}$ will be eigenstates of Q. The electric charge operator is an immediate example.

In section 5.5.2 we argued that the states of a system with symmetry group G are classified according to irreducible representations of the Lie algebra $\mathfrak{g}$. The eigenstates $|\Psi_i\rangle, i = 1, \ldots, r$ of the Hamiltonian, belonging to an r−dimensional representation of $\mathfrak{g}$ have the same energy, i.e. $H |\Psi_i\rangle = E |\Psi_i\rangle$, with E independent of i, and under the action of the group element with parameters $\theta^a, a = 1, \ldots, N$, they transform according to

$$|\Psi'\rangle = e^{iT^a\theta^a} |\Psi\rangle \tag{5.192}$$

where T^a, $a = 1, \ldots, N$, are $r \times r$ matrices representing the generators Q^a of G in the r−dimensional representation. Similarly, an observable $\mathcal{O}_i$ which belongs to an r-dimensional representation of $\mathfrak{g}$ transforms as

$$\mathcal{O}' = e^{iT\cdot\theta}\mathcal{O}e^{-iT\cdot\theta} \tag{5.193}$$

A final remark: Let us consider the example of the electric charge operator Q. It is the generator of infinitesimal phase transformations. Let q_A and q_B be the electric charges of the particle states $|A\rangle$ and $|B\rangle$, respectively. The electric charge of the two particle state $|A, B\rangle = |A\rangle \otimes |B\rangle$ is $q_A + q_B$, since $Q |A, B\rangle = (Q |a\rangle) \otimes |B\rangle + |A\rangle \otimes (Q |B\rangle) = (q_A + q_B) |A, B\rangle$. We say, "$Q$ is *additive*" or that it follows an *additive law*.

Let us now look at discrete transformations, such as the charge conjugation $\mathcal{C}$ and assume that $\mathcal{C} |A\rangle = \eta_A |\overline{A}\rangle$ and $\mathcal{C} |B\rangle = \eta_B |\overline{B}\rangle$. The tensor product state obeys the relation $\mathcal{C} |A, B\rangle = \eta_A \eta_B |\overline{A}, \overline{B}\rangle$, in other words, charge conjugation follows a *multiplicative law*.

5.7.5 Action of Poincaré transformations on $\mathcal{F}$

Poincaré invariance implies that, to each state Ψ and to each Poincaré transformation $\{a, \Lambda\}$, there corresponds a state $\Psi_{\{a, A(\Lambda)\}}$ such that $\Psi_{\{0,1\}} = \Psi$ and such that the transition probabilities are invariant

$$|(\Phi, \Psi)|^2 = |(\Phi_{\{a,A\}}, \Psi_{\{a,A\}})|^2 \tag{5.194}$$

It results from the above identity that there exists a linear mapping $U(a, A)$ on $\mathcal{H}$ which is either unitary

$$(U(a, A)\Phi, U(a, A)\Psi) \equiv (\Phi_{\{a,A\}}, \Psi_{\{a,A\}}) = (\Phi, \Psi) \tag{5.195}$$

or anti-unitary

$$(U(a, A)\Phi, U(a, A)\Psi) = (\Phi, \Psi)^* \tag{5.196}$$

If we are only interested in the restricted Poincaré group, $U(a, A(\Lambda))$ is unitary, unique up to a phase, and such that

$$U(a_1, A(\Lambda_1))U(a_2, A(\Lambda_2)) = \pm U(a_1 + \Lambda_1 a_2, A(\Lambda_1\Lambda_2)) \tag{5.197}$$

These are the so-called representations up to a phase.

For the full Poincaré group we must include the transformations of space inversion and time reversal. In Chapter 11 we shall show that consistency with the quantum mechanical commutation relations imposes that the representation of space inversion must be chosen to be unitary and that of the time reflection anti-unitary.

Let us take the example of space inversion and call $\mathcal{P}$ the corresponding unitary operator acting on $\mathcal{F}$. Since we use representations up to a phase, we will define $\mathcal{P}$ such that $\mathcal{P}^2 = \mathbf{1}$.[28]

• We assume that the vacuum state is invariant under $\mathcal{P}$: $\mathcal{P}\,|0\rangle = |0\rangle$.

• The state of a massive particle at rest is denoted by $|m,s,s_z\rangle$. Like angular momentum, spin is a pseudo-vector and therefore under parity, s_z does not change sign. It follows that $\mathcal{P}\,|m,s,s_z\rangle = \eta\,|m,s,s_z\rangle$ with $\eta = \pm 1$. We call η *the intrinsic parity* of the state. By convention we assign the value $\eta = +1$ to protons and neutrons and we shall discuss the assignment to other particles in Chapter 11. For a one-particle state we often use the notation J^P to denote its spin and parity. For example, the proton is a $\frac{1}{2}^+$ state.

• The state of a massless particle is characterised by its momentum $\boldsymbol{p}$ and its helicity h. Under parity $\boldsymbol{p} \to -\boldsymbol{p}$ and, since helicity can be viewed as the projection of spin in the direction of motion, h goes also into $-h$. It follows that $\mathcal{P}\,|\boldsymbol{p},h\rangle = \eta\,|-\boldsymbol{p},-h\rangle$ with $\eta = \pm 1$ an intrinsic parity. For example, we shall see in Chapter 11 that the one-photon state has $\eta = -1$, so the photon is a 1^- state.

In Problem 8.1 we will compute the transformation properties under parity of multiparticle states.

We leave the discussion concerning the time reversal operator for Chapter 11.

According to the general rules of quantum mechanics, the action of a Poincaré group transformation $U(a, A)$ on the states Ψ implies well-defined transformation properties of observables. Let us consider a linear operator $\mathcal{O}(x)$ acting on the states of $\mathcal{F}$ and belonging to an irreducible representation of the Lorentz group. The Poincaré transformation $x'^\mu = \Lambda^\mu_{\ \nu} x^\nu + a^\mu$ is implemented in the Hilbert space by the operator $U(a, A)$, under which

$$\mathcal{O}'(x') = U(a, A)\mathcal{O}(x)U^{-1}(a, A) \qquad (5.198)$$

Some examples:

(i) scalar operator $\phi(x)$: $U(a, A)\phi(x)U^{-1}(a, A) = \phi\left(\Lambda^{-1}(x - a)\right)$

(ii) spinor operator $\psi_\alpha(x)$: $U(a, A)\psi_\alpha(x)U^{-1}(a, A) = S_\alpha^\beta(A)\psi_\beta\left(\Lambda^{-1}(x - a)\right)$

(iii) Vector operator $V^\mu(x)$: $U(a, A)V^\mu(x)U^{-1}(a, A) = \Lambda^\mu_{\ \nu}(A)V^\nu\left(\Lambda^{-1}(x - a)\right)$

If the transformation U includes parity we must allow for a phase corresponding to an intrinsic parity. For example, a scalar operator transforms under parity as: $\mathcal{P}\phi(x^0, \boldsymbol{x})\mathcal{P}^{-1} = \eta_\phi \phi(x^0, -\boldsymbol{x})$ and similarly for the others.

---

[28]Following equations (5.171) and (5.172), we could give an alternative definition of $\mathcal{P}$ such that $\mathcal{P}^2 = (-)^F$ with F being the fermion number (the number of fermions minus the number of anti-fermions) of the state in which we apply $\mathcal{P}$. See Problem 16.5.

5.8 Problems

Problem 5.1 Using the Clebsch–Gordan decomposition (5.58) compute the following products of $SU(2)$ representations and interpret the results in terms of the quantum mechanical spin composition rules: $\mathbf{2} \otimes \mathbf{2}$, $\mathbf{3} \otimes \mathbf{3}$, $\mathbf{3} \otimes \mathbf{4}$.

Problem 5.2 Prove the following properties related to $O(3)$ representations:

1. The fully antisymmetric symbol ϵ_{ijk} $i, j, k = 1, 2, 3$, is an invariant $SO(3)$ pseudo-tensor.

2. The three quantities $(\boldsymbol{V} \wedge \boldsymbol{U})_i \equiv \epsilon_{ijk} V_j U_k$, built out of the components of two vectors $\boldsymbol{V}$ and $\boldsymbol{U}$, form a pseudo-vector. It is the well-known vector product of two three-dimensional vectors.

3. The quantity $\boldsymbol{V} \cdot \boldsymbol{U} = \delta_{ij} V_i U_j \equiv V_i U_i$, built out of the components of two vectors $\boldsymbol{V}$ and $\boldsymbol{U}$, is a scalar. It is the inner product of two vectors, also called the *scalar product*.

4. The quantity $M \equiv \boldsymbol{A} \cdot (\boldsymbol{B} \times \boldsymbol{C}) \equiv \epsilon_{ijk} A_i B_j C_k$, formed out of the three vectors $\boldsymbol{A}, \boldsymbol{B}$ and $\boldsymbol{C}$ is a pseudo-scalar. The volume of the parallelepiped formed by these three vectors is the absolute value of M and, as expected, is invariant under the full group $O(3)$.

Problem 5.3 Prove by explicit computation that $SU(2)$ and $O(3)$ have the same Lie algebra.

Problem 5.4 1. Show that the generator of spatial rotations around $\boldsymbol{n}$ is $Q = \boldsymbol{n} \cdot \boldsymbol{J}$ with matrix elements $Q_{ij} = -i\epsilon_{ijk} n_k$.
Show that $(Q^2)_{ij} = \delta_{ij} - n_i n_j$ and $Q^3 = Q$.

2. Show that the elements of the rotation matrix $R(\boldsymbol{\alpha})$ are $R_{ij} = n_i n_j + (\delta_{ij} - n_i n_j) \cos \alpha + \epsilon_{ijk} n_k \sin \alpha$

3. Prove the relations (5.100).

Problem 5.5 Prove the relation (5.179)

Problem 5.6 Prove that $SU(2) \times SU(2)$ and $O(4)$ (the group of rotations in a four-dimensional Euclidean space) have the same Lie algebra.

Problem 5.7 1. Use the tensor method to find the decomposition into direct sums of irreducible representations for the following products of $SU(3)$ representations: $\mathbf{3} \otimes \bar{\mathbf{3}}$, $\mathbf{3} \otimes \bar{\mathbf{6}}$

2. Use the Young tableaux to decompose the product $\mathbf{8} \otimes \mathbf{10}$

3. Use the combination of (5.109) and (5.110) to decompose the product $\overline{\mathbf{15}} \otimes \mathbf{10}$, with $\mathbf{15} = (2, 1)$.

Problem 5.8 *Subgroups of $SU(3)$*.

1. Show that $SO(3)$ is a subgroup of $SU(3)$. Write a set of common generators. How do the $\mathbf{3}$ and the $\mathbf{8}$ of $SU(3)$ transform under $SO(3)$?

2. Show that $SU(3)$ has a $SU(2) \times U(1)$ subgroup. Write its generators in the Gell-Mann basis. Show that the **3** of $SU(3)$ decomposes according to $\mathbf{3} \to \mathbf{2}_{1/2\sqrt{3}} + \mathbf{1}_{-1/\sqrt{3}}$ under $SU(2) \times U(1)$, with the subscript denoting its $U(1)$ charge, i.e. the eigenvalue of the corresponding generator. How does the **6** of $SU(3)$ transform under $SU(2) \times U(1)$?

Problem 5.9 1. Find the weight diagram of the representation **6** of $SU(3)$.

  Hint: Start from the product $\mathbf{3} \otimes \mathbf{3} \cong \mathbf{6} \oplus \mathbf{\bar{3}}$. *From the vector sum of the weights of the two* **3***s, you take out the roots of the* **$\bar{3}$** *and what is left is the weight diagram of the* **6**.

  2. Find the weight diagram of the representation **10** of $SU(3)$.

Problem 5.10 Show that complex conjugation changes the (j_1, j_2) representation of the Lorentz group to (j_2, j_1), i.e. that $\overline{(j_1, j_2)} = (j_2, j_1)$.

Problem 5.11 This is a continuation of Problem 3.4. Consider the invariance of the Lagrangian density under a group G of internal global transformations and let Q_a, $a = 1, \dots, N$, be the corresponding conserved charges. Show that they satisfy the algebra $\{Q_a, Q_b\} = f_{abc}Q_c$, with f_{abc} the structure constants of G. Thus, the Poisson brackets of the Q_as generate the Lie algebra $\mathfrak{g}$.

6
Particle Physics Phenomenology

6.1 Introduction

It is perhaps natural to expect a book on elementary particles to begin with a definition of what is an elementary particle. The trouble is that we have no such definition. Our ideas on the subject have evolved along with the increasing resolution power of our "microscopes", the instruments we use to study the microscopic structure of matter. So the only working definition we will use is that an elementary particle is a particle for which we have not been able to detect the presence of any constituents.

The idea that the structure of matter is discontinuous is very old and goes back to Democritus, a Greek philosopher who, around 400 BC, postulated the existence of fundamental building blocks of all matter which he called "atoms", "unbreakables". For Democritus the basic constituents of matter are the "atoms" and the "vacuum", empty space between the atoms.[1] Today we know that the objects we call "atoms" are not elementary but have instead a complex internal structure and can be broken. However, they are "unbreakable" in the sense that the pieces of an iron atom once broken, are no longer iron. The physical existence of atoms was established only by the beginning of last century and during the last hundred years we have uncovered deeper and deeper layers of the subatomic structure:

atoms → nuclei + electrons → protons + neutrons + electrons → quarks + electrons → ??

There is no reason to believe that there exists such a thing as "an innermost layer" and, even less, that we have already reached it.

6.2 Rutherford and the Atomic Nucleus

Our story will start in 1909. By that time radioactivity had been discovered and people had identified three kinds of "radiation" emitted by certain materials. In the absence of a real understanding of the underlying physics, these radiations were assigned the first three letters of the greek alphabet. These names have remained ever since, and even today we talk about α, β and γ radiation, although we now know that they correspond to emission of helium nuclei, the "alpha" particles; electrons and neutrinos, the "beta" decays; and photons, the "gamma" rays. It is remarkable that each one of these processes involves a different type of interaction among elementary particles, so

---

[1] "Νόμω γαρ χροιή, νόμω γλυκύ, νόμω πικρόν, ετεή δ' άτομα καί κενόν", which in free translation says, "Laws determine the tone, the sweetness or the bitterness, but everything consists of atoms and empty space".

Elementary Particle Physics. John Iliopoulos and Theodore N. Tomaras, Oxford University Press.
© John Iliopoulos and Theodore N. Tomaras (2021). DOI: 10.1093/oso/9780192844200.003.0006

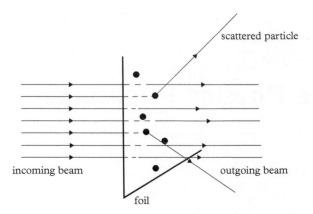

Fig. 6.1 A schematic representation of the Rutherford–Geiger–Marsden experiment. A beam of incoming α-particles hits a target consisting of a thin foil of gold. Most particles go through the foil unscattered, but a few get deflected even at large angles.

their study has contributed to our understanding of all fundamental forces. It is this understanding that we shall present, in a phenomenological way, in this book.

In 1911, E. Rutherford, along with H. Geiger and E. Marsden, established experimentally the existence of the atomic nucleus. The prevailing conception of atomic structure at that time presented atoms as small balls full of soft "matter", with no precise idea of the nature of the latter. Rutherford was inspired to investigate this question using a simple method which became classic: he bombarded, with a beam of α-particles, a thin foil of gold. He had previously established that α-particles interact strongly with matter and that their penetration length is short. Rutherford asked Geiger and Marsden to look at scattered α-particles at large deflection angles, something not expected from the theory of diffusion of particles from a soft medium. To their surprise they found that such scattering events did occur. More precisely, the picture they obtained was the following (see Figure 6.1): most α-particles were going through the foil unaffected, but occasionally, some were deflected even at very large angles. In fact, there were particles that were even scattered in the backwards direction. Rutherford interpreted these results as meaning that the atoms were mostly empty space containing some small hard grains. Starting from this idea he formulated the first realistic atomic model with a hard heavy nucleus and electrons moving in orbits around it. He thus initiated the field of nuclear physics, but also, he became the father of a new method of studying the structure of matter by making use of scattering experiments.

The α-particles used by Rutherford had kinetic energy of order 5 MeV and, consequently, a space resolution on the order of 10^{-12} cm, high enough to resolve the structure of atoms of 10^{-8} cm, but much too low to study the details of a nucleus whose typical size is a few times 10^{-13} cm. Through his experiments Rutherford showed that most of the mass of an atom is concentrated in the nucleus; so a simple assumption could be that there exist "elementary" heavy particles carrying charge opposite to that of the electron – the protons as such were identified by Rutherford only in

1917 – and a nucleus is a bound state of those. Two experimental facts contradicted this simple picture: First, if A is the mass number and Z the atomic number, (the charge of a nucleus), A was measured to be considerably larger than Z for most elements. Second, some elements were known to be unstable against β-decay, a process that involves the emission of an electron. In Rutherford's times, this decay was thought to be of the form

$$A(p_A) \to B(p_B) + e^-(p_e) \tag{6.1}$$

where A and B denote the initial and final nuclei, p_A and p_B their respective momenta and p_e the momentum of the electron. Today, we know that, in reality, the electron is accompanied by a third, neutral particle which we shall call an *anti-neutrino*, but this was discovered much later. Since it was unthinkable for the physicists of early twentieth century that a particle could be emitted from a nucleus without being already inside it, the first nuclear model assumed that nuclei were bound states of massive positively charged particles, to be identified soon as the protons and electrons. This answered also the mass-charge puzzle because the electrons could neutralise part of the charge without contributing much to the mass.

6.3   β-Decay and the Neutrino

The study of beta-decay played a very important role in the development of the entire field of subatomic physics, so we shall briefly review the main steps. We start with a simple kinematical analysis. Consider a particle A which decays into two particles: $A(p_A) \to B(p_B) + C(p_C)$. The conservation of energy and momentum applied in the rest frame of the decaying particle implies

$$M_A = p_B^0 + p_C^0, \quad 0 = \boldsymbol{p}_B + \boldsymbol{p}_C \tag{6.2}$$

and gives

$$p_B^0 = \frac{M_A^2 + M_B^2 - M_C^2}{2M_A}, \quad p_C^0 = \frac{M_A^2 + M_C^2 - M_B^2}{2M_A} \tag{6.3}$$

i.e. the energy of each one of the final particles is fixed. In nuclear decays, both the electron mass $m_e \equiv M_C \simeq 0.5$ MeV, and the mass difference $M_A - M_B$ between the initial and the final nuclei, typically of order a few MeV, are negligible compared to the nuclear masses which are on the order of several GeV. Therefore equation (6.3) tells us that, if β-decay is of the form given in (6.1), in the rest frame of the decaying nucleus the electrons are monoenergetic with energy given by

$$E_e \equiv p_C^0 \simeq \Delta M = M_A - M_B \tag{6.4}$$

The experiments performed up to 1914 did not show any appreciable contradiction with this result. We know today that this was due to lack of sufficient accuracy in the determination of the electron energy. The latter was estimated by the penetration length of the emitted electrons in various materials and the results were only approximate. The best measurements were performed in the chemistry department of the University of Berlin by O. Hahn, a former assistant of Rutherford, and L. Meitner.

The breakthrough occurred in 1914 with the arrival in Berlin of J. Chadwick, a student of Rutherford's who came to work with H. Geiger, Rutherford's former assistant in the α-particle scattering experiment and a world expert in particle counters. Chadwick had the bright idea to use a magnetic spectrometer to determine the electron energy, a method allowing for a much more accurate determination. The result was astonishing: the electrons were emitted with a continuous spectrum of kinetic energies ranging from almost zero to a maximum value consistent with the mass difference (6.4). Let us emphasise here that formula (6.4) is a straightforward consequence of energy and momentum conservation, so Chadwick's continuous spectrum seemed to be in contradiction with one of the most fundamental laws of physics. It was the first *energy crisis* in β-decay.

Apart from a second experiment in 1916 which confirmed Chadwick's result, nothing much happened with the electron spectrum until 1920. There was a world war going on. The years after 1920 were those of great confusion. Hahn and Meitner performed a new series of measurements, still insisting on the assumption of a two-body decay (6.1). They proposed an ingenious model trying to reconcile the theoretical formula (6.4) with the continuous spectrum suggested by the experiments. The idea was roughly the following: β-decay is a two-body process, therefore the electron is emitted with a fixed energy. However, nuclei were assumed to be bound states of protons and electrons, so the latter were part of nuclear matter. The emitted electron could suffer re-scattering and lose energy before getting out of the nucleus, so the observed energy was not the "primary" energy given by ΔM of equation (6.4), but rather the difference $\Delta M - \delta E$, where δE denotes the energy loss. Since there is no reason to assume that the latter has a fixed value, one could easily explain the continuous spectrum. If this explanation is correct, the electrons are always emitted initially with energy given by ΔM, which is *the maximum energy* $E_{\max}$ in the spectrum.

How can one measure the "primary" energy of the electron, if one has only access to its final energy? C.D. Ellis, another young assistant of Rutherford in the Cavendish, found the answer; in doing so he introduced into nuclear and particle physics a new measuring device, *the calorimeter*. Here is the principle: For a given initial nucleus we could measure independently the decay rate, which means that we could determine the number of decays N_D in a time interval ΔT. Let us assume that we can also measure the *total* energy E_T released during that time. If Hahn and Meitner are right and the primary electrons are mono-energetic, we must find $E_T = N_D E_{\max}$. If Chadwick is right and the original spectrum is indeed continuous, the result should be $E_T = N_D E_{\mathrm{av}}$, where E_{av} denotes the *average* energy in the spectrum. Therefore, what we need is not a spectrometer, because we are not interested in the energy of each individual electron, but a calorimeter, an instrument capable of measuring the total energy released in ΔT. For this we can surround the source with a material absorbing all electrons, thermally isolate the apparatus and measure the increase in temperature.

In this experiment Ellis was assisted by W.A. Wooster, a graduate student in Cambridge. Ellis and Wooster took two years to build and operate their calorimeter, but the results, obtained in 1927, left no room for argument. They were using as beta-ray source a rare earth ^{210}Bi with $E_{\max} = 1.05$ MeV and $E_{\mathrm{av}} = 390$ keV, with an uncertainty on the order of 30 keV. Their measurement corresponded to $E = 344 \pm 34$

keV, in agreement with E_{av}, but in complete disagreement with E_{max}. The spectrum was, indeed, continuous! This was the second, and more severe, *energy crisis*.

What was the reaction of the scientific community to this result? Meitner declared that she felt "a great shock!". She repeated the experiment and confirmed the result with better accuracy, but in her paper, in which she fully acknowledges the work of Ellis and Wooster, she offers no explanation. Resolving the energy crisis was now left to the theorists.

Already in the early twenties, N. Bohr was playing with the idea that in quantum mechanics, all conservation laws, including those of energy and momentum, were valid only on the average, and substantial deviation could occur. Together with H.A. Kramers and J.C. Slater, they published a paper in 1924 with a rather sketchy form of such a theory. Even Einstein and Dirac considered such possibilities for a while. The only one among the leading theorists who was consistently critical to all these attempts was Pauli. In the meantime, the crisis was enlarged: not only was energy conservation apparently violated, but so was the conservation of angular momentum, as well as the electron's statistical properties: if electrons were among the constituents of nuclei, they did not seem to obey the Pauli exclusion principle.

This confusion brings us to the end of 1930. A conference was organised in Tübingen in December to debate all the relevant issues. Pauli was invited, but he decided not to go. He sent a short letter to the organisers, dated 4th December, in which he made one of the boldest, but at the same time simplest, suggestions to solve the problem. We reproduce an English translation of the main parts of the letter here. It goes as follows:

Dear radioactive ladies and gentlemen,

I have come upon a desperate way out regarding the "wrong" statistics of the N- and the Li6-nuclei, as well as to the continuous β-spectrum, in order to save the "alternation law" of statistics [*today this is known as the "spin-statistics" theorem according to which integer spin particles are bosons and half-integer ones are fermions*], and the energy law. To wit, the possibility that there could exist in the nucleus electrically neutral particles, which I shall call neutrons, which have spin-$\frac{1}{2}$ and satisfy the exclusion principle and which are further distinct from light-quanta in that they do not move with the speed of light. The mass of the neutrons should be of the same order of magnitude as the electron mass and in any case not larger than 0.01 times the proton mass...

For the time being I dare not publish anything about this idea and address myself confidentially first to you, dear radioactive ones, with the question how it would be with the experimental proof of such a neutron, if it were to have a penetrating power equal to or about ten times larger than a γ-ray...

Your most humble servant, W. Pauli.

Pauli suggests that the correct form of β-decay is not (6.1) but instead:

$$N_1(p_1) \to N_2(p_2) + e^-(k_e) + \bar{\nu}(k_\nu) \tag{6.5}$$

where, anticipating later developments, we denoted the new particle as an *antineutrino*.

Concerning the suggestion itself, we note that it does solve all the problems, in fact better than what Pauli himself was thinking. If β-decay is a three-body process, the kinematics does not fix the energies of each one of the three final particles and the spectrum is predicted to be continuous. If the third particle has spin one-half, both the angular momentum and the exclusion principle problems can also be solved. Note that Pauli does not challenge the prevailing idea according to which particles coming out from a nucleus are inside it; this new particle is assumed to be part of nuclear matter.

Pauli called his new particle a *neutron*, but in January 1932 Chadwick discovered another particle, which he also called *neutron*, with mass slightly heavier than a proton's. In a few years time the proton–neutron model for the nucleus replaced the old proton–electron model. Nuclear structure took its present form. The times were ripe. We saw in Chapter 2 that a few years earlier, in 1928, Dirac had formulated the theory of spontaneous photon emission using the concept of a quantised electromagnetic field. The photon which is emitted in the decay of an excited state "does not come out of the atom"; it is created the moment of the emission. It took a few years before scientists could accept that what was true for light-quanta was also true for all other particles, including electrons. We shall present later the theory of quantum fields corresponding to the various particle species, but the principle is not different from what we saw for photons. In β-decay the electron, and Pauli's new particle, are created during the decay process. The theory was formulated by E. Fermi in 1933 and we shall present it in a later section. It is also Fermi who baptised the new particle. Reporting Pauli's suggestion at a conference in Varenna in 1933, he introduced the name *neutrino*, in Italian "little neutron", a name which has remained ever since. Although some theorists, including Niels Bohr, continued to consider energy violating models for some years, the energy crisis was definitely resolved.[2]

Having come this far, let us anticipate and give here the next step. Neutrinos went undetectable for many long years. Indeed, we shall explain soon that they have only weak interactions with matter and the probability for a neutrino to interact in a detector is extremely small.[3] For years, physicists thought it would be impossible to observe them. The picture changed only after the Second World War and the development of nuclear reactors. They produce a large number of neutrons which undergo β-decay

$$n \to p + e^- + \bar{\nu} \tag{6.6}$$

It follows that nuclear reactors produce a very high flux of anti-neutrinos, making their detection possible. In 1956, F. Reines and C.L. Cowan, working at the Savannah River nuclear reactor,[4] succeeded in observing a reaction caused by an anti-neutrino. It was the reverse of (6.6)

$$\bar{\nu} + p \to n + e^+ \tag{6.7}$$

in which an anti-neutrino interacts with a proton and gives a neutron and a positron. Cowan and Reines observed, in coincidence, the positron annihilation with an electron

---

[2]Bohr continued, for some time, his attempt on energy non-conserving interactions. It seems that he had even considered the emission of gravitational waves carrying the missing energy and momentum. As A. Pais in his book "Inward Bound" notes: "It is clear ... that the theory of particles and fields belongs to the post-Bohr era".

[3]In Problem 20.1 we ask the reader to estimate the thickness of a lead wall necessary to ensure an interaction of a 1 GeV neutrino. It turns out that it is on the order of a few 10^8 km!

[4]The $\bar{\nu}$ flux was 5×10^{13} $\bar{\nu}/\text{cm}^2\text{s}$.

Table 6.1 The complete table of elementary particles in the year 1932

Particle	Symbol	Mass (MeV)	Spin (s)	Electric charge (Q)
Proton	p	938.272	$\frac{1}{2}$	1
Neutron	n	939.565	$\frac{1}{2}$	0
Neutrino	ν	?	$\frac{1}{2}$	0
Electron	e^-	0.511	$\frac{1}{2}$	-1
Antiparticles	$\bar{p}, \bar{n}, e^+, \bar{\nu}$	same	same	opposite
Photon	γ	0	1	0

of the detector giving two photons and the neutron capture. More precisely, the target they used was tanks filled with water and liquid scintillator enriched with a cadmium compound. The chain of reactions was

$$\bar{\nu} + p \to n + e^+$$

followed by

$$e^+ + e^- \to 2\gamma \quad \text{and} \quad n +\, ^{108}Cd \to\, ^{109}Cd^\star \to\, ^{109}Cd + \gamma \tag{6.8}$$

We find three photons in the final state which can be detected in the scintillators. The signature of the event consists of the two photons from the electron–positron annihilation which are detected promptly, followed by the single photon produced by the decay of the cadmium excited state. This photon has a well-defined energy, the energy difference of the two atomic states, and it is detected with a time delay relative to the other two of about 5 microseconds, the mean lifetime of the excited state. Their first announcement of the discovery was a telegram to Pauli.

6.4 1932: The First Table of Elementary Particles

6.4.1 Everything is simple

Table 6.1 contains all the particles considered to be elementary in 1932. We saw previously that the year is not chosen randomly. It is the year of the discovery of the neutron and the formulation of the correct nuclear model. Our basic ideas on the structure of matter date from that year. The table is separated into two parts. The upper part contains two doublets, the proton–neutron and the electron–neutrino which, as we shall explain shortly, can be considered as the constituents of matter. We shall call them *matter particles.* The lower part contains only one entry, the photon, which is the quantum manifestation of the electromagnetic field. We shall call it the *quantum of radiation.*

Let us have a first cursory look at each entry separately.

• The protons and the neutrons form the nuclei. We shall call them collectively *nucleons.*[5] We shall see in one of the following sections that this is more than just a common name. They both have spin one-half and obey the Pauli exclusion principle.

---

[5]This name was introduced by C. Møller in 1941.

Compared to the other particles in Table 6.1, they appear to be heavy, with masses of order 1 GeV and this fact explains why the mass of the atoms is contained almost entirely in the nuclei.

• The electron is the oldest-known elementary particle. It has a long history but its discovery is attributed to J.J. Thomson who, in 1897, established that cathode rays consist of corpuscles carrying negative electric charge. Furthermore, he measured the ratio of charge over mass, e/m, quite accurately and determined that these properties are independent of the chemical composition of the cathode. In 1900, H. Becquerel proved that the same particles are emitted in β-decay. The ratio e/m was measured more accurately by R. Millikan in 1909. The electrons also have spin one-half and obey the Pauli exclusion principle.

• We have already presented the story of the neutrino.[6] In Table 6.1 we see a question mark in the entry of its mass. It shows our ignorance concerning its precise value. We shall come back in more detail to the problem of the neutrino mass in Chapter 20.

• At the end of the matter particles in the table we see a line under the title "anti-particles". When Dirac proposed his equation in 1928, it was meant to be a relativistic wave equation for the electron. It was, however, soon realised by Dirac that the same equation admits solutions describing a positively charged particle with the same mass and spin as the electron. We shall introduce and study the Dirac equation in Chapter 7. Dirac thought for a while to identify this new solution with the proton and he published a paper along these lines in 1930, but he soon understood that it is in fact a new particle. In 1931, he predicted the existence of such an *anti-electron*, which we call the *positron*,[7] and the following year C.D. Anderson detected this particle[8] in the cosmic rays using *a cloud chamber*, which was invented by C.T. Wilson two decades earlier.[9] This successful prediction was a triumph of the Dirac theory. Concerning the antiparticle of the neutrino, we saw previously that, by convention whose origin will be clear later, we called anti-neutrino the particle conjectured by Pauli. So, in 1932 this was the only one that people knew. The existence of what we call today the "neutrino" was established in the same indirect way in 1934 when F. and I. Joliot-Curie discovered a reaction that looked like β-decay but the emitted particle was a positron rather than an electron. In the notation of nuclear physics (A, Z) changes into $(A, Z - 1)$ according to

$$(A, Z) \rightarrow (A, Z - 1) + e^+ + \nu \tag{6.9}$$

---

[6]The terms "anti-particle" and "anti-neutrino" were used for the first time in 1934 by L. de Broglie. He had a theory in which the photon was a $\nu - \bar{\nu}$ bound state. The theory was soon abandoned but the terms remained.

[7]In the early literature it was often called a *positon*.

[8]Inconclusive evidence for its existence had been found earlier.

[9]Called also a Wilson chamber, it is a device to detect electrically charged particles. It is a sealed chamber containing a supersaturated vapour of water, alcohol, or other easily evaporated liquid. When a charged particle passes through the chamber it ionises the molecules which cause condensation of the vapour. This leaves a visible track along the particle's trajectory which can be recorded on a photographic plate. We can also add an external magnetic field which bends the trajectory according to the sign of the electric charge, thus making possible the distinction between positively and negatively charged particles. In addition, measuring the curvature of the trajectory in the magnetic field, allows the determination of the corresponding charge/mass or charge/energy ratio.

In other words, here a proton becomes a neutron with the emission of a positron and a neutrino. Such a reaction is energetically forbidden for free nucleons because the neutron is heavier than the proton. It becomes possible only for some isotopes of relatively light elements with a high proton content. Examples are $P(A = 30, Z = 15)$, which is the element discovered by F. and I. Joliot-Curie, $C(A = 11, Z = 6)$ etc. Since the reaction (6.9) involves the emission of a positron, which is the anti-particle of the electron, it is natural to expect also the emitted neutrino to be the anti-particle of the one emitted in neutron decay. Note, however, that in 1934 there was no evidence that these two were distinct particles. The conjecture about the existence of anti-particles was soon extended to all particles, including the nucleons, although in 1932 none of them could have been discovered. The anti-particles have the same mass, spin and lifetime as the corresponding particles, but carry the opposite of all charges, such as the electric charge. It follows that particles and anti-particles annihilate each other. We saw an example of such an annihilation in the reaction (6.8) between an electron and a positron. It took much longer to establish experimentally the existence of the anti-nucleons, because of their heavy masses and their very fast annihilations with ordinary nucleons. It was only in the years 1955 and 1956, with the advent of the first high energy particle accelerators, that anti-protons and anti-neutrons could be produced and detected. It was done in the Bevatron accelerator of the Berkeley National Laboratory through the reactions $p+p \rightarrow p+p+p+\bar{p}$ and $p+p \rightarrow p+p+n+\bar{n}$. We shall see in Chapter 8 that this particle–anti-particle correspondence is one of the most profound consequences of relativistic quantum theories.

• The last particle in Table 6.1 is our familiar photon, the quantum of the electromagnetic field. It has zero mass,[10] helicity one and no electric charge. Like the electromagnetic field in classical electrodynamics, the photon is supposed to mediate the interactions among electrically charged particles. We called it *a quantum of radiation* and we shall see that such mediators exist in fact for all interactions. In 1932, the photon was the only known quantum of radiation. A final remark: the photon has no electric charge, or any other additive quantum number. It follows that nothing can distinguish it from its anti-particle, so photons and "anti-photons" are in fact identical.

• Looking again at Table 6.1 we can formulate three empirical rules:

1. All matter particles have spin one-half. Radiation quanta have helicity one.

2. The matter particles can be separated into two symmetric groups: the two nucleons (proton–neutron) and the other two (neutrino–electron).

3. The role of each one of the elementary particles in the structure of matter is clear and well understood.

The structure of the world looked simple with a small number of fundamental constituents. Although matter appeared in an enormous variety of forms and properties, physicists believed that they could understand them following simple rules.

---

[10]No measurement of a particle mass can possibly give the value zero. Experimentally we can only obtain upper bounds. They are of various sorts: A bound of 6 ×10^{-16} eV is obtained from satellite studies of planetary magnetic fields. A laboratory Cavendish-type experiment gave a better bound of about 7×10^{-17} eV, while the much more stringent, though less reliable and model dependent, bound of order 10^{-27} eV is obtained from studies of galactic magnetic fields. All these bounds are already much smaller than the mass of any other elementary particle, and, as it will become evident later, there are good theoretical reasons to believe that the photon mass is actually exactly equal to zero.

6.4.2 Conservation laws – baryon and lepton number

We know from experiment that there exist quantities whose values remain constant in all physical processes. In classical physics we have encountered conserved quantities such as energy, momentum or angular momentum. We have seen that they are related to the assumed invariance of the equations of motion under translations and rotations. There exist also conserved quantities that are not related to any symmetry transformations of space-time. The most common example is the electric charge. This conservation law has many observable consequences. Let us mention two which apply, in particular, to particle physics:

• In a reaction of the form $A_1 + A_2 + \ldots \to B_1 + B_2 + \ldots$, in which a set of initial particles, the As, gives, through their interactions, a set of final particles, the Bs, the algebraic sum of the electric charges of the As is equal to that of the Bs.

• The electron, being the lightest of all electrically charged particles, is absolutely stable. The current limit is $\tau_e \geq 6.6 \times 10^{28}$ years.

Already in classical electromagnetic theory the conservation of the electric charge is a consequence of the equations of motion. Experiments in particle physics show the existence of additional conserved charges, although they are not sources of any macroscopic external field. We shall present two examples in this section and we shall have a more complete discussion later.

The baryon number.[11] In 1932 we could define this conserved charge as corresponding to the number of the particles we called "nucleons" in Table 6.1. They are the proton and the neutron. Although, as we saw already, in nuclear reactions a neutron can turn into a proton, or vice versa, we have never observed a reaction in which the number of nucleons is not conserved. For example, we never observed a proton decaying as $p \to e^+ + \gamma$, although this would have been allowed by electric charge conservation.[12] We conclude that we can assign to each particle in the table a new *charge*, which can take positive, negative or zero values, such that in every reaction the algebraic sum of the charges remains the same in the initial and the final state. In 1932, we could have called this new charge the *nucleon number*, but, anticipating what will come next, we call it the *baryon number* B. Among the particles in Table 6.1, nucleons are assigned the value $B = 1$, (we shall call them *baryons*), anti-nucleons the value $B = -1$, while all other particles have $B = 0$. Baryon number has the same properties as the electric charge, namely that it is conserved, baryons can be produced in baryon–anti-baryon pairs and the lightest baryon, the proton, is stable. The difference is that, as we have said already, baryon number is not the source of a classical field.

The lepton number.[13] In complete analogy we can define a new quantum number L with $L = 1$ for the electron and the neutrino, $L = -1$ for the positron and the anti-neutrino and $L = 0$ for all other particles. All experiments show that L is conserved in every reaction.

---

[11]From the Greek word "βαρύς" which means "heavy".

[12]The current lower limit on the lifetime of the proton is on the order of 10^{32} years.

[13]From the Greek "λεπτός" which means "thin", "fine", "delicate". This term was introduced by C. Møller and A. Pais in 1946.

Before closing this section, let us come back to the property of the quantum mechanical states we called *superselection rule* in section 5.7.2. We noted there that the operator of electric charge Q defines such a rule in the sense that only eigenstates of Q correspond to physically realisable states. Here we want to generalise this concept in the form of a general rule:

For any physical system, there exist operators $\mathcal{O}$ which commute with the Hamiltonian of the system, $[\mathcal{O}, H] = 0$, and are, therefore, constants of the motion, such that only their eigenstates are physically realisable states.

Experiments show that besides the electric charge, superselection rules are also defined by the other two conserved quantities we introduced, namely the baryon and lepton numbers. We have never observed a state of the form $C_1 |p\rangle + C_2 |e^+\rangle$, although such a state is an eigenstate of Q. The superselection rules split the Hilbert space into subspaces corresponding to the eigenvalues of the conserved charges. The vectors belonging to these subspaces form the set of physically realisable states. A final remark: Superselection rules are established experimentally and they are valid as long as the corresponding conservation laws remain valid. For example, there have been theoretical speculations according to which baryon and lepton numbers may not be absolutely conserved. In this case, states like $C_1 |p\rangle + C_2 |e^+\rangle$ would be physically realisable. Let us also note that superselection rules are always related to conserved quantities, but the inverse is not always true. Not all operators that commute with the Hamiltonian form such a rule. Angular momentum is a conserved quantity, but the physical states are not necessarily eigenstates of it.

6.4.3 The four fundamental interactions

In order to understand the structure of matter, a knowledge of the elementary constituents is not enough. We should supplement it with that of the interactions among them. Although our ideas on the nature of the elementary particles evolved considerably during the twentieth century, those on the fundamental interactions have remained remarkably unchanged. In all scales, from that of the elementary particles to that of the universe as a whole, the structure of matter is due to four fundamental interactions. One of the main subjects of this book is to describe their detailed properties, but, for the purposes of this discussion, it will be sufficient to classify them according to two simple parameters, whose understanding does not require any knowledge of quantum field theory. They are: (i) the *strength* of the interaction and, (ii) its *range*. These concepts were introduced in Chapter 4, when we discussed scattering from a non-relativistic potential, but we can also adopt a purely phenomenological viewpoint. In a decreasing strength order, we can list the interactions as follows:

• *The strong interactions.* They are mainly responsible for nuclear structure. Indeed, we have explained already that the various nuclei are bound states of protons and neutrons. Each proton carries one unit of positive electric charge and, therefore, protons are subject to electrostatic repulsion. The remarkable cohesion of nuclei shows the existence of attractive forces among protons and neutrons, which must be much stronger to outweigh the electrostatic repulsion. We know experimentally that strong interactions have a short range, typically on the order of a few times 10^{-13} cm, and their effects are not manifest in everyday life. Not all known particles are subject to strong interactions. Those that are, are called *hadrons*.[14] Among the particles of Table 6.1, the nucleons are, naturally, hadrons but all other particles, namely electrons, neutrinos and photons, are not.[15] Following the terminology we used in the last section, we will call the matter particles that are not hadrons i.e. electrons and neutrinos, *leptons*. Understanding the nature of the strong interactions has been a long-lasting problem in high energy physics. Its solution has offered deep insight into the laws of nature, insight that far exceeds the domain of nuclear forces. We shall give a brief description of these ideas later in this book.

• *The electromagnetic interactions.* They are responsible for the atomic and molecular structure. They have long range and their effects are observable at macroscopic scales. We mentioned in Chapter 2 that the electromagnetic interactions, expressed order by order in perturbation theory, can be viewed as the results of the exchange of virtual photons, the quanta of the electromagnetic field, between electrically charged particles. We also saw in Chapter 4 that the long range is attributed to the zero mass of the photon.

• *The weak interactions.* They are responsible for nuclear β-decay as well as the decays of other unstable particles. To give a measure of how weak the weak interactions are, we note that, experimentally, neutrinos, which have no strong or electromagnetic interactions, are produced at the centre of stars and escape from them without being absorbed. For a very massive star, neutrino radiation is the only known cooling mechanism. We know today that weak interactions share many common features with the electromagnetic ones. They are also due to the exchange of virtual quanta, the weak vector bosons W^+, W^- and Z. However, these quanta are massive and, as a result, the weak interactions are of short range. Naturally, the existence of these weak interaction intermediaries was unknown in 1932. In this book we shall explore this weak-electromagnetic analogy much further.

• *The gravitational interactions.* Their importance in both terrestrial and cosmic phenomena is well known but, at the microscopic level, their effects are too weak to be observable. We shall not discuss them much further in this book.

6.5   Heisenberg and the Symmetries of Nuclear Forces

As we have noted already, 1932, the year of the discovery of the neutron, is the year when nuclear physics started taking its present form. Of course, the transition from the old nuclear model of protons and electrons to the modern one of protons and neutrons was not instantaneous. It took several years for the new paradigm to be generally accepted. What is less known is that the same year marks another revolution in our understanding of the fundamental interactions, which is closely related to the new nuclear model. It is the introduction of the concept of *internal symmetries*, a

---

[14]From the Greek word "αδρός" which means "strong", "hard", "rough".

[15]The reader may be puzzled by the fact that we have introduced three different terms to denote apparently the same particles. We called the proton and the neutron, first "nucleons", then "baryons" and now "hadrons". Indeed, if we restrict ourselves to the 1932 elementary particles, the three names are synonymous. However, we shall see soon, with the discovery of new particles, that each of the terms denotes a different set of particles. The proton and the neutron are the only ones which belong to all three sets.

concept that introduced into theoretical physics a degree of abstraction not easy to grasp intuitively. The original idea is due to W. Heisenberg but, as we shall explain briefly later, the actual history is more complicated. We shall present the idea using our present understanding and we shall comment on the historical developments at the end.

We are used to the fact that particles may carry internal degrees of freedom. An electron at rest is described by two orthogonal states in the Hilbert space, the $|\uparrow\rangle$ with spin projection $+\frac{1}{2}$ and the $|\downarrow\rangle$ with spin projection $-\frac{1}{2}$. They satisfy the orthogonality relation $\langle\uparrow|\downarrow\rangle = 0$. Normally, we could have invented two different names to describe these two states. Nevertheless, we talk about one electron and the reason why this is correct is that there exist transformations which leave the equations of motion invariant and transform one state to the other. Indeed, if the spin projections are taken along the z-axis, a rotation of 180° around any axis in the x-y plane interchanges $|\uparrow\rangle$ and $|\downarrow\rangle$.

The study of the energy levels in various nuclei has shown that, as far as the nuclear forces are concerned, the protons and the neutrons play a very symmetrical role. In 1932, the evidence for this symmetry was very sketchy, but today it is very well established. Experiments show that nuclei obey a kind of approximate *exchange symmetry*, namely a symmetry under the interchange $p \leftrightarrow n$. It is often called *charge independence*, in the sense that the two-body nuclear forces are the same for any pair p–p, p–n, or n–n. Of course, this symmetry ceases to be valid when the electromagnetic interactions are taken into account because the proton is charged and the neutron is not.

It is tempting to try to enlarge this discrete symmetry into a continuous one. Since in quantum mechanics protons and neutrons are described by complex-valued wave functions, the straightforward generalisation of the exchange symmetry should be a complex two-dimensional rotation. It can be formulated as follows:

In non-relativistic quantum mechanics the wave function of an electron is represented by a two-component complex spinor

$$\psi(x) = \begin{pmatrix} \psi_\uparrow(x) \\ \psi_\downarrow(x) \end{pmatrix} \tag{6.10}$$

whose components transform among themselves under a rotation of the three-dimensional space. Here $\psi_\uparrow(x)$ and $\psi_\downarrow(x)$ denote the wave functions of an electron in the $|\uparrow\rangle$ and $|\downarrow\rangle$ spin state, respectively. Following the notation introduced by W. Heisenberg in 1932, we introduce another two-component complex spinor, analogous to the one of equation (6.10), of the form

$$\Psi(x) = \begin{pmatrix} \Psi_p(x) \\ \Psi_n(x) \end{pmatrix} \tag{6.11}$$

As indicated, here the components $\Psi_p(x)$ and $\Psi_n(x)$ denote the wave functions of a proton and a neutron, respectively. Note that, since protons and neutrons have also spin equal to $\frac{1}{2}$, each component $\Psi_p(x)$ and $\Psi_n(x)$ is itself a complex two-component spinor of the form of (6.10). So, (6.11) should be read of as an object with four components

$$\Psi(x) = \begin{pmatrix} \Psi_{p\uparrow}(x) \\ \Psi_{p\downarrow}(x) \\ \Psi_{n\uparrow}(x) \\ \Psi_{n\downarrow}(x) \end{pmatrix} \tag{6.12}$$

Obviously, rotations in our three-dimensional space rotate the components of each one of the spinors Ψ_p and Ψ_n among themselves, but do not mix Ψ_p and Ψ_n. A rotation in our space does not result in a state which is a superposition of a proton and a neutron, so the notation introduced in (6.11) appears to be meaningless. Here comes the big step: we shall assume that there exists a second three-dimensional space, isomorphic to, but distinct from, the one in which we live. A rotation in this new space mixes Ψ_p and Ψ_n the same way that a rotation in our space mixes the up and down components of a spinor.[16] E.P. Wigner introduced in 1937 the term *isotopic spin*, later simplified to *isospin* for this new concept, so we say that Ψ of equation (6.11) is a two-component spinor in isospin space,[17] each component of which is a two-component spinor in ordinary space. The assumption is that, up to electromagnetic corrections, the strong nuclear forces are invariant under rotations in isospin space, under which a proton and a neutron behave like the two components of an isospinor. We call such an invariance under transformations that do not affect the space-time point *an internal symmetry*. With this assumption it is justified to use the common name "nucleons", which we introduced earlier for protons and neutrons. They are two different forms of the same "particle", the nucleon.

As we shall see later, with the discovery of new hadrons, this invariance under isospin transformations has been extended and covers not only the nuclear forces, but also the entire domain of hadronic physics. The conceptual change has been very important: for the first time, non-trivial internal symmetries have been considered in particle physics.[18]

Heisenberg's isospin space was three-dimensional and the transformations form a group $O(3)$, or $SU(2)$, like our familiar rotations. However, the concept was subsequently enlarged as new particles were discovered and larger internal symmetry groups were brought into evidence. In order to appreciate, in its full significance, the conceptual change in our way of thinking brought about by Heisenberg, let us point out that in mathematics it is possible to characterise a space by studying the group of symmetry transformations which act on it. From this point of view, to the question, "Which is the space of elementary particle physics?" we should answer that it is a multi-dimensional manifold with complicated geometrical and topological properties, and that only a sub space of it, the four-dimensional Minkowski space, is directly accessible to our senses.

---

[16]We emphasise here that all these considerations are valid only in the absence of the electromagnetic interactions. In this case the electric charge is not defined and there is no superselection rule forbidding the linear superposition of a proton and a neutron state.

[17]We often use the term *isospinor*.

[18]Strictly speaking this is not absolutely correct. Even in 1926 people had realised that both the Schrödinger equation and the normalisation condition are invariant under a multiplication of the wave function by a constant phase. This is an internal symmetry since the phase transformations do not affect the space-time point x. Nevertheless they had never been viewed this way and, from the group theory point of view, the phase transformations are kind of trivial.

It took six years, as well as the work of many physicists, for Heisenberg's original suggestion of 1932 to become the full isospin symmetry of hadronic physics we know today. That is a remarkably short time, given the revolutionary nature of the concept.

Heisenberg's 1932 papers are an incredible mixture of the old and the new. He does not propose a clearly new nuclear model with the proton and the neutron as elementary constituents. For many people at that time the neutron was a new bound state of a proton and an electron, rather like a "small" hydrogen atom. Heisenberg does not reject this idea. Although for his work he considers the neutron as a spin one-half Dirac fermion, something incompatible with a proton–electron bound state, he notes that "...under suitable circumstances the neutron will break up into a proton and an electron in which case the conservation laws of energy and momentum probably do not apply." On the β-decay controversy he does not take any clear stand, but he sides more with his master Bohr than with his friend Pauli:[19] "...The admittedly hypothetical validity of Fermi statistics for neutrons as well as the failure of the energy law in β-decay proves the inapplicability of present quantum mechanics to the structure of the neutron." In fact, Heisenberg's fundamental contribution should be appreciated not *despite* these shortcomings, but precisely *because* of them. We should remember that in 1932 experimental data on nuclear forces were almost entirely absent. The deuteron had just been discovered, but its spin, or the precise value of its binding energy, were not known. Quantitative measurements on nucleon–nucleon scattering did not exist. Heisenberg had to guess the values of the nuclear attractive forces between nucleon pairs by using a strange analogy with molecular forces. He postulated a p–n and an n–n nuclear force,[20] but not a p–p one, so his theory was not really isospin invariant. Nevertheless, he made the conceptual step of describing the nucleon wave functions in terms of the four-component spinor we introduced in equation (6.12) and the nuclear potential in the form of two-by-two Pauli matrices.

In the following years three important developments allowed Heisenberg's initial suggestion to become a complete isospin invariant theory.

The first, and probably the most important, was the progress in experimental techniques, which brought more detailed and more precise data. They concerned both new determinations of nuclear energy levels and measurements of nucleon–nucleon scattering. They showed the need for the introduction of a p–p force and confirmed the charge independence of all nuclear forces. By 1936, this result was firmly established. In a paper by B. Cassen and E.U. Condon we find the full theoretical formulation of charge independence but also the extension of the Pauli exclusion principle to include both spin and isospin. In the notation we used before, if Ψ is the wave function of two nucleons, which depends on their positions x_a, their spin and isospin indices s_a and t_a, it is assumed to satisfy the full antisymmetry property, namely

$$\Psi(x_1, s_1, t_1; x_2, s_2, t_2) = -\Psi(x_2, s_2, t_2; x_1, s_1, t_1) \tag{6.13}$$

---

[19]We know from a letter he sent to Pauli dated 1 December 1930 that the latter had discussed his neutrino hypothesis with him prior to its announcement at the Tübingen conference.

[20]In fact, after Fermi published his paper on β-decay, which we present in the next section, Heisenberg played with the idea of the p–n nuclear force being the result of a virtual exchange of an electron–antineutrino pair between the two nucleons.

The second was the formulation by Fermi in 1933–1934 of a theoretical model for the amplitude of neutron β-decay, which we shall present in the next section. The influence of this work by far exceeds this initial problem and covers subjects even beyond the theory of elementary particles. In Chapter 2 we introduced the notion of the quantised electromagnetic field and showed that it describes a particle, which we called the *photon*. Fermi extended this idea and introduced for the first time the concept of quantised fermion fields and showed that this was the correct description of nucleons and leptons. Since that time, quantum field theory has become the universal language and one of the cornerstones of modern theoretical physics. To every particle corresponds a quantum field which, when expanded in creation and annihilation operators the way we explained in Chapter 2, describes the excitations we call particles. We shall study this theory in Chapter 10.

The third important development was Yukawa's introduction of the *meson* as an intermediary for the nuclear forces. Although his initial suggestion implied a particle transforming under Lorentz transformations as the zero component of a vector, it was soon replaced by a spin zero one. In 1938, N. Kemmer had the idea to incorporate Yukawa's meson into Heisenberg's isospin formalism and write the first fully isospin invariant pion–nucleon interaction. He assumed the existence of three pions with charges $+1$, -1 and 0, which he grouped together in an isospin triplet $\boldsymbol{\pi}$. Like the two nucleons, the proton and the neutron, the charged and the neutral pions should be nearly degenerate in mass, with a mass difference sufficiently small to be attributable to electromagnetic effects. Kemmer used the formalism of quantum field theory which means that he introduced quantum fields to describe nucleons and pions. In terms of these fields he wrote an interaction Lagrangian density of the form

$$\mathcal{L}_I = g\overline{\Psi}(x)\boldsymbol{\tau} \cdot \boldsymbol{\pi}(x)\Psi(x) \tag{6.14}$$

where $\Psi(x)$ is an isospin doublet field describing the nucleons and $\boldsymbol{\pi}(x)$ an isospin triplet describing the pions. $\boldsymbol{\tau}$ are the three Pauli matrices. We will not explain the notation in detail here. This is done in Chapters 10 and 11.

Although the particle, which was believed in 1938 to be the Yukawa meson, turned out, in fact, to be the μ lepton, the formalism remained unchanged when the real pion was discovered in 1947. The only difference is that, as was determined experimentally, it was a 0^- pseudo-scalar and the interaction (6.14) became what we know today:

$$\mathcal{L}_I = g\overline{\Psi}(x)\boldsymbol{\tau} \cdot \boldsymbol{\pi}(x)\gamma_5\Psi(x) \tag{6.15}$$

where again the precise definition of the various symbols used will be explained later.

When Kemmer wrote this interaction in 1938 there was evidence for the existence of a charged Yukawa meson, although, as we shall see soon, it was the wrong one, but there was no evidence at all for a neutral one. Kemmer understood that isospin symmetry required such a neutral meson, and in 1940 S. Sakata and Y. Tanikawa remarked that it could have escaped detection because it could decay into two photons with a very short lifetime. It is precisely what turned out to be the case. It was discovered at the Berkeley electron synchrotron in 1950 by H.J. Steinberger, W.K.H. Panofsky and J.S. Steller. So, π^0 is the first particle whose existence was predicted as a requirement for an internal symmetry[21] and the first to be discovered in an accelerator. We shall find more examples later on.

6.6 Fermi and the Weak Interactions

Already in 1926, before the introduction of the Schrödinger wave equation, Fermi had published two papers with the statistical rules which established Fermi quantum statistics and gave fermions their name. In 1933, he came back with one of the most influential papers in particle physics in which he proposed a field theory model for the β-decay of neutrons. Even today, when this theory has been superseded by the Standard Model of electroweak interactions, which we shall develop in this book, Fermi's theory is still used as a good low energy approximation.

This paper contains many revolutionary ideas. Fermi was one of the first physicists who believed in the physical existence of the neutrino. Contrary to Heisenberg, in the Bohr–Pauli controversy Fermi sided clearly with the second. But he went further and broke completely with the prevailing philosophy, according to which particles that come out from a nucleus ought to be present inside it.[22] In his paper he formulated the full quantum field theory for fermion fields and introduced the formalism of creation and annihilation operators, the analogue of the ones we used for the electromagnetic field in equations (2.23)and (2.24). It was the first time that quantised fermion fields appeared in particle physics. We shall give a full account of this formalism in Chapter 10 where we shall also indicate the novel features that Fermi introduced in order to incorporate Fermi statistics. The paper appeared at the beginning of 1934 in Italian[23] under the title *Tentativo di una teoria della emissione di raggi β*.

Fermi's starting point was an analogy with the electromagnetic interactions, in which the current j_μ produced by the charged particles acts as the source of the electromagnetic potential A^μ. We remind the reader that the process of neutron β-decay is $n \rightarrow p + e^- + \bar{\nu}$. Fermi considered the interaction as one between the nucleon pair and the lepton pair. He assimilated the first with the source j and the second with the potential A. As we noted before, his novel idea was to introduce quantised fermion fields for the electron, $\psi_e(x)$, and the neutrino, $\psi_\nu(x)$. He did not do the same for the nucleons because there was still confusion regarding the magnetic moment of the proton and it was not clear what kind of wave equation a proton field should satisfy. He bypassed this difficulty by considering a static density for the nucleons and, since he wanted it to describe transitions between a proton and a neutron, he used Heisenberg's $\tau_\pm$ isospin operators, which describe precisely the $p \leftrightarrow n$ transitions. The result was the following expression for the β-decay interaction Hamiltonian density[24]

---

[21]Kemmer has not received the credit he deserves for his contribution. Although his equation (6.15) can be found in practically every book on particle physics, his name is rarely mentioned.

[22]Similar ideas had been expressed before by D. Iwanenko in 1932 and Francis Perrin in 1933. The latter wrote: "...The neutrino ...does not preexist in atomic nuclei, it is created when emitted, like the photon.", but Fermi was the first to show how such a thing could actually happen.

[23]An English version had been submitted earlier in *Nature*, but it was rejected "because it contained speculations too remote from reality to be of interest to the reader".

[24]As we did in equations (6.14) and (6.15), we anticipate here a notation for the fermion fields which will be explained in Chapter 10.

$$\mathcal{H}_I = \frac{G_F}{\sqrt{2}} \left[\tau_- \psi_e^\dagger(x) \psi_\nu(x) + \tau_+ \psi_\nu^\dagger(x) \psi_e(x) \right] \tag{6.16}$$

where † means "Hermitian adjoint" and $G_F/\sqrt{2}$ is a coupling constant. F stands for Fermi.[25]

Using this simple form, Fermi derived already in 1934 many important physical results. Let us mention some of them, a more detailed discussion being left for Chapter 18:

• The value of the coupling constant. In Chapter 11, we will see that simple dimensional analysis shows that G_F has dimensions $[M]^{-2}$. By comparison with the observed decay rates, we deduce the approximate value

$$\frac{G_F}{\sqrt{2}} \simeq 10^{-5} M_p^{-2} \tag{6.17}$$

Fermi's initial estimate was off by a factor of two.

• The neutrino mass. Fermi computed the dependence of the energy spectrum of the emitted electron on the assumed neutrino mass. We shall repeat this calculation later. We note here that, since the decay is a three-body process, the spectrum is predicted to be continuous and it is this experimental observation which led Pauli to conjecture the existence of the neutrino. Fermi's important observation was that the spectrum is most sensitive to the neutrino mass at the end, when the electron takes almost all available energy and the neutrino is emitted with almost zero kinetic energy. It is called *the end-point spectrum* and, even today, it provides the best method for a direct measurement of the neutrino mass. By comparison with the available data at the time, and including an estimate of the electromagnetic corrections, Fermi concluded that the β-decay data were compatible with a zero value of the neutrino mass, a result still valid today. We shall devote a special section to the neutrino masses in Chapter 20.

• Most results on β-decay do not come from free neutrons but from ones bound in nuclei. Fermi observed that the result should depend on the nuclear matrix element between the initial and the final nucleus. This led him to present a first classification of the decays, which opened a new chapter in nuclear physics. We shall mention some results very briefly later.

As we see, the effort to understand the nuclear forces went in parallel with that of the weak interactions and they influenced each other considerably. By 1936, the use of Dirac fields quantised *à la* Fermi for the nucleons became common in describing all nuclear interactions, and the β-decay Hamiltonian of equation (6.16) took the form

$$H_I = \frac{G_F}{\sqrt{2}} \sum_{i=1}^{5} \overline{\Psi}_{(p)}(x) O_i \Psi_{(n)}(x) \overline{\psi}_{(e)}(x) O_i \psi_{(\nu)}(x) \tag{6.18}$$

where again the notation will be explained in Chapter 10. We just point out here that the sum runs over five operators which are bilinear in the fermion fields and transform

---

[25]This is the modern notation. Fermi used simply the symbol g.

under Lorentz transformations as scalar (S), pseudo-scalar (P), vector (V), pseudo-vector (or axial vector) (A) and two-index tensor (T). It is the form under which the Fermi theory is known.[26] As we shall see later, following an impressive series of experimental and theoretical investigations, the form (6.18) was gradually reduced to a superposition of only the vector and pseudo-vector parts in the V-A combination which violates maximally the invariance under space inversions.

Another important property of the Fermi Hamiltonian was discovered in 1936. M. Fierz computed the cross section for neutrino scattering and found that, at high energies, it increases with the neutrino energy

$$d\sigma(\bar{\nu} + p \to n + e^+) = \frac{G_F^2}{2\pi^2} \, \boldsymbol{p}_\nu^2 \, d\Omega \qquad (6.19)$$

where $\boldsymbol{p}_\nu$ is the neutrino momentum in the centre-of-mass system and $d\Omega$ is the element of the solid angle in the direction of the positron momentum. Similar conclusions were also reached by Heisenberg for the inelastic cross sections. One could guess these results by a simple dimensional analysis argument, taking into account the fact that the coupling constant G_F has the dimensions of inverse mass squared. It became immediately obvious that such a behaviour is unacceptable. Indeed, at order G_F^3 we would expect a $\boldsymbol{p}_\nu^4$ term and, at sufficiently high energies, the higher order terms would exceed the lowest order result making an expansion in powers of G_F meaningless. This problem haunted weak interactions for many years and led to the formulation of the new theory, whose exposition is one of the main subjects of this book.

6.7 The Muon and the Pion

The apparent simplicity of Table 6.1, with a small number of elementary particles fully explaining the structure of matter, did not last long. In the years 1936−1937 a new particle was discovered in cosmic rays. C.D. Anderson and S.H. Neddermeyer used the same Wilson chamber, which made possible positron's discovery, in order to clear up a peculiar feature which had appeared in cosmic rays. The latter seemed to have two components, a "soft" one, consisting of electrons and positrons, and a "hard" one, much more penetrating. They improved the apparatus with the insertion of a platinum absorber, which stopped the soft particles and isolated the hard component. They concluded that the data indicated the existence of a new particle, having the same electric charge as the electron and a mass much larger than that of an electron, but much smaller than that of a proton. They called it a "heavy electron".[27] It was R. Oppenheimer and R. Serber who first suggested in 1937 that this particle could be the Yukawa meson which was proposed to mediate the nuclear forces.[28] We saw that Kemmer followed this suggestion and wrote the first isospin invariant pion–nucleon interaction in 1938. It took ten years, largely owing to the intervening Second World

---

[26]It is not clear whether Fermi ever wrote this form. It is possible that he had introduced it in one of his lectures. It appeared for the first time in a review article by H.A. Bethe and R.F. Bacher in 1936.

[27]The same result was found by J.C. Street and E.C. Stevenson at Harvard and Y. Nishina, M. Takeuchi and T. Ichimaya in Tokyo.

[28]E.C.G. Stückelberg, from Geneva, as well as Yukawa himself, made the same suggestion.

War, for people to realise that the initial Anderson and Neddermeyer intuition to call the particle a "heavy electron" was in fact correct and the Yukawa meson was still to be discovered. The particle was finally given the name *muon* and the symbol μ, but for many years people continued to call it "mu-meson", despite the fact that it was known to be a new heavy lepton. Today we know that it has a mass of 106 MeV and it is unstable, decaying according to

$$\mu^- \to e^- + \nu_\mu + \bar{\nu}_e \tag{6.20}$$

with a mean lifetime of order 2×10^{-6} s. ν_μ is a new kind of neutrino associated to the muon. Of course, the early results showed only that it was a three-body decay, from the continuous spectrum of the emitted electron, and the identification of these distinct types of neutrinos came later. We shall discuss these developments soon.

In retrospect, the discovery of the muon marks a turning point in the physics of elementary particles: it was the first particle whose role in the structure of the world was, and still remains to the present day, unknown.[29]

So, by 1937 there was a candidate for the Yukawa meson, and in the following years several of its properties were extracted from cosmic ray data. Its mass and its lifetime were approximately determined and it was known to decay into electrons. However, with the accumulation of more data, this interpretation started facing serious problems. Today they have only historical interest, so we will not present them in any detail. We just point out that the assignment of the muon as the carrier of the nuclear forces was soon found to be incompatible with the measured values of its lifetime and its absorption cross section in nuclear matter.

In a 1947 conference in Shelter Island (we shall have occasion to talk about it in more detail later), R. Marshak proposed a radical solution to all these problems: The Yukawa meson is not the one you think! In fact, there are two mesons, one of which is strongly coupled to nucleons and mediates the nuclear forces. It is unstable and decays into a second one, weakly coupled to nucleons and therefore not strongly absorbable in nuclear matter. As a result, it traverses the atmosphere and is the one found by Anderson and Neddermeyer at low altitudes.[30] The prediction was very timely: A few weeks earlier C.F. Powell and his group in Bristol (C.M.G. Lattes, H. Muirhead, G.P.S. Occhialini and C.F. Powell) published evidence coming from emulsion plates exposed to high altitude cosmic rays showing the track of a charged particle, stopping in the emulsion and a new track emerging, forming a kink with respect to the first one, see Figure 6.2. The obvious interpretation is that the original particle, the pion, decays into a μ-lepton, which in turn decays as shown in (6.20). The masses of the parent and the produced particle were roughly estimated and they were of the right order to be mesons. The complete cascade decay, as it was finally established, is

$$\pi^+ \to \mu^+ + \nu_\mu, \; \mu^+ \to e^+ + \bar{\nu}_\mu + \nu_e \tag{6.21}$$

---

[29]It was I. Rabi who first asked the question, "Who ordered that?"

[30]Similar suggestions had been made previously in Japan by Y. Tanikawa (1942) and S. Sakata and T. Inoue (1943), but, because of the war, these works became known later. In fact, Sakata and Inoue made the correct guess assigning spin zero to the parent meson.

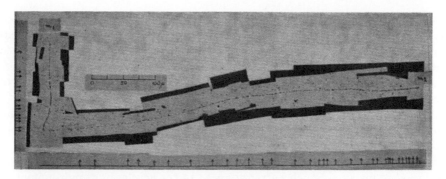

Fig. 6.2 Pion discovery in nuclear emulsions. The vertical track on the left is that of a pion which stops and decays giving a muon. *(Source: C.M.G. Lattes et al., Nature, **159**, 694 (1947)).*

The pions are abundantly produced in hadronic collisions and their properties have been studied in great detail. They come in three charge states, π^+, π^- and π^0 and, as Kemmer had predicted, they form a triplet of isospin. The two charged ones are each other's antiparticles, π^0 is its own antiparticle. Their masses are very similar, about 140 MeV for the charged ones and 135 MeV for the neutral, but they differ considerably in their decay properties and lifetimes. The main decay mode of the charged ones is the one shown in (6.21) with a mean lifetime on the order of 2.6×10^{-8} s, while the neutral one decays as

$$\pi^0 \to \gamma + \gamma \qquad (6.22)$$

with a much shorter lifetime on the order of 8.5×10^{-17} s. The presence of the two photons shows that the decay mechanism involves the electromagnetic interactions and we shall see in section 6.10 that this explains the great difference in the lifetimes. We shall also see that the fundamental strong interactions are not mediated by π-mesons, nevertheless, the latter play a very important role.

6.8 From Cosmic Rays to Particle Accelerators

6.8.1 Introduction

The charged pions were the last particles whose discovery can be entirely credited to cosmic rays. Starting in the 1950s the emphasis was shifted towards particle accelerators. The advantages were obvious: they could be built and operated in the laboratory and they could provide much higher fluxes at high energies. They allowed particle physics experiments to be performed under controlled conditions. Research in cosmic rays continued but, quite soon, the two communities, particle physicists and cosmic ray physicists, split.[31] Results from cosmic rays were not announced in the high energy physics conferences. This split has been reversed in recent years with the emergence of a new field under the combined name "astro-particle" physics. It

---

[31]This split can be placed in 1953. A conference held in July of that year at Bagnères-de-Bigorre was probably the last joint conference of the two communities. C.F. Powell, the discoverer of the pion and a leading figure in cosmic ray research, said in his summary talk: "Gentlemen, we have been invaded. ... The accelerators are here."

involves the use of detection techniques, which were developed in particle physics, to study phenomena related to cosmic radiation. The scientific interest is not restricted to the properties of elementary particles, but covers topics mainly in astrophysics and cosmology. It shows a growing methodological convergence between the quest for the microscopic properties of matter and the study of the large scale structure of the universe.

All particle accelerators are based on the simple principle that an electric charge is accelerated under the action of an electric field. It follows that only charged particles can be accelerated.

In parallel with the rise of particle accelerators, completely new technologies of detection were invented, much better adapted to the accelerator environment. In this and the following section we shall give a brief account of these two developments which changed profoundly the landscape of high energy physics, set up a new pace for its growth and led to what is known as "Big Science".

6.8.2   Electrostatic accelerators

The earliest electrostatic accelerators were just large capacitors with the ground as one of the electrodes and atmospheric air under normal pressure as a dielectric. They suffered from a limitation known as "spark", namely the maximum field $V_{\max}$ air can sustain before becoming ionised. For dry air under normal pressure this maximum field is experimentally determined to be of order 30 kV/cm. To increase the energy, early accelerators had to increase their linear size in order to increase the distance between the positive electrode and the ground, requiring larger and larger buildings to house the device. It was soon realised that a more economic solution was to increase $V_{\max}$ and this could be achieved by increasing the pressure of the gas and using inert gases with higher ionisation threshold. This implied to enclose the main parts of the accelerator – the ion source, placed next to the negative electrode, the high voltage generator and the acceleration tube – inside a pressurised tank. Modern Van de Graaff accelerators follow this principle. A further improvement came with the introduction of the "tandem" machine which essentially doubles the total energy. It is an ingenious device which consists of combining two conventional accelerators with a common high voltage positive electrode placed in the middle of a pressurised tube with two negative electrodes, one at each end. A negative ion is injected at the top, it is accelerated until the centre, where it undergoes a stripping reaction and turns into a positive ion. This way it can be further accelerated until it reaches the bottom. This simple principle gets more complicated in practice because of several technical problems, such as the need to have a pressure tube capable of standing the entire voltage drop from the positive electrode to the ground, to avoid the appearance of sparks, as well as to keep the beam focused, a problem which is of the utmost importance in all accelerators. Modern devices accelerate ions this way and reach energies exceeding 100 MeV. They are used both for fundamental research, such as the study of nuclear structure, and for various applications in the production of radioactive elements for use in medicine, or for research in material sciences. Their rather low maximum energy does not allow them to have any significant impact in particle physics.

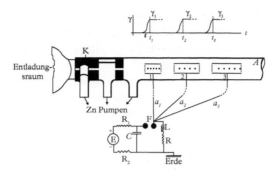

Fig. 6.3 The first drawing for a linear accelerator taken from Ising's 1924 paper.

6.8.3 Linear accelerators

Linear accelerators, or *linacs*, share with the electrostatic ones the property of a straight line trajectory of the accelerated particles, but the principle of acceleration is quite different. As we just saw, the problem with the electrostatic accelerators is that, in order to achieve high energies, they require a high voltage over large distances. This limits their performance because of the spark effect in the gas and the corona formation near the electrodes. The solution, which allowed the construction of all modern accelerators, consisted in applying a relatively moderate voltage repeatedly in a large number of steps. For a linear accelerator the idea was first proposed as early as 1924 by the Swede G.A. Ising.[32] In 1928, the first semi-realistic device was built along these lines by the Norwegian R. Widerøe, a leading figure in the field of accelerator technology. Figure 6.3 shows Ising's original design. It has a two-stage accelerator consisting of a vacuum tube containing three cylindrical metal drift tubes. An alternating voltage $V(t) = V\sin\omega t$ is applied between the first and the second, as well as between the second and the third tube. Obviously, the timing is very important. Widerøe used a radio frequency oscillator in order to apply synchronised voltages to the drift tubes. The particles going down the vacuum tube should arrive at the first gap when the voltage has the right sign to accelerate them. Then they drift inside the metal tube which acts as a Faraday cage and shields them from any external field. Arriving at the second gap they should feel the accelerating field again and the process can be repeated, theoretically, an arbitrary number of times without the potential exceeding anywhere the value V.

Assuming non-relativistic kinematics, we see that after the nth stage the particles have an energy given by

$$E_n = \frac{mv_n^2}{2} = nQV\sin\theta \tag{6.23}$$

where v_n is the particle velocity, Q their electric charge and θ is the average phase of the alternating voltage the particles feel in the gap. One would have thought that we could maximise the energy by arranging a value as close as possible to $\theta = \pi/2$, but this ignores the problem of stability which we shall mention presently.

---

[32]Not to be confused with the German E. Ising of the *Ising model* known in statistical mechanics.

Equation (6.23) shows that the non-relativistic velocity grows as $\sqrt{n}$, and this brings a new problem: since the particles go faster, the length of the drift tubes should grow accordingly for the particles to remain in phase with the alternating voltage.[33] This is shown in Ising's design in Figure 6.3. It is only when the particles reach speeds close to the speed of light that this length remains constant, on the order of one-half the wavelength of the applied field. In the early days the power supplies available were in the domain of radio frequencies with wavelengths of several hundred metres and this limited linear accelerators to very low energies. The development after the Second World War of large *klystrons*, power supplies with frequencies of several GHz, brought the wavelength down to a few centimetres and allowed the building and commissioning of large linacs. The design for protons and heavy ions follows Ising's original idea except that the drift tubes are replaced by a series of cavities (known as the *Alvarez structure*) acting as waveguides which house also the focusing devices. These accelerators reach energies on the order of tenths of MeV per nucleon and, for particle physics, are used as injectors to larger systems.

Ising's original proposal was supposed to accelerate electrons, but this idea had to wait for the development of powerful klystrons. Owing to their low mass electrons are very soon ultra-relativistic and this simplifies the design, at least conceptually. Electrons will go so close to the speed of light that the distance they travel in every accelerating cycle will barely change. As a result one can avoid having drift tubes of varying length and consider instead a waveguide carrying the electric field kept always in phase with the particles. At any given point in space the field oscillates and reverses its direction, but the electrons, which ride on the wave, always see an accelerating field.

Electron linear accelerators played an important role in particle physics and they will probably continue to do so. The largest such machine was built at Stanford in 1962 under the name of SLAC *(Stanford Linear Accelerator Center)*. It was three kilometres long[34] and the electrons reached an energy of 20 GeV. It was with this machine that the quark structure of the nucleon was first discovered.

Before leaving this section we want to come back to the issue of beam stability we mentioned earlier. A machine is never perfect and the particles of the beam will have a certain spread of velocities. We must have a built-in mechanism ensuring that a particle faster than the average will be slowed down and a slower one will be accelerated more. The general problem is very complex but there is a simple way to achieve this, which is easy to understand. It is the subject of Problem 6.1 at the end of this chapter. This phenomenon of phase focusing is crucial in accelerator design.

6.8.4 Cyclotrons

Cyclical accelerators follow the same principle of applying an accelerating field repeatedly several times, but the particles, instead of moving in a straight line, follow a circular orbit, bent by a perpendicular magnetic field. The father of cyclotrons is

---

[33]We assume that the frequency of the applied voltage remains constant.

[34]Often called *The Monster*.

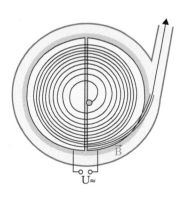

The first circular accelerator
(Berkeley 1930)

Fig. 6.4 The principle of the cyclotron (left) *(Source: Wikipedia)*. The Laurence–Livingston cyclotron (right) *(Source: newscenter.lbl.gov)*.

the American E.O. Lawrence after whom the *Lawrence Berkeley National Laboratory* (LBL) was named.[35]

The physical principle is very simple: a non-relativistic particle with mass m and electric charge q, moving in a plane under the influence of a constant homogeneous magnetic field B perpendicular to the plane, revolves on a circle with frequency f given by

$$f = \frac{qB}{2\pi m} \tag{6.24}$$

This value is independent of the speed of the particle and this observation is the basis of the cyclotron. Figure 6.4 (left) shows the principle: the device can be seen as two D-shaped electrodes (they are often called *dees*). Along their edge we apply the Radio Frequency (RF) accelerating voltage. The entire system is placed in a constant magnetic field perpendicular to the plane of the dees. A charged particle is injected near the centre with an initial velocity in the plane. It goes on a circle and every time it crosses the edges it is accelerated by the voltage. Its speed increases together with the radius of the circle (we assume non-relativistic kinematics), but the frequency remains constant. If the RF voltage is suitably synchronised, the charged particle will continue to be accelerated every time it goes through the edge and it will follow a spiral trajectory as shown in the figure. When it reaches the outer circle it is magnetically extracted and directed to the target for the experiment.

Figure 6.4 (right) shows the first cyclotron. Livingston had succeeded in obtaining resonance between the value of the magnetic field which determines the cyclotron frequency and that of the RF power which accelerates the particles for 41 cycles, thus reaching an energy of 80 keV.

---

[35]Wideröe was instrumental in bringing both these ideas: he was among the first to describe an accelerator with a circular orbit; he called it *a ray transformer* and, as we saw, he was the first to apply repeated accelerations in the linac. But he did not put the two together. In fact, it seems that Lawrence had the idea of the cyclotron while trying to decipher Wideröe's paper. The first working cyclotron was built in 1930 under Laurence's guidance by his graduate student M.S. Livingston; it was the basis of his PhD thesis.

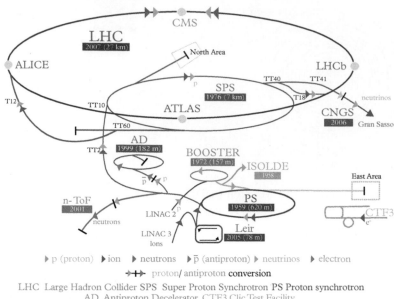

Fig. 6.5 The CERN accelerator complex. It is built near Geneva and it has operated continuously since 1957. It consists of several components, each one representing the frontier in accelerator technology the time it was built: Proton Synchro-Cyclotron (SC) in 1957, Proton Synchrotron (PS) in 1960, Super Proton Synchrotron (SPS) in 1971, Large Electron-Positron Collider (LEP) in 1989 followed by the Large Hadron Collider (LHC) in 2008. As we see in the design, each accelerator is used as part of the injector system of the next one. *(Source: CERN).*

For comparison, Figure 6.5 shows the largest cyclotron ever built, the one at CERN near Geneva. The main ring has a circumference of almost 27 km and the protons in it reach an energy of 7 TeV. In 80 years we have gained a factor of almost 10^8. This gain is the result of a large number of ingenious improvements and technological breakthroughs, whose presentation would have made a passionate story, but would fill an entire book. We shall only mention here some of the most significant ones.

• As with the linacs, an important element has been the building of very high frequency power sources. It was largely a by-product of the radar technology developed during the Second World War.

• The early cyclotrons followed the design shown in Figure 6.4. This limited their performance for several reasons: first, their operation was based on equation (6.24), which is non-relativistic. In the relativistic regime the frequency is given by

$$f = \frac{qB}{2\pi E} \tag{6.25}$$

and so it decreases as the particle energy increases. For this reason the early cyclotrons were proton or ion machines. Second, and most important, the very principle of the accelerator required that the increase in energy be obtained by circulating the particles a large number of times. As a result the size of the dees, and hence that of the machine, increased with energy. The first Laurence–Livingston cyclotron shown in Figure 6.4 had a diameter of only 4 inches, while the last machine Laurence built at Berkeley was 184 inches across. If the same technique had been used for the LHC of Figure 6.5, it would have needed D-shaped electrodes of 8.5 km in diameter. Having a uniform magnetic field everywhere in that space is obviously inconceivable.[36] The solution sounds easy: force all particles, irrespectively of their energy, to follow the outer trajectory, thus concentrating the magnetic field in a narrow ring around this trajectory. It seems that this suggestion was first made by M.L.E. Oliphant, an Australian physicist working at the University of Birmingham in the UK. Oliphant moved to the USA during the war to join the scientists working on nuclear weapons and in 1943, while he was at Oak Ridge, in his spare time he developed the principle of the *synchrotron*. The name comes from the required synchronisation: a particle passing through the accelerating electric field should see a higher magnetic field in order to keep the same trajectory. In a note Oliphant wrote: *"Particles should be constrained to move in a circle of constant radius thus enabling the use of an annular ring of magnetic field... which would be varied in such a way that the radius of curvature remains constant as the particles gain energy through successive accelerations by an alternating electric field applied between coaxial hollow electrodes."* Notice that the technique of synchronising the accelerating electric field and the bending magnetic field allowed also to adjust the machine to relativistic energies. Synchrotrons were built after the war in many places (the first operational one was at Brookhaven), and it is this technology which allowed the construction of large accelerators. A further ingredient has been the development and use of superconducting magnets in order to obtain higher fields. Figure 6.6 shows a part of the tunnel housing the LHC. The magnetic field of 8.4 T is produced by 1232 superconducting dipole magnets kept at a temperature of 1.9 K.

• The final element in this series of innovations, which we would like to mention, has to do with the stability and the focusing of the beam. It is a large and very active domain of research and there exist several specialised books on it. We have already presented the issue of phase stability, which was understood in the very early days of accelerators. Here we shall go a bit further. In a beam of particles there are many sources of instability. The bending magnetic field is never perfect, especially near the edges of the magnets, and the resulting trajectories are not the ideal ones. In the early cyclotrons one would lose as much as 90% of the particles, which would hit the iron of the machine. The beam itself is unstable. The particles in it have their own transverse motion and, furthermore, having the same electric charge, they repel each other. This phenomenon is accentuated in modern synchrotrons in which the beam is not in the form of a continuous current but consists of a series of bunches of particles. In the LHC design we have around 2800 bunches of 10^{11} protons each. If no correcting measures are taken, the beam will diverge and will be lost in the walls of the pipe. Adjusting

---

[36]The magnet of the largest Laurence cyclotron had the steel required to build a frigate. To go much further you would need the steel for a whole fleet!

Fig. 6.6 Part of the LHC tunnel at CERN. We see the succession of the large 15 m long cylindrical cryo-dipole bending magnets. *(Source: CERN).*

the field at the edges of the bending magnets is not enough and one introduces special focusing magnets. They are quadrupole, sextuple, etc. magnets placed along the beam between the dipole bending magnets. An important and counter-intuitive discovery made this method effective: It is *the alternating gradient* or *strong focusing* design.[37] It is counter-intuitive because it does not attempt to focus the beam continuously. We can understand the problem by considering Liouville's theorem which states that the volume in phase space of a conservative dynamical system is constant in time. Focusing, on the other hand, implies a reduction in the phase space volume, and the theorem tells us that this cannot be achieved in a simple way. The idea was already known in optics, where, in order to focus a light beam, you use an alternating sequence of convex and concave lenses. Here this role was played by the magnets, hence the name *Alternating Gradient Synchrotron (AGS)*. Since the late 1950s strong focusing has been used in all high energy accelerators.[38]

6.8.5   Colliders

All early machines were using the accelerated beam to hit a material target, hence the name *laboratory frame* for the reference frame in which one of the colliding particles is at rest. Another useful frame in describing collisions is the *centre-of-mass frame* in which the two colliding particles have equal and opposite three-momenta. It is a simple exercise in relativistic kinematics (see Problem 6.2) to relate the two. If $E_{\rm L}$ is

---

[37]The history of the discovery is interesting. It was first made by N.C. Christofilos. He was born in Boston, MA, but his parents went back to Greece when he was seven. He studied electrical and mechanical engineering in Greece and was working as an elevator engineer in Athens. As an autodidact he learned nuclear and plasma physics and discovered the strong focusing principle in 1950. He patented the idea and wrote an article which he sent to many places, including Berkeley. His work went unnoticed until E.D. Courant, M.S. Livingston and H.S. Snyder, working at the Brookhaven Cosmotron, rediscovered the principle by trial and error two years later. Christofilos' priority was nevertheless recognised and he was invited to work in the US. He spent most of his career in Livermore, where he contributed many very original and seminal ideas in accelerator and plasma physics.

[38]The last machine with weak focusing was a 10 GeV synchro-cyclotron built in Dubna in the Soviet Union in 1955 under the name of "Synchro-Phasotron".

the beam energy in the lab frame and M_t and M_b the masses of the target and the beam particles respectively, we find the relation

$$E_{cm}^2 = M_t^2 + M_b^2 + 2E_L M_t \tag{6.26}$$

where E_{cm} is the total energy in the centre-of-mass frame. In this simple formula we can understand the drive for colliders.

A collider consists in fact of two accelerators (up to now they were mostly circular, but future plans involve also linear ones), whose beams collide head-on. At the LHC at CERN, the largest collider ever built, we have two beams of protons, each one with an energy of 7 TeV. The total energy in the centre-of-mass frame is 14 TeV. It is the energy available to explore new physics, for example to produce new particles. If we want to create the same conditions with a fixed target experiment, formula (6.26) shows that we would need a beam with energy about 10^5 TeV!

The early accelerators were all fixed target machines because they are easier to design and operate. They offered great versatility by the possibility to obtain several secondary beams from a given target and run several experiments in parallel. All the early discoveries were made with such machines. However, soon the quest for higher and higher energies forced physicists to consider colliders because of the simple kinematical relation of equation (6.26). The main technological problem which had to be solved has to do with the density of particles in a beam. In an experiment the number of events, and hence the discovery potential, is proportional to the number of collisions per unit time, which in turn is proportional to the densities of the beams, see for example equation (4.8). For a fixed target experiment the target density is huge, given essentially by the Avogadro number. We can get enough collisions even with a relatively poor beam. This advantage is lost in a collider, which therefore requires very dense and very well collimated beams. It took some years before this technology was mastered. The first collider for particle physics was an electron–positron machine built by B. Touschek, an Austria-born physicist working at the University of Rome. It was commissioned in Frascati, near Rome, in 1961 but it was moved to Orsay, near Paris, where there was a linac with an intense positron beam. The first hadron collider was *the Intersecting Storage Rings* or ISR, built at CERN in 1969 with two proton beams of 31.4 GeV each. It consisted of two 150 m diameter interlaced rings and operated between 1971 and 1984. It was built to discover quarks and in fact it contributed to their discovery, albeit in an indirect way, as we shall see later. Many of the techniques used in subsequent colliders were invented for ISR.

A conceptually new step was the design and building of the first proton–antiproton collider at CERN in 1980. Using antiprotons rather than protons in one of the beams presented an initial advantage, but also several formidable technical challenges. The advantage was that both beams could circulate in the same ring. Protons and antiprotons, having opposite electric charges, would bend in opposite directions in the same magnetic field. In comparison, the proton beams of ISR or LHC require each one its own ring because they must be subject to opposite magnetic fields. On the other hand, the challenges are many. The first is that, unlike protons, antiprotons do not exist on Earth; we must produce them with the accelerator working as a fixed target machine. At CERN a 26 GeV proton beam was directed to a copper target. The production probability is not large and we must accumulate them over a period of time before we have enough to be able to inject them into the first acceleration ring. But storing antiprotons is not simple because they annihilate instantly in contact with ordinary matter, so at CERN they built a special *Antiproton Accumulator* or AA, a roughly circular ring, later improved to a double ring, with a diameter of 40–60 m.[39] The most important technical challenge was that the antiprotons were produced from the target with an average momentum of 3.5 GeV but with a large dispersion in both longitudinal and transverse components. We have seen already that a synchrotron works efficiently if there is a synchronisation between the frequency of the circulating particles and the one which controls the magnetic field. Injecting a beam with large momentum dispersion would lead to beam loss. Reducing this dispersion became known as *stochastic cooling* and it is a technique invented by S. van der Meer.

Probably the simplest way to explain the process of stochastic cooling is to explain the use of the terms: why *stochastic* and why *cooling?* We start with the second. A beam consists of an ensemble, in the sense of statistical mechanics, with a very large number of particles. Therefore, it is tempting to assign thermodynamic properties to it. In particular, we can introduce an *entropy* as a measure of the disorder, i.e. the momentum dispersion in the beam. High disorder corresponds to a hot gas. Reducing the momentum dispersion results in a more ordered ensemble to which we should assign a lower temperature, hence the term "cooling". Before going to the term "stochastic" we want to present the method for the case of a single particle, see Figure 6.7. We limit the presentation to the transverse momentum dispersion. A corresponding cooling process should be applied also to the longitudinal dispersion. Any initial error in the injection will cause a deviation of the particle from the ideal trajectory, represented by the blue curve in the figure. The machine focusing system will try to correct it and, as a result, the particle will follow an oscillatory curve which we assume to be sinusoidal around the ideal trajectory with a frequency known as *betatron frequency*. We call the corresponding half wavelength λ_B. The cooling system is a feedback electronic device whose purpose is to damp this oscillation. Let us place at a point A a pick-up sensor which measures the deviation of the particle's position at that point. It sends a signal to a corrector, known as kicker, placed at a distance equal to $\lambda_B/2 + n\lambda_B$ with n integer. It follows that an error in position at the pick-up will be corrected by an angular deflection at the kicker. Figure 6.7 shows the ideal situation in which the particle crosses the pick-up sensor at the maximum deviation of its trajectory. In practice, this will not be always the case and it will take more revolutions for the trajectory to be fully corrected. Note also that, in order for the system to work, we must achieve a synchronisation between the times needed for the particle and the signal to move from the pick-up to the kicker. The particle moves essentially with the speed of light while the signal is delayed in passing through the cables and the amplifier. This is the reason why, as we show in the Figure, the signal should follow a path cutting across the ring.

So far we have considered a single particle. Stochasticity enters in considering also the other particles. Contrary to the standard methods of beam focusing which view

---

[39]The initial AA has now been replaced by an Antiproton Decelerator (AD), which is shown in Figure 6.5.

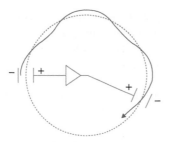

Fig. 6.7 An oversimplified image of momentum dispersion reduction. The ideal trajectory is shown in blue. The red curve represents a real particle oscillating around the ideal trajectory. The feedback system has two parts. A detector, or "pick-up", represented by two electrodes at the left of the figure. It measures the deviation from the centre of the particle position. If it is found to be off centre, it sends a signal to the corrector, or "the kicker", shown on the right. For illustrative purposes the betatron wavelength is largely exaggerated as compared to the ring dimensions.

the beam as a whole, stochastic cooling addresses an individual particle, but randomly chosen. In this case, it is not at first clear that the argument we presented for the single particle applies. Let us assume that our test particle, passing through the pick-up, has a position deviation x. It creates an electric signal which is transmitted to the kicker. The transmission line has a certain passing bandwidth W and, as a response, the kicker generates a correcting pulse centred at time t_0, the arrival of the test particle, but having a time duration $T \sim 1/2W$ (see Problem 6.3) Typical values for W are on the order of a few hundred MHz which gives $T \sim \mathcal{O}(1 \text{ ns})$. When the kicker corrects the trajectory of the chosen particle, it simultaneously affects the trajectories of all the other particles which come with it in the interval $t_0 \pm T/2$ and it is not a priori obvious that the average dispersion will be reduced. In fact, a simple-minded argument would tend to show it will not.

Let us push the statistical mechanics analogy further and view the beam as a classical fluid. To the extent that this analogy is valid, Liouville's theorem tells us that the phase space volume remains constant, in other words "cooling" cannot be achieved by applying electromagnetic or any other conservative forces. The reason why this argument fails is that the fluid analogy is not correct for stochastic cooling for which the beam is a collection of individual particles. As van der Meer pointed out in his Nobel Prize lecture, a beam, although it contains a very large number of particles, is mostly empty space so, if we want to view it as a fluid, empty space should be part of it. In this sense "cooling" amounts into a rearrangement during which more particles move in trajectories close to the centre and more empty space is pushed outside. In fact, it is more appropriate to say that the fluid analogy breaks down.

The first p–$\bar{p}$ collisions at the CERN SpS (*Super-proton Synchrotron*), converted to a proton–anti-proton collider (Sp$\bar{p}$S) using the van der Meer cooling method, occurred in July 1981 and the first physics run started in December of the same year. The centre-of-mass energy was 540 GeV. The scientific yield was the discovery of the intermediate vector bosons of the weak interactions $W^{\pm}$ and Z in 1983, as we will discuss in Chapter 22. The Sp$\bar{p}$S operated until 1990 when it was shut down to make room for a new project of an electron–positron collider, the LEP, but also because it was no longer competitive against the p–$\bar{p}$ collider which started operating at Fermilab in 1987 under the name of Tevatron and reached an energy of 1.8 TeV. The Tevatron, which was the last major particle physics installation in the USA, is credited with the firm discovery of the top quark. It operated until 2011 when it was superseded by the 14 TeV Large Hadron Collider of CERN. In Europe the last two colliders in particle physics, both at CERN, were the Large Electron-Positron machine (LEP), where most of the precision measurements for the electromagnetic and weak interactions were made, and the LHC, the most powerful accelerator ever built (and still in operation), where the Brout–Englert–Higgs boson was discovered.

6.9 The Detectors

6.9.1 Introduction

Parallel to the progress in accelerators, new detection techniques were developed in order to allow a full use of the new machines. The discoveries made in cosmic rays were made using either Wilson cloud chambers, whose principle we presented in section 6.4.1, or photographic emulsions. The latter were solid blocks, or stacks of plates, containing light-sensitive crystals, usually silver halides. A particle passing through this medium ionises some atoms along its path which in turn interact with the silver crystals. In order to visualise the image, one should develop it like an ordinary picture. The particle track is reconstructed by taking slices of the block and following the points in the microscope. Anderson discovered the positron and the muon using a Wilson chamber and Powell discovered the pion using emulsions. The development of accelerators changed the situation drastically. The data were arriving at a high rate and none of the previous detection techniques could cope with it. In fact, pions were abundantly produced in the cyclotrons built in the 1930s and early 40s, but nobody knew how to detect them. It is also true that Laurence was primarily interested in building larger and larger accelerators and he did not devote enough time and resources to developing detection techniques.

6.9.2 Bubble chambers

The first breakthrough took place in 1952 with the invention of the *bubble chamber* by the American physicist D.A. Glaser. The principle is similar to that of the cloud chamber in the sense that a charged particle's track is recorded as a trail of bubbles caused by ionisation. It consists of a vessel filled with some liquid which is kept under pressure in a superheated state, i.e. at a temperature just above its normal boiling point. The chamber is equipped with a piston and when charged particles enter, it expands causing the liquid to enter a metastable state. A charged particle leaves behind a trail of ions which nucleate bubbles. This way the track can be photographed. One could add a magnetic field in order to bend the tracks and determine the charge-over-mass ratio of the particles. The great advantage is that the chamber can be used repeatedly in a rapid cycle of compression-expansion, with the expansion phase synchronised with the arrival of the beam.

Glaser presented his first pictures in 1953 using a chamber of 2 cm³. He was also in the first team to use this new technique for particle physics. In 1957, a group of four physicists, working at the Brookhaven Cosmotron, published the cross section measurements of hyperon production[40] using a 12 inch propane bubble chamber exposed to a beam of 1.1 GeV charged pions. After these first results the pace increased very fast with larger and larger bubble chambers built and operated both in Europe and the USA. This also brought a change in the way data were studied and analysed. Bubble chamber pictures were accumulated by the thousands and there was no way for physicists to look at each one of them individually. Semi-automatic scanning techniques were developed and the size of the collaborations increased. We had entered the era of *Big Science*. The Berkeley group under L. Alvarez built a 72 inch hydrogen bubble chamber put in a 15000 gauss magnetic field, the first large detector in physics. It was commissioned in 1959 and gave a rich harvest of new particles. To give an idea of the data size, we mention that in 1968 the Berkeley group alone was able to analyse 1.5 million bubble chamber pictures. The style of the experimental work changed as well. In the old days an experimentalist would design, build and perform the experiment and analyse the results in the lab. With the millions of pictures provided by the bubble chambers, these functions were to a large extent disconnected. An experimental group could borrow a set of pictures and analyse them at home. For example, the η meson we shall see in the next section was discovered by the experimental group at Johns Hopkins in a set of 100,000 pictures borrowed from Berkeley. As a matter of fact Alvarez, who was held responsible for this evolution, was heavily criticised by the old-timers, but the evolution turned out to be inevitable.

An important feature of the bubble chambers was that the liquid they contained was used both as target for the incoming beam and as detector for the final particles. As we have mentioned already, there were two main kinds of bubble chambers according to the liquid they were using: hydrogen, or a heavy liquid such as propane. Each type had advantages and disadvantages. Hydrogen had the advantage of providing cleaner signals because the interaction was on free protons, the nuclei of hydrogen, while on a heavy liquid chamber the interaction occurred on complex nuclei. Hydrogen's disadvantage, on the other hand, was the lower density implying larger volumes for the same number of nucleons. Furthermore they operated at much lower temperatures to keep hydrogen liquid, thus requiring more complex and expensive cryogenics.

The largest bubble chambers were built in Europe in the late 1960s and early 70s. *Gargamelle*, a 12,000 litre, heavy liquid chamber, started operating at CERN in 1970. It is credited with the discovery of the weak neutral currents. Figure 6.8 shows one of its historic pictures. An even bigger one, BEBC for *Big European Bubble Chamber*, with 35,000 litre of hydrogen, was commissioned at CERN in 1975. But they were the dinosaurs of their kind. With the increase in energy and luminosity[41] of the accelerators, even the fastest bubble chambers could not respond anymore.

---

[40]Some new particles we will introduce in section 6.10

[41]The *luminosity* of an accelerator is defined as the number of scattering events per unit time divided by the total cross section. For a fixed target accelerator it is proportional to the flux of the beam.

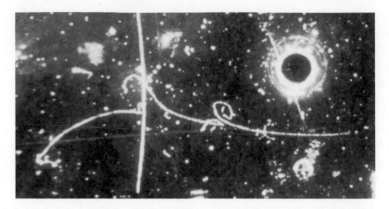

Fig. 6.8 The first picture of a neutral current event corresponding to the elastic scattering $\nu + e^- \rightarrow \nu + e^-$. The neutrino beam comes from the right. The electron track is visible in the chamber. *(Source: CERN, Gargamelle collaboration).*

6.9.3   Counters

All present-day detectors consist of various kinds of electronic counters. Strictly speaking *a counter* is a device which counts the particles passing through it. In a rather primitive form counters were developed at the beginning of last century. A great name in counter technology is J. (Hans) W. Geiger (1882–1945), Rutherford's assistant in the atomic nucleus experiment. In the first measurements they were using a fluorescent screen to determine the trajectory of the α particles. Each impact produced a tiny flash of light. Geiger worked in full darkness counting these flashes using a microscope. It was the first "counter". In subsequent years he kept on improving the experimental set-up used to count the particles, but it was only in 1928, in collaboration with his student W. Müller, that he produced the device which became known as *the Geiger counter*. It is based on a conducting tube filled with a low-pressure inert gas. There is a metal wire along the axis of the tube with a high voltage of several hundred volts between it and the tube. When some sort of ionising radiation hits the tube it produces an avalanche of ion–electron pairs in the gas which are collected in the anode and cathode, respectively and create a visible, or audible, pulse. This counter, with various improvements, is still used today.

The principle of particle detection through the effects of ionisation has become a standard method in the field. The materials used are gases, as in the Geiger counter, but also liquids, crystals or scintillating plastics. An important invention, made in 1944, allowed the rapid expansion of this technique: it is *the photomultiplier*, an electronic device capable of producing an electrical pulse out of a very faint photonic signal consisting of even a single photon. The early photomultipliers used vacuum-tube diodes, which were later replaced by transistors and finally microchips. They are part of almost all detector systems today. The output pulse can be very short, on the order of a nanosecond, and this increased the data collection rate enormously.

The standard Geiger counter cannot identify the kind of the particle that hits it. An array of counters, placed in a magnetic field, can be used to measure the momentum from the bending of the trajectory. In addition, for non-relativistic particles, one could measure the velocity by measuring the time of flight between two counters. A better particle identification can be achieved by using a different kind of counter, based on the *Cherenkov effect*, discovered in 1934 by the Soviet physicist P.A. Cherenkov. It is the phenomenon of radiation emitted when a charged particle is moving in a medium faster than the speed of light in that medium. It is emitted in a cone whose symmetry axis is the particle trajectory and its angle is determined by the particle velocity. A more sophisticated version of the counter, named RICH for *Ring Imaging Cherenkov*, built by T. (Tom) Ypsilantis, allows to extend the identification capabilities to relativistic particles.

In the early 1960s some electronic devices started being used in order to speed up data taking as compared to bubble chambers. The first of them were called *spark chambers* [42] and consisted of a stack of parallel metal plates with a high voltage between adjacent plates. Some inert gas under pressure filled the gaps and was ionised when a charged particle passed through. This in turn created sparks between oppositely charged plates and the trail of such sparks could be recorded in a picture. Although the path of the particle was shown less accurately than in a bubble chamber, the spark chamber had the advantage of having a much faster duty cycle, since one needed only to charge the plates, rather than moving a heavy compressing and expanding piston. Furthermore, the spark chamber could be made highly selective with the addition of external counters which would screen unwanted events. On the other hand, at least in their early versions, spark chambers still required development and scanning of the pictures.

An important breakthrough came in 1968 with the invention by G. Charpak of *the multiwire proportional chamber*. The principle is again ionisation creating an avalanche, but the discharge occurs between planes formed by fine wires. The signal is proportional to the initial number of ions, hence the name, and the position of the spark is determined by the time it takes the signal to reach the end of the wire. One can further improve the accuracy by taking into account the time for the ions to *drift* to the nearest wire (*drift chambers*). Everything is computed electronically and no pictures are taken.[43]

This initial design evolved into a multitude of special-purpose detectors which make full use of the recent progress in micro-electronics and fast computing. Rather than describing each one separately, we present, as an example, ATLAS,[44] one of the two large LHC all-purpose detectors, which is a fully integrated system. Figure 6.9 shows a

---

[42]Early versions of spark chambers had been used in cosmic ray searches already in the late 1940s. A more advanced design, capable of observing multiple particles at once, was proposed in 1959 by two Japanese scientists, M.S. Fukui and S. Miyamoto. It was one of those instances in which a new idea was immediately applied. Upon reading their paper, J.W. Cronin built a desktop model of the chamber and the first full use of spark chambers in a major particle physics experiment was done in 1962 in the Brookhaven neutrino experiment, which we shall present in Chapter 20

[43]It took some time for physicists to accept this absence of a visual image. In the CERN proton–antiproton collider they had developed some software which showed a three-dimensional image of every event. This has not been used in the subsequent colliders, the Tevatron, LEP or LHC.

[44]A Toroidal LHC AparatuS

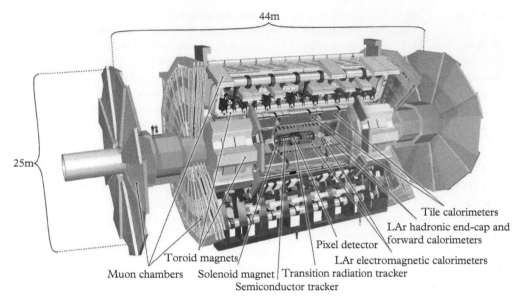

44m

25m

Tile calorimeters
LAr hadronic end-cap and forward calorimeters
Pixel detector
LAr electromagnetic calorimeters
Toroid magnets
Muon chambers   Solenoid magnet | Transition radiation tracker
Semiconductor tracker

Fig. 6.9 A computer generated image of the ATLAS detector. For illustration purposes two tiny human figures are shown in the left part of the figure on the axis of the detector. *(Source: ATLAS collaboration).*

rather schematic view. We shall go through its main parts very briefly. A more detailed description can be found in the web page of the ATLAS experiment.

The detector takes the form of a big cylinder, closed at the two ends, with the two proton beams of the LHC moving in opposite directions inside a thin vacuum tube along its axis and arranged to collide at its centre. Obviously, no detector can be placed directly on the beam, because it would be immediately saturated and burned. The sensitive elements of the detector are all around the collision point. An important property is hermeticity. The detector should record, identify and measure as many of the particles produced in a collision as possible. No detector can be absolutely hermetic, since particles going along the beams in the forward or backward directions will necessarily go undetected, but hermeticity should be as tight as possible, at least in the transverse directions. The reason is that, as we shall see later, an important signature of possible new particles could be missed, i.e. undetected, transverse energy and momentum. The general geometry of the detector is largely determined by this requirement. It consists of a series of co-axial cylinders around the beam axis closed at the two ends by detector systems called *end-caps*, which are mounted on discs perpendicular to the axis. We go from the innermost layer to the outer ones:

• *The central detector.* It surrounds the collision region and its purpose is to determine the trajectories of the particles produced with great accuracy, so that: (i) the primary collision point can be found by extrapolation, (ii) possible secondary vertices away from the collision region, due to the decays of short-lived particles, can be identified. The electronics must be *hardened, i.e.* capable of withstanding the high radiation

Fig. 6.10 At the LHC a typical crossing of two proton bunches results in an average of 25 collisions. *(Source: ATLAS collaboration).*

level around the collisions. This last requirement was a serious technological challenge. In Figure 6.9 we see that the central detector consists of three parts: Closest to the collision region we find the highest granularity semiconductor pixel detectors followed by a silicon microstrip detector. This part is surrounded by a barrel containing semiconductor trackers (SCT) which, in turn, is inside a cylinder of transition radiation trackers (TRT). We find the same elements on the discs of the forward and backward end-caps. The outer radius of the central detector is 1.15 m and the total length 7 m. Figure 6.10 shows the reconstruction of a real event involving the decay of a Z into a $\mu^+\mu^-$ pair, represented by the two yellow straight lines, together with 24 additional collisions, and this gives an idea of the complexity of the task.

- *The solenoid magnet.* The central detector is contained in a solenoid magnet which provides a field of 2 T.

- *The calorimeters.* A calorimeter is a device which absorbs the particles and measures the energy they deposit. We have seen an early version in the Ellis–Wooster β-decay experiment of 1927 (see section 6.3). At that time, the total energy was measured by the rise in temperature. Today's calorimeters are more sophisticated and show the energy deposited locally at each point. In the ATLAS detector we have two kinds, again placed in co-axial cylinders with end-caps. They consist of metal plates acting as absorbers and sensing elements that detect and measure the energy by measuring the shower of particles produced when a particle hits an absorber. In ATLAS the inner part is an *electromagnetic calorimeter* that stops and counts essentially electrons and photons. The sensing element is liquid argon. The outer part is a *hadronic calorimeter* which does the same for hadrons, with tiles of scintillating plastic as sensors. The electromagnetic calorimeter was the main element used in the recent discovery of the Brout–Englert–Higgs particle, which was identified through its decay mode into two photons.

- *The toroidal magnet.* This is one of the main features of the ATLAS detector. The central element is a barrel-shaped part with eight superconducting coils with separate cryostats. In Figure 6.9 we see clearly one of them. Each coil has a length of 25.3 m and a radial extent from 9.4 to 20.1 m. They are assembled radially and symmetrically around the beam axis. They produce a toroidal field of 3.9 T. As with the other elements, this system is also repeated in the forward and backward end-caps.

- *The muon chambers.* Of all the known particles which a detector is expected to identify, the most penetrating are the muons. We find them in the decay products of many short-lived particles. They are themselves long-lived with a mean lifetime on the order of 2.2×10^{-6} s, and have a small absorption cross section because they have no strong interactions. They can traverse the calorimeters without being stopped. On the other hand, it is crucial to detect and measure them because they may signal the presence of new physics. In ATLAS, this is achieved with the help of muon chambers, which, as we can see in Figure 6.9, form the outermost element of the detector. It surrounds the calorimeter and measures muon paths to determine their momenta with high precision. It consists of thousands of charged particle sensors placed in the toroidal magnetic field.

• *The trigger.* Data recording at the LHC is not simple for two reasons. First, proton–proton cross section is very large, since it is a strong interaction process. Second, the machine was designed to have a high luminosity in order to increase its discovery potential. As a result the event rate is very high, on the order of 10^9 per second. It is impossible to record and store all of them.[45] In practice, only about 100 events per second are recorded. In fact, we can estimate, knowing the strong interaction cross section, that the enormous majority are just ordinary, well-known hadronic events and there is no scientific loss in not recording them. So, the challenge that the experimentalists face is to record only a tiny fraction, of order 10^{-7}, of the events and still not miss any discovery. This is achieved using a very sophisticated two-level logical trigger, which uses parts of the information from the detector and decides very rapidly[46] whether the event is worth recording. It is conceivable that in future detectors, with the continuous increase of computing and storage power, a significant part of this decision mechanism will be performed on line by the computer, thus easing the task of the trigger.

6.10   New Elementary Particles and New Quantum Numbers

6.10.1   Unstable particles

As mentioned earlier, the charged π-meson, discovered in 1947, was first seen in cosmic rays.[47] Since the early 1950s accelerators took over. They provided high incident fluxes under controlled energy and momentum conditions. Their extensive use in higher and higher energies yielded a rich harvest of new "elementary" particles. They were all unstable with lifetimes ranging from 2×10^{-6} s (the μ-lepton) to $\sim 10^{-23}$ s for the very short-lived hadrons. Obviously, these last ones cannot leave a visible track in the detector and were discovered as resonances, in the way we indicated in Chapter 4. At present the tables published by the *Particle Data Group* (PDG) contain several hundred entries.

In Chapters 2 and 4 we pointed out that the mean lifetime of an unstable state is the inverse of the width Γ of the state, the latter being equal to its decay probability per unit time. The enormously wide span of the observed lifetimes indicates that all

---

[45]It is often said that recording all the events would fill 100,000 CDs every second and the resulting pile of CDs would rise to the moon every three months!

[46]The first level trigger reaches a decision in 2 µs and the second in around 10 ms.

[47]In fact, charged pions were already produced copiously in the first cyclotrons but people were unable to detect them. It took a while before physicists got used to the accelerator environment and developed the appropriate detection techniques.

these decays are not due to one and the same underlying interaction. Indeed, we shall see in this book that we can distinguish roughly three classes of unstable particles according to the order of magnitude of their lifetimes.

- Particles with "long" lifetimes, longer than roughly 10^{-10} s. We saw that the muons and the charged pions belong to this class. This means that they have a relatively small decay probability and the interactions responsible are the ones we called "weak" in section 6.4.3. An immediate consequence of this weak probability is that these particles live long enough to leave a measurable trace in the detector. A useful quantity is the product $c\tau$, where c is the speed of light in the vacuum and τ the lifetime in the particle rest frame. For the muon it is almost 660 m and for the charged pion 7.8 m. It follows that we can build beams out of these particles and use them as projectiles for scattering experiments.

- Particles with intermediate lifetimes of order 10^{-15} s. These decays are due to the electromagnetic interactions. The neutral pion ($\tau \simeq 10^{-16}$ s) belongs to this class. The corresponding $c\tau$ value is on the order of 25 nm, much too short to build a beam of π^0s.

- Very short-lived particles with lifetimes shorter than 10^{-20} s. The interactions responsible for their decays are the strong interactions. They leave no measurable track in the detector and they are observed as resonances. In fact, if we look at the tables, we see that it is more convenient for these particles to talk about their width Γ, rather than their lifetime τ.[48]

6.10.2 Resonances

In Chapter 4 we introduced the concept of *resonance* in scattering as a sharp increase of the cross section around a certain value of the energy. This should be understood as an increase with respect to a certain background which is the expected cross section in the absence of such a resonance. In Chapter 12 we shall develop precise methods to estimate this background, but in most cases it is given by the phase space integral we introduced in equation (4.7). The first hadronic resonance was discovered in the late 1940s, but it took some time before the correct interpretation was given. It is a pion–nucleon resonance of mass 1232 MeV and width around 115 MeV. It has spin and isospin equal to 3/2, positive parity and in the particle tables appears today as $\Delta(1232)$ $3/2^+$. It was first seen as a sharp increase in the pion–nucleon cross section. It was soon followed by several other resonances of the pion–nucleon system, a fact which greatly confused people in their efforts to build a simple dynamical model for the interaction of pions and nucleons. In Figure 6.11, we show a 1961 plot with the evidence of resonances made out of three pions. The statistics is still poor and the

---

[48]A word of caution: this correspondence between the value of the lifetime and the underlying interaction responsible for the decay is not at all rigorous. It is valid as long as other reasons, dynamical or kinematical, do not play a dominant role. We know already in quantum mechanics that certain nuclei which decay with the emission of α particles may have extremely long lifetimes, even though they are due to the strong nuclear force. The reason there is dynamical. In order to escape, the α particle has to tunnel through a potential barrier. In other cases it may be kinematical. For example, a free neutron has a lifetime of order 880 s, much longer than the typical values we mentioned above, although it decays through the weak interactions. The reason is the very small proton–neutron mass difference, which results in a reduced phase space for the decay products. We shall come back to these points when we develop quantitative tools to compute the various lifetimes.

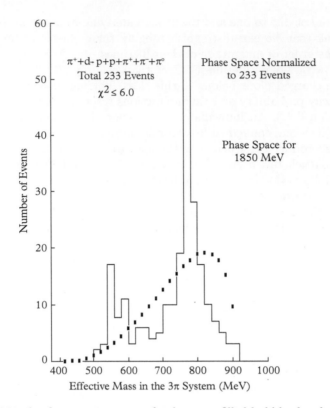

Fig. 6.11 A 1961 plot from an exposure of a deuteron-filled bubble chamber to a 1.23 GeV beam of positive pions at Berkeley. The plot shows the number of events versus the three pion invariant mass, i.e. the variable $\sqrt{s_{3\pi}}$ with $s_{3\pi} = (k_1 + k_2 + k_3)^2$, where k_i is the momentum of the ith pion in the final state of the reaction $\pi^+ + d \rightarrow p + p + \pi^+ + \pi^- + \pi^0$. Two resonances are clearly visible. The one around 780 MeV was discovered by the Berkeley team and corresponds to the spin 1 meson ω^0. The one with a lower mass of about 550 MeV was found by a collaboration of Johns Hopkins and Northwestern Universities and signalled the discovery of a new resonance, named η, which turned out to have spin zero and negative intrinsic parity. The dotted curve is the expected background in the absence of any resonance and is given by the phase space integral of the system $(p + \pi^+ + \pi^- + \pi^0)$ normalised to a centre-of-mass energy of 1850 MeV. *(Source: A. Pevsner et al. PRL **7**, 421, (1961)).*

errors quite large, but the resonance phenomena are clear. For comparison, we show a very precise resonance curve of the neutral weak gauge boson Z, which was obtained at LEP (Figure 6.12). Finally, Figure 22.11 shows the discovery of the last elementary particle, the Brout–Englert–Higgs boson, at the LHC. In this case the background is not simply phase space but the result of a very elaborate calculation of the expected two photon events as a function of their invariant mass. As we see, the signal-to-background ratio is very small, but it was compensated by the very accurate resolution in the invariant mass measurements combined with a very precise determination of the

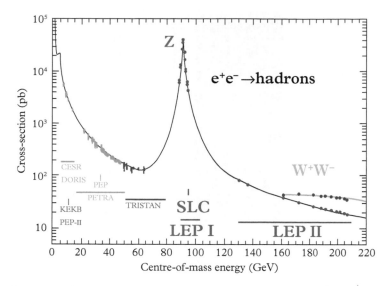

Fig. 6.12 The neutral spin-one boson Z seen as resonance in the $e^+e^- \rightarrow$ hadrons cross section by the LEP collaborations. *(Source: CERN-LEP collaborations).*

expected background. The excess of a few hundred events consistent with the decay of a resonance can be clearly seen.

6.10.3 $SU(2)$ as a classification group for hadrons

In section 6.5 we saw that the concept of isospin was introduced initially as a property of nuclear forces but it was gradually extended to a symmetry of strong interactions. The discovery in 1950 of π^0, whose existence was predicted in 1938 by Kemmer, was the first confirmation of this idea. The new particles which were discovered in the accelerators were all found to form multiplets of $SU(2)$ whose members had approximately the same mass. Some examples are:

Baryons: The nucleons (p, n) with $I = 1/2$, the Δ (Δ^{++}, Δ^{+}, Δ^{0}, Δ^{-}) with $I = 3/2$.

Mesons: The pions (π^+, π^0, π^-) with $I = 1$, the ρ (ρ^+, ρ^0, ρ^-) with $I = 1$, the η^0 with $I = 0$.

In section 5.7.4 we introduced the operator $\mathcal{C}$ which maps particles to antiparticles. π^0, ρ^0 and η^0 are eigenstates of $\mathcal{C}$, e.g. $\mathcal{C} \left| \pi^0 \right\rangle = \left| \pi^0 \right\rangle$,[49] but their charged partners in the isospin multiplet are not. This prompted L. Michel, as well as T.D. Lee and C.N. Yang, in 1956, to extend the transformation of charge conjugation to an isospin multiplet. They defined the operator $\mathcal{G}$ ($\mathcal{G}$-*parity*) as

$$\mathcal{G} = \mathcal{C}\mathrm{e}^{\mathrm{i}\pi I_2} \tag{6.27}$$

---

[49]In Chapter 8 we shall show that the one-photon state is an eigenstate of $\mathcal{C}$ with eigenvalue -1. Since π^0 decays into two photons, its $\mathcal{C}$ eigenvalue must be $+1$.

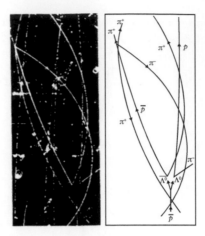

Fig. 6.13 A picture showing the reaction $\bar{p} + p \rightarrow \Lambda + \bar{\Lambda}$. Both Λs decay in the chamber giving $p + \pi^-$ and $\bar{p} + \pi^+$. They form V-shaped traces, hence the old name *V-particles*. The anti-proton from the $\bar{\Lambda}$ decay hits a proton in the chamber, seen at the upper-left corner, and produces four charged pions, all visible in the chamber. *(Source: Laurence Berkeley Laboratory)*.

with I_2 the generator of rotations around the second axis in isospin space. It is easy to check that all pions are eigenstates of $\mathcal{G}$ with eigenvalue -1. In Problem 6.7 we deduce the $\mathcal{G}$-parity of the ρ and η multiplets.

6.10.4 Strange particles

Among the many particles discovered in the 1950s, there were some which showed a strange behaviour. The first evidence came again from cosmic rays.[50] They appeared as relatively long-lived particles, leaving a trace in the detector, cloud chamber or emulsion, with masses which did not fit either with those of the nucleons, or with those of the pions. Some were heavier than the nucleon, while others were lighter.[51] Some were charged, with a visible decay vertex in the detector and others were neutral, decaying into two opposite charged particles. They were called *V-particles*. In Figure 6.13 we show a bubble chamber picture of the reaction $\bar{p} + p \rightarrow \Lambda + \bar{\Lambda}$.

In the 1953 Conference at Bagnères-de-Bigorre the following terminology was adopted, which is still in use today: *K-mesons* for particles intermediate between the pion and the nucleon and *hyperons* for those intermediate between the proton and the deuteron. In 1954, the term *baryon* was introduced to denote collectively nucleons and hyperons.

The study of these particles turned out to be among the most exciting and most rewarding enterprises in particle physics. It forced physicists to forge new concepts and new theoretical tools and to question some of their well accepted doctrines. We shall

---

[50] A single event of this kind was reported by L. Leprince-Ringuet and M. Lhéritier in 1944. G.D. Rochester and C.C. Butler published two more events in 1947. By the early 1950s there were a few dozen reported by various groups. Then the accelerators took over.

[51] In the early days masses were measured as multiples of the electron mass. The first estimates for the new particles were very uncertain, between 770 and 1600 m_e. Heavier ones were discovered later.

not follow the historical development, even though it is passionate and instructive, but we shall present directly the final picture, as it emerged in the years 1953 to 1955.

The first thing that was clear for these particles was that they were copiously produced in reactions like $\pi^- + p \to K^0 + \Lambda$, which meant that their production was governed by the strong interactions.[52] But this seemed to be incompatible with the fact that their decays such as $K^+ \to \pi^+ + \pi^0$ and $\Lambda \to p + \pi^-$ to pions and nucleons were very slow and their lifetimes were of order 10^{-8} to 10^{-10} s, which are characteristic of the weak interactions. The explanation of this strange behaviour was given, independently, by M. Gell-Mann as well as by T. Nakano and K. Nishijima in 1953, further elaborated in 1955 by Gell-Mann and Nishijima. It involves the introduction of a new, partially conserved quantum number, called *strangeness* by Gell-Mann.[53]

Until that time people knew of three conserved quantum numbers: the electric charge Q, the nucleon number which, in anticipation, we called baryon number B, and the lepton number L.[54] The only known hadrons were the nucleons and the pions (together with their resonances). They were assigned to multiplets of isospin, a doublet for the first and a triplet for the second, so we had the relation:

$$Q = I_3 + \frac{B}{2} \tag{6.28}$$

where I_3 is the third component of isospin. The strangeness idea consisted in decoupling these three quantities for the new particles, thus allowing the introduction of integer isospin hyperons and half-integer isospin mesons. If we call S the new quantum number with $S = 0$ for nucleons and pions, the relation (6.28) becomes:

$$Q = I_3 + \frac{S}{2} + \frac{B}{2} \tag{6.29}$$

The combination $Y = S + B$ is often called *hypercharge*. The nucleons, proton and neutron, have $S = 0$ and $Y = +1$ and the pions $S = Y = 0$. If we assume that strangeness is conserved by the strong and electromagnetic interactions but can be violated by the weak ones, we can explain the apparent strange behaviour of the new particles. In π–N and N–N collisions they can be produced, provided the algebraic sum of S of all particles in the final state is zero, hence their strong interaction production probability, but the lightest strange particles could only decay to ordinary particles through weak interactions, hence their long lifetimes. In Table 6.2 we give the first strange particles and their main decay modes.

The introduction of strangeness resolved the puzzle of production and decay of the new particles. But soon further puzzles appeared whose resolution brought novel and revolutionary ideas in particle physics. Let us look in more detail into the neutral K-meson system, the K^0 and its anti-particle $\overline{K^0}$. In the table we give no indication concerning their lifetime and decay products. Let us start by addressing this question. In order to do so, it will be convenient to evoke the simple concept of a *superselection*

---

[52]Experimentally, the production rate is roughly 1% that of pions.

[53]Nishijima had called it η-*charge*.

[54]The separate conservation of the electron-type (L_e) and the muon-type (L_μ) lepton numbers was considered later.

Table 6.2 The first strange particles. Their antiparticles have the same masses and lifetimes, but the opposite of all additive quantum numbers. According to (6.29), Λ, Σ and Ξ are baryons and the Ks are mesons. For the lifetime and decay products of the neutral K-mesons, see in the text

Particle	M(MeV)	τ(sec)	Q	S	T_3	Y	Decay modes
Λ	1115.683	$2.6 10^{-10}$	0	-1	0	0	$p\pi^-$, $n\pi^0$
Σ^+	1189.37	$0.8 10^{-10}$	1	-1	1	0	$p\pi^0$, $n\pi^+$
Σ^0	1192.642	$7.4 10^{-20}$	0	-1	0	0	$\Lambda\gamma$
Σ^-	1197.449	$1.48 10^{-10}$	-1	-1	-1	0	$n\pi^-$
Ξ^0	1314.86	$2.9 10^{-10}$	0	-2	$+\frac{1}{2}$	-1	$\Lambda\pi^0$
Ξ^-	1321.71	$1.64 10^{-10}$	-1	-2	$-\frac{1}{2}$	-1	$\Lambda\pi^-$
K^+	493.677	$1.24 10^{-8}$	1	1	$+\frac{1}{2}$	1	$\mu^+\nu_\mu$, $\pi^+\pi^0$
K^0	497.614		0	1	$+\frac{1}{2}$	1	

rule which we introduced in section 6.4.2. Were the strangeness operator S a conserved quantity, it would define such a superselection rule. In that case, the one particle states K^0 and $\overline{K^0}$ would have been the physical states, i.e. the states with well-defined mass and lifetime

$$S\left|K^0\right\rangle = \left|K^0\right\rangle , S|\overline{K^0}\rangle = -|\overline{K^0}\rangle , H\left|K^0\right\rangle = E_0\left|K^0\right\rangle , H|\overline{K^0}\rangle = E_0|\overline{K^0}\rangle \qquad (6.30)$$

Notice that, even if strangeness was conserved, a K^0 would still be unstable because it could undergo the analogue of β-decay, $K^0 \to K^+ + e^- + \bar\nu$ and the eigenvalue E_0 would be complex, as we explained in equation (2.59).

Strangeness non-conservation by the weak interactions changes this picture in two ways: First, it opens many more channels for K^0 decay, such as $K^0 \to \pi + \pi$ and $K^0 \to \pi + \pi + \pi$. This offers a much larger phase space and the lifetime becomes shorter. Second, and more important, strangeness violation allows virtual transitions $K^0 \leftrightarrow \overline{K^0}$, so the Hamiltonian describing this system has non-vanishing off-diagonal matrix elements in the K^0 and $\overline{K^0}$ basis. A similar situation appears in, for example, the quantum mechanical study of the ammonia molecule. It follows that the energy eigenstates of the system will be linear superpositions of K^0 and $\overline{K^0}$ and, in order to find them, we must diagonalise the 2×2 matrix Hamiltonian. We have no reliable theory to compute this matrix, but we can find a physical solution, following a reasoning introduced by M. Gell-Mann and A. Pais in 1955.

We know that K^0 and $\overline{K^0}$ are anti-particles of each other and, furthermore, that they are pseudo-scalars regarding space inversions. Let us assume that the product of the two operations $\mathcal{CP}$, charge conjugation ($\mathcal{C}$) times parity ($\mathcal{P}$), is conserved by all interactions.[55] Therefore we can diagonalise simultaneously $\mathcal{CP}$ and the Hamiltonian.

---

[55]This is incorrect in two ways. First, historically. In 1955 people did not know yet that weak interactions strongly violate the separate conservation of $\mathcal{C}$ and $\mathcal{P}$ and the argument of Gell-Mann and Pais was formulated using $\mathcal{C}$ alone. Second, physically. We know today that even $\mathcal{CP}$ is not conserved. We shall present the modern version of the argument in sections 18.11 and 22.4

This is done easily. Let us define[56] $CP \left| K^0 \right\rangle = \left| \overline{K^0} \right\rangle$. Then we can write the eigenstates of CP as

$$\left| K_L^0 \right\rangle = \frac{1}{\sqrt{2}} \left(\left| K^0 \right\rangle - \left| \overline{K^0} \right\rangle \right), \left| K_S^0 \right\rangle = \frac{1}{\sqrt{2}} \left(\left| K^0 \right\rangle + \left| \overline{K^0} \right\rangle \right) \tag{6.31}$$

$$CP \left| K_S^0 \right\rangle = \left| K_S^0 \right\rangle, CP \left| K_L^0 \right\rangle = - \left| K_L^0 \right\rangle \tag{6.32}$$

where the S and L subscripts used to denote the states stand for "short-lived" and "long-lived", respectively, and their meaning will be explained presently.

As we pointed out, strangeness violation by the weak interactions makes it possible for K-mesons to decay into two or three pions. Both kaons and pions have zero spin and negative intrinsic parity. It follows that a two-pion state, coming from the decay of a neutral kaon, must have zero orbital angular momentum and as a result it will be an eigenstate of CP with eigenvalue $+1$. This is true for both $\pi^+ \pi^-$ as well as $\pi^0 \pi^0$ states. We conclude that *only K_S^0 can decay into two pions*, while K_L^0 must decay into three pions. Since the phase space for the latter is quite small (the pion mass is about 140 MeV), we conclude that K_S^0 will have a much shorter lifetime. Indeed, the observed lifetimes are 0.9×10^{-10} s and 5×10^{-8} s, respectively.

A further point concerns the mass difference between these two states. In the initial K^0–$\overline{K^0}$ basis the Hamiltonian has the form

$$H = \begin{pmatrix} M & h \\ h & M \end{pmatrix} \tag{6.33}$$

where M is the value of the common mass of K^0 and $\overline{K^0}$ in the absence of strangeness violation and h is the part of the Hamiltonian which is responsible for the virtual transitions $K^0 \leftrightarrow \overline{K^0}$. We shall argue in Chapter 11 that our assumption of CP conservation is in fact equivalent to assuming invariance under time reversal; therefore, the two off-diagonal elements are equal.[57] Diagonalising H we find that the mass difference between the K_S^0 and K_L^0 states is equal to $2h$. Experimentally we find $\Delta m = 3.5 \times 10^{-12}$ MeV. Although we cannot compute h, we see from this value that the transition cannot be of first order in the weak interactions, whose strength is characterised by the Fermi coupling constant given in equation (6.17).[58] We conclude that the transition $K^0 \leftrightarrow \overline{K^0}$, which changes the value of strangeness by two units, must be of second order in the weak interactions. This is consistent with the observation from Table 6.2 according to which the Ξ hyperons decay into $\Lambda + \pi$ and not into $N + \pi$.[59] It seems that first order weak interactions violate strangeness by at most one unit.

A final point about the $K^0 - \overline{K^0}$ system: In 1955, A. Pais and O. Piccioni predicted the phenomenon of *regeneration*. Consider a beam of neutral kaons produced at $t = t_0$

---

[56]The relative phase between $\left| K^0 \right\rangle$ and $\left| \overline{K^0} \right\rangle$ is not fixed by the strong interactions because they conserve strangeness. The definition we adopt here is a convention.

[57]Under time reversal we obtain for the K^0–$\overline{K^0}$ transition at rest: $H_{12} = \left\langle K^0 \right| H \left| \overline{K^0} \right\rangle \rightarrow \left\langle \overline{K^0} \right| H \left| K^0 \right\rangle = H_{21}$.

[58]By dimensional analysis, a first order contribution to Δm would be of order $G_F M^3 \simeq 10^{-3}$ MeV, much larger than the experimental value.

[59]In the early literature Ξ hyperons were called *cascades*, precisely because they decay through the cascade reaction $\Xi \rightarrow \Lambda + \pi \rightarrow N + 2\pi$.

by hadronic collisions. Baryon and strangeness conservation of the strong interactions implies that, if the energy is low enough, only K^0s can be produced in association with Λs or Σs, but not $\overline{K^0}$s. The particles which propagate in the vacuum are the combinations of equation (6.31), so that at $t = t_0 + \epsilon$ the beam consists of an equal mixture of K_S^0s and K_L^0s. Owing to the large difference in their lifetimes, at $t = t_0 + \delta$ with $\tau_S \ll \delta \ll \tau_L$, the K_S^0s have all decayed and we are left with a beam consisting almost entirely of K_L^0s, i.e. an equal mixture of K^0s and $\overline{K^0}$s. This can be checked experimentally by the absence of the two-pion decay mode. Let us now send this beam to hit a material plate. For the same reason as before, the plate will absorb predominantly the $\overline{K^0}$s through the reaction $\overline{K^0} + N \to \Lambda$ (or Σ) $+ \pi$. Such a reaction is not possible for K^0. So, after traversing the plate, the beam will be almost pure K^0, i.e. almost equal mixture of K_S^0 and K_L^0; in other words we have succeeded in regenerating K_S^0, phenomenon which will manifest itself by the reappearance of the two pion decays. All this has been verified experimentally–beautiful manifestations of the superposition principle, one of the most fundamental principles of quantum mechanics.

The study of the K-mesons revealed many more surprises, but we shall leave the story here and pick it up again in Chapter 18.

6.11   The Eightfold Way and the Quarks

6.11.1   From $SU(2)$ to $SU(3)$

With the discovery of strange particles and the identification of strangeness as a new conserved charge of the strong and electromagnetic interactions, it became clear that the initial isospin symmetry based on the group $SU(2)$ should be enlarged to a bigger group. The first attempt in this direction is due to A. Pais in 1953, but the scheme which finally emerged dates from 1961 and is based on $SU(3)$. Contrary to what one might think, going from $SU(2)$ to $SU(3)$ was not at all an obvious step. $SU(2)$ is realised in a direct way with the two nucleons forming a doublet. The generalisation would imply the addition of a third baryon to form a triplet, which is the fundamental representation of $SU(3)$. Such a model, using the triplet (p, n, Λ), was indeed proposed in 1959 by S. Sakata. It was disproved by the data, in particular because it could not accommodate all the other strange particles.

The solution was found independently in 1961 by M. Gell-Mann and Y. Ne'eman. It is an $SU(3)$ symmetry which contains the isospin $SU(2)$ as a subgroup. We know that the algebra of $SU(3)$ has rank equal to 2. Thus, two of its generators can be diagonalised simultaneously. They were called T_3 and T_8 in Chapter 5. Gell-Mann and Ne'eman chose to identify them with the third component $I_3 = T_3$ of isospin and with the hypercharge $Y = (2/\sqrt{3})T_8$, respectively. With this identification the positions in the I_3–Y plane of the *eight spin-$\frac{1}{2}$ positive parity baryons* $(p, n, \Lambda$, the three Σ s and the two Ξ s) are shown in Figure 6.14 (left). The same hexagon is formed by the *eight pseudo-scalar mesons* (the pions, the kaons and a neutral one called η with mass 548 MeV and zero isospin and hypercharge), as well as by the *eight vector mesons* with spin and parity 1^-.[60] In Problem 6.10 we ask the reader to place them in the I_3–Y plane. The $3/2^+$ baryons, like the Δ resonance, we saw earlier, form a decouplet, see

---

[60]In fact, we know experimentally that there are nine pseudo-scalar and nine vector mesons. In $SU(3)$ they form octets and singlets. We will discuss the situation shortly.

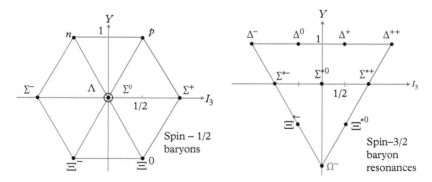

Fig. 6.14 Left: The octet of spin-$\frac{1}{2}$ baryons. By definition, it is identical to the weight diagram of the adjoint representation $\mathbf{8} = (1,1)$ of $\mathfrak{su}(3)$. Right: The spin-$\frac{3}{2}$ baryons. The figure is identical to the weight diagram of the $(3,0)$ of $\mathfrak{su}(3)$. The existence of the Ω^- state was a prediction of $SU(3)$ symmetry.

Figure 6.14 (right). These figures were not all complete at the time. Nevertheless, they inspired Gell-Mann and Ne'eman to conjecture the presence in the strong interactions of an $SU(3)$ symmetry and assign the $1/2^+$ baryons and the 0^- and 1^- mesons to the eight-dimensional **8** representation of $SU(3)$ and the $3/2^+$ baryons to the **10**. An important feature of the scheme is that it leaves the **3** representation empty. Because of the central role played by the eight-dimensional representation, Gell-Mann called this scheme *The Eightfold Way*.

6.11.2 The arrival of the quarks

As we have noticed in section 6.8, in the years around 1960 the strong focusing technique allowed the construction of more powerful accelerators, the alternating gradient synchrotrons, which greatly increased the discovery potential for new "elementary" particles. Faced, though, with a proliferation of new particles several physicists started wondering whether we were not uncovering a new layer of the onion, after those of atoms and nuclei. The introduction of $SU(3)$ made possible a more precise formulation of this question. Three elements played an important role: (i) $SU(3)$ was very successful as a classification scheme for all these new hadrons, in the sense that they could all be placed in appropriate representations. (ii) Among all the $SU(3)$ representations mesons formed only octets (and singlets) and baryons octets and decouplets. (iii) In particular, the fundamental triplet representation remained empty.

These facts motivated the introduction in 1964 by M. Gell-Mann, and independently by G. Zweig, of a constituent model based on the following simple principles: (i) All hadrons are bound states of elementary spin-$\frac{1}{2}$ fermions. Gell-Mann called them *quarks*[61] and Zweig *aces*. (ii) There exist three distinct types of quarks forming a triplet representation of $SU(3)$. Their anti-particles form the conjugate $\bar{\mathbf{3}}$ representation. (iii) Mesons are bound states of a quark–anti-quark pair and baryons are bound states of three quarks.

---

[61]Gell-Mann found that name in *Finnegans wake*, the last book of James Joyce, where we can read the obscure verse *Three quarks for Muster Mark*.

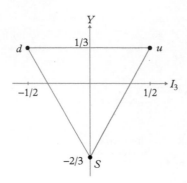

Fig. 6.15 The quark triplet of $SU(3)$.

Table 6.3 The quantum numbers of the quarks

Quark	Q	B	I_3	S
u	$+\frac{2}{3}$	$\frac{1}{3}$	$+\frac{1}{2}$	0
d	$-\frac{1}{3}$	$\frac{1}{3}$	$-\frac{1}{2}$	0
s	$-\frac{1}{3}$	$\frac{1}{3}$	0	-1

Table 6.3 shows the quantum numbers of the three quarks in the Gell-Mann and Zweig scheme. The symbols which are used today are u for "up", d for "down" and s for "strange" (see Figure 6.15). Notice that they carry non-integer values for the electric charge and baryon number.

With these rules at hand it is straightforward to determine the quark content of the low-lying hadrons by matching quantum numbers. For example, the proton is a (uud) bound state, since this combination has $Q = 1$, $B = 1$, $S = 0$, $I_3 = +1/2$ and is consistent with spin $= \frac{1}{2}$ and isospin $I = 1/2$. Similarly, the neutron is a (udd) bound state. Tables 6.4 and 6.5 show the quark composition of various hadrons.

All known baryons are classified according to the tensor product $\mathbf{3} \otimes \mathbf{3} \otimes \mathbf{3}$ in octets and decouplets shown in Figure 6.14,[62] with quark content shown in Table 6.4.

Similar considerations apply to the light 0^- and 1^- mesons. Their quark composition is shown in Table 6.5. We have not indicated the quark content of the isospin zero states. In the limit of exact $SU(3)$ symmetry there would be two orthogonal states with $I = Y = 0$, one which is a member of an octet, and another one of a singlet. They are given by the linear superpositions of $\bar{q}q$ states

$$\psi_8 \sim (\bar{u}u + \bar{d}d - 2\bar{s}s)/\sqrt{6}, \quad \psi_1 \sim (\bar{u}u + \bar{d}d + \bar{s}s)/\sqrt{3} \tag{6.34}$$

However, the mass differences between the members of an $SU(3)$ multiplet show that $SU(3)$ must be broken and this breaking induces a mixing among the octet and singlet

---

[62]In the product there is also a singlet. Such a baryon will be singlet under colour (see next subsection), as well as under flavour $SU(3)$, while its spin state will have either full or mixed symmetry. Thus, Fermi statistics implies that the relative angular momenta among the three quarks cannot all be equal to zero. In a potential model this implies a heavier mass. The first candidate for such a state is a $1/2^+$ Λ resonance of 1600 MeV.

Table 6.4 The quark composition of the $1/2^+$ and the $3/2^+$ baryons. The value in parenthesis gives the average mass of the corresponding isospin multiplet

N (939 MeV)	Λ (1116 MeV)	Σ (1193 MeV)	Ξ (1318 MeV)
uud, udd	$1/\sqrt{2}(ud+du)s$	$dds, 1/\sqrt{2}(ud-du)s, uus$	dss, uss
Δ (1232 MeV)	Σ* (1385 MeV)	Ξ* (1530 MeV)	Ω⁻ (1672 MeV)
$ddd, udd,$ uud, uuu	$dds, uds,$ uus	dss, uss	sss

Table 6.5 The quark composition of the 0^- and the 1^- mesons. For the $I=0$ states, read the text

J^P	$I=1$ $u\bar{d}, 1/\sqrt{2}(u\bar{u}-d\bar{d}), \bar{u}d$	$I=\frac{1}{2}$ $u\bar{s}, d\bar{s}; \bar{d}s, \bar{u}s$	$I=0$	$I=0$
0^-	π (138 MeV)	K (496 MeV)	η (549 MeV)	η' (958 MeV)
1^-	ρ (770 MeV)	K^* (892 MeV)	ϕ (1020 MeV)	ω (782 MeV)

states. It follows that the mass eigenstates are linear combinations of the form

$$\psi = -\cos\theta\,\psi_8 + \sin\theta\,\psi_1 \quad \psi' = \sin\theta\,\psi_8 + \cos\theta\,\psi_1 \tag{6.35}$$

We call "ideal mixing" the one with $\tan\theta = 1/\sqrt{2}$ because it yields a pure $\bar{s}s$ and a pure $(\bar{u}u+\bar{d}d)/\sqrt{2}$ state. Phenomenologically, the mixing turns out to be rather small between η and η' in the 0^- mesons, but it is very close to ideal for the 1^- mesons with ϕ being an almost pure $\bar{s}s$ state.

6.11.3 The breaking of $SU(3)$

We know that even isospin is slightly broken in nature because the proton and the neutron do not have the same mass. However, this breaking was attributed to electromagnetic effects and it was assumed that the strong interactions were $SU(2)$ invariant.[63] For $SU(3)$ the mass difference between the nucleons and the Ξ hyperons is of order 30%, much too large to be of electromagnetic origin. It follows that any realistic model of higher symmetry must include a symmetry breaking pattern. The philosophy was expressed already in 1958, long before the introduction of $SU(3)$, by B. d'Espagnat and J. Prentki.[64] For the first time the whole hierarchy of symmetries and interactions was clearly presented. It involves four levels of interactions, all of which conserve the electric charge Q and the baryon number B, but they differ with respect to $SU(3)$ as follows:

- The *very strong* interactions conserve the full $SU(3)$.
- The *medium strong* interactions break $SU(3)$, but conserve hypercharge and isospin.

[63]We shall come back to this point in Chapter 19.
[64]Their model used an $O(4)$ symmetry group.

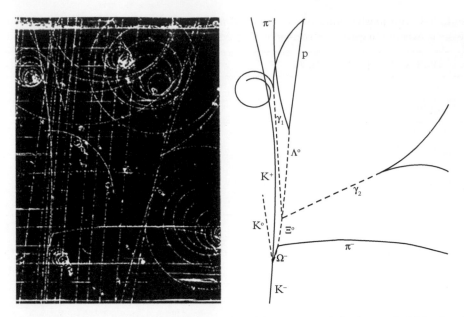

Fig. 6.16 The first Ω^- picture taken in the Brookhaven 80-inch hydrogen bubble chamber by the Nicholas Samios team. The chamber was exposed to a K^- beam. On the right we see the cascade of events: $K^- + p \to \Omega^- + K^0 + K^+$, $\Omega^- \to \Xi^0 + \pi^-$, $\Xi^0 \to \Lambda + \pi^0$, $\Lambda \to \pi^- + p$. *(Source: V.E. Barnes et al., Phys. Rev. Lett. **12**, 204 (1964)).*

• The electromagnetic interactions break isospin, but conserve hypercharge and, hence, the third component of isospin.

• The weak interactions break hypercharge by one unit.

The total Hamiltonian of elementary particles could thus be written in the form

$$H = H_{VS} + H_{MS} + H_{em} + H_{W} \tag{6.36}$$

with H_{VS} an $SU(3)$ singlet operator. Although, as we shall see later in this book, our ideas have considerably evolved concerning especially the strong interaction symmetries, the basic principles remain the same.

A model following this pattern was proposed by M. Gell-Mann in 1961 and S. Okubo in 1962. They postulated that H_{MS} belongs to an $SU(3)$ octet and transforms like the hypercharge generator. The fact that H_{MS} preserves isospin and hypercharge means that it should belong to a representation containing a state with $I = Y = 0$. The octet is the simplest choice and the corresponding state corresponds to the generator Y. Therefore, H_{MS} has the form: .

$$H_{MS} = Y_l^k \hat{\mathcal{O}}_k^l \tag{6.37}$$

with $\hat{\mathcal{O}}$ an operator in the **8** of $SU(3)$ and Y_l^k is the (kl) matrix element of the hypercharge 3×3 matrix, which when contracted with $\hat{\mathcal{O}}$ projects out the Y-component of the latter.[65] In the quark model the octet choice of equation (6.37) means that

---

[65]Under the action of the generator $\hat{T}_a$ of $SU(3)$, H_{MS} changes by $\delta_a H_{MS} = Y_l^k \delta_a \hat{\mathcal{O}}_k^l = Y_l^k [\hat{T}_a, \hat{\mathcal{O}}_k^l] = Y_l^k (T_{a\,n}^{\ l} \hat{\mathcal{O}}_k^n - T_{a\,k}^{\ n} \hat{\mathcal{O}}_n^l) = [Y, T_a]_l^n \hat{\mathcal{O}}_n^l$. It vanishes for $a = 1, 2, 3, 8$.

the strong interaction Hamiltonian $H_{\mathrm{VS}} + H_{\mathrm{MS}}$ transforms under $SU(3)$ as a $\bar{q}q$ pair. We will see in Chapter 21 that this fits very well with our current ideas on strong interactions. Here let us stress that equation (6.37) has far-reaching consequences for particle physics, all of which were confirmed experimentally. First, the model made predictions about the relative strength of strong, as well as weak interaction amplitudes, some of which will be derived in the following chapters. Second, it predicted the existence of particles, which were needed to fill the $SU(3)$ representations and which, as mentioned already, had not yet been observed, when it was formulated in 1961. The most spectacular of those predictions was the particle we call Ω^- in Figure 6.14. It was discovered in a bubble chamber experiment by the team of N. Samios at Brookhaven in 1964 (see Figure 6.16) and gave the most convincing evidence that the Eightfold Way was the correct symmetry scheme.

6.11.3.1 The Gell-Mann–Okubo mass formula. The specific form of the symmetry breaking Hamiltonian (6.37) makes it possible to estimate the mass differences among the members of an $SU(3)$ multiplet. We will assume that these effects are small so that we can compute them using first order perturbation theory. It follows that we should compute the matrix elements of H_{MS} between hadron states of the same multiplet. Note however that, since we ignore the isospin breaking, which may be attributed to electromagnetic interactions, the accuracy of these calculations cannot be expected to be better than a few per cent.

Let us consider a D-dimensional irreducible representation. We must compute the matrix elements $\langle D_1 | (H_{\mathrm{VS}} + H_{\mathrm{MS}}) | D_2 \rangle$ with $|D_1\rangle$ and $|D_2\rangle$ two states belonging to the representation D. The $SU(3)$ symmetric part of the Hamiltonian H_{VS} will give a common mass M_D to all members. Using the Young tableaux or the tensor method discussed in the previous chapter, we can establish the following result:
  If $D \ncong \overline{D}$ the product $D \otimes \overline{D}$ contains the octet only once.
  If $D \cong \overline{D}$ the product $D \otimes \overline{D}$ contains the octet twice.

Applying the Wigner–Eckart theorem in the computation of the matrix elements $\langle D_1 | H_{\mathrm{MS}} | D_2 \rangle$, we conclude that in the first case we shall need one reduced matrix element and in the second we shall need two.

• Let us start with the masses in the baryon decouplet. Since $\mathbf{10} \ncong \overline{\mathbf{10}}$, all matrix elements of H_{MS} are proportional to a single one, which we choose to be the matrix element of the hypercharge operator Y. We conclude that the symmetry breaking splits the levels according to the value of hypercharge and we obtain equal spacing for the various levels

$$M_\Delta = M_D + \delta M_D, \quad M_{\Sigma^*} = M_D, \quad M_{\Xi^*} = M_D - \delta M_D, \quad M_{\Omega^-} = M_D - 2\delta M_D \quad (6.38)$$

where δM_D is the value of the reduced matrix element. Using the experimental values we obtain

$$(M_{\Sigma^*} - M_\Delta)/(M_{\Xi^*} - M_{\Sigma^*})/(M_{\Omega^-} - M_{\Xi^*}) = 153\,\mathrm{MeV}/145\,\mathrm{MeV}/142\,\mathrm{MeV} \quad (6.39)$$

with uncertainties on the order of a few MeV in each mass difference, in excellent agreement with equal spacings. Note also that when this formula was written the Ω^- state had not yet been identified. Knowing its mass through this formula, certified the identification shown in Figure 6.16.

• As a second application of the Gell-Mann–Okubo mass formula let us look at the mass splittings among the members of the $1/2^+$ baryon octet. The formalism remains the same with an important difference: in the product of two **8**s we find twice the representation **8**, see equation (5.114). As a result, the Wigner–Eckart theorem tells us that the matrix elements $\langle B_1|H_{\text{MS}}|B_2\rangle$ with $|B_1\rangle$ and $|B_2\rangle$ two states belonging to the **8** representation, depend on two reduced matrix elements. We conclude that the mass differences will depend on two arbitrary constants. We can perform the computation following the general method we used to prove the Wigner–Eckart theorem (see Problem 6.10), but there exists a general derivation that applies to all representations.

By assumption, H_{MS} conserves hypercharge and isospin. The only term with these properties which is linear in the generators is Y. For the octet we need a second operator $\mathcal{O}$ and we choose it to be quadratic in the generators. It must be a superposition of the form (see Problem 6.10)

$$\mathcal{O} = Y^2 + A\boldsymbol{I}^2 \tag{6.40}$$

where $\boldsymbol{I}$ are the three isospin generators and A a constant. We determine A by the following condition: Had we used a quadratic operator for the mass splittings of the decouplet, we should have obtained the results of equation (6.38). This gives $A = -4$ so the general mass formula valid for any representation is of the form

$$M_{I,Y}^{(j)} = M^{(j)} + c_1 Y + c_2[Y^2 - 4I(I+1)] \tag{6.41}$$

where $M_{I,Y}^{(j)}$ is the mass of the member with spin-j, isospin I and hypercharge Y. $M^{(j)}$ is a common mass for the representation and c_1 and c_2 are constants depending on the representation. This is the *Gell-Mann–Okubo mass formula*. As for the baryon octet, ignoring the possibility of a singlet-octet mixing, it leads to the relation

$$2(M_N + M_\Xi) = 3M_\Lambda + M_\Sigma \tag{6.42}$$

where the masses denote the average masses in each isospin multiplet. The experimental values give: $4.514 \simeq 4.538$, in very good agreement with (6.42).

• We can continue and apply this general formula (6.41) to the meson octets, but the agreement is less impressive. First, as we have pointed out already, the singlet-octet mixing among the 1^- vector mesons is known to be important, so this formula can be used to determine the $\omega - \phi$ mixing angle. Second, there is an empirical fact that for mesons the Gell-Mann–Okubo formula works better if we replace all masses by their squares. We find sometimes in the literature the argument that this is due to the fact that Lagrangian field theories are linear in fermion masses and quadratic in boson masses, but there is no obvious logical connection between the two.

6.11.4   Colour

It was soon realised that this simple picture with three quarks as the sole constituents of all hadronic matter should be made a little more complex. The reason was quantum statistics. The quarks have spin one-half, so they must be fermions. If we consider the wave function of a baryon, it must be totally antisymmetric under the exchange of any two of its three constituent quarks.[66] For example, let us take Ω^-. It has strangeness

[66]In fact, we can extend the requirement of antisymmetry of the wave function to $SU(3)$, the way we did it for $SU(2)$ in equation (6.13)

$S = -3$, so it is made out of three s-quarks. It must correspond to the lightest bound state of this system, so we expect the quarks to be in relative $l = 0$ states. This is corroborated by calculations using potential models as well as by extended symmetry arguments. In addition, since Ω^- has spin equal to $3/2$, the spin state is symmetric. But this leads to a wave function which is fully symmetric in both space and spin variables, in contradiction with Fermi statistics. Several solutions were proposed, but the one which proved to be the right one was proposed by O.W. Greenberg, in 1964. It amounts, first, to the introduction of an extra index to label each quark, which can take three values[67] and was given the rather uninspired name *colour* (no connection, of course, with the common sense of the word) and, second, to the assessment that the wave functions of the baryons should be totally antisymmetric in the colour indices. As a result, the baryon wave functions were now consistent with the fermionic nature of the quarks. It should be pointed out that even though each quark appears in three forms differing in their colour, the coloured quark model is not equivalent to a nine-quark model, because it is supplemented with the extra assumption that every hadron contains all three colours in equal proportions, so that matter appears colourless.

As we shall study in detail in Chapter 21, the concept of colour turned out to be fundamental in describing the dynamics of the strong interactions.

6.12 The Present Table of Elementary Particles

With quarks we moved one step deeper into the structure of matter. In 1964, the known constituents were the three quarks u, d and s, each one coming in three colours, and four leptons, the electron, the muon and their associated neutrinos. Soon this number increased. A third leptonic doublet was discovered, the τ and its neutrino ν_τ and three more quark species, c for *charm*, b for *bottom* and t for *top*. As we shall see in Chapter 19, the existence of these additional states was predicted for theoretical reasons, for most of them before their experimental discovery. So, now we have six lepton and six quark species,[68] each one of the latter coming in three colours. The complete list of elementary particles as we know them today – 2021 – is shown in Table 6.6, while in Table 6.7 we give the measured, or deduced, values of the masses of all the particles whose existence has been established.

A few words to explain each of the entries.

• *The quanta of radiation* denote the particles whose exchanges generate the interactions. They generalise the concept of the photon to the other three fundamental interactions. We see that we have eight mediators of the strong interactions at the level of the quarks; they are called *gluons* in the table, the photon for the electromagnetic interactions, and three vector bosons, W and Z for the weak interactions. We believe that there exists also a mediator for the gravitational interactions, the *graviton*, but it has not been identified. For the first three interactions we know that their associated

---

[67]Greenberg's original model used the concept of *parastatistics*, first introduced in 1940. A particle is said to obey parastatistics of order n if we allow at most n identical particles to occupy the same state. $n = 1$ corresponds to fermions and $n = \infty$ to bosons. For the Gell-Mann–Zweig quarks it was proven in 1966 by Greenberg and D. Zwanziger that parastatistics of order $n=3$ is equivalent to the colour model.

[68]An equally uninspired terminology has been introduced: the six quark species are called *flavours*.

Table 6.6 This table shows our present ideas on the structure of matter. Quarks and gluons do not exist as free particles and the graviton has not been observed

TABLE OF ELEMENTARY PARTICLES	
QUANTA OF RADIATION	
Strong Interactions	Eight gluons
Electromagnetic Interactions	Photon (γ)
Weak Interactions	Bosons W^+ , W^- , Z
Gravitational Interactions	Graviton (?)

MATTER PARTICLES		
	Leptons	Quarks
1st Family	ν_e , e^-	u_a , d_a , $a = 1, 2, 3$
2nd Family	ν_μ , μ^-	c_a , s_a , $a = 1, 2, 3$
3rd Family	ν_τ , τ^-	t_a , b_a , $a = 1, 2, 3$
BEH BOSON		

Table 6.7 Since the quarks do not exist as free particles, their masses are deduced from theoretical considerations. The masses of the neutrinos will be discussed in a later chapter

MASSES OF ELEMENTARY PARTICLES					
QUANTA OF RADIATION				BEH BOSON	
g_a	γ	$W^\pm$	Z	H	
0	0	80.4 GeV	91.2 GeV	125.1 GeV	
QUARKS					
u	d	c	s	t	b
2.2 MeV	4.7 MeV	1.27 GeV	96 MeV	173.2 GeV	4.2 GeV
LEPTONS					
ν_e	e	ν_μ	μ	ν_τ	τ
?	0.511 MeV	?	105.7 MeV	?	1.777 GeV

quanta of radiation are vector (spin-1) particles, while the graviton is believed to be a helicity-2 massless particle.

• *Matter particles.* The constituents of matter appear to be all spin one-half particles. They are the six quarks in three colours each and the six leptons we mentioned before.

Quarks and gluons do not appear as free particles. They form a large number of bound states, the hadrons.

Quarks and leptons seem to fall into three distinct groups, or "families". This family structure is one of the great puzzles in elementary particle physics. We believe we understand the importance of the first family. It is composed of the electron and its associated neutrino, as well as the up and down quarks. These quarks are the constituents of protons and neutrons. The role of each member of this family in the structure of matter is obvious. In contrast, the role of the other two remains obscure.

The muon and the tau leptons seem to be heavier versions of the electron, but they cannot be viewed as excited states of it, because they seem to carry their own quantum numbers. The associated quarks with exotic names such as charm, strange, top and bottom, form unstable hadrons, which are not present in ordinary matter. Why nature needs three similar copies of apparently the same structure remains a mystery.

• An observation, whose importance will not be discussed in any detail here, is that the sum of all electric charges inside each family is equal to zero.

• The final entry in Table 6.6 is a spin zero particle, which was discovered in 2012 at CERN. Its significance will be explained in Chapter 14.

In the following chapters we shall describe in some detail the fundamental theory of elementary particles – the so-called Standard Model – together with the rules and techniques of computing scattering cross sections and decay rates of unstable particles. In the process, we will elaborate further on several open questions.

6.13 Problems

Problem 6.1 *Stability of the beam.* Consider a linear accelerator. Prove that if we choose the phase θ of equation (6.23) to be smaller than its "optimal" value $\pi/2$, we lose a little of accelerating power, but the machine itself acts as a beam stabiliser, i.e. a particle faster than the average is slowed down and a slower one is accelerated.

Problem 6.2 *Relativistic kinematics.* Consider a beam of relativistic particles of mass M_b and momentum $\boldsymbol{p}_b$ hitting a target of particles of mass M_t at rest. Compute the square of the total energy E_{tot}^2 of a pair of colliding particles in the centre-of-mass frame.

Problem 6.3 *The bandwidth–duration relation of a pulse.* Verify the relation $\Delta\omega\Delta t \geq 2\pi$ between the bandwidth $\Delta\omega$ and the duration Δt of a pulse in the following simple example: Consider a square pulse $\phi(t)$, which is nonzero only in a time interval Δt, and whose value in that interval is $\phi(t) = 1/\Delta t$. (i) Show that with appropriate choice of the origin $t = 0$ of the time axis, its Fourier amplitude $\tilde{\phi}(\omega)$ can be written in the form $\tilde{\phi}(\omega) = (1/\pi) \int_{-\infty}^{\infty} dt\,\phi(t) \cos\omega t$. (ii) Compute $\tilde{\phi}(\omega)$. (iii) Take as natural definition of the bandwidth $\Delta\omega$ the range of ω from its minimum value $\omega = 0$ to the first zero of $\tilde{\phi}(\omega)$ and show that $\Delta\omega\Delta t = 2\pi$. (iv) What is $\tilde{\phi}(\omega)$ in the limit $\Delta t \to 0$?

Problem 6.4 *Isospin-3/2 dominance in π–N scattering (Fermi, 1951).* The low energy pion–nucleon scattering data show that the ratios of the cross sections for $\pi^+ p \to \pi^+ p(\sigma^{++})$, $\pi^- p \to \pi^0 n(\sigma^{-0})$ and $\pi^- p \to \pi^- p(\sigma^{--})$ are given, approximately, by $\sigma^{++} : \sigma^{-0} : \sigma^{--} = 9 : 2 : 1$. Assuming isospin invariance of the strong interaction Hamiltonian, prove that this result implies that in pion–nucleon scattering the $I=3/2$ channel dominates.

Problem 6.5 Compute the relations between the total cross sections of the processes: (a) $\pi^- p \to K^0 \Sigma^0$, (b) $\pi^- p \to K^+ \Sigma^-$ and (c) $\pi^+ p \to K^+ \Sigma^+$.

Problem 6.6 *The Σ-baryons.*

1. In Table 6.2 we see that the lifetimes of $\Sigma^{\pm}$ are on the order of 10^{-10} s, while that of Σ^0 is $\sim 10^{-20}$ s. Explain why.

2. The lifetime of Σ^- is almost twice as large as that of Σ^+. Why?

Problem 6.7 *The mesons ρ and K.*

1. The ρ-meson we find in Table 6.5 has the quantum numbers 1^- and isospin $I = 1$. It decays mainly into two pions: $\rho \to 2\pi$. (i) Give the result of the operation $C\left|\rho^0\right\rangle$. (ii) Using symmetry arguments prove that the decay $\rho^0 \to 2\pi^0$ is forbidden. (iii) Show that a one-ρ state is an eigenstate of $\mathcal{G}$ with eigenvalue $+1$. (iv) Give the G-parity of η.

2. The K^+ meson has quantum numbers 0^-, isospin $\frac{1}{2}$ and decays through the weak interactions. Prove that in the decay mode $K^+ \to \pi^+ + \pi^0$ the two pions are in an isospin equal to 2 state.

Problem 6.8 What is the isospin of a resonance X, which decays via the strong interaction to $p\pi^-$ and $n\pi^0$ with branching ratios 36% and 18% respectively?

Problem 6.9 A resonance $\Lambda^\star$, which is an isosinglet with electric charge 0 and mass 1405 MeV, decays to $\Sigma\pi$ via the strong interaction. Show that the partial widths $\Gamma(\Lambda^\star \to \Sigma^+\pi^-)$, $\Gamma(\Lambda^\star \to \Sigma^0\pi^0)$ and $\Gamma(\Lambda^\star \to \Sigma^-\pi^+)$ are equal.

Problem 6.10 *The flavour group $SU(3)$.*

1. The spin-$\frac{1}{2}$ baryons form an octet of $SU(3)$. Therefore, they can be described collectively by a wave function which is a 3×3 traceless matrix B with elements B_j^i. Using the quantum numbers of the members of the **8** determine their position in the matrix B.

2. Use the result in order to derive the Gell-Mann−Okubo mass formula for the baryon octet, equation (6.42).

Problem 6.11 Place the octets of 0^- and 1^- mesons of Table 6.5 on the $I_3 - Y$ plane.

7
Relativistic Wave Equations

7.1 Introduction

The Lorentz transformations were discovered as an invariance of the theory of electromagnetism described by Maxwell's equations. The formulation of special relativity was based on Maxwell's equations, which in empty space were shown to be relativistically covariant wave equations. In modern terminology they describe the propagation of a vector field with zero mass parameter. In the present chapter we will derive the relativistically covariant differential equations for fields with spins equal to zero, one-half and one, which are used in elementary particle physics.

7.2 The Klein–Gordon Equation

We start with the simplest case, the equation for a real, scalar field. In our notation of Chapter 5, it belongs to the trivial one-dimensional (0,0) representation of the Lorentz algebra. In this case the elements which are at our disposal are the field itself ϕ and the four-vector operator of derivation ∂_μ. It is clear that the lowest order, non-trivial, relativistically covariant equation, which can be built with these quantities is

$$\left(\partial_\mu \partial^\mu + m^2\right)\phi(x) = 0 \tag{7.1}$$

This is the Klein–Gordon equation. In our usual system of units $\hbar = c = 1$ the parameter m^2 has the dimensions of $[\mathrm{M}]^2$. We shall often call this equation *the massive Klein–Gordon equation*, although at this stage we have no real justification for this name. The m^2 is just a parameter which can take any real value.[1] This equation can be derived by the variational principle applied to the action

$$S[\phi] = \int \mathrm{d}^4x\, \mathcal{L}(x) = \frac{1}{2}\int \mathrm{d}^4x \left(\partial_\mu \phi(x)\partial^\mu \phi(x) - m^2 \phi^2(x)\right) \tag{7.2}$$

where

$$\mathcal{L} = \frac{1}{2}\left(\partial_\mu \phi(x)\partial^\mu \phi(x) - m^2 \phi^2(x)\right) \tag{7.3}$$

is the Lagrangian density. The canonical momentum associated to $\phi(x)$ is given by

$$\pi(x) = \frac{\partial \mathcal{L}}{\partial(\partial_0 \phi(x))} = \partial_0 \phi(x) \tag{7.4}$$

---

[1] In the classical theory with $c = 1$ the parameter m has dimensions $1/L$ and in equation (4.15) was interpreted as the attenuation length of the resulting Yukawa potential. We also noted in Chapter 4 that, upon quantisation, the particle mass is given by $m\hbar$, which indeed has dimensions of mass.

Elementary Particle Physics. John Iliopoulos and Theodore N. Tomaras, Oxford University Press.
© John Iliopoulos and Theodore N. Tomaras (2021). DOI: 10.1093/oso/9780192844200.003.0007

and the Hamiltonian density by

$$\mathcal{H} = \frac{1}{2}\left[\pi^2(x) + (\partial_i\phi(x))^2 + m^2\phi^2(x)\right] \tag{7.5}$$

Equation (7.1) admits plane wave solutions

$$e^{-i(p_0\,t - \boldsymbol{p}\cdot\boldsymbol{x})}, \quad p_0 = \pm\omega_p\,, \; \omega_p = \sqrt{\boldsymbol{p}^2 + m^2} \tag{7.6}$$

In Fourier transform, the Klein–Gordon equation can be written as $(p^2 - m^2)\widetilde{\phi}(p) = 0$ whose most general solution is $\widetilde{\phi}(p) = F(p)\delta(p^2 - m^2)$ with $F(p)$ an arbitrary function of p provided it is sufficiently regular at $p^2 = m^2$. We see that the solution has, as support, the two branches of the hyperboloid $p^2 = m^2$. As a result, the most general x-space solution of the Klein–Gordon equation can be written as an expansion in plane waves, i.e.

$$\phi(x) = \int \mathrm{d}\Omega_m \left[a(p)e^{-ip\cdot x} + a^*(p)e^{ip\cdot x}\right] \tag{7.7}$$

where we used the invariant measure on the positive energy branch of the mass hyperboloid given in equation (2.19), and a^* denotes the complex conjugate of a. The measure $\mathrm{d}\Omega_m$ contains the δ function $\delta(p^2 - m^2)$ which implies that the complex function $a(p)$ depends only on three of the components of the four-vector p. We make the choice to eliminate p^0 and we determine $a(\boldsymbol{p})$ by the initial data. Indeed, since the Klein–Gordon equation is a second order differential equation, the initial data at $t = 0$ involve the value of the function and its first time derivative $\phi(0, \boldsymbol{x})$ and $\dot{\phi}(0, \boldsymbol{x})$. Inverting the expression (7.7) we obtain

$$a(\boldsymbol{p}) = \int \mathrm{d}^3x\, e^{-ip\cdot x}\left[\omega_p\phi(0, \boldsymbol{x}) + i\dot{\phi}(0, \boldsymbol{x})\right] = i\int_{x_0=t} \mathrm{d}^3x \left[e^{i(\omega_p t - \boldsymbol{p}\cdot\boldsymbol{x})}\overleftrightarrow{\partial_0}\phi(x)\right] \tag{7.8}$$

where we have introduced the notation $f\overleftrightarrow{\partial_0}g = f\,\partial_0 g - (\partial_0 f)\,g$.

The Hamiltonian H expressed in terms of $a(\boldsymbol{p})$ and $a^*(\boldsymbol{p})$ is

$$H = \int \mathrm{d}\Omega_m \frac{1}{2}\,\omega_p\left[a(\boldsymbol{p})\,a^*(\boldsymbol{p}) + a^*(\boldsymbol{p})\,a(\boldsymbol{p})\right] \tag{7.9}$$

For later convenience we have kept the peculiar combination $aa^* + a^*a$, even though the two terms are equal in the classical theory we are discussing here.

7.2.1 The Green's functions

The solution of a linear homogeneous wave equation is always rather trivial. However, in practice we are often interested in the dynamics of the field $\phi(x)$ coupled to a given external source described by a function $j(x)$. The corresponding equation of motion is

$$\left(\Box + m^2\right)\phi(x) = j(x) \tag{7.10}$$

It is an equation of hyperbolic type. As a second order differential equation, its solutions are determined by the Cauchy data, the value of the function and its first derivatives on a surface, called the Cauchy surface. In practice, every time we use local coordinates, we will take as a Cauchy surface the hyperplane $\mathbb{R}^3 = \{(t, \boldsymbol{x}) \in M^4 | t = 0\}$ or its time translations. By relativistic invariance, the properties of the solutions are independent of this particular choice and we can choose any space-like hypersurface.

Since the equation (7.10) is linear, the solution for a general $j(x)$ will be given by the superposition principle starting from the solution of the equation corresponding to a point source

$$\left(\Box + m^2\right) G(x, y) = \delta^4(x - y) \tag{7.11}$$

In physics, these solutions are the so-called elementary solutions or *Green functions*. $G(x, y)$ is the field produced by a point source, which appears at the point $\boldsymbol{y}$ instantaneously at time y_0. We are particularly interested in those solutions that are translationally invariant. The general solution of (7.10) will then be of the form

$$\phi(x) = \phi_0(x) + \int \mathrm{d}^4 y \; G(x - y) j(y) \tag{7.12}$$

where $\phi_0(x)$ is a solution of the homogeneous equation $\left(\Box + m^2\right) \phi_0(x) = 0$, which is fixed by the Cauchy data.

By Fourier transform, the equation for the elementary solution becomes

$$\left(-p^2 + m^2\right) \widetilde{G}(p) = 1 \tag{7.13}$$

In other words, the general Green function is given by the inverse of the differential operator on the left-hand side. This is very general and we shall use it often in what follows. Let us consider a linear, homogeneous differential equation of the form

$$\mathcal{D} \, \Psi(x) = 0 \tag{7.14}$$

where the function $\Psi(x)$ may be an n-component vector and the differential operator $\mathcal{D}$ an $n \times n$ matrix. The associated general Green function is given by $[\mathcal{D}]^{-1}$, where the inverse is understood both in the sense of the differential operator and as that of the matrix.

In equation (7.13), we see that if $p^2 \neq m^2$, $\widetilde{G}(p) = (-p^2 + m^2)^{-1}$. It follows that $\widetilde{G}$ is completely fixed, up to a distribution proportional to $\delta(-p^2 + m^2)$. We conclude that we can obtain several Green functions, depending on the choice of this distribution. We shall describe them in this section.

In order to satisfy the simple-minded causality principle – that is, to ensure that the value of the solution at a space-time point depends only on its values in the past light-cone of that point – we will look at the so-called *retarded Green function* $G_\mathrm{R}(x - y)$ whose support is in the future cone of y, i.e. for $x^0 \geq y^0$. It is given by

$$\begin{aligned} G_\mathrm{R}(x - y) &= -\frac{1}{(2\pi)^4} \lim_{\varepsilon \to 0+} \int \mathrm{d}^4 p \, \frac{\mathrm{e}^{-\mathrm{i} p \cdot (x - y)}}{(p_0 + \mathrm{i}\varepsilon)^2 - \boldsymbol{p}^2 - m^2} \\ &= -\frac{1}{(2\pi)^4} \int \mathrm{d}^3 p \, \mathrm{e}^{\mathrm{i} \boldsymbol{p} \cdot (\boldsymbol{x} - \boldsymbol{y})} \left\{ \int_{-\infty}^{+\infty} \mathrm{d} p^0 \, \frac{\mathrm{e}^{-\mathrm{i} p^0 (x^0 - y^0)}}{(p^0 + \mathrm{i}\varepsilon)^2 - \omega_p^2} \right\} \end{aligned} \tag{7.15}$$

In what follows, the limit $\varepsilon \to 0+$ will be implicit.

Using the expression of the Heaviside θ-function

$$\int_{-\infty}^{+\infty} dp^0 \frac{e^{-ip^0 x^0}}{p^0 \pm i\varepsilon + a} = \mp 2i\pi e^{iax^0} \theta(\pm x^0) \tag{7.16}$$

which can be easily proved using the Cauchy theorem, the expression in the brackets in (7.15) can be written as $(-i\pi/\omega_p)\theta(x^0 - y^0)\left(e^{-i\omega_p(x^0 - y^0)} - e^{i\omega_p(x^0 - y^0)}\right)$ so that, finally,

$$G_{\mathrm{R}}(x - y) = \frac{i\theta(x^0 - y^0)}{(2\pi)^3} \int \frac{d^3\boldsymbol{p}}{2\omega_p} e^{i\boldsymbol{p}\cdot(\boldsymbol{x}-\boldsymbol{y})} \left\{ e^{-i\omega_p(x^0 - y^0)} - e^{i\omega_p(x^0 - y^0)} \right\} \tag{7.17}$$

Let $V^+ = \{x \in \mathbb{R}^4 | x^0 \geq 0, x^2 \geq 0\}$ and $V^- = -V^+$ be the future and past half-cones with respect to the origin of the coordinates. The function $G_{\mathrm{R}}(x - y)$ has its support in the variable x in $\{x \in \mathbb{R}^4 | x \in y + V_y^+\}$ where we define $V_y^{\pm}$ the future and past cones, respectively, with the origin taken at the point y. In fact, the Heaviside function limits its support to the x's in the future cone of y and it is Lorentz invariant. We see that G_{R} propagates both positive and negative energy solutions towards the future.

It is straightforward to repeat this analysis and obtain the advanced Green function given by the expression

$$G_{\mathrm{A}}(x - y) = -\frac{1}{(2\pi)^4} \lim_{\varepsilon \to 0+} \int d^4p \frac{e^{-ip\cdot(x-y)}}{(p_0 - i\varepsilon)^2 - \boldsymbol{p}^2 - m^2} \tag{7.18}$$

whose support for x is in the past light-cone of y. $G_{\mathrm{A}}(x - y)$ propagates both positive and negative energy solutions towards the past.

These two Green functions are given by real distributions and they are the ones often used in classical electromagnetic theory.

Another elementary solution will play a fundamental role in relativistic quantum field theory, the Feynman Green function G_{F}. It is a complex distribution which we shall define as

$$G_{\mathrm{F}}(x) = \frac{i}{(2\pi)^4} \int d^4p\, e^{-ip\cdot x} \frac{1}{p^2 - m^2 + i\varepsilon} \tag{7.19}$$

and we can show that the positive energy solutions propagate towards the future, while the negative energy solutions propagate towards the past. These last ones will later be interpreted as the propagation of antiparticles.[2]

It is useful to define the function $D(x)$ by

$$D(x) = \int d\Omega_m\, e^{-ip\cdot x} \tag{7.20}$$

In Problem 7.1 we ask the reader to express all Green functions in terms of D and the function θ. These expressions show clearly their propagation properties.

---

[2]In the literature there is no established convention regarding the phases of the Green functions. Here we choose to define G_{R} and G_{A} as real distributions satisfying equation (7.11). They are the ones used in classical electrodynamics. For G_{F} we adopted the definition (7.19) which will be more convenient later. It satisfies the equation $(\Box + m^2)\, G_{\mathrm{F}}(x, y) = -i\delta^4(x - y)$.

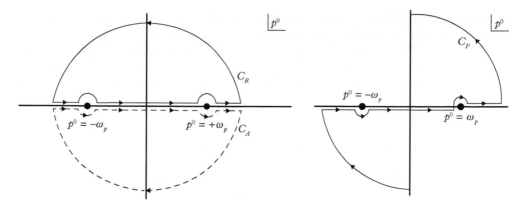

Fig. 7.1 Left: Two of the integration contours in the complex p^0 plane. For the retarded (advanced) Green function, we close the contour in the upper (lower) half plane and find that G_R (G_A) vanishes if $x^0 < 0$ ($x^0 > 0$). Right: The integration contour in the complex p^0 plane which shows the propagation properties of the Feynman Green function.

In the special case with $m = 0$, $\omega_p = |\boldsymbol{p}| = p$, and for $y^\mu = 0$ we obtain[3]

$$
\begin{aligned}
G_{\mathrm{R}}(x) &= \frac{\theta(x^0)}{(2\pi)^3} \int \mathrm{d}^3\boldsymbol{p} \, \frac{\sin p x^0}{p} \mathrm{e}^{\mathrm{i}\boldsymbol{p}\cdot\boldsymbol{x}} = \frac{2\theta(x^0)}{(2\pi)^2} \int_0^{+\infty} \mathrm{d}p \, \frac{\sin p x^0 \sin p|\boldsymbol{x}|}{|\boldsymbol{x}|} \\
&= \frac{\theta(x^0)}{(2\pi)^2} \int_{-\infty}^{+\infty} \frac{\mathrm{d}p}{2|\boldsymbol{x}|} \left(\mathrm{e}^{\mathrm{i}p(x^0-|\boldsymbol{x}|)} - \mathrm{e}^{\mathrm{i}p(x^0+|\boldsymbol{x}|)} \right) \\
&= \frac{\theta(x^0)}{2\pi} \delta(x^2)
\end{aligned}
\tag{7.22}
$$

It follows that in the massless case the function G_{R} has its support on the boundary of the future light-cone.

Before closing this section, we want to note that all three Green functions we introduced, G_{R}, G_{A} and G_{F}, have a simple geometrical interpretation. We saw that they are boundary values of the same analytic function $f = 1/(p^2 - m^2)$. Considered as a function of p^0, f has two poles in the complex plane corresponding to the values $p^0 = \pm \omega_p$. In taking the Fourier transform we must integrate over p^0 from $-\infty$ to $+\infty$. Since the singularities are on the real axis we must deform the contour, going above or below the singularities, the way it is shown in Figure 7.1. It is clear that there exist four possible deformations, denoted in the Figure as C_{R}, C_{A} and C_{F} together with its complex conjugate $C_{\mathrm{F}}^\star$. It is easy to verify that they describe propagation properties of G_{R}, G_{A}, G_{F} and $G_{\mathrm{F}}^\star$, respectively.

The particular choice of the contour C_{F} suggests the following interpretation for G_{F}: if we ignore for the moment the pieces of the circle at infinity, C_{F} contains an

---

[3]We recall that for a function $f(x)$ with a finite number of distinct zeros x_a

$$
\delta(f(x)) = \sum_a \frac{1}{|f'(x_a)|} \delta(x - x_a)
\tag{7.21}
$$

integration over p^0 from $-\infty$ to $+\infty$ as well as one from $+i\infty$ to $-i\infty$. The two are formally related by the substitution $p^0 \to -ip^0$. But this is precisely the substitution that would bring the Minkowski space to the corresponding Euclidean space, since $p^2 = (p^0)^2 - \boldsymbol{p}^2 \to -((p^0)^2 + \boldsymbol{p}^2)$. So, let us imagine that we start from the Klein–Gordon equation in a four-dimensional Euclidean space with coordinates $x_E = (x_1, x_2, x_3, x_4)$, i.e.

$$\left(\sum_{i=1}^{4} \partial_i \partial_i - m^2\right) \phi(x_E) \equiv (\Box_E - m^2)\phi(x_E) = 0 \tag{7.23}$$

It is an equation of elliptic type, and the corresponding equation for the Green function in Euclidean momentum space can be written as

$$(\sum_{i=1}^{4} p_E^i p_E^i + m^2)\, \widetilde{G}_E(p_E) \equiv (p_E^2 + m^2)\, \widetilde{G}_E(p_E) = -1 \tag{7.24}$$

It has a unique solution given by $\widetilde{G}_E(p_E) = -(p_E^2 + m^2)^{-1}$ or, by Fourier transform,

$$G_E(x_E) = \frac{-1}{(2\pi)^4} \int \mathrm{d}^4 p_E \, \frac{e^{ip_E \cdot x_E}}{p_E^2 + m^2} \tag{7.25}$$

This integral can be performed without ambiguity because in Euclidean space the denominator never vanishes. The resulting Green function admits an analytic continuation in the variable $(x_E)_4$ such that, for $(x_E)_4 \to ix_0$, $G_E(x_E) \to G_F(x)$ as defined in equation (7.19).

7.2.2   Generalisations

An obvious generalisation is to consider the equations for N real scalar fields $\phi_a(x)$, $a = 1, ..., N$. If we have N different mass parameters m_a, the theory is just N copies of the one we have studied already. More interesting is the special case with all mass parameters equal. We thus obtain the $O(N)$ invariant Lagrangian density

$$\mathcal{L} = \frac{1}{2} \sum_{i=1}^{N} \left(\partial_\mu \phi_a(x) \partial^\mu \phi_a(x) - m^2 \phi_a(x)\phi_a(x)\right) \tag{7.26}$$

The theory is invariant under the transformations $\Phi \to \mathcal{O}\Phi$, with $\Phi = (\phi_1, \phi_2, ..., \phi_N)^T$ and $\mathcal{O}$ an orthogonal $N \times N$ matrix. The corresponding conserved current is a special case of the one we computed in Problem 3.3.

Of particular interest will turn out to be the case $N = 2$. Taking into account the fact that $O(2)$ is locally isomorphic to $U(1)$, we can rewrite the two-component real scalar field as a complex field $\phi = (\phi_1 + i\phi_2)/\sqrt{2}$. The Lagrangian becomes

$$\mathcal{L} = \partial_\mu \phi(x) \partial^\mu \phi^*(x) - m^2 \phi(x)\phi^*(x) \tag{7.27}$$

and it is invariant under phase transformations of the field: $\phi \to e^{i\theta}\phi$. We can compute the corresponding conserved current to obtain

$$j_\mu(x) = \phi(x)\partial_\mu\phi^*(x) - \phi^*(x)\partial_\mu\phi(x) \tag{7.28}$$

We can still expand the complex field ϕ in terms of plane waves, but now the coefficients of the positive and negative frequency terms will not be complex conjugates of each other. To keep the notation as close as possible to the real field case, we will write the field $\phi(x)$ as

$$\phi(x) = \int d\Omega_m \left[a(\boldsymbol{p})e^{-ip\cdot x} + b^*(\boldsymbol{p})e^{ip\cdot x} \right] \tag{7.29}$$

and similarly for $\phi^*(x)$ with coefficient functions $a^*(\boldsymbol{p})$ and $b(\boldsymbol{p})$.

Although we obtained the Klein–Gordon equation from purely group theory considerations, we can think of possible physical applications. In classical physics this equation, or its non-linear generalisations, can be applied to problems of relativistic fluid dynamics. Another application, which we shall encounter in this book, involves the addition of a cubic term in the field and it is known as *the Landau–Ginzburg* equation. It describes the order parameter in various problems of phase transitions. In fact, as we will see later on in this book, a generalised Klein–Gordon equation will be used precisely in order to describe a particular phase transition, the *Brout–Englert–Higgs mechanism*, which is an important ingredient of the theory we call "The Standard Model" of particle physics.

7.3 The Dirac Equation

7.3.1 Introduction

Historically, the Dirac equation was formulated as a response to a physics problem, namely the need to write a relativistic equation for the wave function of the electron. In doing so, Dirac was the first to discover, in 1928, by trial and error, the spinorial representations of the Lorentz group. Today we do not have to follow this historic route. We can use the results of our analysis on the Lorentz group and write directly the covariant equations for spinor fields.

In Chapter 5 we saw that there exist two inequivalent, irreducible, two-dimensional representations of the restricted Lorentz group, the $(\frac{1}{2}, 0)$ corresponding to spinors ξ^α with undotted indices and the $(0, \frac{1}{2})$ corresponding to spinors $\eta^{\dot\alpha}$ with dotted indices. These representations do not mix under the transformations of the restricted Lorentz group but transform into each other by space inversion which, in the particle physics jargon, is often called *parity operation* $\mathcal{P}$. We also saw that the derivative operator p_μ, which transforms as a four-vector, according to the $(\frac{1}{2}, \frac{1}{2})$ representation of the Lorentz group, is represented in the space of spinors by a 2×2 matrix, having one undotted and one dotted index, by[4]

$$\tilde{P}^{\alpha\dot\alpha} = p_\mu\bar\sigma^\mu = p^0\,\mathbf{1} + \boldsymbol{\sigma}\cdot\boldsymbol{p} = \begin{pmatrix} p^0 + p^3 & p^1 - ip^2 \\ p^1 + ip^2 & p^0 - p^3 \end{pmatrix} \tag{7.30}$$

where $\sigma^\mu = (\mathbf{1}, \sigma^i)$ and $\bar\sigma^\mu = (\mathbf{1}, -\sigma^i)$ with $\mathbf{1}$ the unit matrix and σ^i the three Pauli matrices. It follows that, in order to write Lorentz covariant wave equations, we have

---

[4]We shall often omit the unit matrix $\mathbf{1}$, when there is no danger of confusion.

at our disposal the derivative operator $\widetilde{P}^{\alpha\dot\alpha}$ and the spinor fields $\xi^\alpha(x)$ or $\eta^{\dot\alpha}(x)$, each being a two-component complex spinor.

7.3.2 Weyl and Majorana equations

The simplest linear, homogeneous, Lorentz covariant, differential equations we can write are[5]

$$\widetilde{P}_{\alpha\dot\alpha}\xi^\alpha(x) = 0\,,\quad \widetilde{P}_{\alpha\dot\alpha}\eta^{\dot\alpha}(x) = 0 \tag{7.31}$$

They have the following properties: (i) They are not invariant under parity since the latter interchanges ξ^α with $\eta^{\dot\alpha}$. (ii) By multiplying with $\widetilde{P}$ and using equation (5.173), we see that the spinors ξ and η satisfy the Klein–Gordon equation with $m^2 = 0$, i.e.

$$\Box\,\xi^\alpha(x) = 0\,,\quad \Box\,\eta^{\dot\alpha}(x) = 0$$

(iii) Both equations (7.31) are invariant under the $U(1)$ group of phase transformations, for example $\xi \to \mathrm{e}^{i\theta}\xi$. As a result, each one implies the existence of a conserved current, which yields a conserved charge.

As we shall see later, such equations with no mass parameter are good candidates to describe particles with spin $\frac{1}{2}$ and zero mass. These equations were first written by H. Weyl in 1929 and they were dismissed because, as we explained above, they were not invariant under parity, which, at that time, was believed to be an exact symmetry of nature.

A slightly more complicated equation was discovered by E. Majorana in 1936. In comparison with the Weyl equations it conserves parity and yields a massive Klein–Gordon equation, but it does not have a conserved charge. Majorana starts from the observation (see Chapter 5) that for the representations of the Lorentz group, the operation of complex conjugation is not always an equivalence. In particular, a member of the $(\frac{1}{2}, 0)$ representation is transformed by complex conjugation into a member of the $(0, \frac{1}{2})$. We pointed out this property in Chapter 5, where we noted that $(\xi_\alpha)^*$ transforms as $\eta^{\dot\alpha}$. Therefore, the equation

$$\widetilde{P}_{\alpha\dot\alpha}\xi^\alpha(x) = m(\xi^*)_{\dot\alpha} \tag{7.32}$$

with m a real parameter with dimensions of mass is Lorentz covariant. Combining equation (7.32) with its complex conjugate we conclude that the spinor ξ satisfies the Klein–Gordon equation $(\Box + m^2)\,\xi^\alpha(x) = 0$. On the other hand, since equation (7.32) contains both ξ and ξ^*, it does not remain invariant under a phase transformation of ξ. This becomes more transparent if we combine the two-component complex spinors ξ and ξ^* into a four-component bi-spinor

$$\psi(x) = \begin{pmatrix} \xi(x) \\ \xi^\star(x) \end{pmatrix} \tag{7.33}$$

---

[5]Spinor indices are raised, lowered and contracted by the antisymmetric ϵ-matrix. For example, $\xi_\alpha = \epsilon_{\alpha\beta}\,\xi^\beta$, $\widetilde{P}_{\alpha\dot\alpha}\eta^{\dot\alpha} = \widetilde{P}_{\alpha\dot\alpha}\epsilon^{\dot\alpha\dot\beta}\eta_{\dot\beta}$, see equations (5.162) and (5.163).

It is now clear that by a change of basis in the spinor space of the form $\psi' = U\psi$ with U a unitary matrix, we can make all four components of ψ' real. In later chapters we shall have the opportunity to use the Weyl and Majorana equations in particle physics.

7.3.3 The Dirac equation

If, following Dirac, we are looking for a relativistic wave equation to describe an electron in an electromagnetic field, we need an equation that contains a parameter with dimensions of mass, is invariant under parity and admits a conserved charge. At the linearised level, group theory leads to a unique answer.

We saw in Chapter 5 that the four-dimensional direct sum representation $(\frac{1}{2}, 0) \oplus (0, \frac{1}{2})$ of the Lorentz group is reducible under the transformations of the restricted Lorentz group, but it becomes irreducible if we include parity. This leads us to consider two spinor fields $\xi^\alpha(x)$ and $\eta_{\dot\alpha}(x)$ and write the following system of two coupled wave equations, which generalise the Weyl equations (7.31)

$$\widetilde{P}^{\alpha\dot\beta}\eta_{\dot\beta} = m\,\xi^\alpha\,, \quad \widetilde{P}_{\alpha\dot\beta}\xi^\alpha = m\,\eta_{\dot\beta} \tag{7.34}$$

where $\widetilde{P}^{\alpha\dot\beta} = (p_\mu\bar\sigma^\mu)^{\alpha\dot\beta} = (p^0 + \boldsymbol{p}\cdot\boldsymbol{\sigma})^{\alpha\dot\beta}$ and $\widetilde{P}_{\alpha\dot\beta} = \epsilon_{\alpha\gamma}\epsilon_{\dot\beta\dot\delta}\widetilde{P}^{\gamma\dot\delta}$. The spinors ξ and η with components ξ^α and $\eta_{\dot\alpha}$, respectively, satisfy separately the Klein–Gordon equation

$$\left(\Box + m^2\right)\xi^\alpha(x) = 0\,, \quad \left(\Box + m^2\right)\eta_{\dot\beta}(x) = 0 \tag{7.35}$$

Equations (7.34) can be written in the form

$$(p_0 + \boldsymbol{p}\cdot\boldsymbol{\sigma})\,\eta = m\,\xi\,, \quad (p_0 - \boldsymbol{p}\cdot\boldsymbol{\sigma})\,\xi = m\,\eta \tag{7.36}$$

or, be combined in

$$\begin{pmatrix} 0 & p_0 + \boldsymbol{p}\cdot\boldsymbol{\sigma} \\ p_0 - \boldsymbol{p}\cdot\boldsymbol{\sigma} & 0 \end{pmatrix} \begin{pmatrix} \xi \\ \eta \end{pmatrix} = m \begin{pmatrix} \xi \\ \eta \end{pmatrix} \tag{7.37}$$

It will be convenient to define the bi-spinor

$$\psi(x) = \begin{pmatrix} \xi(x) \\ \eta(x) \end{pmatrix} \tag{7.38}$$

We see how simple group theory implies that the spinor field with the desired properties to describe an electron should have four components, a result first obtained by Dirac by trial and error.

We introduce the 4×4 matrices

$$\gamma^\mu = \begin{pmatrix} 0 & \bar\sigma^\mu \\ \sigma^\mu & 0 \end{pmatrix} \tag{7.39}$$

and we write the matrix equation (7.37) in its usual form, the celebrated *Dirac equation*,

$$(\not{p} - m)\,\psi = 0 \quad \text{or} \quad (\mathrm{i}\gamma^\mu\,\partial_\mu - m)\psi(x) = 0 \tag{7.40}$$

We introduced the symbol $\not{a} \equiv \gamma^\mu a_\mu$ for any four-vector a and the correspondence $p_\mu \leftrightarrow \mathrm{i}\partial_\mu$ to write $\not{p} = \gamma^\mu p_\mu = \gamma\cdot p \leftrightarrow \mathrm{i}\gamma^0\partial_0 + \mathrm{i}\boldsymbol{\gamma}\cdot\boldsymbol{\nabla} = \mathrm{i}\gamma^\mu\,\partial_\mu$

The matrices $\gamma^\mu = (\gamma^0, \gamma^i)$ in (7.39) are called *gamma* (or *Dirac*) *matrices* and satisfy the anti-commutation relations

$$\{\gamma^\mu, \gamma^\nu\} = \gamma^\mu \gamma^\nu + \gamma^\nu \gamma^\mu = 2\,\eta^{\mu\nu}\,\mathbf{1} \tag{7.41}$$

7.3.4 The γ matrices

We obtained the Dirac equation (7.40) with the γ matrices (7.39) from our knowledge on spinors in a representation which naturally exhibits its relativistic invariance. It is called a *spinorial* or *Weyl representation*. However, in applications, it may be more useful to change the representation, that is to apply unitary linear transformations on the four components of the bi-spinor ψ, $\psi \to \psi' = U\psi$, where U is a 4×4 unitary matrix.[6] It is easy to show that if ψ satisfies the Dirac equation $(\gamma \cdot p - m)\psi = 0$, ψ' satisfies the Dirac equation $(\gamma' \cdot p - m)\psi' = 0$, with $\gamma'^\mu = U\gamma^\mu U^\dagger$, which also obey the algebra $\{\gamma'^\mu, \gamma'^\nu\} = 2\eta^{\mu\nu}\mathbf{1}$. Conversely, we can also prove the opposite, namely, given a set of four 4×4 matrices satisfying (7.41), there exists a unitary 4×4 matrix U, such that the transformed matrices $\gamma'^\mu = U\gamma^\mu U^\dagger$ are equal to our first choice (7.39). It follows that any such set can be used to write the Lorentz covariant Dirac equation in a particular basis. In other words, (7.41) is equivalent to the property of Lorentz covariance.

The algebra defined by the anti-commutation relation (7.41) is called *Clifford algebra*. It is the necessary and sufficient condition for a set of γ matrices to give, through (7.40), a Lorentz covariant Dirac equation. It is also the necessary relation for the Dirac equation $(\gamma \cdot p - m)\Psi = 0$ to give the Klein–Gordon equation by left multiplication by the operator $\gamma \cdot p + m$.

The γ matrices satisfy the hermiticity relations

$$(\gamma^0)^\dagger = \gamma^0, \; \gamma^{i\dagger} = -\gamma^i \;\text{ and }\; \gamma^{\mu\dagger} = \gamma^0\,\gamma^\mu\,\gamma^0 \tag{7.42}$$

It is easy to verify that these relations are independent of the chosen representation. It will be useful to introduce a fifth 4×4 matrix, called γ^5. It is defined by

$$\gamma^5 = \gamma_5 = \mathrm{i}\gamma^0\gamma^1\gamma^2\gamma^3 = \frac{-\mathrm{i}}{4!}\,\varepsilon_{\mu\nu\varrho\sigma}\,\gamma^\mu\gamma^\nu\gamma^\varrho\gamma^\sigma \tag{7.43}$$

and satisfies

$$(\gamma_5)^2 = 1 \quad \text{and} \quad \{\gamma_5, \gamma^\mu\} = 0 \tag{7.44}$$

We also define the six 4×4 matrices

$$\sigma^{\mu\nu} = \frac{-\mathrm{i}}{4}\,[\gamma^\mu, \gamma^\nu] \tag{7.45}$$

They represent the generators of the Lorentz algebra in the $(\frac{1}{2},0) \oplus (0,\frac{1}{2})$ representation. The transformation of a Dirac spinor under the Lorentz transformation $x'^\mu = \Lambda^\mu{}_\nu x^\nu$

---

[6]The Dirac equation is linear and homogeneous like the Schrödinger equation. Therefore, we must supplement it with a normalisation condition. The unitarity property of the matrix U guarantees that the norm of ψ remains the same.

with parameters $\omega_{\mu\nu} = -\omega_{\nu\mu}$ is given by $\psi(x) \to \psi'(x) = S(\omega)\psi(\Lambda^{-1}x)$, with the 4×4 matrix $S(\omega)$ being

$$S = \exp\left(\frac{\mathrm{i}}{2}\omega_{\mu\nu}\sigma^{\mu\nu}\right)$$

Finally, note that

$$\frac{1}{4}\gamma^\mu\,\gamma_\mu = \mathbf{1}\,, \quad \mathrm{Tr}\,\gamma^\mu = 0\,, \quad \mathrm{Tr}\,\mathbf{1} = 4\,, \quad \mathrm{Tr}\,\gamma_5 = 0 \tag{7.46}$$

The following identities are often used with $\slashed{a} = a_\mu\gamma^\mu$:

$$\slashed{a}^2 = a^2\,\mathbf{1}\,, \qquad \slashed{a}\,\slashed{b} + \slashed{b}\,\slashed{a} = 2\,a\cdot b\,\mathbf{1} \tag{7.47}$$

A useful exercise is the computation of the trace of a product of γ matrices. Using the Clifford algebra and the cyclic property of the trace, we can easily obtain the following results: If $T^{\mu_1\cdots\mu_n} \equiv \mathrm{Tr}(\gamma^{\mu_1}\gamma^{\mu_2}\ldots\gamma^{\mu_n})$ and $T_5^{\mu_1\cdots\mu_n} \equiv \mathrm{Tr}(\gamma_5\gamma^{\mu_1}\gamma^{\mu_2}\ldots\gamma^{\mu_n})$, we have

$$T^{\mu_1\mu_2} = 4\,\eta^{\mu_1\mu_2}\,, \quad T^{\mu_1\cdots\mu_{2n+1}} = T_5^{\mu_1\cdots\mu_{2n+1}} = 0$$

$$T_5^{\mu_1\cdots\mu_n} = 0 \ \text{ for } \ n < 4\,, \quad T_5^{\mu_1\cdots\mu_4} = -4\mathrm{i}\epsilon^{\mu_1\cdots\mu_4}$$

$$T^{\mu_1\cdots\mu_{2n}} = \sum_{k=1}^{2n-1}(-)^k\eta^{\mu_1\mu_k}T^{\mu_2\cdots\mu_{k-1}\mu_{k+1}\cdots\mu_{2n}} \tag{7.48}$$

The last relation expresses the $2n$-trace in terms of the $2n - 2$ ones. With the help of these relations we can compute, by induction, the trace of the product of any number of γ matrices.

7.3.5 The conjugate equation

Consider the complex conjugate of (7.40). The definition $p_\mu = \mathrm{i}\,\partial_\mu$ implies $p_\mu^* = -p_\mu$, so that from $(\slashed{p} - m)\psi = 0$ we obtain

$$0 = (\slashed{p}^* - m)\psi^* = \left((\gamma^0)^T p_0^* - (\gamma^i)^T p_i^* - m\right)\psi^* \tag{7.49}$$

or $(\psi^*)^T\left(-\gamma^0 p_0 + \gamma^i p_i - m\right) = 0$, the differential operators acting on the left. Here ψ^* is the four-component spinor whose elements are the complex conjugates of the elements of ψ. Multiplying this equation on the right by γ^0 and using $\gamma^0\gamma^0 = \mathbf{1}$ and $\gamma^0\gamma^i\gamma^0 = -\gamma^i$ we obtain $(\psi^*)^T\gamma^0(-\gamma^0 p_0 - \gamma^i p_i - m) = 0$, i.e.

$$\overline{\psi}\,(\slashed{p} + m) = 0 \tag{7.50}$$

where we have introduced the *Dirac conjugate* spinor[7]

$$\overline{\psi} = (\psi^*)^T\gamma^0 = \psi^\dagger\gamma^0 \tag{7.51}$$

---

[7]We use the "star" ($*$) to denote complex conjugation. A "bar" over a Dirac spinor symbolises Dirac conjugation.

7.3.6   The standard representation

The spinorial representation is the most natural one from the mathematical point of view and it is the one we found when we first constructed the equation using essentially only group theoretic considerations. We also saw that we can transform this representation by applying a unitary transformation $\psi \to U\psi$. Depending on the physical application we have in mind, some particular forms of these transformed spinors will turn out to be useful. We shall present here the ones which are most commonly used.

In order to study the non-relativistic approximation we use the so-called *standard,* or *Dirac representation.* It can be deduced from the spinorial representation by the following unitary transformation on (ξ, η)

$$\phi = \frac{1}{\sqrt{2}}(\xi + \eta), \quad \chi = \frac{1}{\sqrt{2}}(\xi - \eta). \tag{7.52}$$

The corresponding γ matrices are

$$\gamma_D^i = \begin{pmatrix} 0 & \sigma^i \\ -\sigma^i & 0 \end{pmatrix}, \quad \gamma_D^0 = \begin{pmatrix} 1 & 0 \\ 0 & -1 \end{pmatrix} \tag{7.53}$$

In this representation, parity transforms ϕ and χ into themselves, according to the formula

$$\phi \to i\phi, \quad \chi \to -i\chi \tag{7.54}$$

The γ^5 matrix becomes

$$\gamma_D^5 = \begin{pmatrix} 0 & 1 \\ 1 & 0 \end{pmatrix} \tag{7.55}$$

an expression to be compared with the corresponding one in the spinorial representation in which γ^5 is diagonal, namely

$$\gamma_{sp}^5 = \begin{pmatrix} 1 & 0 \\ 0 & -1 \end{pmatrix} \tag{7.56}$$

Correspondingly, in the Dirac representation ψ is given by

$$\psi(x) = \begin{pmatrix} \phi(x) \\ \chi(x) \end{pmatrix} \tag{7.57}$$

and the Dirac equation (7.40) implies for ϕ and χ the equations

$$\begin{aligned} p_0\,\phi - \boldsymbol{p} \cdot \boldsymbol{\sigma}\,\chi &= m\,\phi \\ p_0\,\chi - \boldsymbol{p} \cdot \boldsymbol{\sigma}\,\phi &= -m\,\chi, \end{aligned} \tag{7.58}$$

or

$$\begin{aligned} i\,\partial_0\,\phi &= -i\boldsymbol{\sigma} \cdot \boldsymbol{\nabla}\chi + m\,\phi \\ i\,\partial_0\,\chi &= -i\boldsymbol{\sigma} \cdot \boldsymbol{\nabla}\phi - m\,\chi \end{aligned} \tag{7.59}$$

Solutions with positive energy ($p_0 > 0$) and zero momentum ($\boldsymbol{p} = 0$) have $\chi = 0$, and the four-component Dirac spinor $\psi = (\phi, \chi)^T$ reduces to the two-component spinor

ϕ. In the non-relativistic approximation, we will see that the norm of χ remains small compared to that of ϕ. For this reason, we call χ the *small* components of the four-dimensional spinor and ϕ the *large* ones.

7.3.7   Lagrangian, Hamiltonian and Green functions

The Lagrangian formulation of the Dirac equation presents some subtleties, because the latter is a first order differential equation. Recall the results we obtained for the case of a complex scalar field. The Lagrangian density (7.27) depends on the fields ϕ *and* ϕ^* as well as their first derivatives. This is consistent with the fact that the equations of motion are second order differential equations and we must assign initial values for the fields *and* their first derivatives. When we vary with respect to the field ϕ in order to obtain the Euler–Lagrange equations, we must also take into account the variation of $\partial\phi$. For the Dirac equation, however, which is a first order equation, we can vary only with respect to ψ and $\overline{\psi}$, not to their derivatives. This implies in turn that the standard way to obtain the Hamiltonian through a Legendre transformation should be reformulated. In this section we want to present the rules which will make it possible for us to use the Lagrangian and Hamiltonian formalisms, without attempting a mathematically rigorous justification.

We choose the Lagrangian density corresponding to the Dirac equation in the form

$$\mathcal{L}_{\mathrm{D}} = \frac{\mathrm{i}}{2}(\overline{\psi}\gamma^\mu \partial_\mu \psi - \partial_\mu \overline{\psi}\gamma^\mu \psi) - m\overline{\psi}\psi \qquad (7.60)$$

This is justified by the fact that we obtain the Dirac equations for ψ and $\overline{\psi}$ as a consequence of the stationarity requirement of the action $S = \int \mathcal{L}_{\mathrm{D}}\, d^4x$ under independent variations of $\overline{\psi}$ and ψ, respectively. Because of the linear dependence of $\mathcal{L}_{\mathrm{D}}$ on ψ or $\overline{\psi}$, the action has neither a minimum nor a maximum. Thus, the overall sign of the action can be chosen at will. Provided that the field vanishes at infinity, we can rewrite the action as

$$S = \int \mathrm{d}^4x \left(\mathrm{i}\,\overline{\psi}\,\slashed{\partial}\,\psi - m\overline{\psi}\psi \right) \qquad (7.61)$$

The action is invariant under the $U(1)$ group of phase transformations $\psi \to \mathrm{e}^{\mathrm{i}\theta}\psi$. The corresponding conserved current is

$$j_\mu(x) = \overline{\psi}(x)\gamma_\mu \psi(x), \quad \partial^\mu j_\mu(x) = 0 \qquad (7.62)$$

We follow a similar approach in order to obtain the Hamiltonian. In the canonical formalism we first compute the conjugate variable to ψ

$$\pi_D(x) = \frac{\partial \mathcal{L}}{\partial \dot{\psi}} = \mathrm{i}\psi^\dagger \qquad (7.63)$$

Similarly, the conjugate momentum to $\overline{\psi}$ is proportional to $\psi(x)$. But this sounds strange. The independent variables turn out to be the same with their conjugate momenta. This is again a direct consequence of the fact that we are dealing with a first order differential equation. We could ignore this problem and proceed with the formalism. Alternatively, we can bypass it by the following trick: We first derive Hamilton's equations by multiplying the Dirac equation $(i\gamma^0 \partial_0 + i\gamma^i \partial_i - m)\psi = 0$ on the left by γ^0, to obtain[8]

$$i\hbar \frac{\partial}{\partial t}\psi = \hat{H}\psi \qquad (7.64)$$

with

$$\hat{H} = c\,\boldsymbol{\alpha}\cdot(-i\hbar\boldsymbol{\nabla}) + \beta\,m\,c^2 = c\,\boldsymbol{\alpha}\cdot\boldsymbol{p} + \beta\,m\,c^2. \qquad (7.65)$$

where

$$\alpha^i = \gamma^0\gamma^i = \begin{pmatrix} \sigma^i & 0 \\ 0 & -\sigma^i \end{pmatrix} \quad \text{and} \quad \beta = \gamma^0 = \begin{pmatrix} 0 & 1 \\ 1 & 0 \end{pmatrix} \qquad (7.66)$$

We can verify, see Problem 7.7, that, applying blindly the canonical formalism, we obtain the same expression for the Hamiltonian, namely

$$H = \int d^3x\psi^\dagger(x)\left(c\,\boldsymbol{\alpha}\cdot\boldsymbol{p} + \beta\,m\,c^2\right)\psi(x) = \int d^3x\psi^\dagger(x)\hat{H}\psi(x) \qquad (7.67)$$

We will not attempt to give a correct justification of these steps here and we will briefly come back to this problem in section 10.3.

The Green functions of the Dirac equation are obtained following the same steps we used for the Klein–Gordon theory. We must find the inverse of the differential operator in the quadratic part of the Lagrangian (7.60). Therefore, the general Green function is proportional to $(i\,\partial\!\!\!/ - m)^{-1}$, where the inverse is understood with respect to the differential operator, as well as the four-by-four matrix. As usual, the expression simplifies in momentum space. Taking into account the fact that $(k\!\!\!/ + m)(k\!\!\!/ - m) = k^2 - m^2$, we can write the general expression for the Green function as

$$\widetilde{S} \propto \frac{k\!\!\!/ + m}{k^2 - m^2} \qquad (7.68)$$

Depending on the $i\epsilon$ prescription that we will choose for the singularities in the complex k^0 plane, we will obtain the retarded, the advanced or the Feynman Green function. For the phase convention, we shall follow the prescription we adopted for the Green functions of the Klein–Gordon equation. In particular, the Feynman Green function is given by

$$\widetilde{S} = i\frac{k\!\!\!/ + m}{k^2 - m^2} \qquad (7.69)$$

7.3.8   The plane wave solutions

We have shown that the solutions of the Dirac equation are solutions of the Klein–Gordon equation $(\Box + m^2)\psi = 0$ as well. Consequently, a plane wave solution $\psi(x) \sim \exp(-ik\cdot x)$ of the Dirac equation has to satisfy the condition $k^2 = k_0^2 - \boldsymbol{k}^2 = m^2$, which means that its energy k_0 can have either sign. We shall be interested in the full set of plane wave solutions of both positive and negative energies, since only their union forms

---

[8]The constants c and $\hbar$ were restored by dimensional analysis.

a basis. We fix the zero component of the wave vector k^μ to $k_0 = +\sqrt{\mathbf{k}^2 + m^2} \equiv E_k$. Then, we denote the positive energy solution of wave vector $\mathbf{k}$ by

$$\psi^{(+)}(x) = \mathrm{e}^{-\mathrm{i}k \cdot x}\, u(\mathbf{k}) \tag{7.70}$$

and the negative energy one by

$$\psi^{(-)}(x) = \mathrm{e}^{\mathrm{i}k \cdot x}\, v(\mathbf{k}) \tag{7.71}$$

where u and v are four-component spinors, whose components are labelled u_r and v_r, $r = 1, \dots, 4$.

From $(\mathrm{i}\gamma^\mu \partial_\mu - m)\psi^{(\pm)}(x) = 0$, we obtain

$$(\not{k} - m)u(\mathbf{k}) = 0 \quad \text{and} \quad (\not{k} + m)v(\mathbf{k}) = 0 \tag{7.72}$$

Let us choose the γ matrices in the standard representation. For $\mathbf{k} = 0$, the equations (7.72) simplify to

$$(\gamma^0 - 1)u(\mathbf{0}) = 0 \quad \text{and} \quad (\gamma^0 + 1)v(\mathbf{0}) = 0 \tag{7.73}$$

and lead to $u_3 = u_4 = v_1 = v_2 = 0$. A possible basis of the solutions is

$$\hat{u}^{(1)}(m, \mathbf{0}) = \begin{pmatrix} 1 \\ 0 \\ 0 \\ 0 \end{pmatrix}, \quad \hat{u}^{(2)}(m, \mathbf{0}) = \begin{pmatrix} 0 \\ 1 \\ 0 \\ 0 \end{pmatrix}, \quad \hat{v}^{(1)}(m, \mathbf{0}) = \begin{pmatrix} 0 \\ 0 \\ 1 \\ 0 \end{pmatrix}, \quad \hat{v}^{(2)}(m, \mathbf{0}) = \begin{pmatrix} 0 \\ 0 \\ 0 \\ 1 \end{pmatrix} \tag{7.74}$$

To obtain the solutions for arbitrary wave vector $\mathbf{k}$, we just have to boost the above spinor solutions, by applying on them the Lorentz transformation, which takes the four-vector $(m, \mathbf{0})$ to $(\pm\sqrt{\mathbf{k}^2 + m^2}, \mathbf{k})$. We shall short circuit the process by a simple trick.

Let us consider, for example, the spinor u. We observe that in order for $u(\mathbf{k})$ to satisfy $(\not{k} - m)u(\mathbf{k}) = 0$, since $k^2 - m^2 = (\not{k} - m)(\not{k} + m) = 0$, it is enough to look for $u(\mathbf{k})$ of the form $u(\mathbf{k}) = (\not{k} + m)f(k)\hat{u}(m, \mathbf{0})$ with $f(k) \to$ constant when $k \to (m, \mathbf{0})$. A convenient choice resulting from the normalisation conventions we shall adopt[9] is $1/f(k) = \sqrt{m + E_k}$. A similar argument is applied to $v(\mathbf{k})$, with the end result

$$u^{(\alpha)}(\mathbf{k}) = \frac{\not{k} + m}{\sqrt{m + E_k}} \hat{u}^{(\alpha)}(m, \mathbf{0}) \quad \text{and} \quad v^{(\alpha)}(\mathbf{k}) = \frac{-\not{k} + m}{\sqrt{m + E_k}} \hat{v}^{(\alpha)}(m, \mathbf{0}) \tag{7.75}$$

or, explicitly

---

[9]The conventions regarding the normalisation of the spinor solutions are not standard and they are dictated by the physical applications one has in mind. In some treatments the choice $f(k) \to 1/2m$ is used, which makes the non-relativistic limit simpler. In contemporary high energy physics experiments, however, the zero-mass limit is often more appropriate and our normalisation makes it simpler. Naturally, the final results of physically measurable quantities are independent of normalisation conventions.

$$u^{(1)}(\boldsymbol{k}) = \sqrt{E_k + m}\begin{pmatrix} 1 \\ 0 \\ \dfrac{\boldsymbol{\sigma} \cdot \boldsymbol{k}}{E_k + m}\begin{pmatrix} 1 \\ 0 \end{pmatrix} \end{pmatrix}$$

$$u^{(2)}(\boldsymbol{k}) = \sqrt{E_k + m}\begin{pmatrix} 0 \\ 1 \\ \dfrac{\boldsymbol{\sigma} \cdot \boldsymbol{k}}{E_k + m}\begin{pmatrix} 0 \\ 1 \end{pmatrix} \end{pmatrix} \tag{7.76}$$

and similar expressions for the $v^{(\alpha)}(\boldsymbol{k})$'s. They can be rewritten collectively as

$$u^{(\alpha)}(\boldsymbol{k}) = \sqrt{E_k + m}\begin{pmatrix} \phi^{(\alpha)} \\ \dfrac{\boldsymbol{\sigma} \cdot \boldsymbol{k}}{E_k + m}\phi^{(\alpha)} \end{pmatrix}$$

$$v^{(\alpha)}(\boldsymbol{k}) = \sqrt{E_k + m}\begin{pmatrix} \dfrac{\boldsymbol{\sigma} \cdot \boldsymbol{k}}{E_k + m}\kappa^{(\alpha)} \\ \kappa^{(\alpha)} \end{pmatrix} \tag{7.77}$$

with an explicit use of the non-vanishing two-component spinors $\phi^{(\alpha)}$ and $\kappa^{(\alpha)}$, with

$$\phi^{(1)} = \kappa^{(1)} = \begin{pmatrix} 1 \\ 0 \end{pmatrix}, \quad \phi^{(2)} = \kappa^{(2)} = \begin{pmatrix} 0 \\ 1 \end{pmatrix} \tag{7.78}$$

as they appear in $\hat{u}^{(\alpha)}(m, \boldsymbol{0})$ and $\hat{v}^{(\alpha)}(m, \boldsymbol{0})$.

For the conjugate spinors we obtain

$$\bar{u}^{(\alpha)}(\boldsymbol{k}) = \bar{\hat{u}}^{(\alpha)}(m, \boldsymbol{0})\frac{\not{k} + m}{\sqrt{m + E_k}} \quad \text{and} \quad \bar{v}^{(\alpha)}(\boldsymbol{k}) = \bar{\hat{v}}^{(\alpha)}(m, \boldsymbol{0})\frac{-\not{k} + m}{\sqrt{m + E_k}} \tag{7.79}$$

It is straightforward to verify the following orthogonality relations

$$\bar{u}^{(\alpha)}(\boldsymbol{k})u^{(\beta)}(\boldsymbol{k}) = 2m\delta^{\alpha\beta}, \quad \bar{u}^{(\alpha)}(\boldsymbol{k})v^{(\beta)}(\boldsymbol{k}) = 0$$

$$\bar{v}^{(\alpha)}(\boldsymbol{k})v^{(\beta)}(\boldsymbol{k}) = -2m\delta^{\alpha\beta}, \quad \bar{v}^{(\alpha)}(\boldsymbol{k})u^{(\beta)}(\boldsymbol{k}) = 0 \tag{7.80}$$

Let us consider the "three-vector" matrix $\boldsymbol{\Sigma}$ of components

$$\Sigma^i \equiv -\varepsilon^{ijk}\sigma_{jk} = \begin{pmatrix} \sigma^i & 0 \\ 0 & \sigma^i \end{pmatrix} \tag{7.81}$$

The spinors $\hat{u}^{(1)}(m, \boldsymbol{0})$ and $\hat{v}^{(1)}(m, \boldsymbol{0})$ are eigenvectors of Σ^3 with eigenvalue $+1$, while $\hat{u}^{(2)}(m, \boldsymbol{0})$ and $\hat{v}^{(2)}(m, \boldsymbol{0})$ are eigenvectors with eigenvalue -1. They correspond to a given direction of the z-axis in the $\boldsymbol{k} = 0$ frame. It will be convenient for many applications to choose them instead as eigenvectors of $\boldsymbol{\Sigma} \cdot \hat{\boldsymbol{k}}$. Then, (see Problem 7.13) the action of $\not{k} + m$ on them leads to eigenstates of the *helicity* operator $h \equiv \frac{1}{2}\boldsymbol{\Sigma} \cdot \hat{\boldsymbol{k}}$, with eigenvalues $+\frac{1}{2}$ and $-\frac{1}{2}$, respectively. Furthermore, in the ultra-relativistic limit, $E_k \to |\boldsymbol{k}|$, these latter spinors become eigenvectors also of the *chirality* operator γ_5. Specifically, the $h = \pm\frac{1}{2}$ spinors have eigenvalue of $\gamma_5 = \mp 1$ and are called *right-* and *left-handed*, respectively (see also section 8.3.8).

The matrices

$$Q_\pm(k) = \frac{1}{2}\left(\mathbf{1} \pm \mathbf{\Sigma} \cdot \hat{\mathbf{k}}\right) \tag{7.82}$$

are the projectors on the two helicity states.

The matrices

$$P_+(k) = \frac{\slashed{k} + m}{2m} \quad \text{and} \quad P_-(k) = \frac{-\slashed{k} + m}{2m} \tag{7.83}$$

are the projectors respectively on the positive and negative energy components of a solution with momentum $\mathbf{k}$. They satisfy

$$P_\pm^2(k) = P_\pm(k), \quad P_+ + P_- = \mathbf{1}, \quad \mathrm{Tr}P_\pm(k) = 2 \tag{7.84}$$

It is instructive to compute the fermion density $\varrho = j^0 = \bar{\psi}\gamma^0\psi$ for a plane wave of positive energy and momentum $\mathbf{k}$

$$\varrho = \bar{\psi}^{(+)}\gamma^0\psi^{(+)} = \bar{u}(\mathbf{k})\gamma^0 u(\mathbf{k}) = \frac{1}{2m}\bar{u}(\mathbf{k})\{\slashed{k},\gamma^0\}u(\mathbf{k}) = \frac{E_k}{m}$$

As expected, it is positive.

Finally, as we did with the Klein–Gordon equation, we can expand an arbitrary solution of the Dirac equation in the basis of plane waves. By analogy with equation (7.29) for a complex scalar field, we have

$$\psi(x) = \int \frac{\mathrm{d}^3k}{(2\pi)^3}\frac{1}{2E_k}\sum_{\alpha=1}^{2}\left[a_\alpha(\mathbf{k})u^{(\alpha)}(\mathbf{k})\mathrm{e}^{-\mathrm{i}k\cdot x} + b_\alpha^*(\mathbf{k})v^{(\alpha)}(\mathbf{k})\mathrm{e}^{\mathrm{i}k\cdot x}\right] \tag{7.85}$$

$$\bar{\psi}(x) = \int \frac{\mathrm{d}^3k}{(2\pi)^3}\frac{1}{2E_k}\sum_{\alpha=1}^{2}\left[a_\alpha^*(\mathbf{k})\bar{u}^{(\alpha)}(\mathbf{k})\mathrm{e}^{\mathrm{i}k\cdot x} + b_\alpha(\mathbf{k})\bar{v}^{(\alpha)}(\mathbf{k})\mathrm{e}^{-\mathrm{i}k\cdot x}\right] \tag{7.86}$$

where a_α and b_α, $\alpha = 1, 2$ are two pairs of arbitrary complex functions and * means "complex conjugation". With our normalisation conventions the integration measure in the expansions (7.85) and (7.86) is the invariant measure $d\Omega_m$ on the mass hyperboloid.

7.4 Relativistic Equations for Vector Fields

Up to now we have obtained relativistic wave equations for scalar and spinor fields. In addition we have written Maxwell's equations for the electromagnetic field. In this section we want to write the general relativistically covariant wave equation for fields belonging to the $(\frac{1}{2}, \frac{1}{2})$ representation of the Lorentz group.

Let $A_\mu(x)$ denote a general vector field. In writing a differential equation we can use the vector differential operator ∂_μ and the field itself. So, the most general Lorentz covariant, linear, second order differential equation for A_μ is of the form

$$\mathcal{D}_{\mu\nu}A^\nu = \Box A_\mu(x) - \frac{1}{\alpha}\partial_\mu\partial_\nu A^\nu(x) + m^2 A_\mu(x) = 0 \tag{7.87}$$

It involves two arbitrary parameters, m^2, which has dimensions of mass squared and α, which is dimensionless. Maxwell's equations in the vacuum are obtained for the particular choice $m^2 = 0$ and $\alpha = 1$. For these values the equation acquires an extra invariance which, in Chapter 2, we called *gauge invariance*. It is the invariance under the transformation $A_\mu(x) \to A_\mu(x) + \partial_\mu \theta(x)$. In general, this transformation applied to (7.87) gives an extra term of the form $[(1 - \alpha^{-1})\Box + m^2]\partial_\mu\theta$, which, for arbitrary θ, vanishes only when $m^2 = 0$ and $\alpha = 1$.

The equation (7.87) can be obtained from the Lagrangian density

$$\mathcal{L} = -\frac{1}{2}\left(\partial_\mu A^\nu \partial^\mu A_\nu - \frac{1}{\alpha}\partial_\mu A^\mu \partial_\nu A^\nu - m^2 A^\mu A_\mu\right) \tag{7.88}$$

It is straightforward to obtain the corresponding Green functions. They are given by the inverse of the differential operator $\mathcal{D}_{\mu\nu}$, and, consequently, their Fourier transform satisfies the equation

$$\widetilde{\mathcal{D}}_{\mu\nu}(k)\widetilde{G}^{\mu\rho}(k) = \left(k^2 \eta_{\mu\nu} - \frac{1}{\alpha}k_\mu k_\nu - m^2 \eta_{\mu\nu}\right)\widetilde{G}^{\mu\rho}(k) = \eta_\nu^\rho \tag{7.89}$$

which gives, see Problem 7.8,

$$\widetilde{G}^{\mu\rho}(k) = \frac{1}{k^2 - m^2}\left(\eta^{\mu\rho} - \frac{k^\mu k^\rho}{k^2(1-\alpha) + \alpha m^2}\right) \tag{7.90}$$

with, as always, the appropriate $i\epsilon$ prescription. Regarding the phase convention we follow the choice we made for the Klein–Gordon field: $\widetilde{G}_R^{\mu\rho}(k)$ and $\widetilde{G}_A^{\mu\rho}(k)$ are real and $\widetilde{G}_F^{\mu\rho}(k)$ is given by

$$\widetilde{G}_F^{\mu\rho}(k) = \frac{-i}{k^2 - m^2 + i\epsilon}\left(\eta^{\mu\rho} - \frac{k^\mu k^\rho}{k^2(1-\alpha) + \alpha m^2}\right) \tag{7.91}$$

As expected, the limit of (7.90) for $m^2 = 0$ and $\alpha = 1$ does not exist. The reason is gauge invariance. For this choice of the values of the parameters the wave equation differential operator satisfies the condition $\mathcal{D}_{\mu\nu}\partial^\mu\theta = 0$ (or equivalently $\widetilde{\mathcal{D}}_{\mu\nu}(k)k^\mu = 0$), which means that $\mathcal{D}_{\mu\nu}$ has a zero mode and, consequently, it has no inverse.

Let us analyse the two cases $m^2 = 0$ and $m^2 \neq 0$ separately.

7.4.1   The Maxwell field

Maxwell's equation for the electromagnetic field in the vacuum is obtained from the general equation (7.87) setting $m^2 = 0$ and $\alpha = 1$. So, it is the most general equation for a vector field, which is invariant under both Lorentz and gauge transformations. In the presence of an external source $j_\mu(x)$ the equation is obtained by the variational principle, from the action

$$S[A] = \int d^4x \left(-\frac{1}{4}F_{\mu\nu}(x)F^{\mu\nu}(x) - A^\mu(x)j_\mu(x)\right) \tag{7.92}$$

where, as we have noted already, consistency of Maxwell's equations requires that the externally prescribed current $j_\mu(x)$ is conserved.

Because of gauge invariance, not all components of the vector potential can be considered as independent dynamical variables. As a result, the variational method will not give a unique answer. For every solution satisfying the equations, we can find an infinity of others by adding the gradient of an arbitrary scalar function $\theta(x)$. Even if we were willing to ignore this problem, we would face its consequences when attempting to compute the Green functions of the Maxwell theory. Indeed, as we just explained, gauge invariance implies the existence of a zero mode of the differential operator $\mathcal{D}_{\mu\nu}$, which means that $\mathcal{D}_{\mu\nu}$ has no inverse.

In order to overcome this problem and find well-defined Green functions we want to impose a condition among the four components of A_μ to remove this arbitrariness. Therefore, *a gauge fixing condition* is a condition of the form $G(A) = 0$, where G is some suitably chosen functional of A. The obvious requirement on G is that the equation

$$G[A + \partial\theta] = 0 \tag{7.93}$$

considered as an equation for $\theta(x)$, should yield a unique solution for every given vector potential $A_\mu(x)$. It is clear that this condition implies that G is not gauge invariant. The *Lorenz condition*[10] $G = \partial_\mu A^\mu(x)$ satisfies this criterion if we impose on θ to vanish at infinity. Indeed, for $G = \partial_\mu A^\mu(x)$, (7.93) implies $\Box\theta = 0$, which has a unique solution, $\theta(x) = 0$, for functions which vanish at infinity. In fact, an apparently much simpler option is to choose one component of A_μ to vanish, for example to impose $A_0(x) = 0$. It is easy to check that this condition also satisfies the criterion expressed by (7.93). The same applies to the "Coulomb gauge" condition given by the choice $G = \nabla \cdot A = 0$. The advantage of these non-covariant gauge conditions is that they can be solved explicitly and eliminate the redundant degrees of freedom. The disadvantage is that they break manifest Lorentz covariance and this makes explicit calculations quite complicated. For Lorentz covariant conditions we are led to use the method introduced by Lagrange in order to solve constrained problems in classical mechanics. Let us see how it works.

Maxwell's equations together with the subsidiary gauge condition $G = 0$, can be obtained from the Lagrangian density

$$\mathcal{L} = -\frac{1}{4}F_{\mu\nu}(x)F^{\mu\nu}(x) + b(x)G[A] - A^\mu(x)j_\mu(x) \tag{7.94}$$

where, $b(x)$ is an auxiliary field, called "Lagrange multiplier". Indeed, varying the action independently with respect to A_μ and b we obtain the system of equations

$$\partial^\mu F_{\mu\nu}(x) + \frac{\delta}{\delta A^\nu(x)} \int d^4y\, b(y)\, G[A(y)] = j_\nu(x) \quad \text{and} \quad G[A] = 0 \tag{7.95}$$

For the particular example of the Lorenz gauge condition in which $G = \partial A$, this system is equivalent to

$$\Box A_\nu(x) - \partial_\nu b(x) = j_\nu(x) \quad \text{and} \quad \partial_\mu A^\mu(x) = 0 \tag{7.96}$$

---

[10]Introduced in 1867 by the Dane L.V. Lorenz, not to be confused with the Dutch H.A. Lorentz of the Lorentz transformations.

Since $j_\nu(x)$ is conserved, these equations imply $\Box b(x) = 0$ even in the presence of the external current, in other words, the field $b(x)$ is a free field and does not participate in the dynamics. If it vanishes at infinity, the system (7.96) is equivalent to

$$\Box A_\nu(x) = j_\nu(x) \quad \text{and} \quad b(x) = 0 \tag{7.97}$$

For more complicated gauge conditions, for example if G depends non-linearly on A, the Lagrange multiplier field does not decouple and we should study the coupled system of A_μ and b.

We can obtain a slightly more general formulation[11] by adding to the Lagrangian density (7.94) a term proportional to $b^2(x)$. We obtain

$$\mathcal{L} = -\frac{1}{4} F_{\mu\nu}(x) F^{\mu\nu}(x) + b(x) G[A] - \frac{\lambda}{2} b^2(x) - A^\mu(x) j_\mu(x) \tag{7.98}$$

where λ is an arbitrary constant. It is straightforward to show that, for $G = \partial A$, b is still a free field and the Lagrangian takes the form

$$\mathcal{L} = -\frac{1}{4} F_{\mu\nu}(x) F^{\mu\nu}(x) + \frac{1}{2\lambda} G^2[A] - A^\mu(x) j_\mu(x) \tag{7.99}$$

which is equivalent to the one of equation (7.88) with $m^2 = 0$. We thus obtain a family of gauge fixed Lagrangians depending on the gauge parameter λ.

The presence of the gauge fixing term in (7.99) eliminates the zero mode from the gauge field equation of motion and leads to the Green functions of the gauge field given by equation (7.90) with $m^2 = 0$. For the Feynman Green function we find

$$\widetilde{G}_{\mathrm{F}}^{\mu\rho}(k) = \frac{-\mathrm{i}}{k^2 + \mathrm{i}\epsilon} \left(\eta^{\mu\rho} - \frac{k^\mu k^\rho}{k^2(1 - \lambda)} \right) \tag{7.100}$$

Obviously, the Green functions do depend on λ but, as we ask the reader to verify in Problem 7.9, physically measurable quantities are λ-independent. This is the justification of the gauge fixing procedure.

7.4.2   Massive spin-1 field

For $m^2 \neq 0$ the equation (7.87) describes a theory which, as we shall show later, is appropriate to describe the fields of massive particles with spin equal to one. Of particular interest is the choice $\alpha = 1$ for which the equation is known as *the Proca equation*

$$\Box A_\mu(x) - \partial_\mu \partial_\nu A^\nu(x) + m^2 A_\mu(x) = 0 \tag{7.101}$$

We will use this equation extensively in this book.

Taking the divergence ∂^μ of this equation we obtain

$$m^2 \, \partial^\mu A_\mu(x) = 0 \tag{7.102}$$

which, for $m^2 \neq 0$, implies a condition among the four components of the vector field. Two remarks are in order here: First, for the Proca equation with the term proportional

---

[11]All these steps appear to be arbitrary and, in some sense, they are. Their final justification is based on the fact that, as we can show, they lead to a mathematically consistent and physically correct theory.

to m^2, this condition is a consequence of the equation of motion, in contradistinction to what happens in the $m = 0$ case in which this, or any analogous condition, had to be imposed by hand. Second, note that this condition remains valid even if we introduce a source term $j_\mu(x)$ on the right-hand side of equation (7.101), provided the source is a conserved current. It follows that out of the four components of the field, only three are independent dynamical variables, which, upon quantisation, will correspond to the three spin states of massive spin-1 particles.

The Feynman Green function of the Proca equation takes the form

$$\widetilde{G}_{\mathrm{F}}^{\mu\rho}(k) = \frac{-\mathrm{i}}{k^2 - m^2 + \mathrm{i}\epsilon}\left(\eta^{\mu\rho} - \frac{k^\mu k^\rho}{m^2}\right) \tag{7.103}$$

7.4.3 Plane wave solutions

As we did for the Dirac equation, we can find plane wave solutions, which we can use as basis to expand any field configuration.

We start by defining, for every wave vector $k^\mu = (k^0, \boldsymbol{k})$, four basic vectors $\epsilon_\mu^{(\lambda)}(\boldsymbol{k})$, with λ running from 0 to 3. We can use any system of linearly independent vectors, but a particularly convenient choice is the following: We choose $\epsilon_\mu^{(0)}(\boldsymbol{k})$ to be a unit vector in the time direction with $\epsilon_0^{(0)}(\boldsymbol{k}) = 1$. The remaining three vectors are of the form $\epsilon_\mu^{(\lambda)}(\boldsymbol{k}) = (0, -\boldsymbol{e}^{(\lambda)}(\boldsymbol{k}))$, with unit three-vectors $\boldsymbol{e}^{(\lambda)}(\boldsymbol{k})$, $\lambda = 1, 2, 3$. Then we take the three-axis along the chosen vector $\boldsymbol{k}$. For the $m = 0$ case in particular, the natural choice for $\boldsymbol{k}$ is the direction of propagation. We take $\boldsymbol{e}^{(3)}(\boldsymbol{k})$ to be the unit vector in this direction. The vectors $\boldsymbol{e}^{(1)}(\boldsymbol{k})$ and $\boldsymbol{e}^{(2)}(\boldsymbol{k})$ are chosen in the plane perpendicular to that formed by the first two and orthogonal to each other. In our particular reference frame we have

$$\epsilon^{(0)} = \begin{pmatrix} 1 \\ 0 \\ 0 \\ 0 \end{pmatrix} \qquad \epsilon^{(1)} = \begin{pmatrix} 0 \\ 1 \\ 0 \\ 0 \end{pmatrix} \qquad \epsilon^{(2)} = \begin{pmatrix} 0 \\ 0 \\ 1 \\ 0 \end{pmatrix} \qquad \epsilon^{(3)} = \begin{pmatrix} 0 \\ 0 \\ 0 \\ 1 \end{pmatrix} \tag{7.104}$$

The four four-vectors $\epsilon_\mu^{(\lambda)}(k)$ are called *polarisation vectors* and, when $\boldsymbol{k}$ is chosen along the direction of propagation, we call $\epsilon^{(3)}$ *longitudinal*, $\epsilon^{(1)}$ and $\epsilon^{(2)}$ *transverse* and $\epsilon^{(0)}$ *scalar*.[12] In an arbitrary reference frame they satisfy the orthonormality relations

$$\sum_{\lambda,\lambda'=0}^{3} \eta^{\lambda\lambda'} \epsilon_\mu^{(\lambda)}(\boldsymbol{k})\epsilon_\nu^{(\lambda')*}(\boldsymbol{k}) = \eta_{\mu\nu}\,, \qquad \epsilon^{(\lambda)\rho}(\boldsymbol{k})\epsilon_\rho^{(\lambda')*}(\boldsymbol{k}) = \eta^{\lambda\lambda'} \tag{7.105}$$

With the help of these unit vectors, an arbitrary solution of the wave equation for a real vector field can be expanded in plane waves. For the simple case of $m = 0$ in a linear gauge, the expansion reads

$$A_\mu(x) = \int \frac{\mathrm{d}^3 k}{(2\pi)^3 2\omega_k} \sum_{\lambda=0}^{3} \left(a^{(\lambda)}(\boldsymbol{k})\,\epsilon_\mu^{(\lambda)}(\boldsymbol{k})\,\mathrm{e}^{-\mathrm{i}k\cdot x} + a^{(\lambda)*}(\boldsymbol{k})\,\epsilon_\mu^{(\lambda)*}(\boldsymbol{k})\,\mathrm{e}^{\mathrm{i}k\cdot x} \right) \tag{7.106}$$

---

[12]We can show that the terminology corresponds to the usual polarisations of electromagnetic waves.

with $a^{(\lambda)}(\boldsymbol{k})$ four complex functions. Here $\omega_k = k^0 = |\boldsymbol{k}|$. As we shall see in a later chapter, depending on the choice of gauge, the number of independent polarisation vectors can be reduced. Note the Minkowski metric $\eta^{\lambda\lambda'}$ in the second of the relations (7.105) which is necessary to reproduce the correct transformation properties of the field A_μ.

We can obtain a similar expansion for the solutions of the Proca equation. The four-dimensional transversality condition (7.102) can be used to eliminate the polarisation in the zero direction. For the general $m \neq 0$ equation, the expansion involves all four polarisation vectors.

We can obtain relativistically covariant equations for fields belonging to higher representations of the Lorentz group, but we shall not use any of them in this book.

7.5 Problems

Problem 7.1 Show that the retarded, advanced and Feynman Green functions of the Klein–Gordon equation can be expressed as

$$G_R(x) = \mathrm{i}\,\theta(x^0)[D(x) - D(-x)], \quad G_A(x) = -\mathrm{i}\,\theta(-x^0)[D(x) - D(-x)]$$

$$G_{\mathrm{F}}(x) = \theta(x^0)D(x) + \theta(-x^0)D(-x) \tag{7.107}$$

Problem 7.2 *Properties of the Pauli matrices.* Show that
  1. $\sigma_i\sigma_j = \delta_{ij} + \mathrm{i}\epsilon_{ijk}\sigma_k$
  2. $\sigma_i\sigma_j\sigma_k = \sigma_i\delta_{jk} - \sigma_j\delta_{ik} + \sigma_k\delta_{ij} + \mathrm{i}\epsilon_{ijk}$
  3. $\sigma_i\sigma_j\sigma_k\sigma_l = \delta_{ij}\delta_{kl} - \delta_{ik}\delta_{jl} + \delta_{il}\delta_{jk} + \mathrm{i}[\delta_{ij}\epsilon_{klm} - \delta_{ik}\epsilon_{jlm} + \delta_{jk}\epsilon_{ilm}]\sigma_m + \mathrm{i}\epsilon_{ijk}\sigma_l$

Problem 7.3 *Properties of the Dirac matrices and of Dirac bilinears.*
  1. Show that a general 4×4 matrix can be written as a linear superposition of the matrices $\mathbf{1}$, γ^5, γ_μ, $\gamma_\mu\gamma^5$ and $\sigma^{\mu\nu}$.
  2. Let $\psi(x)$ be a Dirac spinor. Compute the transformation properties under an extended Lorentz transformation (which includes space inversion) $x'^\mu = \Lambda^\mu_{\ \nu}x^\nu$ of the following bilinears and show that they transform as indicated: *(The factor i in the definition of the pseudo-scalar has been put in order to make the bilinear Hermitian.)*

$\overline{\psi}(x)\psi(x)$ scalar (S)
$\overline{\psi}(x)\gamma^\mu\psi(x)$ vector (V)
$\overline{\psi}(x)\sigma^{\mu\nu}\psi(x)$ antisymmetric tensor (T)
$\overline{\psi}(x)\gamma_5\gamma^\nu\psi(x)$ pseudo-vector (A)
$\mathrm{i}\overline{\psi}(x)\gamma_5\psi(x)$ pseudo-scalar (P)

Problem 7.4 Show that for $k^2 = m^2$

$$(\slashed{k} + m)\gamma^0(\slashed{k} + m) = 2k^0(\slashed{k} + m) \tag{7.108}$$

Problem 7.5 Consider the spinors $u^{(\alpha)}(\boldsymbol{k})$ and $v^{(\alpha)}(\boldsymbol{k})$ given in equations (7.75). Show that they satisfy the matrix relations

$$\sum_{\alpha=1,2} u^{(\alpha)}(\boldsymbol{k}) \otimes \bar{u}^{(\alpha)}(\boldsymbol{k}) = \frac{2m}{m+E_k}(\not{k}+m), \quad \sum_{\alpha=1,2} v^{(\alpha)}(\boldsymbol{k}) \otimes \bar{v}^{(\alpha)}(\boldsymbol{k}) = \frac{2m}{m+E_k}(\not{k}-m)$$

(7.109)

Problem 7.6 Prove the identities

$$\gamma_\nu \gamma_\mu \gamma^\nu = -2\gamma_\mu, \quad \gamma_\nu \gamma_\mu \gamma_\rho \gamma^\nu = 4\eta_{\mu\rho}, \quad \gamma_\nu \gamma_\mu \gamma_\rho \gamma_\lambda \gamma^\nu = -2\gamma_\mu \eta_{\rho\lambda} + 4i\sigma_{\rho\lambda}\gamma_\mu \quad (7.110)$$

Problem 7.7 Apply the canonical formalism and derive the expression for the Hamiltonian given in equation (7.67).

Problem 7.8 Using that $\widetilde{G}^{\mu\rho}(k)$ is a two-index symmetric Lorentz tensor, prove equation (7.90).

Problem 7.9 Consider the gauge fixed Lagrangian of equation (7.99) with a general external source $j_\mu(x)$ satisfying $\partial^\mu j_\mu(x) = 0$. Prove that the resulting field $A_\mu(x)$ is independent of the gauge parameter λ, and therefore the resulting electric and magnetic fields are also λ-independent.

Problem 7.10 Consider the Cauchy problem of the homogeneous Maxwell's equations with initial data given by a field configuration at $t = 0$ which is a pure gauge, namely $A_\mu(0, \boldsymbol{x}) = \partial_\mu\theta(x)|_{t=0}$ and $\dot{A}_\mu(0, \boldsymbol{x}) = \partial_\mu\dot\theta(x)|_{t=0}$ with $\theta(x)$ an arbitrary differentiable scalar function. Show that the field at time $t > 0$ is also a pure gauge.

Problem 7.11 Show that if $\psi(x)$ satisfies the Dirac equation $(i\gamma^\mu\partial_\mu - m)\psi(x) = 0$, so does its transform under a Lorentz transformation with parameters $\omega_{\mu\nu} = -\omega_{\nu\mu}$

$$\psi'(x) = S(\omega)\psi(\Lambda^{-1}x) \tag{7.111}$$

where

$$x'^\mu = \Lambda^\mu{}_\nu(\omega)x^\nu \text{ and } S(\omega) = \exp\left(\frac{i}{2}\omega_{\mu\nu}\sigma^{\mu\nu}\right), \text{ with } \sigma^{\mu\nu} = -\frac{i}{4}[\gamma^\mu, \gamma^\nu] \tag{7.112}$$

Problem 7.12 Consider the solution of the Dirac equation $\psi_0(x)$ in the frame in which the wave vector has vanishing space components and the spin is oriented in the $+z$ direction, i.e.

$$\psi_0(x) = A\,e^{-imt}\hat{u}^{(1)}(m, \boldsymbol{0}), \quad \hat{u}^{(1)}(m, \boldsymbol{0}) = \begin{pmatrix} 1 \\ 0 \\ 0 \\ 0 \end{pmatrix} \tag{7.113}$$

where A is a normalisation constant.

1. Act on ψ_0 with the appropriate Lorentz boost to write the solution $\psi(x)$ in the frame in which the wave vector is $k^\mu = (k^0, k^1, 0, 0)^T$.

2. Check that your answer agrees with the one we found using the shortcut described in the text.

Problem 7.13 *Helicity and chirality of spinors.*

1. Write the four independent solutions of the Dirac equation in the frame in which the wave vector is $(m, \mathbf{0})$, which are eigenvectors of $\boldsymbol{\Sigma} \cdot \hat{\boldsymbol{k}}$.

2. Show that the commutator $[\not{k} + m, \boldsymbol{\Sigma} \cdot \hat{\boldsymbol{k}}] = 0$. Consequently, the four solutions we obtain after multiplication of the above with $\not{k} + m$ are still eigenvectors of the *helicity* $h = \frac{1}{2}\boldsymbol{\Sigma} \cdot \hat{\boldsymbol{k}}$, which is the projection of the spin in the direction of the wave vector, and takes values $\pm\frac{1}{2}$.

3. Show that in the ultra-relativistic limit $E \to |\boldsymbol{k}|$, the solutions with $h = +\frac{1}{2}$ $(-\frac{1}{2})$ become also eigenvectors of the *chirality* γ_5 with eigenvalues -1 $(+1)$, respectively.

8
Towards a Relativistic Quantum Mechanics

8.1 Introduction

In the previous chapter we derived relativistically covariant wave equations for fields of low spin. Although the derivation was purely mathematical, these equations have important physical applications. The obvious case is Maxwell's equation, which describes the dynamics of the classical electromagnetic field.

In the present chapter we wish to use these wave equations in order to formulate an extension of Schrödinger's equation for the description of the quantum mechanics of a relativistic particle.

Actually, we may wonder why Schrödinger, who wrote his equation in 1926, 21 years after the discovery of special relativity, chose to write a non-relativistic equation for the electron. In fact, historically, it seems that he first tried to find a relativistic one and considered the analogue of the Klein–Gordon equation, since, at that time, the degrees of freedom associated to the spin of the electron were not known. We will see here that this approach leads to serious difficulties and we will present the extent to which they can be overcome.

8.2 The Klein–Gordon Equation

We start with the simplest Lorentz covariant wave equation, namely the Klein–Gordon equation. From our study of the Lorentz group, we expect it a priori to be relevant to the description of the quantum mechanics of a spinless particle. Since the wave function in quantum mechanics is complex valued, it is natural to start with the Klein–Gordon equation (7.1) for a complex function $\Phi(x)$

$$\left(\frac{1}{c^2} \frac{\partial^2}{\partial t^2} - \triangle + m^2 \right) \Phi(t, \boldsymbol{x}) = 0 \tag{8.1}$$

It is straightforward to verify that in the non relativistic limit, $c \to +\infty$, the Klein–Gordon equation reduces to the Schrödinger equation. To show this it is enough to extract the mass from the energy dependence of $\Phi(x)$. This is obtained by parametrising Φ in the following way:

$$\Phi(t, \boldsymbol{x}) = \exp\left(-\frac{imc^2}{\hbar} t \right) \Psi(t, \boldsymbol{x}).$$

Elementary Particle Physics. John Iliopoulos and Theodore N. Tomaras, Oxford University Press.
© John Iliopoulos and Theodore N. Tomaras (2021). DOI: 10.1093/oso/9780192844200.003.0008

Then, the Klein–Gordon equation becomes

$$\left(\frac{1}{c^2}\left(\frac{\partial}{\partial t}\right)^2 - \frac{i}{\hbar}2m\frac{\partial}{\partial t} - \triangle \right) \Psi(x) = 0.$$

In the non-relativistic approximation, $c \to +\infty$, the first term can be neglected, and we recover the free Schrödinger equation

$$i\hbar\frac{\partial}{\partial t}\Psi(x) = -\frac{\hbar^2}{2m}\triangle\Psi(x)$$

So, at first sight, the Klein–Gordon equation seems to have the right properties to give the relativistic generalisation of the Schrödinger equation. We suspect though that this could not be right because, if it were that simple, Schrödinger would have written directly this more general relativistic equation. Let us show that, indeed, interpreting $\Phi(x)$ as the particle wave function, does lead to physical inconsistencies.

Let us choose m^2 to be real. The functions Φ and Φ^* satisfy the equation (8.1) and its complex conjugate, respectively. If Φ is to be considered as a wave function, there must exist a conserved probability current, as for the Schrödinger equation. In the previous chapter we constructed this current. We saw that the two equations derive from the variational principle applied to the real action (7.27). The invariance of this action under phase transformations of Φ implies the conservation of the current (7.28). Let us separate the zero component and the three spatial components

$$\varrho(x) = \frac{i\hbar}{2m} \left(\Phi^*\left(\frac{\partial}{\partial t}\Phi\right) - \left(\frac{\partial}{\partial t}\Phi^*\right)\Phi \right) \tag{8.2}$$

$$\boldsymbol{j}\,(x) = -\frac{i\hbar}{2m}\left(\Phi^*(\boldsymbol{\nabla}\Phi) - (\boldsymbol{\nabla}\Phi^*)\Phi\right)$$

We see that ϱ satisfies the conservation equation, which is the necessary condition for its interpretation as a probability density. However, such an interpretation is impossible because the expression (8.2) is not positive definite. Indeed, the Klein–Gordon equation has solutions of the form $\Phi = \exp(-iEt/\hbar)u(\boldsymbol{x})$ for all $E \geq mc^2$ and $E \leq -mc^2$. It follows that $\varrho = (1/m)E|u|^2$ can be negative and therefore it cannot be interpreted as a probability density. Technically, the problem arises from the fact that the energy dependence of the Klein–Gordon equation is quadratic. This is the clue that led Dirac to the formulation of the equation that bears his name.

A second problem, as fundamental as the first one, which will be discussed below in relation to the Dirac equation, is that the energy spectrum E is not bounded from below, since $E = \pm c\sqrt{\boldsymbol{p}^2 + m^2c^2}$. We conclude that a complex function $\Phi(x)$, which satisfies the Klein–Gordon equation, cannot be interpreted as the wave function of a relativistic particle.

8.3 The Dirac Equation

8.3.1 Introduction

Dirac derived his equation in 1928 as a relativistic generalisation of Schrödinger's equation for the electron. Schrödinger's equation is first order in time derivatives, but second order in derivatives with respect to the spatial coordinates. It does have a conserved probability current with positive definite density. We just saw that this property is lost in the obvious relativistic extension, the Klein–Gordon equation, which has second order derivatives with respect to both time and space. It is this difference, which motivated Dirac to look for an equation with first order derivatives. For a single scalar function there is no such non-trivial Lorentz covariant first order differential equation, so Dirac assumed a multi-component wave function and looked for a matrix equation. In doing so, he discovered the spinorial representations of the Lorentz group. In the notation we used in section 7.3 the equation reads

$$(i\partial\!\!\!/ - m)\Psi(x) = 0 \tag{8.3}$$

As we have shown already, this equation is Lorentz covariant, provided the γ matrices satisfy the Clifford algebra relation (7.41). It is obtained from the Lagrangian density given in equation (7.60).

8.3.2 The conserved current

The free Dirac Lagrangian density of equation (7.60) is invariant under the $U(1)$ group of phase transformations

$$\Psi(x) \to e^{i\theta}\Psi(x) \tag{8.4}$$

By Noether's theorem, to this invariance corresponds a conserved current

$$j_\mu(x) = \overline{\Psi}(x)\gamma_\mu\Psi(x); \quad \partial^\mu j_\mu(x) = 0 \tag{8.5}$$

whose zero component is given by

$$j_0(x) = \overline{\Psi}(x)\gamma_0\Psi(x) = \Psi^\dagger(x)\Psi(x) \geq 0 \tag{8.6}$$

Contrary to what happened in the Klein–Gordon equation (8.2), here we have a positive definite density, which admits, a priori, a probabilistic interpretation.

8.3.3 The coupling with the electromagnetic field

If we want to use the Dirac equation to describe electrons bound in atoms, we must specify the coupling to the electromagnetic field. According to our discussion in Chapter 2, this is achieved by the so-called *minimal substitution* which amounts to the replacement of the ordinary derivatives ∂_μ by the *covariant derivatives*

$$D_\mu = \partial_\mu + i\,e\,A_\mu \tag{8.7}$$

where A_μ is the vector potential of the electromagnetic field and e the charge of the electron.

Thus, the Dirac equation for the electron coupled to electromagnetism is

$$(i\,\partial\!\!\!/ - e\,A\!\!\!/ - m)\Psi(x) = 0 \tag{8.8}$$

It is invariant under the local, or *gauge transformation*

$$\Psi(x) \to e^{i\,e\,\theta(x)}\Psi(x), \quad A_\mu(x) \to A_\mu(x) - \partial_\mu\theta(x) \tag{8.9}$$

8.3.4   The non-relativistic limit of the Dirac equation

It is instructive to verify, as a first check, that (8.8) has the correct non-relativistic limit.

As we did with the Klein–Gordon equation, we separate the rest energy by defining

$$\psi = \Psi \exp\left(\mathrm{i}\frac{mc^2}{\hbar}t\right)$$

which satisfies the equation (we have reintroduced the dependence on $\hbar$ and c)

$$\left(\mathrm{i}\hbar\frac{\partial}{\partial t} + mc^2\right)\psi = \left[c\,\boldsymbol{\alpha}\cdot\left(-\mathrm{i}\boldsymbol{\nabla} - \frac{e}{c}\boldsymbol{A}\right) + \beta mc^2 + eA_0\right]\psi \qquad (8.10)$$

We work in the standard representation most appropriate for the study of the non-relativistic limit. We define

$$\psi = \begin{pmatrix} \phi \\ \chi \end{pmatrix}$$

in which case (8.10) splits into

$$\left(\mathrm{i}\hbar\frac{\partial}{\partial t} - eA_0\right)\phi = c\boldsymbol{\sigma}\cdot\left(\boldsymbol{p} - \frac{e}{c}\boldsymbol{A}\right)\chi$$

$$\left(\mathrm{i}\hbar\frac{\partial}{\partial t} - eA_0 + 2mc^2\right)\chi = c\boldsymbol{\sigma}\cdot\left(\boldsymbol{p} - \frac{e}{c}\boldsymbol{A}\right)\phi \qquad (8.11)$$

Let us consider the non-relativistic approximation, in which the electron self-energy is much greater than all other terms with dimensions of energy. This means, in particular, that in the second equation, the leading term on the left-hand side is $2mc^2$ and therefore that[1]

$$\chi = \frac{1}{2mc}\boldsymbol{\sigma}\cdot\left(\boldsymbol{p} - \frac{e}{c}\boldsymbol{A}\right)\phi$$

which, upon substitution in the first equation, leads to

$$\left(\mathrm{i}\hbar\frac{\partial}{\partial t} - eA_0\right)\phi = \frac{1}{2m}\left(\boldsymbol{\sigma}\cdot\left(\boldsymbol{p} - \frac{e}{c}\boldsymbol{A}\right)\right)^2\phi$$

Using the fact that

$$\boldsymbol{\sigma}\cdot\boldsymbol{a}\,\boldsymbol{\sigma}\cdot\boldsymbol{b} = \boldsymbol{a}\cdot\boldsymbol{b} + \mathrm{i}\boldsymbol{\sigma}\cdot\boldsymbol{a}\wedge\boldsymbol{b}$$

we obtain

$$\left(\boldsymbol{\sigma}\cdot\left(\boldsymbol{p} - \frac{e}{c}\boldsymbol{A}\right)\right)^2 = \left(\boldsymbol{p} - \frac{e}{c}\boldsymbol{A}\right)^2 + \mathrm{i}\boldsymbol{\sigma}\cdot\left(\boldsymbol{p} - \frac{e}{c}\boldsymbol{A}\right)\wedge\left(\boldsymbol{p} - \frac{e}{c}\boldsymbol{A}\right)$$

and since $\boldsymbol{p}$ and $\boldsymbol{A}$ do not commute, the last term is non-zero and equal to

$$-\frac{e\hbar}{c}\boldsymbol{\sigma}\cdot(\boldsymbol{\nabla}\wedge\boldsymbol{A} + \boldsymbol{A}\wedge\boldsymbol{\nabla}) = -\frac{e\hbar}{c}\boldsymbol{\sigma}\cdot(\boldsymbol{\nabla}\wedge\boldsymbol{A}) = -\frac{e\hbar}{c}\boldsymbol{\sigma}\cdot\boldsymbol{B}$$

---

[1] Note that $\chi \sim \mathcal{O}(v/c)\phi$.

so that ϕ obeys the Pauli equation

$$i\hbar\frac{\partial}{\partial t}\phi = \left[\frac{1}{2m}\left(\boldsymbol{p} - \frac{e}{c}\boldsymbol{A}\right)^2 + eA_0 - \frac{e\hbar}{2mc}\boldsymbol{\sigma}\cdot\boldsymbol{B}\right]\phi \tag{8.12}$$

This equation differs from the Schrödinger equation by the last term, which represents the coupling between the electromagnetic field and a magnetic moment $\boldsymbol{\mu}$ given by

$$\boldsymbol{\mu} = \frac{e\hbar}{2mc}\boldsymbol{\sigma} = \frac{g\mu_B\boldsymbol{S}}{\hbar}$$

where $\mu_B = e\hbar/2mc = 9,27 \times 10^{-24}\,\mathrm{J/T}$, $\boldsymbol{S} = \hbar\boldsymbol{\sigma}/2$ is the electron spin-1/2 and $g = 2$ its *gyromagnetic ratio*.

The Dirac equation is therefore the theory of a massive particle of charge e, spin-1/2 and gyromagnetic ratio[2] predicted to be equal to 2. This prediction was the first great success of the Dirac theory.

We can continue the expansion and obtain the next order relativistic corrections. In Problem 8.3 we ask the reader to compute the second order terms and show that we obtain the effects of a spin-orbit coupling.

8.3.5 Negative energy solutions

We saw that the Dirac equation solves the positivity problem of the probability density, which was the first difficulty we encountered with the Klein–Gordon equation. We want here to discuss the second one, namely the solutions of negative energy. In the previous chapter we saw that, as a consequence of Lorentz invariance, the spectrum of the free Dirac Hamiltonian is given by $(-\infty, -mc^2] \cup [mc^2, +\infty)$. It follows that it is continuous and unbounded from below, characteristics which remain unchanged even in the presence of a potential that vanishes at spatial infinity. This is, in particular, the case of the electromagnetic potential of a nucleus, therefore, an electron bound in an atom will be able to lower its energy by the emission of electromagnetic radiation. Since its spectrum is not bounded from below, there is no limit for such a process, which means that all atoms are unstable.

In order to tackle this problem, Dirac made a bold proposal: According to Pauli, electrons obey the exclusion principle, which forbids two electrons to occupy the same quantum state. Dirac assumed that under normal conditions all states with negative energies are occupied, which of course forbids the unbounded fall of the energy of an electron. This hypothetical infinite set of negative energy electrons has been called the *Dirac sea*, and determines the ground state of the Dirac theory. It can be interpreted as a Fermi gas with infinite density. All observables of the full system (energy, electric charge etc.) must take into account all possible interactions between the Dirac sea and the electrons with positive energy. In Dirac's point of view, the energy of an electron must be computed as the energy of the system [one electron+Dirac sea] minus the

---

[2]Experimentally $g = 2.00233184$. The value of the "anomalous" magnetic moment $(g-2)/2$ is known up to 10^{-7}. The theoretical computations predict $g = 2 + \alpha/\pi + \ldots$, with $\alpha = e^2/(4\pi\hbar c)$ the *fine structure constant*. The deviation from the Dirac value $g = 2$ is due to quantum electrodynamics and coincides with the measured value and with the above precision. We will discuss this in Chapter 16.

energy of the system [zero electron+Dirac sea]. In the Dirac theory, when a photon with energy $\hbar\omega > 2mc^2$ is absorbed by an electron of the sea, this electron can become an electron with positive energy, leaving behind a vacancy, *a hole* in the sea. The electron and the hole become an observable system. The energy of the hole is positive, its electric charge is the opposite of that of an electron and its mass is identical to that of the electron. Dirac thought for a while that he could identify the holes with the protons, but it was quickly understood that, were that the case, the hydrogen atom would decay into two photons with an unrealistic lifetime on the order of 10^{-10} s. Therefore the holes must correspond to *new* particles. The second great triumph of the Dirac theory, after the prediction of electron's gyromagnetic ratio, came with the experimental discovery of the *positron* by Anderson in 1932. This discovery was the cornerstone of the establishment of the existence of *anti-particles*. After that, instead of being a mathematical artifact, a hole was to be understood as a particle with positive energy, and all puzzles could be resolved by interpreting the Dirac equation as describing in a unified way either a positron or an electron, each of them having positive energy larger than, or equal to, their rest energy mc^2. In this picture the absorption of a photon by a negative energy electron,which we discussed previously, is interpreted as an electron–positron pair creation by the photon: $\gamma \to e^-_{E>0} + e^+_{E>0}$.[3] This process is not possible for free photons in the vacuum because it would violate energy-momentum conservation, but it becomes possible, and in fact frequently observed, in the field of a nucleus which absorbs the recoil momentum.

We see that, although the Dirac equation was originally constructed to describe a single charged particle in a straightforward generalisation of Schrödinger's picture, its thorough analysis enforces the introduction of another particle, called the *anti-particle*. Eventually, it implies a multi-particle interpretation, due to the possibility of creating pairs of electrons and positrons, provided there is energy and momentum conservation.[4] This new paradigm, in which the number of particles is not conserved, was originally called second quantisation. Nowadays, it is just called quantum field theory and it is simply the quantum mechanics of a system with an indefinite number of particles. We just saw that it is the logical consequence of putting together quantum mechanics and special relativity and it will be described in this book. Nevertheless, although the traditional one-particle interpretation of quantum mechanics is not consistent for a relativistic theory, the Dirac equation can still be used as an approximation in situations where the effects of electron–positron pair creation can be neglected.

8.3.6 Charge conjugation

It is possible to give a precise mathematical meaning to the positron interpretation of the negative energy solutions. To this effect, we will next establish a symmetry of the

[3]At the level of the Dirac equation this interpretation is not fully satisfactory. It relies on the hypothesis that electrons obey the Pauli exclusion principle, although this property does not follow from the equation and it must be imposed by hand. We will see in Chapter 10 that, in the framework of relativistic quantum field theory, this problem too is resolved.

[4]It is possible to show that insisting on a single particle interpretation of the Dirac equation leads to inconsistencies. In Problems 8.5 and 8.6 we exhibit a few of these problems.

Dirac equation called *charge conjugation*. We want the positrons to play a role symmetric to that of electrons, therefore the former should satisfy the same Dirac equation as the latter, but with the opposite electric charge $-e$. Thus, we are after a transformation $\Psi \rightarrow \Psi^C$ such that

$$(i \not{\partial} - e \not{A} - m)\Psi = 0 \tag{8.13}$$

$$(i \not{\partial} + e \not{A} - m)\Psi^C = 0 \tag{8.14}$$

According to the basic principles of quantum mechanics, in order to preserve the norm, this transformation must be local and involutive up to a phase.

Assuming that the spinor Ψ is a solution of the Dirac equation (8.13), its complex conjugate satisfies

$$\left(-i \left(\gamma^0 \gamma^{\mu\dagger} \gamma^0 \right)^T (\partial_\mu - ieA_\mu) - m \right) \overline{\Psi}^T = 0 \tag{8.15}$$

Suppose there exists a matrix C such that

$$C \left(\gamma^0 \gamma^{\mu\dagger} \gamma^0 \right)^T C^{-1} = -\gamma^\mu \tag{8.16}$$

Then,

$$\Psi^C \equiv C \overline{\Psi}^T \tag{8.17}$$

satisfies the Dirac equation (8.14), which differs from (8.13) only in the sign of the electric charge.

Condition (8.16) is generic for any representation of the Dirac matrices. Using $\gamma^{\mu\dagger} = \gamma^0 \gamma^\mu \gamma^0$ it simplifies to

$$C\gamma^{\mu T} C^{-1} = -\gamma^\mu \tag{8.18}$$

In the standard representation the solution for C is

$$C = i\gamma^2\gamma^0 = \begin{pmatrix} 0 & -i\sigma_2 \\ -i\sigma_2 & 0 \end{pmatrix} \tag{8.19}$$

and satisfies

$$-C = C^{-1} = C^T = C^\dagger \tag{8.20}$$

Note, in particular, that for the spinor solution representing a negative energy electron at rest, we have:

$$\Psi = e^{imt} \begin{pmatrix} 0 \\ 0 \\ 0 \\ 1 \end{pmatrix} \quad \text{and} \quad \Psi^C = e^{-imt} \begin{pmatrix} 1 \\ 0 \\ 0 \\ 0 \end{pmatrix} \tag{8.21}$$

This shows that what Dirac called an electron with negative energy and spin projection $-\frac{1}{2}$ is transformed by charge conjugation into an electron with positive energy, positive electric charge and spin projection $+\frac{1}{2}$. The charge conjugated state has the quantum numbers of the "absence" from the Dirac sea of the initial negative energy electron. This is identified with the positron.

8.3.7 CPT symmetry

The Dirac equation is by construction invariant under space inversions, or parity transformations ($\mathcal{P}$). We just showed the existence of a second symmetry transformation, that of charge conjugation ($\mathcal{C}$). We wish to investigate here the consequences of time reversal $\mathcal{T}$, with $(\mathcal{T}x)^\mu = (-t, \boldsymbol{x})$. We can easily check that $\gamma^0\gamma^5\Psi^C(\mathcal{T}x)$, which is an anti-unitary operation, also satisfies the Dirac equation. It is interesting to consider the product of these three symmetries, namely the product of [5]

$$\mathcal{P} : \Psi(x) \to i\gamma^0\Psi(\mathcal{P}x)$$

$$\mathcal{C} : \Psi(x) \to \Psi^C(x) = C\gamma^0\Psi^*(x) \tag{8.22}$$

and

$$\mathcal{T} : \Psi(x) \to \gamma^0\gamma^5\Psi^C(\mathcal{T}x) \tag{8.23}$$

where $(\mathcal{P}x)^\mu = (t, -\boldsymbol{x})$ denotes space inversion. We can verify that the product CPT of the three operators taken in any order is an invariance of the theory. Indeed, start with

$$\begin{aligned}
PCT\,\Psi(x) &= i\gamma^0 \left(C\gamma^0[\gamma^0\gamma^5 C\gamma^0\Psi^*(\mathcal{PT}x)]^* \right) \\
&= i\gamma^0 \left(C\gamma^0(\gamma^0\gamma^5 C^{-1}\gamma^0\Psi(-x)) \right) \\
&= i\gamma^5\Psi(-x)
\end{aligned} \tag{8.24}$$

and ask the following question: Given a wave function $\Psi(x)$, which satisfies the Dirac equation in an external electromagnetic field $A_\mu(x)$, can we find the equation satisfied by its CPT-transformed wave function $i\gamma^5\Psi(-x)$? The answer is simple:

$$\begin{aligned}
0 &= [i\,\partial\!\!\!/_x - e\,A\!\!\!/(x) - m]\,\Psi(x) \\
&= i\gamma^5 [i\,\partial\!\!\!/_x - e\,A\!\!\!/(x) - m]\,\Psi(x) \\
&= [-i\,\partial\!\!\!/_x + e\,A\!\!\!/(x) - m]\,i\gamma^5\Psi(x) \\
&= [i\,\partial\!\!\!/_x + e\,A\!\!\!/(-x) - m]\,CPT\,\Psi(x)
\end{aligned} \tag{8.25}$$

where, in the last step, we have performed the change of variable $x^\mu \to -x^\mu$.

This equation shows that the CPT-transformed wave function $i\gamma^5\Psi(-x)$ satisfies the same Dirac equation *provided* at the same time we replace the external electromagnetic field $A_\mu(x)$ by $CPTA_\mu(x) = -A_\mu(-x)$. This is not surprising. We have already seen that the operation of charge conjugation relates an electron to a positron, whose coupling to the electromagnetic field has the opposite sign.

At this level the invariance under CPT looks like a definition. It is the consequence of our assumption that the electromagnetic potential changes sign under this transformation. However, we shall see later that, in fact, this is a much more general property. It is possible to prove, using only general principles based on locality and Lorentz invariance, that every relativistically invariant quantum theory with local interactions is invariant under the product, taken in any order, of three appropriately defined operators, $\mathcal{P}$ for space inversion, or parity, $\mathcal{T}$ for time reversal and $\mathcal{C}$

---

[5]Here we choose to work with the standard representation of the Dirac matrices.

for particle–anti-particle, or charge, conjugation. This "CPT-theorem" is one of the few very general results, which test the most basic axioms of any local quantum field theory. Among its immediate consequences is that particles and anti-particles have exactly the same mass.

8.3.8 Chirality

In section 7.3.6 we introduced the so-called "standard" representation, which separates the components of a Dirac spinor that vanish in the $c \to \infty$ non-relativistic limit. At the other end, at the ultra-relativistic limit where the mass is negligible, we expect, at least in the absence of any external potential, the Dirac equation to split into a pair of Weyl equations. We shall introduce a formal way to demonstrate this fact.

We define two orthogonal projectors P_L and P_R, L and R standing for "left" and "right" respectively, by

$$P_L = \frac{1+\gamma^5}{2}, \quad P_R = \frac{1-\gamma^5}{2} \tag{8.26}$$

They are Hermitian operators and satisfy the projection, orthogonality and completeness relations: $P_{L,R}^2 = P_{L,R}$, $P_L P_R = P_R P_L = 0$ and $P_L + P_R = 1$. With the help of these projectors we define the "left" and "right" components of a general Dirac spinor $\Psi(x)$ by

$$\Psi_{L,R}(x) = P_{L,R}\Psi(x), \quad \Psi(x) = \Psi_L(x) + \Psi_R(x) \tag{8.27}$$

In the Weyl basis we used in equation (7.38), in which γ^5 is diagonal, P_L and P_R project onto the ξ and η spinors, respectively. Using Ψ_L and Ψ_R, the free Dirac action (7.61) becomes

$$S = \int d^4x \left[\overline{\Psi}_L i\partial\!\!\!/\Psi_L + \overline{\Psi}_R i\partial\!\!\!/\Psi_R - m(\overline{\Psi}_L\Psi_R + \overline{\Psi}_R\Psi_L) \right] \tag{8.28}$$

We see that the kinetic energy part is diagonal in the "left"-"right" basis, while the mass term mixes "left" and "right" components. This is as expected, because we know that the two Weyl equations decouple in the massless limit. In section 7.3.8 and in Problem 7.13 the reader may find the connection between the helicity and the chirality of Dirac spinors. In the ultra-relativistic limit, the eigenstates of the helicity h of a Dirac fermion become also eigenstates of the chirality operator γ_5. As a result, in the ultra-relativistic limit, $\Psi_L(x)$ and $\Psi_R(x)$ become the wave functions of an electron with spin projections $-\frac{1}{2}$ and $+\frac{1}{2}$, respectively, on the axis defined by its momentum. For this reason these two states are called *left* and *right chirality states*.[6] Note also that, since γ^0 anti-commutes with γ^5, the charge conjugate of a left-handed spinor is right-handed: $(\Psi_L)^C = (\Psi^C)_R$.

8.3.9 Hydrogenoid systems

The extremely precise analysis of the hydrogen spectrum played an essential role in the development of quantum mechanics in the first half of the last century. A major success

---

[6]From the Greek word "$\chi\varepsilon\acute{\iota}\rho$" which means "hand".

of the Dirac equation was the accurate determination of the relativistic corrections to the Schrödinger theory. This came in two steps, first at the level of the one-particle approximation and, second, at that of the complete picture, where one takes into account the effects of particle–anti-particle creation and annihilation.

8.3.9.1 *Hydrogenoid atoms in non-relativistic quantum mechanics.*   In non-relativistic quantum mechanics the states of an electron bound to a nucleus with Z protons are described by the Schrödinger equation

$$\left(-\frac{\hbar^2}{2M}\triangle - \frac{Ze^2}{4\pi r} - E\right)\psi(\boldsymbol{r}) = 0 \tag{8.29}$$

where M is the reduced mass

$$M = \frac{m_e m_N}{m_e + m_N} \simeq m_e$$

of the system. The m_e and m_N stand for the electron and the nucleus masses, respectively. Using the rotational symmetry of the Coulomb potential, one can parameterise the solutions in terms of the eigenfunctions of the angular momentum operator $\psi_{l,m}(\boldsymbol{r}) = Y_l^m(\theta, \phi)\Phi(r)$. Then, the Schrödinger equation reads

$$\left(-\frac{\hbar^2}{2M}\triangle - \frac{Ze^2}{4\pi r} - E\right)\psi_{l,m}(\boldsymbol{r}) = 0 \tag{8.30}$$

while $\psi_{l,m}$ satisfies in addition

$$\boldsymbol{L}^2\,\psi_{l,m} = l(l+1)\,\psi_{l,m} \qquad l = 0, 1, 2, \ldots$$
$$L_z\,\psi_{l,m} = m\,\psi_{l,m} \qquad\quad m = -l, -l+1, \ldots, +l$$

With the Laplacian in spherical polar coordinates given by

$$\triangle = \frac{\partial^2}{\partial r^2} + \frac{2}{r}\frac{\partial}{\partial r} - \frac{\boldsymbol{L}^2}{r^2}$$

we obtain the radial Schrödinger equation in the form

$$\left(-\frac{\hbar^2}{2M}\left(\frac{\partial^2}{\partial r^2} + \frac{2}{r}\frac{\partial}{\partial r} - \frac{l(l+1)}{r^2}\right) - \frac{Ze^2}{4\pi r} - E\right)\Phi(r) = 0 \tag{8.31}$$

Its normalisable solutions are

$$\Phi(\boldsymbol{r}) = r^l e^{-\sqrt{2ME}\,r/\hbar} P_{n'}(r) \tag{8.32}$$

where $P_{n'}$ is a polynomial of degree n', while the corresponding energy eigenvalue is

$$E = -\frac{Mc^2(Z\alpha)^2}{2(n'+l+1)^2} \tag{8.33}$$

with α the fine structure constant

$$\alpha \equiv \frac{e^2}{4\pi\hbar c} \simeq \frac{1}{137}$$

Defining the quantum number $n = n' + l + 1$, we now see that the stationary solutions of the Schrödinger equation are functions $\psi_{n,l,m}(\boldsymbol{r})$, depending also on the principal quantum number n. They are eigenfunctions of $\boldsymbol{L}^2$, L_z and the Hamiltonian with energy

$$E_n = -\frac{Mc^2(Z\alpha)^2}{2\,n^2}\,,\qquad n = 1, 2, \ldots \tag{8.34}$$

The quantum numbers n and l are such that $n' = n - (l+1)$ is a non-negative integer. For a given value of n, this fixes the allowed values of l to be $l = 0, 1, 2, \ldots, n - 1$. Correspondingly, the degeneracy of the energy eigenvalue corresponding to n is given by

$$\sum_{l=0}^{n-1}(2l + 1) = n^2 \tag{8.35}$$

8.3.9.2 Hydrogenoid atoms in relativistic quantum mechanics. We next proceed with the study of the spectrum of hydrogenoid atoms in the framework of relativistic quantum mechanics. Specifically, we shall use the Dirac equation to derive the energy spectrum of a spin-$\frac{1}{2}$ particle with mass M and charge e in the Coulomb field of an infinitely heavy nucleus of charge $-Ze$. It is convenient to start with (8.13), which under the standard substitution $i\partial_\mu \to p_\mu$ takes the form

$$(\not{p} - e\,\not{A} - M)\psi = 0 \tag{8.36}$$

and multiply it by $\not{p} - e\,\not{A} + M$ to obtain in steps

$$\begin{aligned}
(\not{p} - e\,\not{A} + M)\ (\not{p} - e\,\not{A} - M)\,\psi &= \left((\not{p} - e\,\not{A})^2 - M^2\right)\psi \\
&= \left((i\partial - eA)^2 + \frac{1}{4}[\gamma^\mu, \gamma^\nu][i\partial_\mu - eA_\mu, i\partial_\nu - eA_\nu] - M^2\right)\psi \\
&= \left((i\partial - eA)^2 + e\sigma^{\mu\nu}F_{\mu\nu} - M^2\right)\psi
\end{aligned} \tag{8.37}$$

In the problem at hand we have $\boldsymbol{A} = 0$ and $A_0 = -\frac{Ze}{4\pi r}$, so that $F_{0i} = -F_{i0} = -\partial_i A_0 = E^i$. In the Weyl representation, we obtain

$$e\sigma^{\mu\nu}F_{\mu\nu} = \begin{pmatrix} -\mathrm{i}\,e\,\boldsymbol{\sigma}\cdot\boldsymbol{E} & 0 \\ 0 & \mathrm{i}\,e\,\boldsymbol{\sigma}\cdot\boldsymbol{E} \end{pmatrix}$$

where

$$\mathrm{i}\,e\,\boldsymbol{\sigma}\cdot\boldsymbol{E} = -\mathrm{i}\,Z\alpha\,\frac{\boldsymbol{\sigma}\cdot\hat{\boldsymbol{r}}}{r^2}$$

with $\hat{\boldsymbol{r}}$ denoting the unit vector along $\boldsymbol{r}$. Since (8.37) has a block diagonal form, we naturally look for the solution in the form

$$\psi = \begin{pmatrix} \psi_+ \\ \psi_- \end{pmatrix}$$

with ψ_+ and ψ_- two-component spinors, in which case equation (8.37) for stationary solutions with energy E reduces to the pair of independent spinorial equations

$$\left(-\frac{\partial^2}{\partial r^2} - \frac{2}{r}\frac{\partial}{\partial r} + \frac{\boldsymbol{L}^2 - Z^2\alpha^2 \mp \mathrm{i}Z\alpha\,\boldsymbol{\sigma}\cdot\hat{\boldsymbol{r}}}{r^2} - \frac{2Z\alpha E}{r} - (E^2 - M^2)\right)\psi_\pm(\boldsymbol{r}) = 0 \tag{8.38}$$

Spherical symmetry implies that the total angular momentum $\boldsymbol{J} = \boldsymbol{L} + \boldsymbol{\sigma}/2$ commutes with H, the Dirac Hamiltonian in the Coulomb potential, as well as with $\boldsymbol{L}^2$. Thus, the energy eigenstates can be classified with the quantum numbers of the total angular momentum $j = \frac{1}{2}, \frac{3}{2}, \ldots$, of its third component $j_3 = -j, -j+1, \ldots, +j$ and of the orbital angular momentum l, whose possible values for a given j are

$$l_\pm = j \pm \frac{1}{2}$$

Let us compute the energy eigenvalues E in (8.38) for given j and j_3, i.e. in the two-dimensional subspace of the Hilbert space spanned by the eigenvectors $\{|jj_3l_+\rangle, |jj_3l_-\rangle\}$ of $\boldsymbol{L}^2$, satisfying

$$\boldsymbol{L}^2 |jj_3l_\pm\rangle = l_\pm(l_\pm + 1) |jj_3l_\pm\rangle$$

$\boldsymbol{L}^2 - Z^2\alpha^2$ is a diagonal operator in this basis, with diagonal elements $l_+(l_+ + 1) - Z^2\alpha^2$ and $l_-(l_- + 1) - Z^2\alpha^2$. The operator $\boldsymbol{\sigma} \cdot \hat{\boldsymbol{r}}$, on the other hand, is Hermitian and traceless and its square is the unit operator $\boldsymbol{1}$. Moreover, its diagonal elements $\langle jj_3l_\pm | \boldsymbol{\sigma} \cdot \hat{\boldsymbol{r}} | jj_3l_\pm \rangle = 0$, since $\boldsymbol{\sigma} \cdot \hat{\boldsymbol{r}}$ is odd under parity, $\boldsymbol{r} \to -\boldsymbol{r}$. Thus,

$$\boldsymbol{\sigma} \cdot \hat{\boldsymbol{r}} = \begin{pmatrix} 0 & f \\ f^* & 0 \end{pmatrix}$$

where f is a phase. We conclude that in the basis $\{|jj_3l_+\rangle, |jj_3l_-\rangle\}$ the operators $\boldsymbol{L}^2 - Z^2\alpha^2 \mp iZ\alpha\boldsymbol{\sigma} \cdot \hat{\boldsymbol{r}}$ are represented by the 2×2 matrices

$$\boldsymbol{L}^2 - Z^2\alpha^2 \mp iZ\alpha\boldsymbol{\sigma} \cdot \hat{\boldsymbol{r}} = \begin{pmatrix} (j+\frac{1}{2})(j+\frac{3}{2}) - Z^2\alpha^2 & \mp if Z\alpha \\ \mp if^* Z\alpha & (j+\frac{1}{2})(j-\frac{1}{2}) - Z^2\alpha^2 \end{pmatrix}$$

These matrices have identical eigenvalues equal to

$$\left(j + \frac{1}{2}\right)^2 - Z^2\alpha^2 \pm \sqrt{\left(j + \frac{1}{2}\right)^2 - Z^2\alpha^2} \tag{8.39}$$

which can be written in the form $\lambda_a(\lambda_a + 1)$, $a = 1, 2$, with

$$\lambda_1 = \sqrt{\left(j + \frac{1}{2}\right)^2 - Z^2\alpha^2} \quad \text{and} \quad \lambda_2 = \lambda_1 - 1 \tag{8.40}$$

Having solved for the eigenvalues of the angular part of the operator in (8.38), we are left with the four radial equations

$$\left(-\frac{\hbar^2}{2M}\left(\frac{\partial^2}{\partial r^2} + \frac{2}{r}\frac{\partial}{\partial r} - \frac{\lambda_a(\lambda_a + 1)}{r^2}\right) - \frac{Z\alpha E}{Mr} - \frac{E^2 - M^2}{2M}\right) \Phi_\pm(r) = 0 \tag{8.41}$$

with the index $\pm$ being inherited from (8.38) and $a = 1, 2$.

With the substitutions

$$l(l+1) \to \lambda_a(\lambda_a+1), \quad \alpha \to \frac{\alpha E}{M}, \quad E \to \frac{E^2 - M^2}{2M} \tag{8.42}$$

separately for the two values λ_1 and λ_2, the corresponding operators are formally the same as in equation (8.31) of the non-relativistic case. We shall exploit this fact to derive the energy spectrum of interest here.

The parameters λ_a, $a = 1,2$ are not integers. They differ from the integers $l_{\pm}$, respectively, by the same quantity

$$\delta_j = l_+ - \lambda_1 = l_- - \lambda_2 \simeq \frac{Z^2\alpha^2}{2j+1} + \mathcal{O}(Z^4\alpha^4)$$

The substitutions (8.42) into the energy levels (8.33) lead to the two relations for the energy levels of (8.41) [7]

$$\frac{E^2 - M^2}{2M} = -\frac{MZ^2\alpha^2}{2}\frac{E^2}{M^2}\frac{1}{(n'+\lambda_a+1)^2} \tag{8.43}$$

one for each value of $a = 1,2$ and with n' non-negative integer. But, $n' + \lambda_1 + 1 = n' + j + \frac{3}{2} - \delta_j \equiv n - \delta_j$ with $n \equiv n' + j + \frac{3}{2} = 2,3,\ldots$ and $j \leq n - \frac{3}{2}$. Similarly, $n' + \lambda_2 + 1 = n' + j + \frac{1}{2} - \delta_j \equiv n - \delta_j$ with $n \equiv n' + j + \frac{1}{2} = 1,2,\ldots$ and $j \leq n - \frac{1}{2}$.

We substitute in (8.43) $n' + \lambda_a + 1 = n - \delta_j$ and solve for the energy to obtain finally for the positive energy spectrum

$$E_{nj} = \frac{M}{\sqrt{1 + (\frac{Z\alpha}{n-\delta_j})^2}} \simeq M - \frac{MZ^2\alpha^2}{2n^2} - \frac{MZ^4\alpha^4}{n^3(2j+1)} + \frac{3}{8}\frac{MZ^4\alpha^4}{n^4} + O(Z^6\alpha^6) \tag{8.44}$$

for $n = 1,2,\ldots$ and $j = \frac{1}{2}, \frac{3}{2}, \ldots, n - \frac{1}{2}$. Furthermore, for a given n the energy levels with $j \leq n - \frac{3}{2}$ are doubly degenerate, while the one with $j = n - \frac{1}{2}$ is non-degenerate.[8]

The difference between the non-relativistic and the relativistic cases is the occurrence of a *fine structure* that lifts the degeneracy of the j dependent levels, for a given value of n. Using the atomic physics notations, and for $Z = 1$, we have

$$E(2P_{\frac{3}{2}}) - E(2P_{\frac{1}{2}}) \simeq \frac{m\alpha^4}{32} = 4.53 \times 10^{-5} eV \tag{8.45}$$

for $n = 2$, $l = 1$ and $j = \frac{3}{2}$ or $j = \frac{1}{2}$.

One can refine further the relativistic correction by including the effect of the interaction between the spin of the nucleus and the spin of the electron. This means adding to the Hamiltonian a term of the form

$$\Delta H = -\frac{e}{2m_e}\boldsymbol{\sigma}_e \cdot \boldsymbol{B} \tag{8.46}$$

with $\boldsymbol{B}$ the magnetic field due to the proton magnetic moment. Since $\boldsymbol{B}$ is proportional to the magnetic moment of the proton $\mu_p \simeq g_p e/2m_p$, we obtain an additional

---

[7] Obviously, in discussing equations (8.39) or (8.40) we have assumed that $Z\alpha < 1$, which is true for all known stable atoms.

[8] Here we suppress the $2j + 1$ degeneracy due to the possible values of $j_3 = -j, -j+1, \ldots, +j$.

contribution which lifts the double degeneracy for $j_3 = \pm\frac{1}{2}$ of the ground state $1S_{\frac{1}{2}}$ by the amount

$$\Delta E = \frac{4}{3}m_e\alpha^4\frac{m_e}{m_p}g_p = 5,89\times 10^{-6}eV \tag{8.47}$$

This new effect is called the *hyperfine structure*. The new correction is proportional to the mass ratio m_e/m_p, which makes it smaller by an order of magnitude compared to the previous effect.

The following figure illustrates these corrections on the energy levels of the hydrogen atom. The figure also contains a further correction, the *Lamb shift*, due to the quantum field effect, which cannot be computed within the framework of the Dirac equation. It played a crucial role in the development of modern quantum field theory and we shall describe it briefly in Chapter 16.

8.4 Problems

Problem 8.1 We consider the following two-particle states (π is a pion and p is a proton): $|\pi^+\pi^-; l\rangle$, $|\pi^0\pi^0; l\rangle$, $|pp; l, s\rangle$, $|p\bar{p}; l, s\rangle$, $|p\pi; l\rangle$, where l and s denote the orbital angular momentum and the spin of the state, respectively. Give the results of the action of P and C on them.

Problem 8.2 Consider a relativistic particle of charge e and mass m which interacts with an electromagnetic field $A_\mu(x)$, and whose dynamics is described by the Dirac equation.

1. Show that the current density $j^\mu(x) = \bar{\psi}(x)\gamma^\mu\psi(x)$ is conserved.
2. Show that $j_\mu(x) = j_\mu^{(1)}(x) + j_\mu^{(2)}(x)$ with

$$j_\mu^{(1)} = \frac{i}{2m}\bar{\psi}\overleftrightarrow{\partial}_\mu\psi - \frac{e}{m}A_\mu\bar{\psi}\psi = \frac{i}{2m}\bar{\psi}\overleftrightarrow{D}_\mu\psi \quad\text{and}\quad j_\mu^{(2)} = \frac{1}{m}\partial^\nu\left(\bar{\psi}\,\sigma_{\nu\mu}\,\psi\right)$$

This identity is known as the Gordon decomposition of the current.

3. Assume that all the energies are small with respect to the rest energy of the particle. Give the non-relativistic limit of the two currents $j_{(1)}$ and $j_{(2)}$. Interpret the results. *(Hint: Use the fact that, classically, a magnetic dipole M generates an effective density of current $J_{eff} = \nabla\wedge M$).*

4. Express the part of the electromagnetic coupling $\int j_\mu^{(2)}A^\mu\,\mathrm{d}^4x$ in terms of the Maxwell tensor $F_{\mu\nu}$. Give the non-relativistic limit.

Problem 8.3 Compute the next order in the non-relativistic expansion of section 8.3.4, assuming for simplicity only electric external field, and show the appearance of a spin-orbit coupling.

Problem 8.4 *An electron in an external magnetic field. The Landau levels.*

We consider the motion of a relativistic electron in a constant homogeneous magnetic field $B = (0, 0, B_z)$.

I. The classical relativistic case.

I.1: Find the trajectory of the electron corresponding to the initial conditions at $t = 0$ x_0 and v_0.

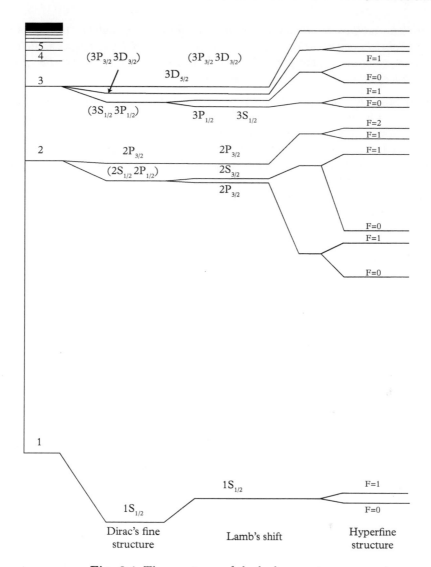

Fig. 8.1 The spectrum of the hydrogen atom.

I.2: Show that its projection on the xy plane is a circle and compute the coordinates of its centre.

II. The quantum relativistic case.

II.1: Solve the Dirac equation for an electron in a constant homogeneous magnetic field $\boldsymbol{B} = (0, 0, B_z)$. Find the energy levels for the relativistic version of the Landau levels we studied in non-relativistic quantum mechanics.

II.2: Find the degree of degeneracy for each one and interpret this degeneracy in terms of the symmetries of the problem.

II.3: Show that the two components of the velocity of the electron v_x and v_y do not commute.

Hint: The method is similar to the one we followed in solving the same problem using the Schrödinger equation. We choose a gauge in which the magnetic field is given by a vector potential of the form $\boldsymbol{A} = (-B_z y, 0, 0)$ and reformulate the problem in terms of an equation for a harmonic oscillator.

Problem 8.5 *The Klein paradox.* The purpose of this and the following problem is to show that it is impossible to interpret the Dirac equation as the quantum mechanical relativistic wave equation for a single particle.

Consider the one-dimensional motion of an electron in a space with a square potential barrier: $V(x) = 0$ for $x < 0$ and $V(x) = V > 0$ for $x > 0$. The incident electron has energy $0 < E < V$.

1. Solve the Dirac equation in the two regions $x < 0$ and $x > 0$ and impose the continuity of the solution at $x = 0$.

2. Compute the transmitted and the reflected waves and check the conservation of probabilities.

3. Assume $V < E + m$. Show that, as we expect intuitively, the transmitted wave decays exponentially in the classically forbidden region $x > 0$.

4. Repeat the calculation for $V \geq E + m$ and show that the behaviour of the transmitted wave becomes oscillatory.

5. In the same regime show also that the reflected flux becomes larger than the incident one.

6. Discuss the physical origin of these paradoxes.

This is a famous example of the problems we encounter when we try to localise an electron at distances comparable to its Compton wavelength. The physical explanation is that by interpreting the Dirac equation as the equation for a single particle, we neglect the effects due to electron–positron pair creation.

Problem 8.6 *Zitterbewegung.*[9] Consider the time evolution of the wave function of an electron represented by a wave packet built out of positive energy solutions of the Dirac equation. Choose at $t = 0$ a wave function in the form of a Gaussian of width d. Study two cases: (i) $d \gg 1/m$, the particle is localised over distances much larger than its Compton wavelength, and (ii) $d \leq 1/m$, the particle is localised over distances smaller than or equal to its Compton wavelength.

Discuss the difference between the two cases.

---

[9]The name comes from the German *zittern*-tremble and *bewegung*-motion

9

From Classical to Quantum Mechanics

9.1 Introduction

In Chapter 2 we introduced the principle of *canonical quantisation* in the form of a rule:

> *Given a classical theory formulated in terms of a set of canonical coordinates $q_a(t)$ and momenta $p_a(t)$, $a = 1, \ldots, N$, we obtain the corresponding quantum theory by postulating that the qs and the ps are operators acting in a certain Hilbert space and satisfying the equal time commutation relations (2.2).*

In Chapter 2 we applied this rule to the classical electromagnetic theory and obtained the quantum theory of radiation. We did not attempt to *justify*, let alone to *prove*, the rule. The rule *defines* the quantum theory. We can apply the same rule to other classical field theories such as those we studied in Chapter 7. Once this rule is imposed, all the results we obtained in Chapter 2 for the quantum electromagnetic field follow. We want to stress at this point that the canonical quantisation rule cannot be understood in any intuitively obvious way. It is justified only a posteriori because, by applying it, we obtain results related to physically measurable quantities which are in agreement with observation. We conclude that any other rule, provided it yields equivalent results, would be equally acceptable. In this chapter we want to present such an alternative quantisation rule. It has a long history but it was raised into the status of a quantum mechanical postulate by R.P. Feynman in 1948,[1] following an earlier proposal by Dirac. Compared to the canonical quantisation rule it offers several advantages: (i) It gives the same results in all cases in which both can be applied. (ii) It is more powerful and can be used in constrained systems in which the canonical rule may be problematic. (iii) It can be made relativistically covariant at any step, while the canonical quantisation, by imposing equal time commutation relations, distinguishes time and space coordinates. (iv) It gives a novel formulation of a quantum field theory suitable to numerical simulations. (v) It offers a new, and probably more profound, perception of quantum mechanics.

---

[1] R.P. Feynman, Rev. Mod. Phys. **20**, 367 (1948).

Elementary Particle Physics. John Iliopoulos and Theodore N. Tomaras, Oxford University Press.
© John Iliopoulos and Theodore N. Tomaras (2021). DOI: 10.1093/oso/9780192844200.003.0009

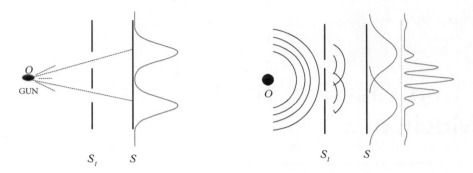

Fig. 9.1 The double slit experiment with bullets and waves. The bullets on the left do not form an interference pattern, the waves on the right do. The figure on the right for the waves with wavelength λ was drawn for slit width approximately equal to 2λ and slit distance about 6λ.

9.2   Quantum Mechanics and Path Integrals

This new rule was first introduced by Feynman in his famous article of 1948. It was further developed in a book published in 1965.[2] In this chapter we shall follow very closely Feynman's pedagogical presentation. Feynman built on the idea that the fundamental principle of quantum mechanics is the particle–wave duality. It is exemplified by the well-known double slit experiment, which he used to present the rule. This experiment is described in practically every textbook on quantum mechanics, so we only recall the results.

The experimental set-up is shown in Figure 9.1. It comprises a source of projectiles, which in our presentation will be successively classical corpuscles, classical waves and electrons, placed at point O. A screen, S_1 in the Figure placed at a distance z_1 from O, is supposed to be totally absorbent but has two holes drilled in it. Another screen S, placed at a distance z, is a "memory screen", one which records the trace of every impact.

• *The double slit experiment with bullets.* The source is a gun firing bullets. We count the impacts at a point A of screen S. We repeat the experiment with hole 1 closed and hole 2 open, the same with 1 open and 2 closed and, finally, with both holes open. The results can be summarised in two properties:

1) Each bullet produces a single impact on the screen S.

2) If we call P_{AO}^i, $i = 1, 2$ the density of impacts at point A when hole i alone is open and P_{AO}^{12} that with both holes open, we obtain

$$P_{AO}^{12} = P_{AO}^1 + P_{AO}^2 \tag{9.1}$$

• *The double slit experiment with waves.* The gun is replaced by a wave emitter. They can be light waves or material waves and lead to the well-known interference phenomena of classical optics. They are summarised as:

[2]R.P. Feynman and H.R. Hibbs, "Quantum Mechanics and Path Integrals", McGraw-Hill.

1) There is no individual impact on the screen S.

2) If we call $\mathcal{A}_{AO}^i$ the *amplitude* of the wave arriving at point A from point O with only the hole i open and $\mathcal{A}_{AO}^{12}$ that with both holes open, the observed interference pattern is described by the amplitude superposition rule:

$$\mathcal{A}_{AO}^{12} = \mathcal{A}_{AO}^1 + \mathcal{A}_{AO}^2 \tag{9.2}$$

- *The double slit experiment with electrons.*

At point O we place a radioactive source emitting electrons one at a time. Otherwise the experiment remains the same. The results are:

1) Each electron produces a single impact on the screen S.

2) At the screen S we observe interference patterns which follow the amplitude rule of equation (9.2).

We stress the point that the electrons are emitted separately one at a time and each time there is a single impact on S. The interference pattern is formed by accumulating a large number of impacts on the screen. Furthermore, if we try to determine which hole each electron went through by placing a detector next to it, we destroy the interference pattern.

If we want to describe the electrons using classical terms, we find that they exhibit a schizophrenic behaviour: when they are materialised as impacts on the screen they behave as classical point-like particles, but their propagation is described by a wave equation.

When Feynman described this experiment in 1948 it was meant to be a *gedanken experiment*, i.e. a thought experiment which cannot be realised in practice. However, with present techniques, the experiment has been performed repeatedly, varying the incident particles, and the measurements are in absolute agreement with the superposition principle of quantum mechanics. In Figure 9.2 we show the results with electrons obtained in a 1976 experiment which followed exactly the procedure described by Feynman. In 2019, a collaboration led by Markus Arndt of the University of Vienna observed interference patterns using molecular beams with molecules containing up to 2000 atoms each, the heaviest objects to date to exhibit matter–wave interference.[3] Beautiful results showing the universal validity of quantum mechanics.

Going back to our experiment, we can complicate the apparatus in several ways. For example, we can imagine drilling more holes on S_1. With k holes equation (9.2) is generalised as

$$\mathcal{A}_{AO}^{1...k} = \sum_{i=1}^{k} \mathcal{A}_{AO}^i \tag{9.3}$$

Drilling more and more holes results into taking away the screen. The sum in equation (9.3) is replaced by a two-dimensional integral over every point in the plane with coordinate z_1. We can also imagine putting a second screen at a point z_2 between z_1 and z with various holes on it. Repeating the argument we arrive at the sum over the amplitudes corresponding to any combination of holes on the successive screens.

---

[3]Yaakov Y. Fein et al. Nature Physics **15** 1242 (2019).

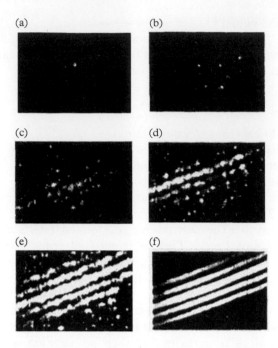

Fig. 9.2 The electron interference experiment. The figure shows a succession of pictures of the screen. We see that: (i) each electron produces a single impact on the screen and (ii) the interference pattern is built up by an accumulation of individual impacts. An impressive illustration of the quantum mechanical rule expressed by the Feynman postulate. *(Source: P.G. Merli et al. Am. J. Phys.* **44**, *306 (1976)).*

At the limit of an infinite number of screens between O and z and an infinite number of holes on each one of them, we express the amplitude superposition rule as follows:

Let us consider a time interval $[t_O, t_A]$. We call a *trajectory* joining the points O and A a function $\boldsymbol{x}_{AO}(t) \in \mathbb{R}^3$, $t_O \leq t \leq t_A$, such that $\boldsymbol{x}_{AO}(t_O) = O$ and $\boldsymbol{x}_{AO}(t_A) = A$. For the purposes of this discussion we consider the trajectories to be continuous and differentiable functions at every point in the interval $[t_O, t_A]$. We call $\mathcal{S}$ the set of all trajectories $\boldsymbol{x}_{AO}(t)$. Let $\mathcal{A}_{AO}[\boldsymbol{x}]$, a functional of $\boldsymbol{x}(t) : \mathcal{S} \to \mathbb{C}$, be the amplitude for the electron to go from O to A following the trajectory $\boldsymbol{x}(t)$. Generalising the superposition rule (9.3), the total probability amplitude can be formally written as

$$\mathcal{A}_{AO} = \int \mathcal{D}[\boldsymbol{x}] \mathcal{A}_{AO}[\boldsymbol{x}] \tag{9.4}$$

with $\mathcal{D}[\boldsymbol{x}]$ some appropriately defined measure in the set of all trajectories. This expression is only formal, first because we have not yet defined precisely the amplitude as a functional of the trajectory and, second, because we have not specified the integration measure. We will answer these questions presently.

9.2.1 The Feynman postulate

Let $L(\boldsymbol{x}, \dot{\boldsymbol{x}})$ be the classical Lagrangian describing the motion of the particle. For example, for a non-relativistic particle of mass m moving in a potential V we have $L = m\dot{\boldsymbol{x}}^2/2 - V$. The classical action corresponding to a trajectory $\boldsymbol{x}(t) = \boldsymbol{x}_{AO}(t)$, $t_O \leq t \leq t_A$, is given by

$$S[\boldsymbol{x}] = \int_{t_O}^{t_A} \mathrm{d}t\, L(\boldsymbol{x}, \dot{\boldsymbol{x}}) \tag{9.5}$$

Feynman postulated the following form for the amplitude $\mathcal{A}_{AO}[\boldsymbol{x}]$:

$$\mathcal{A}_{AO}[\boldsymbol{x}] = C \exp\left[\frac{\mathrm{i}}{\hbar} S[\boldsymbol{x}]\right] \tag{9.6}$$

with $C \in \mathbb{C}$ a constant independent of the trajectory $\boldsymbol{x}(t)$. For simplicity let us assume that the classical equations of motion admit a single solution with trajectory $\boldsymbol{x}^c(t)$. The generalisation to more complex situations is not difficult. Without loss of generality we can assume that the action vanishes for the trajectory $\boldsymbol{x}^c(t)$ since any constant value of S will give a phase which can be absorbed in the constant C. The rule expressed by (9.6) looks strange and counter-intuitive. Before discussing it let us first note that, inserted in equation (9.4), it has the correct classical limit when $\hbar \to 0$. This is heuristically obvious without knowing the precise form of the measure $\mathcal{D}[\boldsymbol{x}]$. Indeed, when $\hbar \to 0$ the contribution of every trajectory goes to zero *except* of those trajectories for which the phase is stationary with vanishing action. But this is precisely the trajectory which satisfies the classical equations of motion (3.1). The rule is nevertheless counter-intuitive, because we would expect, naïvely, trajectories close to the classical trajectory to give a larger contribution to the amplitude. But this is not the case with rule (9.6); the modulus of the amplitude of each trajectory, no matter how far from the classical one, is the same. In the next section we will prove that this rule correctly reproduces quantum mechanics, but here we see already that it offers a new perception of the quantum world. The concept of *a trajectory* makes sense only in the semi-classical approximation when $\hbar$ is very small. In the quantum regime a particle has no trajectory, only the phase of the amplitude depends on it and produces the interference patterns. Although this sounds very strange when we talk about particles, we want to remind the reader that it is the analogue of Huygen's principle which we know and apply in classical optics.

9.2.2 Recovering quantum mechanics

9.2.2.1 The wave function. $\mathcal{A}_{AO}$ has been defined as the amplitude of a conditional probability: the probability for a particle to be at point A at time t_A *provided* it was at point O at an earlier time t_O. Let us change slightly the notation in order to make this dependence explicit: we write $\mathcal{A}_{AO}$ as $\mathcal{A}(\boldsymbol{x}_A, t_A; \boldsymbol{x}_O, t_O)$ with $t_O < t_A$. Let t_B be some time in the interval $[t_O, t_A]$. Since the action (9.5) is additive $S_{AO} = S_{BO} + S_{AB}$, the amplitude satisfies the composition law

$$\mathcal{A}(\boldsymbol{x}_A, t_A; \boldsymbol{x}_O, t_O) = \int \mathrm{d}^3 \boldsymbol{x}_B \mathcal{A}(\boldsymbol{x}_A, t_A; \boldsymbol{x}_B, t_B) \mathcal{A}(\boldsymbol{x}_B, t_B; \boldsymbol{x}_O, t_O) \tag{9.7}$$

In order to simplify the notation, let us consider the limit $t_O \to -\infty$. We will call *the wave function* of the particle $\Psi(\boldsymbol{x}, t)$ the quantity

$$\Psi(\boldsymbol{x}, t) = \lim_{t_O \to -\infty} \int d^3\boldsymbol{x}_O \mathcal{A}(\boldsymbol{x}, t; \boldsymbol{x}_O, t_O) \qquad (9.8)$$

since it is, by construction, the probability amplitude to find the particle at the point $\boldsymbol{x}$ at time t, no matter where it was at some time in the remote past. Furthermore, $\Psi(\boldsymbol{x}, t)$, through the equations (9.4) and (9.6), can be used to predict the future state of the particle for any time larger than t. By the composition law (9.7), the wave function satisfies the integral equation

$$\Psi(\boldsymbol{x}, t) = \int d^3\boldsymbol{x}' \mathcal{A}(\boldsymbol{x}, t; \boldsymbol{x}', t') \Psi(\boldsymbol{x}', t') \qquad (9.9)$$

We will prove next that the wave function (9.8) satisfies Schrödinger's equation.

9.2.2.2 *The Gaussian functional measure.*
In order to proceed we will first adopt an engineering definition of the functional measure in equation (9.4), which means that we will generalise the measure of a Riemann integral without worrying about its rigorous mathematical existence. For simplicity, let us restrict ourselves to a particle moving in one space dimension. The generalisation to three dimensions will be straightforward.

Consider a division of the interval $[t_0, t]$ in n equal parts $t_0 < t_1 < t_2 < \cdots < t_n = t$, $t_i - t_{i-1} = \epsilon$, $n\epsilon = t - t_0$. The composition law (9.7) implies that the amplitude is given by

$$\mathcal{A}(x, t; x_0, t_0) = \int_{-\infty}^{+\infty} dx_1 \ldots dx_{n-1} \mathcal{A}(x, t; x_{n-1}, t_{n-1}) \ldots \mathcal{A}(x_1, t_1; x_0, t_0) \qquad (9.10)$$

We will assume that the Lagrangian is sufficiently regular and n sufficiently large, so that we can approximate the action between two neighbouring points as

$$\int_{t_i}^{t_{i+1}} dt L(x, \dot{x}, t) = \epsilon L\left(\frac{x_{i+1} + x_i}{2}, \frac{x_{i+1} - x_i}{\epsilon}, \frac{t_{i+1} + t_i}{2}\right) \qquad (9.11)$$

Under this approximation the amplitude will be given by

$$\mathcal{A}(x_{i+1}, t_{i+1}; x_i, t_i) = \frac{1}{N} \exp\left\{\frac{i\epsilon}{\hbar} L\left(\frac{x_{i+1} + x_i}{2}, \frac{x_{i+1} - x_i}{\epsilon}, \frac{t_{i+1} + t_i}{2}\right)\right\} \qquad (9.12)$$

N is a normalisation factor which is supposed to take into account the fact that we have an infinite number of possible trajectories. In general it will depend on ϵ and should guarantee the existence of the double limit $\epsilon \to 0$ and $n \to \infty$, thus determining the integration measure. We won't be able to find such an N for an arbitrary Lagrangian and we will proceed in steps. We see in (9.12) that, when $\epsilon \to 0$, the dominant terms come from $\dot{x}$, so we will first restrict ourselves to the case in which L is quadratic in

$\dot{x}$ and the potential is velocity independent. Let us recall some formulae related to Gaussian integrals. We start from the simplest one: For $a > 0$ we have

$$\int_{-\infty}^{+\infty} \mathrm{d}x\, e^{-ax^2} = \sqrt{\frac{\pi}{a}} \tag{9.13}$$

By analytic continuation we define, for $b > 0$,

$$\lim_{\delta \to 0+} \int_{-\infty}^{+\infty} \mathrm{d}x\, e^{(ib-\delta)x^2} = \int_{-\infty}^{+\infty} \mathrm{d}x\, e^{ibx^2} = \sqrt{\frac{i\pi}{b}} \tag{9.14}$$

Later we will give a geometric meaning to this particular continuation. Looking at (9.11) we see that, in order to control the kinetic energy part of the Lagrangian, it is sufficient to define the measure by choosing the normalisation factor N of (9.12) as

$$N(\epsilon) = \sqrt{\frac{2\pi i \hbar \epsilon}{m}} \tag{9.15}$$

The next step is to use this definition in order to derive Schrödinger's equation.

9.2.2.3 Schrödinger's equation for a particle in a potential. We consider the Lagrangian $L = \frac{1}{2}m\dot{x}^2 - V(x,t)$. We want to prove that the integral equation (9.9) implies that the wave function satisfies a partial differential equation. We start by computing the time derivative of $\Psi(x,t)$.

$$\frac{\partial \Psi(x,t)}{\partial t} = \lim_{\epsilon \to 0} \frac{\Psi(x, t+\epsilon) - \Psi(x,t)}{\epsilon} \tag{9.16}$$

Using formula (9.9), we compute $\Psi(x, t+\epsilon) - \Psi(x,t)$ up to $\mathcal{O}(\epsilon^2)$. Let us first rewrite the difference

$$\Psi(x, t+\epsilon) - \Psi(x,t) = \int \mathrm{d}y\, \mathcal{A}(x, t+\epsilon; y, t) \Psi(y,t) - \Psi(x,t) \tag{9.17}$$

$$= \int \mathrm{d}y \left(\frac{1}{N} e^{\frac{im}{2\hbar\epsilon}(x-y)^2} e^{-\frac{i\epsilon}{\hbar}V(\frac{x+y}{2}, t+\frac{\epsilon}{2})} \right) \Psi(y,t) - \Psi(x,t)$$

We have to compute $\frac{1}{N(\epsilon)} \int \mathrm{d}y\, e^{\frac{im}{2\hbar\epsilon}(x-y)^2} f(y)$. We find[4]

$$\frac{1}{N(\epsilon)} \int \mathrm{d}y\, e^{\frac{im}{2\hbar\epsilon}(x-y)^2} f(y) = f(x) - \frac{\hbar\epsilon}{2im} f''(x) + \mathcal{O}(\epsilon^2) \tag{9.18}$$

Applying this formula to the function

$$f(y) = e^{-\frac{i\epsilon}{\hbar}V(\frac{x+y}{2}, t+\frac{\epsilon}{2})} \Psi(y,t) \tag{9.19}$$

and keeping the terms of order at most ϵ, we get

---

[4]This shows that $\lim_{\epsilon \to 0+}(i\pi\epsilon)^{-1/2} \exp(ix^2/\epsilon)$ converges as a distribution to $\delta(x)$.

$$\frac{1}{N} \int \mathrm{d}y \, \mathrm{e}^{\frac{im}{2\hbar\epsilon}(x-y)^2} \mathrm{e}^{-\frac{i\epsilon}{\hbar} V\left(\frac{x+y}{2}, t+\frac{\epsilon}{2}\right)} \Psi(y,t) \tag{9.20}$$

$$= \Psi(x,t) - i\frac{\epsilon}{\hbar} V\left(x, t+\frac{\epsilon}{2}\right) \Psi(x,t) + \frac{i\hbar\epsilon}{2m} \frac{\partial^2}{\partial x^2} \Psi(x,t)$$

Collecting all these results we find

$$i\hbar \frac{\partial}{\partial t} \Psi(x,t) = \left(-\frac{\hbar^2}{2m} \frac{\partial^2}{\partial x^2} + V(x,t)\right) \Psi(x,t) = H\Psi(x,t) \tag{9.21}$$

We thus proved that:

The wave function $\Psi(x,t)$, defined by equation (9.8), is a solution of Schrödinger's equation.

In other words, Feynman's postulate, expressed through equations (9.4) and (9.6), supplemented with the Gaussian measure (9.15), reproduces Schrödinger's equation for the wave function. We conclude that the rule of path integration gives a definition of quantum mechanics which should be equivalent to that of Schrödinger, one postulate replacing another.

9.2.2.4 The harmonic oscillator. In this section we want to go one step further in the path integral formulation of quantum mechanics and show how we can actually compute with it the probability amplitudes. We choose the simple problem of the one-dimensional harmonic oscillator and compare the results obtained using the operator formalism of quantisation and the path integral one.

We start with the operator formalism. The Lagrangian L and the Hamiltonian H are given by

$$L = \frac{m}{2}\dot{q}^2 - \frac{m\omega^2}{2}q^2 \quad \text{and} \quad H = \frac{p^2}{2m} + \frac{m\omega^2}{2}q^2 \tag{9.22}$$

Let $\Psi \in L^2(\mathbb{R})$ be a solution of the Schrödinger equation. Since H does not depend on time, Ψ can be written as $\Psi = \mathrm{e}^{-iEt/\hbar}\Phi$ with $H\Phi = E\Phi$ and $\Phi \in L^2(\mathbb{R})$.

The eigenvalues of the Hamiltonian are $E_n = \hbar\omega(n+\frac{1}{2})$, $n = 0, 1, 2 \cdots$ and the corresponding wave functions Φ_n, of unit norm, are given by

$$\Phi_n(q) = \frac{1}{\sqrt{2^n n!}} \left(\frac{m\omega}{\hbar\pi}\right)^{1/4} \mathrm{e}^{-\frac{m\omega}{2\hbar}q^2} H_n\left(q\sqrt{\frac{m\omega}{\hbar}}\right) \tag{9.23}$$

where H_n are the so-called Hermite polynomials. The energy of the ground state is $E_0 = \hbar\omega/2$ and its wave function is $\Phi_0 = (\frac{m\omega}{\hbar\pi})^{1/4}\mathrm{e}^{-\frac{m\omega}{2\hbar}q^2}$.

Let us introduce the following operators acting on functions of $L^2(\mathbb{R})$:

$$a = \sqrt{\frac{m\omega}{2\hbar}}q + \sqrt{\frac{\hbar}{2m\omega}}\frac{\mathrm{d}}{\mathrm{d}q} , \quad a^\dagger = \sqrt{\frac{m\omega}{2\hbar}}q - \sqrt{\frac{\hbar}{2m\omega}}\frac{\mathrm{d}}{\mathrm{d}q} \tag{9.24}$$

From the preceding expressions, we check that

$$a\Phi_0 = 0 , \quad a\Phi_n = \sqrt{n}\,\Phi_{n-1} , \quad a^\dagger\Phi_n = \sqrt{n+1}\,\Phi_{n+1} \tag{9.25}$$

and we obtain the commutation relations

$$[a, a] = 0 \ , \qquad [a^\dagger, a^\dagger] = 0 \ , \qquad [a, a^\dagger] = 1 \tag{9.26}$$

In the ket formalism, the fundamental state Φ_0 is denoted by $|0\rangle$ and it is called the vacuum, or ground state; the n quantum state, or n excitation state, with wave function Φ_n, is denoted $|n\rangle$ and given by

$$|n\rangle = \frac{(a^\dagger)^n}{\sqrt{n!}} |0\rangle \tag{9.27}$$

with $\Phi_n(q)$ in (9.23).

The operators a and $a^\dagger$ are respectively the *annihilation and creation operators.* The Hamiltonian H and the excitation number operator N can be written as

$$H = \hbar\omega \left(N + \frac{1}{2} \right) \ , \qquad N = a^\dagger a \tag{9.28}$$

The states $|n\rangle$ are the common eigenstates of H and N:

$$N |n\rangle = n |n\rangle \ , \qquad H |n\rangle = \hbar\omega \left(n + \frac{1}{2} \right) |n\rangle$$

They form a complete orthonormal basis in $L^2(\mathbb{R})$. In particular, they satisfy the completeness relation

$$\sum_n |n\rangle \langle n| = 1 \tag{9.29}$$

We want to link the creator and annihilator formalism to the functional integral formalism. In the process we will complete the definition of the Gaussian functional measure we started in equation (9.15). An immediate problem we see is that in the first picture two operators in general do not commute. In the functional integral approach, however, we deal only with classical trajectories, and the notion of non-commutativity is not obvious. We shall explain at the end of this section that this question is related to the definition of the functional measure, but, for the moment, let us proceed intuitively. In the operator approach we introduce the notion of chronological product or T-product. For two operators $\mathcal{O}_1(t_1)$ and $\mathcal{O}_2(t_2)$ we define the product

$$T(\mathcal{O}_1(t_1)\mathcal{O}_2(t_2)) = \theta(t_1 - t_2)\mathcal{O}_1(t_1)\mathcal{O}_2(t_2) + \theta(t_2 - t_1)\mathcal{O}_2(t_2)\mathcal{O}_1(t_1) \tag{9.30}$$

which means that in the T-product we write the operator with the later time first. Let us choose the simple case of $\mathcal{O} = \hat{q}$, the position operator of the harmonic oscillator, and consider the vacuum expectation value of the T-product of two qs taken at different times[5]

---

[5]In the Schrödinger representation the operators $\hat{q}$ are time-independent while the states $|\Psi(\boldsymbol{x}, t)\rangle$ depend on t. We can reverse this dependence and go to the Heisenberg representation in which the opposite is true. The connection between the two is given by the time translation operator $\mathrm{e}^{-\mathrm{i}t\hat{H}/\hbar}$, as we explained in Chapter 2.

$$\langle 0| \, T(\hat{q}(t_1)\hat{q}(t_2)) \, |0\rangle \tag{9.31}$$

We first need to compute the value of the operator $\hat{q}(t)$ in terms of annihilators and creators.[6]

From $\hat{q} = \sqrt{\frac{\hbar}{2m\omega}}(a + a^\dagger)$, we get

$$\begin{aligned}
\hat{q}(t) &= e^{it\hat{H}/\hbar}\hat{q}\,e^{-it\hat{H}/\hbar} \\
&= \hat{q} + \frac{it}{\hbar}[\hat{H}, \hat{q}] + \frac{(it)^2}{2!\hbar^2}[\hat{H}, [\hat{H}, \hat{q}]] + \cdots \\
&= \sqrt{\frac{\hbar}{2m\omega}}(a\,e^{-i\omega t} + a^\dagger e^{i\omega t})
\end{aligned} \tag{9.32}$$

where we used

$$[\hat{H}, a] = -\hbar\omega a \qquad [\hat{H}, a^\dagger] = \hbar\omega a^\dagger \tag{9.33}$$

We thus obtain

$$\langle 0| \, \hat{q}(t_1)\hat{q}(t_2) \, |0\rangle = \frac{\hbar}{2m\omega}e^{-iw(t_1-t_2)} \tag{9.34}$$

Therefore

$$\begin{aligned}
\langle 0| \, T(\hat{q}(t_1)\hat{q}(t_2)) \, |0\rangle &= \frac{\hbar}{2m\omega}\left(\theta(t_1 - t_2)\,e^{-i\omega(t_1-t_2)} + \theta(t_2 - t_1)\,e^{i\omega(t_1-t_2)}\right) \\
&= \frac{i\hbar}{2\pi m}\int \frac{e^{-ik_0(t_1-t_2)}}{k_0^2 - \omega^2 + i\varepsilon}dk_0 \\
&= G_{\mathrm{F}}(t_1, t_2)
\end{aligned} \tag{9.35}$$

This is a Green function similar to the one we found in the study of the Klein–Gordon equation in Chapter 7 and we had called *the Feynman Green function*.[7] We want to reproduce this expression using the path integral formalism.

Up to now, we have defined the path integral by fixing the end points of the path over which we perform the integration, a condition which can be read directly on the Feynman postulate. Suppose now that we consider an integral corresponding to paths going from q at time t to q' at time t'.

Let us consider, formally, the eigenstates of the position operator $\hat{q}$: $\hat{q}\,|q\rangle = q\,|q\rangle$.[8]

---

[6] In order to simplify the notation we do not use the "hat" symbol when there is no danger of confusion.

[7] The variable in the Klein–Gordon equation is the field $\phi(t, \boldsymbol{x})$ which depends on four parameters, t and $\boldsymbol{x}$, the four dimensions of space-time. In quantum mechanics the variable q depends only on t. In this sense we often say that quantum mechanics is a one-dimensional theory.

[8] In standard quantum mechanics the position operator has a continuous spectrum, therefore the eigenstates are not square integrable functions. All formulae remain formally the same if in the orthogonality conditions the Kronecker delta is replaced by the Dirac delta function and summations over indices are replaced by integrals. If we wish to give a precise mathematical meaning to the states $|q\rangle$ we must go through the lattice approximation.

Since the kets $|q\rangle$ form a basis, we can define path integrals between any states; in particular, we can choose the vacuum state as the initial and the final state. Using the fact that

$$\int_{-\infty}^{+\infty} |q\rangle \langle q|0\rangle \, dq = |0\rangle \tag{9.36}$$

together with Feynman's time-splitting method which led to Schrödinger's equation, we define

$$\int \mathcal{D}[q]\mathrm{e}^{\mathrm{i}\int_t^{t'} L d\tau} = \int \mathrm{d}q'\mathrm{d}q\, \langle 0|q'\rangle \int_{\substack{q(t')=q' \\ q(t)=q}} \mathcal{D}[q]\mathrm{e}^{\mathrm{i}\int_t^{t'} L d\tau} \langle q|0\rangle \tag{9.37}$$

where $\langle q|0\rangle$ is the wave function $\Phi_0(q)$ of the harmonic oscillator ground state. With this definition,[9] the path integral in the middle of (9.37) can be written as

$$\int \mathcal{D}[q]\mathrm{e}^{\mathrm{i}\int_t^{t'} L d\tau} = \langle q' \ t'|q \ t\rangle = \langle 0| \, \mathrm{e}^{-\mathrm{i}(t'-t)\hat{H}/\hbar} \, |0\rangle = \mathrm{e}^{-\frac{\mathrm{i}\omega}{2}(t'-t)} \tag{9.38}$$

We now consider

$$\int \mathcal{D}[q]q(t_1)q(t_2)\mathrm{e}^{\mathrm{i}\int_t^{t'} L d\tau} \tag{9.39}$$

for $t_1, t_2 \in [t, t']$. Let us suppose that $t_1 > t_2$. Using the factorisation property of the probability amplitude (9.7), we interpret $q(t_i)$ as the multiplication by q at the time t_i and the preceding expression can be written as

$$\int \mathrm{d}q_1 \mathrm{d}q_2 \left\{ \left(\int_{q(t_1)=q_1} \mathcal{D}[q]\mathrm{e}^{\mathrm{i}\int_{t_1}^{t'} L d\tau} \right) q_1 \right. \tag{9.40}$$

$$\left. \cdot \left(\int_{\substack{q(t_1)=q_1 \\ q(t_2)=q_2}} \mathcal{D}[q]\mathrm{e}^{\mathrm{i}\int_{t_2}^{t_1} L d\tau} \right) q_2 \left(\int_{q(t_2)=q_2} \mathcal{D}[q]\mathrm{e}^{\mathrm{i}\int_t^{t_2} L d\tau} \right) \right\}$$

namely, thanks to the Feynman postulate

$$\int \mathrm{d}q'\mathrm{d}q \langle 0|\mathrm{e}^{-\mathrm{i}(t'-t_1)\hat{H}}|q\rangle q\langle q|\mathrm{e}^{-\mathrm{i}(t_1-t_2)\hat{H}}|q'\rangle \, q'\langle q'|\mathrm{e}^{-\mathrm{i}(t_2-t)\hat{H}} |0\rangle$$

$$= \int \mathrm{d}q'\mathrm{d}q \langle 0|\mathrm{e}^{-\mathrm{i}(t'-t_1)\hat{H}} q \, \mathrm{e}^{-\mathrm{i}(t_1-t_2)\hat{H}} q' \mathrm{e}^{-\mathrm{i}(t_2-t)\hat{H}} |0\rangle$$

$$= \langle 0|\mathrm{e}^{-\mathrm{i}t'\hat{H}} q(t_1) q'(t_2)\mathrm{e}^{\mathrm{i}t\hat{H}} |0\rangle$$

$$= \mathrm{e}^{-\frac{\mathrm{i}\omega}{2}(t'-t)} \langle 0|q(t_1)q(t_2) |0\rangle \tag{9.41}$$

Doing this calculation again but now reversing the time ordering, we get

$$\int \mathcal{D}[q]q(t_1)q(t_2)\mathrm{e}^{\mathrm{i}\int_t^{t'} L d\tau}$$

$$= \mathrm{e}^{-\frac{\mathrm{i}\omega}{2}(t'-t)} \left(\theta(t_1-t_2)\langle 0| \, q(t_1)q(t_2) |0\rangle + \theta(t_2-t_1)\langle 0| \, q(t_2)q(t_1) |0\rangle \right)$$

$$= \mathrm{e}^{-\frac{\mathrm{i}\omega}{2}(t'-t)} \langle 0| \, T(q(t_1)q(t_2)) |0\rangle \tag{9.42}$$

---

[9]We omit to indicate the boundary conditions on the trajectories in the vacuum-to-vacuum amplitude.

under the condition that t_1 and t_2 are between t' and t. We have therefore rigorously proven

$$\langle 0 | T(q(t_1)q(t_2)) | 0 \rangle = \lim_{t' \to +\infty, t \to -\infty} \frac{\int q(t_1)q(t_2)\mathrm{e}^{\mathrm{i} \int_t^{t'} L \mathrm{d}\tau} \mathcal{D}[q]}{\int \mathrm{e}^{\mathrm{i} \int_t^{t'} L \mathrm{d}\tau} \mathcal{D}[q]} \qquad (9.43)$$

the limit being trivial since, as long as $t < t_1, t_2 < t'$, the functional integral is independent of t and t'.

A few remarks before closing this section.

• We see in equation (9.43) that the 2-point function is expressed as a ratio of two path integrals. Naïvely we would expect the denominator to be equal to one since it represents the probability amplitude for the vacuum state to remain unchanged. But this forgets the fact that we are dealing with functional integrals in an infinite-dimensional space. The volume in the functional space of all trajectories connecting the vacuum states of $t = -\infty$ and $t' = +\infty$ is infinite. The denominator, which is a constant independent of t_1 and t_2, should be viewed as a normalisation factor whose presence is essential to absorb this divergence.

• It is obvious that we can extend the notion of the T-product to that of any number of operators. For n operators we define

$$T(\mathcal{O}_1(t_1)\mathcal{O}_2(t_2)\dots\mathcal{O}_n(t_n)) \qquad (9.44)$$

$$= \sum_\sigma \theta(t_{\sigma(1)} - t_{\sigma(2)})\dots\theta(t_{\sigma(n-1)} - t_{\sigma(n)})\, \mathcal{O}_{\sigma(1)}(t_{\sigma(1)})\, \mathcal{O}_{\sigma(2)}(t_{\sigma(2)})\dots\mathcal{O}_{\sigma(n)}(t_{\sigma(n)})$$

where σ is a permutation of $(1, 2, \dots, n)$ such that the times $t_{\sigma(j)}$, $j = 1, \dots, n$, are ordered: $t_{\sigma(1)} > \dots > t_{\sigma(n)}$.

This definition implies that the T-product is a symmetric function of the operators, that is, for any permutation σ

$$T(\mathcal{O}_1(t_1)\mathcal{O}_2(t_2)\dots\mathcal{O}_n(t_n)) = T(\mathcal{O}_{\sigma(1)}(t_{\sigma(1)})\mathcal{O}_{\sigma(2)}(t_{\sigma(2)})\dots\mathcal{O}_{\sigma(n)}(t_{\sigma(n)})) \qquad (9.45)$$

Using either of the two formalisms, the operator or the path integral one, we can compute the vacuum expectation value of the T-product of n operators $\hat{q}(t_1),\dots,$ $\hat{q}(t_n)$ of the harmonic oscillator. The relation between the two formalisms is the direct generalisation of (9.43), namely

$$\langle 0 | T(\hat{q}(t_1),\dots,\hat{q}(t_n)) | 0 \rangle = \frac{\int q(t_1)\dots q(t_n)\, \mathrm{e}^{\mathrm{i} \int_{-\infty}^{+\infty} L \mathrm{d}\tau} \mathcal{D}[q]}{\int \mathrm{e}^{\mathrm{i} \int_{-\infty}^{+\infty} L \mathrm{d}\tau} \mathcal{D}[q]} \qquad (9.46)$$

We call the expression $\langle 0 | T(\hat{q}(t_1),\dots,\hat{q}(t_n)) | 0 \rangle$ *the n-point function* of the operator $\hat{q}$. In Chapter 10 we shall give an explicit formula of this quantity in terms of products of Feynman Green functions.

• We can generalise the computation we just made for the operator $\hat{q}$ to any observable in the functional integral formalism. In the canonical quantisation it is represented by an operator $\hat{A}$ for which we can compute the eigenvalues and eigenfunctions. Using the completeness relation we can expand them in the basis of the eigenfunctions of the operator $\hat{q}$ and then use the formulae we just obtained.[10] It takes a bit more care to

[10]See, for example, the book of Feynman and Hibbs, already cited, for explicit calculations.

recover the notion of creation and annihilation operators whose eigenfunctions, as we have shown in Problem 2.3, are the coherent states which form an over-complete basis, but it can be done. This way we can establish the equivalence of the two pictures. We will not present all these calculations here and, for each particular physical problem, we will use the formalism which appears to be better adapted to it.

• The computation of the 2-point function of the operator $\hat{q}$ allows us to give a geometrical meaning to the particular functional measure we adopted in equation (9.15). The Gaussian integral (9.13) is well defined for real and positive a. Going to (9.14) required the particular limit $ib = \lim_{\delta \to 0+} i(b+i\delta)$. Note that had we chosen the limit from negative values of δ, the integral would diverge. In the functional integral the equivalent analytic continuation is the following: the action S is defined as the integral (9.5) from t_0 to t_A taken along the real time axis. It is an oscillating integral, like the one of (9.14), which should be defined through analytic continuation from a Gaussian one. The action of the harmonic oscillator, in particular, is obtained from the Lagrangian (9.22) which has the form $L = (m/2)(\dot{q}^2 - \omega^2 q^2)$ with positive m and ω^2. We conclude that the right analytic continuation is to set $t \to -i\tau$, in other words to start by integrating over imaginary times. By Fourier transform, this implies the continuation $k_0 \to ik_E$ in the expression (9.35), i.e. to imaginary energy. But this is precisely what we found in section 7.2.1 when we defined the Feynman Green function as the one obtained from Euclidean space through analytic continuation. We conclude that the concept of time ordering in the operator formalism is equivalent in the path integral formalism to the analytic continuation from Minkowski to Euclidean time defined by $t \to -i\tau$. In the physicists' jargon this rotation of the integration contour from real to imaginary times is called *Wick rotation*.

9.3 Problems

Problem 9.1 In section 9.2.2 we established the equivalence between the operator and the path integral formulations of quantum mechanics.

1. Derive the energy spectrum and the wave function of a harmonic oscillator ground state using the path integral formalism.

2. Consider a modification of the Hamiltonian (9.22) with the addition of an unharmonic term λq^4. Find the correction to the ground state energy at first order in λ.

10

From Classical to Quantum Fields. Free Fields

10.1 Introduction

In Chapter 9 we established an equivalence between the canonical quantisation and the one based on Feynman's path integral for a system with one degree of freedom. We saw that the two approaches share a common feature: they start from the corresponding classical formulation. In this chapter we want to extend this equivalence to systems with an infinite number of degrees of freedom, to field theories. We will first restrict ourselves to systems whose classical Lagrangian density is quadratic in its variables and we will leave the discussion of more complicated systems to Chapter 11.

Classical field theory can be thought of as the limit of classical mechanics to an infinite number of degrees of freedom $q_a(t)$. In Chapter 2 we showed a way to visualise this limit by means of the lattice approximation, in which the three-dimensional Euclidean space is replaced by a discrete set of N points. In the appropriate limit when $N \to \infty$ and the lattice spacing goes to zero, we replace $q_a(t)$, $a = 1, \ldots, N$, by a field $\phi(t, \boldsymbol{x})$. At every step of this limiting process we can apply either of the two quantisation rules and obtain the corresponding quantum theory. We see that they both involve subtle mathematical steps. In the canonical quantisation, the fields become operator valued distributions, and operations, such as the multiplication of two fields, are not always well defined. In the path integral quantisation the functional space, over which the integration measure should be defined, becomes that of all possible field configurations, in other words, from the space of the trajectories of a single particle we go to the space of the trajectories of an infinite number of particles. In this chapter we develop this programme for the simple field theories we studied in Chapter 7. We start with a generalisation of the Gaussian integrals that we introduced in section 9.2.

Let $X \in \mathbb{R}^N$ be a N-component real vector and A a $N \times N$ non-singular matrix with positive eigenvalues. The extension of the one-dimensional Gaussian integral (9.13) to N dimensions is

$$\int \mathrm{d}^N X \mathrm{e}^{-\frac{1}{2}(X, AX)} = (2\pi)^{\frac{N}{2}} (\det A)^{-\frac{1}{2}} \tag{10.1}$$

where $(X, AX) = X^T AX$. For notational simplicity let us absorb the factors of 2π into the measure: $(\mathrm{d}X) = \mathrm{d}^N X (2\pi)^{-\frac{N}{2}}$. If $J \in \mathbb{R}^N$ we get also

$$Z(J) \equiv \int (\mathrm{d}X) \, \mathrm{e}^{-\frac{1}{2}(X, AX) + J \cdot X} = \mathrm{e}^{\frac{1}{2}(J, A^{-1} J)} (\det A)^{-\frac{1}{2}} \tag{10.2}$$

Elementary Particle Physics. John Iliopoulos and Theodore N. Tomaras, Oxford University Press.
© John Iliopoulos and Theodore N. Tomaras (2021). DOI: 10.1093/oso/9780192844200.003.0010

where $J \cdot X = J^T X$. Knowing the function $Z(J)$ makes it possible to compute the integral of any polynomial $P(X)$ with the measure given by (10.1)

$$\int (\mathrm{d}X) P(X) \, \mathrm{e}^{-\frac{1}{2}(X,AX)} = P\left(\frac{\partial}{\partial J}\right) Z(J)\Big|_{J=0} \qquad (10.3)$$

When we integrate over vectors Y taking values in a complex space, $Y \in \mathbb{C}^N$, the integral (10.1) becomes

$$\int (\mathrm{d}Y^*)(\mathrm{d}Y) \, \mathrm{e}^{-(Y^*,AY)} = (\det A)^{-1} \qquad (10.4)$$

10.2 The Klein–Gordon Field

We start with the simplest field theory of a free, real, scalar field $\phi(x)$. The classical theory was presented in section 7.2. The dynamics is described by the Lagrangian density of equation (7.3)

$$\mathcal{L} = \frac{1}{2}\left(\partial_\mu \phi(x)\partial^\mu \phi(x) - m^2 \phi^2(x)\right) \qquad (10.5)$$

We shall study this theory using both the canonical and the path integral quantisation.

10.2.1 The canonical quantisation

10.2.1.1 The canonical commutation relations. The canonical variables corresponding to $q(t)$ and $p(t)$ of classical mechanics are $\phi(t, \boldsymbol{x})$ and $\pi(t, \boldsymbol{x}) = \dot{\phi}(t, \boldsymbol{x})$, where, as usual, "dot" means derivative with respect to t. According to the canonical quantisation rule, they are assumed to satisfy the equal time commutation relations (2.4)

$$[\phi(t, \boldsymbol{x}), \pi(t, \boldsymbol{y})] = \mathrm{i}\hbar\delta^3(\boldsymbol{x} - \boldsymbol{y}) \qquad (10.6)$$

with the remaining equal time commutators of two ϕs, as well as of two πs vanishing. As we did with the electromagnetic field, we obtain a simpler system by expanding the field in plane waves, solutions of the equation of motion (7.1), following the equations (7.7) and (7.8). In terms of the operators $a(\boldsymbol{p})$ and its Hermitian conjugate $a^\dagger(\boldsymbol{p})$, corresponding to the coefficient functions in the expansion of $\phi(x)$, the commutation relations (10.6) become

$$\left[a(\boldsymbol{p}), a^\dagger(\boldsymbol{p}')\right] = 2E_p(2\pi)^3\delta^3(\boldsymbol{p} - \boldsymbol{p}') \ , \ \ [a(\boldsymbol{p}), a(\boldsymbol{p}')] = \left[a^\dagger(\boldsymbol{p}), a^\dagger(\boldsymbol{p}')\right] = 0 \quad (10.7)$$

where $E_p = \sqrt{\boldsymbol{p}^2 + m^2}$. We see that in momentum space the commutation relations are diagonal in $\boldsymbol{p}$.

These relations make it possible to interpret the quantum field ϕ as describing an infinite set of harmonic oscillators, one for each value of the momentum $\boldsymbol{p}$, with $a^\dagger(\boldsymbol{p})$ and $a(\boldsymbol{p})$ the corresponding creation and annihilation operators.

10.2.1.2   The Fock space of states.   From this point on, the quantisation proceeds along the lines we set in section 2.2. We define *the vacuum state* $|0\rangle$ as the ground state of the system. It is annihilated by all annihilation operators $a(\boldsymbol{p})$. The generic state with n excitations is obtained by acting on the vacuum with n creation operators (see equation (2.30))

$$|\boldsymbol{p}_1, \boldsymbol{p}_2, \ldots, \boldsymbol{p}_n\rangle = a^\dagger(\boldsymbol{p}_n) \ldots a^\dagger(\boldsymbol{p}_2) a^\dagger(\boldsymbol{p}_1) |0\rangle \qquad (10.8)$$

As they are constructed, these states are not normalisable, but this can be easily remedied by using wave packets instead of plane waves (see equations (2.28) and (2.29)). Each one of the four-momenta $p_a, a = 1, 2, \ldots, n$, satisfies the dispersion relation $(p_a^0)^2 = (\boldsymbol{p}_a)^2 + m^2$, which means that it is concentrated on the hyperboloid $p_a^2 = m^2$. This in turn makes it possible to interpret the excitations as *particles* of mass m. The fact that creation operators commute implies that the n-particle state is completely symmetric under the exchange of any two of them, in other words the corresponding particles are *bosons*.

The Fock space of states is the one given in equation (2.31), with the same notation but without the sum over polarisations, which do not exist for scalar fields, namely

$$\mathcal{F} = \sum_{n=0}^{\infty} \oplus \mathcal{H}_n \qquad (10.9)$$

10.2.1.3   Renormalised energy and momentum operators. Normal ordering.   In computing the energy and momentum operators H and $\boldsymbol{P}$ in terms of creation and annihilation operators we encounter the same difficulty we found in section 2.2, namely that the vacuum expectation values of H and $\boldsymbol{P}$ diverge due to the sum over the zero-point energies of an infinite number of harmonic oscillators. We discussed this problem from both the mathematical and the physical points of view in section 2.2 and the conclusions remain the same: we are led to introduce the *renormalised* expressions for the total energy and momentum operators

$$H_{\text{ren}} = \int d\Omega_m \, E_p \, a^\dagger(\boldsymbol{p}) \, a(\boldsymbol{p}) \; ; \quad \boldsymbol{P}_{\text{ren}} = \int d\Omega_m \, \boldsymbol{p} \, a^\dagger(\boldsymbol{p}) \, a(\boldsymbol{p}) \qquad (10.10)$$

which differ from the original, un-renormalised expressions by infinite additive constants.[1] In section 2.2 we have also introduced the concept of *normal ordering*, which amounts to writing any product of creation and annihilation operators by placing all annihilation operators to the right of all creation operators. It follows that the vacuum expectation value (v.e.v.) of a normal-ordered operator vanishes identically. We had denoted the normal-ordered expression of an operator $\mathcal{O}$ by $:\mathcal{O}:$ and we see that

$$H_{\text{ren}} = :H: \quad \text{and} \quad \boldsymbol{P}_{\text{ren}} = :\boldsymbol{P}: \qquad (10.11)$$

---

[1] For the momentum operator of equation (10.10) this additive constant turns out to be equal to zero by symmetric integration over the momentum $\boldsymbol{p}$, but we prefer to keep the notation of the renormalised expression which will be useful later.

10.2.1.4 Causality. Given two operators $\mathcal{O}_1(x_1)$ and $\mathcal{O}_2(x_2)$, which are functions of the field $\phi(x)$ and its derivatives, we can compute their commutator using the basic commutation relations (10.7). Let us do it explicitly for the simplest example, $[\phi(x), \phi(y)]$. It will be convenient to separate the positive and negative frequency parts of the field by writing respectively

$$\phi^{(+)}(x) = \int d\Omega_m \, a(\boldsymbol{p}) \, e^{-i(E_p t - \boldsymbol{p} \cdot \boldsymbol{x})} \quad \text{and} \quad \phi^{(-)}(x) = \int d\Omega_m a^\dagger(\boldsymbol{p}) \, e^{i(E_p t - \boldsymbol{p} \cdot \boldsymbol{x})} \quad (10.12)$$

$\phi^{(+)}(x)$ contains the annihilation operators and $\phi^{(-)}(x)$ the creation ones. We obtain

$$[\phi(x), \phi(y)] = [\phi^{(+)}(x), \phi^{(-)}(y)] + [\phi^{(-)}(x), \phi^{(+)}(y)] \qquad (10.13)$$
$$= D(x - y) - D(y - x) \equiv i\Delta(x - y)$$

where the function $D(x)$ is defined in (7.20). For the real distribution $\Delta(x)$ we find

$$\Delta(x) \equiv i \int \frac{d^3p}{(2\pi)^3 2E_p} \left(e^{i(E_p t - \boldsymbol{p} \cdot \boldsymbol{x})} - e^{-i(E_p t - \boldsymbol{p} \cdot \boldsymbol{x})} \right) = \frac{-i}{(2\pi)^3} \int d^4p \, \delta(p^2 - m^2) \, \varepsilon(p_0) \, e^{-ip \cdot x}$$
$$(10.14)$$

We can check easily that if $x^0 = 0$ the two terms contributing to $\Delta(x)$ are equal, so that $\Delta(0, \boldsymbol{x}) = 0$. Since Δ is Lorentz invariant, this means that $\Delta(x) = 0$ for space-like four-vectors x, that is to say for $x^2 < 0$. It follows that fields at different space-like separated points commute. This is the manifestation of causality for a scalar quantum field theory.

10.2.1.5 Time ordering. Propagators. We have seen already that the product of two local operators $\mathcal{O}_1(x_1)$ and $\mathcal{O}_2(x_2)$ is often ill defined when $x_1 = x_2$. Normal ordering is a particular prescription to define it by assigning the value zero to its vacuum expectation value. It applies to operators which are functions of the free field and consequently can be expressed in terms of the creation and annihilation operators. In discussing the harmonic oscillator in Chapter 9 we introduced a different ordering called *the time-ordered*, or *T-product* which applies to any time-dependent operators. The same computation which we performed for the harmonic oscillator gives

$$\langle 0 | \, T(\phi(x)\phi(y)) \, | 0 \rangle = \theta(x^0 - y^0) \, \langle 0 | \, \phi(x)\phi(y) \, | 0 \rangle + \theta(y^0 - x^0) \, \langle 0 | \, \phi(y)\phi(x) \, | 0 \rangle$$
$$= G_{\text{F}}(x - y) \qquad (10.15)$$

with $G_{\text{F}}(x-y)$ the Feynman Green function given in equation (7.19). This computation shows also the physical meaning of the time-ordered product. Remember that $\phi^{(-)}$ creates a particle and $\phi^{(+)}$ destroys it. It follows that the two terms in the expression of the 2-point function describe the amplitude for the following process: (i) when $x^0 > y^0$, creation of a particle at the point y and its subsequent destruction at the point x, and (ii) when $y^0 > x^0$, creation of a particle at the point x and its subsequent destruction at the point y. In both cases it gives the probability amplitude for a particle to move from one point to another. For this reason the 2-point function is called *the propagator*.

10.2.1.6 Several scalar fields. Dealing with N real, independent scalar fields presents no novelty. The field operators acquire an index $\phi_a(x), a = 1, 2, \ldots, N$ and the commutation relation (10.7) becomes

$$\left[a_a(\boldsymbol{p}), a_b^\dagger(\boldsymbol{p}') \right] = 2\,\delta_{ab}\, E_p^a (2\pi)^3 \delta^3(\boldsymbol{p} - \boldsymbol{p}') \tag{10.16}$$

with $E_p^a = \sqrt{\boldsymbol{p}^2 + m_a^2}$, where m_a is the mass of ϕ_a, and with all other commutators equal to zero.

From the notational point of view, the case with $N = 2$ with equal masses, written as a complex field $\phi = (\phi_1 + i\phi_2)/\sqrt{2}$, is interesting. The Lagrangian density and the plane wave expansions are given in equations (7.27) and (7.29). The commutation relations become

$$\left[a(\boldsymbol{p}), a^\dagger(\boldsymbol{p}') \right] = 2E_p (2\pi)^3 \delta^3(\boldsymbol{p} - \boldsymbol{p}')\,, \quad \left[b(\boldsymbol{p}), b^\dagger(\boldsymbol{p}') \right] = 2E_p (2\pi)^3 \delta^3(\boldsymbol{p} - \boldsymbol{p}') \tag{10.17}$$

with all other commutators, including those of as with bs, equal to zero. Note that the Lagrangian density $\mathcal{L}$ is invariant under the $U(1)$ transformations $\phi(x) \rightarrow e^{i\theta}\phi(x)$. It follows that the 2-point functions of either ϕ or $\phi^\dagger$ vanish. By explicit calculation we can verify that

$$\langle 0 | \, T(\phi(x)\phi(y)) \, |0\rangle = \langle 0 | \, T(\phi^\dagger(x)\phi^\dagger(y)) \, |0\rangle = 0$$

and

$$\langle 0 | \, T(\phi(x)\phi^\dagger(y)) \, |0\rangle = G_{\mathrm{F}}(x - y) \tag{10.18}$$

10.2.2 The path integral quantisation

Since a free scalar field can be viewed as an infinite set of harmonic oscillators, we expect the methods we developed in section 9.2.2, appropriately generalised, to give the corresponding quantum field theory. The latter is described by variables $\phi(x)$, $x \in \mathbb{M}^4$. Following the discussion of section 9.2.2, we start by performing a Wick rotation in which the four-dimensional Minkowski space is mapped into the four-dimensional Euclidean space $\mathbb{M}^4 \rightarrow \mathbb{E}^4$. Then we consider a lattice of N^4 points in $\mathbb{E}^4$ with lattice spacing a. The fields $\phi(x)$, $x \in \mathbb{E}^4$, become variables ϕ_i, $i = 1, \ldots, N^4$. The functional integral over ϕ becomes an ordinary N^4-dimensional integral. The next step is to study the limit $a \rightarrow 0$ with $Na \rightarrow \infty$. Finally, we must choose a direction x_4 in $\mathbb{E}^4$ and perform an analytic continuation $x_4 \rightarrow it$ to recover the Minkowski theory.

Let us apply this programme to the Klein–Gordon field. The Euclidean action is given by

$$S_E[\phi] = \int \mathrm{d}^4 x_E \left[\frac{1}{2} \phi(x) \square_E \phi(x) - V(\phi) \right] \tag{10.19}$$

where $\square_E = \sum_{\mu=1}^{4} (\partial^2/\partial x_\mu^2)$. In general, we will assume the potential V to be a polynomial in ϕ, but here we will take it to be equal to $\frac{1}{2}m^2\phi^2$. On the lattice the action takes the form

$$S_E(\phi) = a^4 \sum_i \left[\phi_i \sum_{\mu=1}^4 \frac{\phi_{i+\hat{\mu}a} - 2\phi_i + \phi_{i-\hat{\mu}a}}{a^2} - V(\phi_i) \right] \qquad (10.20)$$

where $\hat{\mu}$ denotes the unit vector in the direction μ. On the lattice the functional integral reduces to a multiple ordinary integral

$$\mathcal{D}[\phi] \to \prod_i \mathrm{d}\phi_i \qquad (10.21)$$

where the product extends over all lattice points, so we must perform an N^4-dimensional integral, each one from $-\infty$ to $+\infty$.

If $V(\phi) = \frac{1}{2}m^2\phi^2$ everything we did in section 9.2.2 applies and we can define the continuum limit $a \to 0$ by an appropriate choice of the normalisation factor $N(a)$, the way we did for the one-dimensional case in equations (9.15) and (9.43). Looking at equations (10.1) and (10.2) we see that for the scalar field of equation (10.19) the matrix A is given by $A = (m^2 - \Box_E)$. In $\mathbb{E}^4$ the d'Alembertian has a continuous spectrum, which means that the determinant diverges in the limit $a \to 0$. It is this divergence which we must absorb in the integration measure. The normalisation factor N which we introduced in equation (9.15) becomes now

$$\int e^{-\int \mathrm{d}^4 x_E \frac{1}{2}\phi(x)(m^2 - \Box_E)\phi(x)} \mathcal{D}[\phi] = \left(\det\left(m^2 - \Box_E\right) \right)^{-\frac{1}{2}} \qquad (10.22)$$

This leads to the following definition of the *Gaussian* functional measure $\mathrm{d}\mu_G$ for the scalar field in Euclidean space: given a functional $F[\phi]$ we *define*

$$\int F[\phi]\mathrm{d}\mu_G \equiv \frac{\int F[\phi]\, e^{-\int \mathrm{d}^4 x_E \frac{1}{2}\phi(x)(m^2 - \Box_E)\phi(x)}\, \mathcal{D}[\phi]}{e^{-\int \mathrm{d}^4 x_E \frac{1}{2}\phi(x)(m^2 - \Box_E)\phi(x)}\, \mathcal{D}[\phi]} \qquad (10.23)$$

At the end of the calculation an analytic continuation to Minkowski space can be performed.

From this point on the formulae are analogous to the ones we obtained for the harmonic oscillator. Specifically, for the 2-point function in Minkowski space we obtain

$$G_{\mathrm{F}}(x_1 - x_2) = \langle 0|\, T(\phi(x_1)\phi(x_2))\, |0\rangle = \frac{\int \phi(x_1)\phi(x_2)\, e^{\mathrm{i}S[\phi]}\, \mathcal{D}[\phi]}{\int e^{\mathrm{i}S[\phi]}\, \mathcal{D}[\phi]} \qquad (10.24)$$

In fact, this is expected on the basis of equations (10.2) and (10.3): the 2-point function is the inverse of the operator $\mathrm{i}(\Box + m^2)$ of the quadratic form, which, by definition, is the Green function. Generalising equation (10.2), we can define a generating functional $Z[J]$ as

$$Z[J] = \frac{\int e^{\mathrm{i}S[\phi] + \mathrm{i}\int \mathrm{d}^4 x J(x)\phi(x)}\, \mathcal{D}[\phi]}{\int e^{\mathrm{i}S[\phi]}\mathcal{D}[\phi]} \qquad (10.25)$$

The n-point function $W(x_1, \ldots, x_n) \equiv \langle 0|\, T(\phi(x_1) \ldots \phi(x_n))\, |0\rangle$ is given by

$$W(x_1, \ldots, x_n) = \frac{\int \phi(x_1) \ldots \phi(x_n) e^{iS[\phi]} \mathcal{D}[\phi]}{\int e^{iS[\phi]} \mathcal{D}[\phi]} = (-i)^n \left. \frac{\delta^n Z[J]}{\delta J(x_1) \ldots \delta J(x_n)} \right|_{J=0} \tag{10.26}$$

It is an easy exercise to show the following two results:

- The $(2k+1)$-point function vanishes for all k.[2]
- The $2k$-point function is given by the sum of all possible products of k Feynman propagators.

For example, the explicit form of the 4-point function is

$$\langle 0| \, T(\phi(x_1)\phi(x_2)\phi(x_3)\phi(x_4)) \, |0\rangle = G_F(x_1 - x_2)G_F(x_3 - x_4)$$
$$+ G_F(x_1 - x_3)G_F(x_2 - x_4) + G_F(x_1 - x_4)G_F(x_2 - x_3) \tag{10.27}$$

10.3   The Dirac Field

We now move to the Dirac field which we studied in section 7.3. The field is a four-component complex spinor $\psi(x)$ whose dynamics is described by the free Dirac action (7.61):

$$S = \int d^4x \left(i\overline{\psi} \, \partial\!\!\!/ \, \psi - m\overline{\psi}\psi \right) \tag{10.28}$$

The plane wave solutions are given in equations (7.85) and (7.86).

10.3.1   The canonical quantisation

10.3.1.1   The canonical (anti-)commutation relations.   In the previous section we showed that a free scalar field can be expanded in terms of an infinite set of creation and annihilation operators satisfying canonical commutation relations, see equations (10.7). Two seemingly unrelated properties follow from them: The first is that the corresponding Hamiltonian is positive definite. The second is that a state which contains two excitations, one with momentum $\boldsymbol{k}_1$ and a second with momentum $\boldsymbol{k}_2$, is symmetric under the exchange of the two momenta; in other words, scalar fields describe bosons.

In this section we will extend these considerations when quantising a Dirac field $\psi(x)$. Heisenberg's canonical quantisation recipe tells us that we should interpret the complex functions $a_\alpha(\boldsymbol{k})$ and $b_\alpha(\boldsymbol{k})$, $\alpha = 1, 2$, given in equations (7.85) and (7.86) as two pairs of creation and annihilation operators satisfying canonical commutation relations. At first sight the differences between the Dirac and the scalar field expansions seem to be unimportant. The fact that we have two pairs of creation and annihilation operators is just a consequence of the fact that the Dirac field is complex. But then we see immediately that we are in trouble. We already found a first sign of it in Chapter 7, when we attempted to apply the canonical formalism at the classical level. We recall that the Dirac equation is obtained from a variational principle starting from the Lagrangian density of equation (7.60), with the fields $\psi(x)$ and $\overline{\psi}(x)$ as the

---

[2]This result can be derived immediately as a consequence of the invariance of the action (10.5) under the transformation $\phi \leftrightarrow -\phi$.

independent dynamical variables. We can compute the Hamiltonian and express it in terms of a_α and b_α. A straightforward calculation gives, for the normal ordered form,

$$: H := \int \mathrm{d}\Omega_m E_k \sum_{\alpha=1}^{2} \left[a_\alpha^\dagger(\boldsymbol{k}) a_\alpha(\boldsymbol{k}) - b_\alpha^\dagger(\boldsymbol{k}) b_\alpha(\boldsymbol{k}) \right] \qquad (10.29)$$

The trouble comes from the minus sign. No matter how we arrange an overall sign, this Hamiltonian will be unbounded from below. The two kinds of excitations, those created by the $a_\alpha^\dagger$ and the $b_\alpha^\dagger$ operators, have opposite energies. This is not a new problem. It is the one we encountered in analysing the energy spectrum of the Hamiltonian, when we were considering the Dirac equation as the wave equation for an electron. The solution we adopted there was to assume that electrons are fermions and obey the Pauli exclusion principle. We thus assumed, following Dirac, that all negative energy states are occupied. This was meant to be an ad hoc principle inspired by the phenomenology of atomic spectra. But here we do not have that option. If the a_α and b_α operators satisfy canonical commutation relations the same argument we developed for the scalar field shows that a two-excitation state is always symmetrical; in other words the excitations are bosons.

Can we modify the commutation relations in order to describe fermions? Let us consider a doubly excited state. For fermions we must have

$$|\boldsymbol{k}_1, \alpha; \boldsymbol{k}_2, \beta\rangle = a_\alpha^\dagger(\boldsymbol{k}_1) a_\beta^\dagger(\boldsymbol{k}_2) |0\rangle = -a_\beta^\dagger(\boldsymbol{k}_2) a_\alpha^\dagger(\boldsymbol{k}_1) |0\rangle \qquad (10.30)$$

This suggests that the operators $a^\dagger$ must satisfy an *anti-commutation* relation of the form $a_\alpha^\dagger(\boldsymbol{k}_1) a_\beta^\dagger(\boldsymbol{k}_2) + a_\beta^\dagger(\boldsymbol{k}_2) a_\alpha^\dagger(\boldsymbol{k}_1) = 0$. This implies, in particular, that the square of any such operator vanishes, for example, $a_\alpha^\dagger(\boldsymbol{k}_1) a_\alpha^\dagger(\boldsymbol{k}_1) = 0$. This is in agreement with Pauli's exclusion principle, which forbids two fermions to occupy the same quantum state. We conclude that the entire Heisenberg recipe should be rewritten with anti-commutators replacing commutators. Notice that this also fixes the energy crisis because in defining the normal ordering we obtain an extra minus sign: $: bb^\dagger := -b^\dagger b$. As we said in section 6.3, this was the solution first presented by Fermi in 1933 in his famous article on β-decay, one of the founding papers of quantum field theory. It was the first time that the formalism of creation and annihilation operators was used for particles other than the photon and, as we just explained, it involved a highly non-trivial conceptual step. It was also the first example of what is known as *the spin-statistics theorem* which we will see again in Chapter 11. Positivity of the energy requires that half-odd-integer spin particles obey the Fermi–Dirac statistics. Their quantisation requires the use of anti-commutation relations instead of commutation relations, which are used for integer spin particles. This rule is not obvious either from the point of view of classical physics or from the elementary formalism of quantum mechanics. Indeed, a quantum theory in the limit $\hbar \to 0$ is supposed to give the corresponding classical theory. This is true for scalar fields. When $\hbar \to 0$ the field operators $\phi(x)$ become classical c-number functions $\phi : \mathbb{M}^4 \to \mathbb{R}$. However, if the Dirac fields satisfy anti-commutation relations, at the limit $\hbar \to 0$ they will become "classical anti-commuting numbers", something we never encountered in studying classical physics. There exists a precise mathematical formalism, called *Grassmann algebra*, which correctly describes what we should call *a classical fermion*, but we shall use a very simplified form of it in this book.

For the Dirac field $\psi(x)$ and its conjugate $\overline{\psi}(x)$ whose plane wave expansions are given in equations (7.85) and (7.86), we are thus led to postulate the anti-commutation relations

$$\left\{a_\alpha(\mathbf{k}), a_\beta^\dagger(\mathbf{k}')\right\} = (2\pi)^3 2\, E_k\, \delta^3(\mathbf{k} - \mathbf{k}')\, \delta_{\alpha\beta} \tag{10.31}$$

$$\left\{a_\alpha(\mathbf{k}), a_\beta(\mathbf{k}')\right\} = \left\{a_\alpha^\dagger(\mathbf{k}), a_\beta^\dagger(\mathbf{k}')\right\} = 0 \tag{10.32}$$

$$\left\{b_\alpha(\mathbf{k}), b_\beta^\dagger(\mathbf{k}')\right\} = (2\pi)^3 2\, E_k\, \delta^3(\mathbf{k} - \mathbf{k}')\, \delta_{\alpha\beta} \tag{10.33}$$

$$\left\{b_\alpha(\mathbf{k}), b_\beta(\mathbf{k}')\right\} = \left\{b_\alpha^\dagger(\mathbf{k}), b_\beta^\dagger(\mathbf{k}')\right\} = 0 \tag{10.34}$$

with the anti-commutators of bs with as vanishing. We see from these expressions that the field ψ annihilates an electron and creates a positron, while the field $\overline{\psi}$ creates an electron and annihilates a positron. The field ψ, being linear in the creation and annihilation operators, inherits the same anti-commutation properties. In particular, field operators inside a time-ordered, or normal-ordered product, anti-commute, that is

$$T(\psi(x_1)\psi(x_2)) = -T(\psi(x_2)\psi(x_1)), \quad :\psi(x_1)\psi(x_2): = -:\psi(x_2)\psi(x_1): \tag{10.35}$$

The 2-point function $\langle 0|\, T(\psi(x)\overline{\psi}(y))\, |0\rangle$ is given by the Dirac propagator (7.69), i.e.

$$\langle 0|\, T(\psi(x)\overline{\psi}(y))\, |0\rangle = \widetilde{S}_\mathrm{F}(k) = \mathrm{i}\frac{\slashed{k} + m}{k^2 - m^2} \tag{10.36}$$

Naturally, operators which are bilinears in the field behave as bosonic operators. For example, if $\mathcal{O}(x) = \overline{\psi}(x)\psi(x)$, we have

$$T(\mathcal{O}(x_1)\mathcal{O}(x_2)) = T(\mathcal{O}(x_2)\mathcal{O}(x_1)), \quad :\mathcal{O}(x_1)\mathcal{O}(x_2): =: \mathcal{O}(x_2)\mathcal{O}(x_1): \tag{10.37}$$

Finally, note that the anti-commutation property is important in order to satisfy the causality condition. We find

$$\{\psi_\alpha(x), \overline{\psi}_\beta(y)\} = \frac{1}{(2\pi)^3} \int \frac{\mathrm{d}^3 k}{2\omega_k} \left(\slashed{k} - m)\mathrm{e}^{\mathrm{i}k\cdot(x-y)} + (\slashed{k} + m)\mathrm{e}^{-\mathrm{i}k\cdot(x-y)}\right)_{\alpha\beta}$$
$$= (\mathrm{i}\,\slashed{\partial} + m)_{\alpha\beta}\, \triangle(x - y) \tag{10.38}$$

where $\triangle(x - y)$ is the causal Green function introduced in formula (10.14).

10.3.2 The path integral quantisation

Describing fermions via path integrals presents some novel features. The reason is the peculiar property of fermion fields in the classical limit to take values in a Grassmann algebra and not in an algebra of real or complex numbers. This means that we must redefine the rules of differential and integral calculus. We shall present here a simple way to introduce such operations.

A finite-dimensional Grassmann algebra is defined as the set of all formal power series of n generators $x_1, x_2, ..., x_n$ which anti-commute

$$\{x_i, x_j\} \equiv x_i x_j + x_j x_i = 0 \tag{10.39}$$

The coefficients of the power series are complex numbers. The anti-commutation rule (10.39) ensures that all x_i satisfy $x_i^2 = 0$, therefore, a power series contains at most 2^n terms. In other words, a Grassmann algebra can be viewed as a linear vector space of 2^n dimensions. Similarly, given a function $f(z), z \in \mathbb{R}^n$ which admits a power series expansion around the origin, we can define through this power series the function $f(x_1, x_2, ..., x_n)$. Obviously, any such function is in fact a polynomial.

We shall need two operations on a Grassmann algebra: differentiation from the left and integration. Since any function is a polynomial, it is sufficient to define these operations on an arbitrary monomial.

A left derivative $\mathrm{d}/\mathrm{d}x$ is defined by

$$\frac{\mathrm{d}}{\mathrm{d}x_i} c = 0 \ , \qquad \frac{\mathrm{d}}{\mathrm{d}x_i} x_j = \delta_{ij} - x_j \frac{\mathrm{d}}{\mathrm{d}x_i} \ , \qquad \left\{ \frac{\mathrm{d}}{\mathrm{d}x_i}, \frac{\mathrm{d}}{\mathrm{d}x_j} \right\} = 0 \tag{10.40}$$

where c is a complex number.

The concept of integration we shall need is that of a definite integral from $-\infty$ to $+\infty$ in real numbers. The important property of the latter is translational invariance. Therefore, following F.A. Berezin, we define integrals of functions on a Grassmann algebra by imposing invariance under translations. Again, it is enough to give the definitions for all monomials. For example, for an algebra with only one generator x, the most general function is of the form $f(x) = c_1 + c_2 x$ and translational invariance of the integral means

$$\int f(x) \mathrm{d}x = \int f(x + a) \mathrm{d}x \tag{10.41}$$

and suggests the definitions

$$\int \mathrm{d}x = 0 \ , \qquad \int x \mathrm{d}x = 1 \tag{10.42}$$

The generalisation to arbitrary n is obvious. For a function

$$f(x_1, \ldots, x_n) = c_0 + c_i x_i + c_{ij} x_i x_j + \ldots + c_{1 \ldots n} x_1 \ldots x_n \tag{10.43}$$

we find

$$\int f(x_1, x_2, \ldots, x_n) \mathrm{d}x_1 \ldots \mathrm{d}x_n = c_{1 \ldots n} \tag{10.44}$$

i.e. the integral of a function is the coefficient of the highest term in the power series expansion. It is in this sense that we often say that integration in a Grassmann algebra corresponds, in fact, to differentiation. An immediate consequence of the definitions (10.42) is that when we perform a change of variables in an integral of the form $x' = cx$, with c a number, the differential changes as $\mathrm{d}x = c\mathrm{d}x'$. As a result, after integration of a Gaussian form over fermion fields, the determinant of the operator of the quadratic form appears in the numerator rather than the denominator, i.e.

$$\int \mathrm{d}^N X \mathrm{e}^{-\frac{1}{2}(X,AX)} = (\det A)^{\frac{1}{2}} \qquad (10.45)$$

if the Xs are Grassmannian variables.

A final remark: a given Grassmann algebra with $2n$ generators may admit an operation ( )* (involution) with the properties

$$(f^*)^* = f \quad ; \quad (f_1 f_2)^* = f_2^* f_1^* \quad ; \quad (f_1 + f_2)^* = f_1^* + f_2^* \quad ; \quad (cf)^* = \bar{c} f^* \qquad (10.46)$$

where $\bar{c}$ is the complex conjugate of c. A function f which satisfies $f^* = f$ is called "real". A basis that satisfies $(x_{n+k})^* = x_k \quad k = 1, ..., n$ is called involutory. This means that if $\psi(x)$ is an element of a Grassmann algebra $\overline{\psi}(x)$ can be considered as another, independent element. The Grassmann algebra we use to describe a Dirac field is involutory. Therefore, the operation we performed in section 7.3.7, which consisted in varying the Dirac action independently with respect to $\psi(x)$ and $\overline{\psi}(x)$, is legitimate. For an involutory algebra the integration formula (10.45) can be rewritten as

$$\int \mathrm{d}\overline{X} \mathrm{d}X \mathrm{e}^{-(\overline{X}, MX)} = \det M \qquad (10.47)$$

Taking into account these particular features of Grassmann algebras, we can reproduce, using the Feynman path integral quantisation, all the results of fermionic quantum field theories.

10.4 The Maxwell Field

10.4.1 The canonical quantisation

We have already studied the quantum field description of the electromagnetic field in Chapter 2. We used the non-covariant Coulomb gauge $\boldsymbol{\nabla} \cdot \boldsymbol{A} = 0$, which has the advantage of exhibiting explicitly the physical degrees of freedom. We introduced creation and annihilation operators for the transversely polarised photons and we constructed the corresponding Fock space of states. The drawback of the method is the lack of explicit Lorentz covariance. The transversality condition refers to a particular frame and as long as we do not perform Lorentz transformations, the free quantum electromagnetic field behaves like a pair of massless scalars.

However, as we pointed out in section 7.4, explicit Lorentz invariance is very useful in calculations, so in this section we will develop the quantisation formalism for the covariant Lorenz gauge $\partial_\mu A^\mu = 0$. Using this gauge condition we cannot eliminate the redundant degrees of freedom in a covariant way, so we have to work keeping all four components of A^μ. The Lagrangian density reduces to

$$\mathcal{L} = -\frac{1}{2} \partial_\mu A_\nu \partial^\mu A^\nu \qquad (10.48)$$

Each component of the field satisfies an independent, massless, Klein–Gordon equation, but with one important difference: Because of the Minkowski metric, the term for A_0 in the Lagrangian has sign opposite to that of the three A_i. We ignore this problem for the moment and proceed as we did with the Klein–Gordon field. We expand in plane waves and use the equation we wrote in (7.106) with the normalisation of the four unit vectors $\epsilon_\mu^{(\lambda)}(\boldsymbol{k})$ given by (7.105). The quantisation procedure leads to the canonical commutation relations

$$\left[a^{(\lambda)}(\boldsymbol{k}), a^{(\lambda')\dagger}(\boldsymbol{k}')\right] = -2E_k(2\pi)^3\,\eta^{\lambda\lambda'}\delta^3(\boldsymbol{k}-\boldsymbol{k}') \qquad (10.49)$$

with $E_k = |\boldsymbol{k}|$. The minus sign and the $\eta^{\lambda\lambda'}$ on the right-hand side are necessary in order to satisfy the causality equation (10.54), which we shall derive shortly. The usual convention implies that $a^{(1)\dagger}$ and $a^{(2)\dagger}$ are the creation operators of transverse photons, while $a^{(0)\dagger}$ and $a^{(3)\dagger}$ are the creation operators of what we call *scalar* and *longitudinal* photons, respectively. Correspondingly, we denote the annihilation operators by $a^{(\mu)}, \mu = 0, 1, 2, 3$. The Fock space appears to be that of four types of excitations, twice as many as the ones we obtained in the Coulomb gauge.

So far, so good. However, we see immediately that a new problem arises: The commutation relations for the scalar photons, $\lambda' = \lambda = 0$, as written in equation (10.49), have the opposite sign from those of the other three. This in turn implies that some states have negative norm! Indeed, for the one scalar photon state, we find

$$|\Psi_1\rangle = \int \mathrm{d}\Omega_0 f(\boldsymbol{k}) a^{(0)\dagger}(\boldsymbol{k})\,|0\rangle\ , \qquad \langle\Psi_1|\Psi_1\rangle = -\int \mathrm{d}\Omega_0\,|f(\boldsymbol{k})|^2 \leq 0 \qquad (10.50)$$

We should emphasise that this difficulty is the reflection of the Minkowski metric of space-time and cannot be eliminated by any redefinition of the fields. As a consequence, the renormalised Hamiltonian is

$$H_{\mathrm{ren}} = \int \mathrm{d}\Omega_0 E_k \left[\sum_{\lambda=1}^{3} a^{(\lambda)\dagger}(\boldsymbol{k})a^{(\lambda)}(\boldsymbol{k}) - a^{(0)\dagger}(\boldsymbol{k})a^{(0)}(\boldsymbol{k})\right] \qquad (10.51)$$

and the energy is not positive definite.

A formal way to describe this situation is to say that in the Fock space of states the scalar product is defined with an indefinite metric of the form

$$(\Psi, \eta\Phi)\,, \text{with}\ \eta\,|\Phi\rangle = (-)^s\,|\Phi\rangle \qquad (10.52)$$

where s is the number of scalar photons in the state $|\Phi\rangle$. The metric η is Hermitian, since it satisfies $\eta^2 = 1$ and $\eta = \eta^\dagger$. Furthermore, the creation and annihilation operators for all four polarisations satisfy

$$(\Psi, a^{(\lambda)}(\boldsymbol{k})\Phi) = (a^{(\lambda)\dagger}(\boldsymbol{k})\Psi, \Phi) \qquad (10.53)$$

which means that the field $A_\mu(x)$ is a self-adjoint operator.

This metric was introduced by S.N. Gupta and K. Bleuler in 1950 and their approach is called *the Gupta–Bleuler formalism*. We could give a more detailed mathematical description of the resulting Fock space with this indefinite metric but, in this book, we will never need anything beyond the straightforward application of formula (10.52).

We still have the problem of the existence of too many states in the Fock space. From our experience with the Coulomb gauge we expect some of them to be unphysical. Can we separate them? In other words, can we find a subspace of the large Fock space, which contains only physical, positive norm, states? And, *last but not least*, can we obtain this separation in a Lorentz covariant way? Of course, we could apply the gauge condition $\partial_\mu A^\mu = 0$, which is Lorentz covariant, but here we are facing a new problem: The canonical quantisation yields the commutation relation

$$[A_\mu(x), A_\nu(y)] = \eta_{\mu\nu}\triangle(x - y) \tag{10.54}$$

where $\triangle(x - y)$ is given by the formula (10.14) and exhibits the causal properties of the theory. Can we apply the Lorenz condition $\partial_\mu A^\mu = 0$ as an operator equation? In other words, can we assume that

$$\partial_\mu A^\mu(x) |\Psi\rangle = 0 \tag{10.55}$$

for *all* states $|\Psi\rangle$ in the Fock space? The answer is no, as we can easily convince ourselves by acting with $\partial/\partial x_\mu$ on both sides of the commutation relation (10.54). We obtain

$$[\partial^\mu A_\mu(x), A_\nu(y)] = \partial_\nu\triangle(x - y) \tag{10.56}$$

and the right-hand side is, obviously, non-zero. There is, therefore, an inconsistency in trying to quantise the photon field in a Lorentz covariant way, while at the same time imposing the Lorenz condition. We overcome this problem by choosing to impose the Lorenz condition in a weaker sense, namely we split the operator ∂A into positive and negative frequency parts, and we define the physical states by:
    A state $|\Psi\rangle$ is physically acceptable if

$$\partial^\mu A_\mu^{(+)}(x) |\Psi\rangle = 0 \tag{10.57}$$

where $(+)$ indicates that we only consider the positive frequency part of A_μ.
    In the reference frame we used to define the basic polarisation vectors in section 7.4.3, this condition is equivalent to

$$(a^{(3)}(\boldsymbol{k}) - a^{(0)}(\boldsymbol{k})) |\Psi\rangle = 0 \tag{10.58}$$

on the physical states.[3] Notice that this separation into "physical" and "unphysical" states is Lorentz covariant because it is based on the condition (10.57), which is valid in any frame.
    The condition (10.57) is obviously satisfied by the vacuum state. It is also satisfied by the one-particle states with transverse photons. In fact, it is satisfied by any state

---

[3]One may wonder whether for an interacting field, the splitting of frequencies into positive and negative ones is meaningful. The answer is yes, because, from the equations of motion, if the electromagnetic current is conserved, the divergence of the field $\partial \cdot A$ satisfies the equation

$$\Box\, (\partial \cdot A) = 0$$

and is, therefore, a free field.

which has only transverse photons. So, the states which we expect, intuitively, to be physical, indeed satisfy this condition. However, and this comes as a surprise, there are other states, which also satisfy the condition and are, according to this definition, "physical". A simple example is given by the one-photon state, which is a superposition of the form $|\Psi\rangle = (a^{(0)\dagger}(\boldsymbol{k}) + a^{(3)\dagger}(\boldsymbol{k}))\,|0\rangle\,/\sqrt{2}$, i.e. the superposition of a one scalar and a one longitudinal photon states of the same momentum. In a similar way we can construct superpositions of states with two, three or more scalar and longitudinal photons, which satisfy the condition. It is easy to show that any such state, which has equal numbers of longitudinal and scalar photons has zero norm. We suspect that such states, although they are classified as "physical" according to our definition (10.57), will not correspond to physically realisable states, but we cannot discuss this point before we have a more precise definition of what is "physically realisable".

Let us summarise: We quantised the free electromagnetic field using various gauge conditions. The canonical quantisation recipe applies to all of them. In each case we obtain a system of massless bosons. However, in the field theory picture and the particle picture, we found totally different quantum field theories: the space of states is different, the Green functions are different. We return to the question we have asked several times before: In what sense can we say that all these field theories describe the same underlying physical theory? We expect the answer to be of the following form: They all must give the same result when we compute "physically measurable quantities". So, the question is to define which quantities are physically measurable. We cannot answer this question if we restrict ourselves to free fields, since any physical measurement requires an interaction of the system with the measuring apparatus. Therefore, we should postpone this discussion, until the interacting theory is studied. Here, let us only explain the physical origin of the problem. For the description of spin-1 particles, Lorentz invariance imposes the use of vector fields, which have four components. On the other hand Poincaré invariance tells us that there exist only two independent massless states with helicity equal to one. It is this mismatch which leads to several descriptions with redundant variables and the need to prove their physical equivalence. Let us also mention that, with increasing spin, this mismatch becomes worse and with it, the difficulty to define a physically meaningful theory.

10.4.2 The path integral quantisation

We have learned that the transition from the classical to the quantum theory is conveniently encoded into the calculation of a Euclidean functional integral. For the electromagnetic field this amounts to the calculation of a generating functional of the form

$$Z[J] = \frac{\int \mathcal{D}[A]\,\mathrm{e}^{-S_{\mathrm{inv}}[A]+\int \mathrm{d}^4 x A\cdot J}}{\int \mathcal{D}[A]\,\mathrm{e}^{-S_{\mathrm{inv}}[A]}} \tag{10.59}$$

where, $S_{\mathrm{inv}}[A]$ is the Maxwell gauge invariant action proportional to the integral of $F_{\mu\nu}^2$. However, we see immediately that this expression is ill-defined. The reason is again gauge invariance. If we denote, symbolically, the gauge transformation as $A \to {}^{\Omega}A$, with Ω an element of the gauge group, both the numerator and the denominator of

(10.59) remain invariant. A simple way to visualise the problem is to consider an integral $\int \mathrm{d}x \mathrm{d}y \, Q(x, y)$ from minus infinity to plus infinity of a function Q of two variables x and y. If Q is translational invariant it is, in fact, a function of only the difference $x - y$. Therefore the integral contains an infinite factor given by $\int \mathrm{d}(x + y)$, the volume of the translation group. Similarly, the expressions in (10.59) contain the volume of the gauge group. Recall that, although the underlying group of electromagnetic gauge invariance is $U(1)$, the gauge potentials take values in the Lie algebra of the group which is an infinite-dimensional space. We say often, in a formal sense, that the electromagnetic gauge group is $U(1)^\infty$, since it is as if we had a $U(1)$ factor at every point in space.[4] One could argue that this is not a serious problem, since we expect the factors to cancel between the numerator and the denominator, and here we want to investigate under which conditions this is indeed the case. For this, let us try to impose a gauge fixing condition of the form $G[A] = 0$ where G is a functional of A satisfying the condition we set in equation (7.93). We want to separate the measure of the functional integral into a product of two factors, one $\mathcal{D}[\Omega]$ on the group and a second $\mathcal{D}[\tilde{A}]$ on the space of gauge potentials satisfying the gauge condition. It will be convenient to use a more general gauge condition of the form $G[A] = C(x)$, with C any given function of x, and then average over all functions C with a Gaussian measure. Consider one of the two integrals, for example the denominator D of (10.59). We write

$$D = \int \mathcal{D}[A] \mathrm{e}^{-S_{\mathrm{inv}}[A]} = \int \mathcal{D}[A] \mathcal{D}[C] \mathcal{D}[G] \, \mathrm{e}^{-S_{\mathrm{inv}}[A] - \frac{1}{2} \int \mathrm{d}^4 x C^2(x)} \delta[G - C] \quad (10.60)$$

where we have introduced a functional delta function which has no effect, since we integrate over all gauge conditions given by functions G, and the Gaussian integral over C is just a multiplicative constant. The discussion is only formal, but we could make it more precise by formulating the theory on the points of a finite Euclidean space lattice.

Looking at the expression (10.60) we see that we can use the delta function to get rid of the integration over C. Then we change variables from G to $\Omega = \theta(x)$, the gauge function. This is, in principle, possible, since we assumed that the gauge condition has a unique solution as an equation for the element of the gauge group.[5] Therefore, taking into account the invariance of the action under gauge transformations, we find

$$D = \int \mathcal{D}[\Omega] \, \mathcal{D}[A] \, \mathrm{e}^{-S_{\mathrm{inv}}[A] - \frac{1}{2} \int \mathrm{d}^4 x G^2[A]} \det \left(\frac{\delta G}{\delta \Omega} \right) \quad (10.61)$$

For the class of gauge conditions we are considering, such as $G = \partial A$, the Jacobi determinant is given by $\det \square$. It is a divergent quantity but we shall never have to compute it. The only thing that matters here is that it is a constant, independent of the vector potential A. So it can be taken out of the integral. The final result is that the denominator D is given by

---

[4]Invariance under a compact group of global transformations does not create any difficulties because the corresponding volume is finite. This is the case, for example, with the $O(4)$ invariance in Euclidean space.

[5]We will not discuss what happens when this assumption is not satisfied.

$$D = K \int \mathcal{D}[A] \, e^{-S_{\text{inv}}[A] - \frac{1}{2} \int d^4 x G^2[A]} \qquad (10.62)$$

with K some infinite constant. The same constant will appear in the expression for the numerator of (10.59). So, the net result of all this gauge fixing procedure is that, for the family of such gauges, it is enough to change the Lagrangian density and use an *effective* one in which we have added the term G^2, in accordance with the formula we used in (7.99). What happens when we use more complicated gauge conditions will be discussed in Chapter 13.

We see that the effective field theory is gauge dependent, since the Lagrangian depends on the arbitrary gauge fixing function G. This is not a trivial dependence since, as we have seen already, the Green functions depend on it. Even if we restrict ourselves to the very special family of $G \propto \partial A$, the Green functions are given by (7.100) and they depend on the arbitrary parameter λ. It seems that we are obtaining an infinite family of different field theories and this is indeed the case. In which sense do all these theories describe the quantum version of Maxwell's electrodynamics? We can answer this question in two steps: In section 10.4.1 we introduced the Fock space of states which is associated to each one of these field theories and tried to define a physical subspace. In Chapter 12 we shall argue that observable quantities are indeed gauge independent.

10.5 Massive Spin-1 Fields

They are described by the Proca equation (7.101). As long as the mass m is different from zero, the quantisation of such theories in either the canonical, or the path integral method, presents no difficulties. The space of physical states is the Fock space of particles with mass m, each one coming in three possible spin states.

10.6 Problems

Problem 10.1 Gaussian integrals: Let $\boldsymbol{X}$ and $\boldsymbol{J}$ be vectors in an N-dimensional Euclidean space and A an $N \times N$ positive definite matrix. We define the generating function $Z(\boldsymbol{J})$ and the "n-point function" by

$$Z(\boldsymbol{J}) = \frac{\int d\boldsymbol{X} e^{-(\boldsymbol{X}, A\boldsymbol{X}) + \boldsymbol{J} \cdot \boldsymbol{X}}}{\int d\boldsymbol{X} e^{-(\boldsymbol{X}, A\boldsymbol{X})}} \quad ; \quad G_{i_1 i_2 \ldots i_n} = \frac{\partial^n Z(\boldsymbol{J})}{\partial J_{i_1} \partial J_{i_2} \ldots \partial J_{i_n}} \bigg|_{J=0}$$

1. Show that all the $(2n+1)$-point functions vanish.
2. Express the $2n$-point functions in terms of 2-point functions.

Problem 10.2 Derive the relation between the imaginary part of the two-point function $G_F(x-y)$ of a free massive scalar field and the function $\Delta(x-y)$ given in (10.14).

Problem 10.3 In Problem 7.3 we computed the transformation properties of the five Dirac bilinears under restricted Lorentz transformations and space inversions. In this problem we propose to also establish the transformation properties under charge conjugation $\mathcal{C}$ and time reversal $\mathcal{T}$. The answer for the three discrete transformations $\mathcal{P}$, $\mathcal{C}$ and $\mathcal{T}$ is summarised in Table 10.1.

Table 10.1 The transformation properties of the five Dirac bilinears under the discrete transformations $\mathcal{P}$, $\mathcal{C}$ and $\mathcal{T}$. The i denotes the total number of space indices for the γ matrices: $i = 0$ for γ^0 and $+1$ for the other three. We have included the result for the product $\mathcal{CPT}$

	S	V	T	A	P
$\mathcal{P}$	$+1$	$(-)^i$	$(-)^i$	$(-)^{i+1}$	-1
$\mathcal{C}$	$+1$	-1	-1	$+1$	$+1$
$\mathcal{T}$	$+1$	$(-)^i$	$(-)^{i+1}$	$(-)^i$	-1
$\mathcal{CPT}$	$+1$	-1	$+1$	-1	$+1$

11
Interacting Fields

11.1 Introduction – The Axioms of Quantum Field Theory

The great advantage of the free field theories we have studied so far is their simplicity. They yield linear equations of motion which admit plane wave solutions. Their great disadvantage is again their simplicity. They cannot describe any interesting observable phenomena which necessarily imply some interaction among physical systems. We are thus forced to complicate the picture and study non-linear equations of motion, which do not admit plane waves as solutions.

When we wrote the linear wave equations in Chapter 7 we started by considering the most general form compatible with the symmetries we wanted to impose. Obviously this approach is not sufficient for non-linear equations and we must make assumptions to restrict the class of possible theories. In this section we want to stay as general as possible and make only a minimum set of "physically reasonable" assumptions. For a theory with a single scalar field ϕ (generalisations to include more fields with various spins is straightforward), these assumptions include:[1]

- *The physical states.* As it is usually done in a quantum theory, we assume that they are the unit rays of a Hilbert space $\mathcal{H}$ defined on the complex numbers.
- *The quantum fields* are assumed to be appropriately defined operators acting on $\mathcal{H}$.

By "appropriately defined" we mean the following: we have seen already that even free fields should be viewed as operator valued distributions. This means that their correct definition requires the introduction of a set of appropriate test functions and, in quantum field theory, we choose $f \in \mathcal{S}$, the space of C^∞ rapidly decreasing functions. This axiom extends this property to interacting fields. It states that the operator

$$\phi(f) = \int \phi(x) f(x) \, \mathrm{d}^4 x \tag{11.1}$$

is an unbounded operator, defined on a dense subset D of $\mathcal{H}$ and, for $|\Phi\rangle$ and $|\Psi\rangle \in D$

$$\langle \Phi | \left(\phi(f) | \Psi \rangle \right) = \left(\langle \Phi | \phi(f^*) \right) | \Psi \rangle \tag{11.2}$$

We call the fields $\phi(f)$ *regularised* or *smeared* fields. Since $\phi(f) D \subset D$, products of smeared fields are well-defined.

---

[1] We shall give here only a very sketchy description of the main assumptions and the results without proofs. There exist many references which offer a rigorous presentation of the axiomatic formulation of quantum field theory. See, for example, R.F. Streater and A.S. Wightman, *PCT, Spin and Statistics, and All That* (W.A. Benjamin, New York, 1964.)

Elementary Particle Physics. John Iliopoulos and Theodore N. Tomaras, Oxford University Press.
© John Iliopoulos and Theodore N. Tomaras (2021). DOI: 10.1093/oso/9780192844200.003.0011

• *Poincaré invariance.* We saw in Chapter 5 that a transformation of the Poincaré group consists of a Lorentz transformation $x \to \Lambda x$ and a translation $x \to x + a$. We also derived in equation (5.178) a map of the form $A \in SL(2,\mathbb{C}) \mapsto \Lambda(A)$. To these transformations correspond unitary operators $U(a,A)$ acting on $\mathcal{H}$. The axiom of Poincaré invariance imposes on the field operators $\phi(x)$ the transformation properties, see equation (5.198),

$$U(a,A)\phi(x)U^{-1}(a,A) = S(A)\phi(\Lambda^{-1}(x-a)) \tag{11.3}$$

where $A \to S(A)$ is a finite-dimensional irreducible representation of $SL(2,\mathbb{C})$. For the smeared fields defined in equation (11.1) this implies

$$U(a,A)\phi(f)U^{-1}(a,A) = S(A)\phi(f_{(a,\Lambda)}) \tag{11.4}$$

where $f_{(a,\Lambda)}(x) = f(\Lambda^{-1}(x-a))$; moreover $U(a,A)D \subset D$

• *Stability of the ground state and positivity of the energy.* The simplest way to present this axiom is to describe the spectrum of the energy operator P_0. For a massive theory with mass m we assume that it is similar to the one we found for the renormalised Hamiltonian of the free theory in section 10.2. It consists of a single ground state $|\Omega\rangle$ corresponding to the lowest eigenvalue of P_0: $P_0 |\Omega\rangle = E_0 |\Omega\rangle$, $E_0 \geq 0$,[2] one-particle states with momenta p^μ satisfying $p^2 = m^2$ and a continuum of multi-particle states[3] with energy $E \geq 2m$. There are no states with negative energy. This implies, in particular, the stability of the ground state, often called *the vacuum state*. As we did for free fields, $|\Omega\rangle$ is supposed to be Poincaré invariant: $U(a,A)|\Omega\rangle = |\Omega\rangle$.

• *Locality.* If $|\Psi\rangle$ is any state of $\mathcal{H}$, we assume that

$$[\phi(x),\phi(y)]_\pm |\Psi\rangle = (\phi(x)\phi(y) \pm \phi(y)\phi(x)) |\Psi\rangle = 0 \quad \text{for} \quad (x-y)^2 < 0 \tag{11.5}$$

where the $+$ or $-$ sign depends on whether the field has half-odd integer, or integer spin, respectively. This axiom guarantees the causal structure of the theory, in the sense that operators commute (or anti-commute) for space-like separations.

• *Completeness.* This axiom states that the set of all vectors of the form $P(\phi)|\Omega\rangle$, where $P(\phi)$ is a polynomial in the field, is dense in $\mathcal{H}$. We can say that the fields form a complete set of operators in the Hilbert space. This axiom is by construction obvious for free fields since the entire Fock space is built out of the vacuum state by applying products of fields, but it must be imposed as an axiom in the case of interacting fields. Note that we only need to impose it in a weak form. For free fields *every* vector of the Fock space is of the form $P(\phi)|0\rangle$, while here we only require the set of $P(\phi)|\Omega\rangle$ to be dense in $\mathcal{H}$. This property is also called *cyclicity of the vacuum state*.

These six axioms, written in a mathematically more precise way, were first formulated by Arthur Wightman and are known as *the Wightman axioms*. They encode all the basic properties a quantum field theory is supposed to satisfy and they hardly need

---

[2]Since in the absence of gravity energies are defined up to an additive constant, for the purposes of this discussion we can set $E_0 = 0$.

[3]In case the interaction creates bound states, these states should be added in the space in order to guarantee completeness.

any justification. In fact, they are so general that we could not expect any non-trivial results to come out of them. Yet, the opposite is true: there are two fundamental theorems, with direct physical implications, which follow from these axioms. We shall state them, without proof in the following section.[4]

A final remark. We saw in section 10.4.1 that a massless vector theory, such as electrodynamics, cannot be quantised consistently if we want to keep simultaneously explicit Lorentz covariance *and* positivity of the metric in the space of states. It follows that this axiomatic framework is not directly applicable to such field theories. In section 10.4.1 we saw that special care is required in order to define a subspace of the general vector space containing only positive norm physical states.

11.2 General Results: Invariance under *CPT* and the Spin-Statistics Theorem

11.2.1 The *CPT* theorem

In section 8.3.7 we introduced some important discrete transformations in the framework of the Dirac theory. Here we want to extend and generalise these notions to the case of a general quantum field theory. The symmetries we have in mind are:

- *Space inversion*, often called *Parity* ($\mathcal{P}$). It is the transformation under which $\boldsymbol{x} \to -\boldsymbol{x}$. In classical physics all three-vectors, such as three-momenta or the electric field, change sign under parity, while pseudo-vectors, such as the angular momentum or the magnetic field, do not.
- *Time reversal* ($\mathcal{T}$). It is the transformation which changes $t \to -t$ and thus reverses the sign of velocities.
- *Charge conjugation* ($\mathcal{C}$). This transformation has no analogue in classical physics. It changes particles to anti-particles but leaves space-time unaffected.

Here we want to define the corresponding operators and study some of their properties in quantum field theory. Following the theorem we proved in Chapter 5, these transformations will be represented on the Hilbert space of physical states by operators which are either linear and unitary or anti-linear and anti-unitary.[5]

In quantum field theory we can prove, using the completeness axiom, that it is sufficient to define these transformations through their action on the fundamental fields. Let us consider the example of space inversion $\mathcal{P}$ for a theory with a single complex scalar field $\phi(x)$. ϕ has no Lorentz indices, so we can write

$$U(\mathcal{P})\phi(x)U^{-1}(\mathcal{P}) = \eta_P \, \phi(\mathcal{P}x) = \eta_P \, \phi(x^0, -\boldsymbol{x}) \tag{11.6}$$

---

[4]See footnote 1 of this chapter.

[5]Reminder: The defining property of a linear operator U is that $U(\alpha \, |\phi\rangle + \beta \, |\psi\rangle) = \alpha U \, |\phi\rangle + \beta U \, |\psi\rangle$ for any complex numbers α and β, and any states $|\phi\rangle$ and $|\psi\rangle$. The Hermitian adjoint of a linear operator U is defined by $\langle\phi| \, U^\dagger \, |\psi\rangle = \langle U\phi|\psi\rangle = \langle\psi| \, U \, |\phi\rangle^*$. By definition, an operator A is anti-linear if it satisfies $A(\alpha \, |\phi\rangle + \beta \, |\psi\rangle) = \alpha^* A \, |\phi\rangle + \beta^* A \, |\psi\rangle$. The definition of the Hermitian adjoint of an operator, given above, is not consistent for anti-linear operators. You can check this explicitly by applying the definition for $|\phi\rangle = \alpha_1 \, |\phi_1\rangle + \alpha_2 \, |\phi_2\rangle$. The two sides of the definition give different results. To correct this, for anti-linear operators we define instead: $\langle\phi| \, A^\dagger \, |\psi\rangle = \langle A\phi|\psi\rangle^* = \langle\psi| \, A \, |\phi\rangle$. Symmetries by definition preserve probabilities. An operator U realises a symmetry transformation on the Hilbert space if and only if $| \, \langle\psi'|\phi'\rangle \, |^2 = | \, \langle\psi|\phi\rangle \, |^2$. Symmetry transformations connected continuously to the identity are necessarily represented by linear and unitary operators.

where we have allowed for a phase η_P. For boson fields it is natural to adopt the convention that $\mathcal{P}^2 = 1$, so $\eta_P = \pm 1$. Although the choice ± 1 is, in principle, arbitrary we shall see that relative phases are dictated by physical considerations. We shall call fields with $\eta_P = 1$ *scalars*, while those with $\eta_P = -1$ *pseudo-scalars*.

This discussion can be repeated for the other two discrete transformations, time reversal and charge conjugation. Staying always with the example of a scalar field, we write

$$U(\mathcal{T})\phi(x)U^{-1}(\mathcal{T}) = \eta_T \, \phi(\mathcal{T}x) = \eta_T \, \phi(-x^0, \boldsymbol{x}) \tag{11.7}$$

$$U(\mathcal{C})\phi(x)U^{-1}(\mathcal{C}) = \eta_C \, \phi^{\dagger}(x) \tag{11.8}$$

where we have introduced two new arbitrary phases, each of which we take equal to ± 1. Note also that $U(\mathcal{C})$ in (11.8) is assumed to be a linear unitary operator. For example, if the field ϕ is a free complex field with annihilation and creation operators a, b, $a^{\dagger}$ and $b^{\dagger}$, the action of charge conjugation is defined, up to a phase, as $a \leftrightarrow b$ and $a^{\dagger} \leftrightarrow b^{\dagger}$, leaving complex numbers unchanged. On the contrary, $U(\mathcal{T})$ is assumed to be anti-unitary.[6]

With these definitions we can announce the CPT theorem:

Theorem: For any quantum field theory satisfying the axioms of section 11.1 there exists always a choice for the phases η of the various fields, usually in more than one way, such that the anti-unitary operator Θ, equal to the product $\mathcal{C}PT$, taken in any order, is a symmetry of the theory.

If we denote by $\phi_I(x)$ a field of arbitrary spin with the index I representing a collection of dotted and undotted indices, the action of Θ is given by

$$\Theta\phi_I(x)\Theta^{-1} = \mathrm{i}^F(-1)^j\phi_I^{\dagger}(-x) \tag{11.9}$$

where F equals 0 if the spin of the field is an integer and 1 if it is a half-odd integer, and j is the number of undotted indices in I. In Problems 11.1 and 12.1 we ask the reader to show that this theorem implies the equality of masses and lifetimes for particles and their anti-particles. The same applies to electric and magnetic properties such as the electric charge (q) and the magnetic moment (g). These equalities are in excellent agreement with experiment. The most stringent limits come from a bound in the mass difference between K^0 and its anti-particle $\overline{K^0}$: $|m_{K^0} - m_{\overline{K^0}}|/m_{\mathrm{av}} < 6 \times 10^{-19}$ with 90% confidence level. For electrons and positrons we have $|m_{e^-} - m_{e^+}|/m_{\mathrm{av}} < 8 \times 10^{-9}$, $|q_{e^-} + q_{e^+}|/e < 4 \times 10^{-8}$, $(g_{e^+} - g_{e^-})/g_{\mathrm{av}} = (-0.5 \pm 2.1) \times 10^{-12}$, and for protons and anti-protons $|m_p - m_{\bar{p}}|/m_p < 7 \times 10^{-10}$, $|q_p + q_{\bar{p}}|/e < 7 \times 10^{-10}$.

11.2.2   The spin-statistics theorem

In section 10.3.2 we showed that a free Dirac field must be quantised using anti-commutation relations in order to satisfy the positive energy condition. We can similarly show that for a free Klein–Gordon field we must use commutation relations. In

---

[6]The fact that $U(\mathcal{T})$ must be an anti-unitary operator in quantum mechanics can be shown using the fundamental commutation relation $[x, p] = \mathrm{i}\hbar$. Under $\mathcal{T}$, x is unaffected but p changes sign. It follows that, for this relation to remain invariant, the r.h.s., which is a purely imaginary number, should change sign.

this section we state a general theorem which generalises these results. Let us first recall that this choice is dictated by the statistical properties of the system we wish to describe. A state of N identical bosons is symmetric under interchanges of the particles, while a state of N identical fermions is completely anti symmetric. These properties are directly observable and, as we know, have important physical consequences. The entire atomic structure depends on the property of the electrons to be fermions. Planck's law of radiation requires the photons to be bosons. Similarly for the phenomenon of Bose–Einstein condensation. In fact, the introduction of the corresponding statistical rules, Bose–Einstein and Fermi–Dirac, antedates the formalism of quantum field theory. In the axiomatic framework we are working, these statistical properties enter through the locality axiom which may be written either in terms of commutators or in terms of anti-commutators. The spin-statistics theorem states:

Theorem: For the set of the axioms exposed in section 11.1 to be consistent, the locality property of equation (11.5) should be written in terms of commutators for integer spin fields and anti-commutators for half-odd-integer fields.

Both these theorems, the CPT and the spin-statistics, require for their proofs the use of essentially all the Wightman axioms. This implies that they are deeply rooted in the foundations of our physical theories. On the other hand, as we just explained, they have very important and observable implications. Any discrepancy between these theorems and experiment would imply a radical change in our ideas on the microscopic properties of space-time and matter.

11.3 Examples of Interacting Theories

The very general framework of the Wightman axioms yields a few important results but does not offer the possibility of detailed numerical computations. For this we must appeal to specific field theory models. In this book we will study only theories whose Lagrangian densities are polynomials in the fields and their first derivatives. Surprising as it may sound, this extremely simple assumption will turn out to be sufficient to describe all fundamental interactions among elementary particles. They are all represented by polynomials of low degree.[7] In a later chapter we will try to explain this result and make it plausible. Under this assumption it is easy to give examples of possible interacting theories. Naturally, Lorentz invariance will always be imposed.

• A single scalar field $\phi(x)$. To the free Lagrangian density of equation (7.3) we can add terms of the form $\phi^n(x)$ with n integer larger than 2. In this book we will consider the simplest choices of $n = 3$ and $n = 4$. The corresponding Lagrangian density is

$$\mathcal{L} = \frac{1}{2}\partial_\mu\phi\partial^\mu\phi - \frac{1}{2}m^2\phi^2 - g\phi^3 - \lambda\phi^4 \tag{11.10}$$

where g and λ are arbitrary coefficients which we shall call *coupling constants*. Simple dimensional analysis shows that in four-dimensional space-time g has dimensions of

---

[7]There is one notable exception: the Lagrangian of general relativity which describes classical gravitational interactions is non-polynomial in the metric tensor.

mass and λ is dimensionless. If, in addition to Lorentz invariance, we require invariance under the discrete transformation $\phi \to -\phi$, the ϕ^3 term must be absent.[8]

• With N scalar fields the $O(N)$ invariant Lagrangian density which generalises equation (7.26) is

$$\mathcal{L} = \frac{1}{2} \sum_{a=1}^{N} \left(\partial_\mu \phi_a \partial^\mu \phi_a - m^2 \phi_a \phi_a \right) - \frac{\lambda}{4} \left(\sum_{a=1}^{N} \phi_a \phi_a \right)^2 \tag{11.11}$$

For $N = 2$ and using the notation of a complex field, which we introduced in section 7.2.2, this becomes

$$\mathcal{L} = \partial_\mu \phi^* \partial^\mu \phi - m^2 \phi^* \phi - \lambda (\phi^* \phi)^2 \tag{11.12}$$

• Similarly, if we consider a spin-$\frac{1}{2}$ Dirac field ψ we can add to the free Dirac Lagrangian density (7.60) terms given by the squares of the bilinear expressions we studied in Problem 7.3

$$\mathcal{L} = \bar{\psi} \, i \partial\!\!\!/ \, \psi - m \, \bar{\psi} \psi + \sum_a G_a (\bar{\psi} \, O_a \psi)^2 \tag{11.13}$$

where the sum extends over the five bilinears of Problem 7.3, O_a stands for $\mathbf{1}$, γ^μ, $\sigma^{\mu\nu}$, $\gamma_5 \gamma^\mu$ and γ_5 and G_a are five arbitrary coupling constants. The Lagrangian density (11.13) is invariant under parity. If we do not impose parity conservation, we can include non-diagonal products of bilinears mixing $\mathbf{1}$ with γ_5, or γ^μ with $\gamma_5 \gamma^\mu$. We saw in Chapter 6 that, in fact, this kind of interaction, involving the four spinor fields of the neutron, the proton, the electron and the neutrino, was first considered by Fermi to describe β-decay.

Using dimensional analysis we can determine the dimensions of the various coupling constants. In the system of units, $\hbar = c = 1$, $[L] = [T] = 1/[M]$ and the action is dimensionless. Thus, in a four-dimensional space-time the Lagrangian density must have dimensions $[M]^4$, so boson fields have dimensions $[M]^1$ and fermion fields $[M]^{3/2}$. It follows, for example, that the coupling constants G_a have dimensions $[M]^{-2}$.

• We can consider fields with different spins. A simple example is a Dirac field ψ and a scalar field ϕ. An example of a Lorentz invariant Lagrangian is

$$\mathcal{L} = \bar{\psi} \, i \partial\!\!\!/ \, \psi - M \bar{\psi} \, \psi + \frac{1}{2} \partial_\mu \phi \, \partial^\mu \phi - \frac{1}{2} m^2 \phi^2 + g \, \bar{\psi} \psi \phi \tag{11.14}$$

The $\bar{\psi}\psi\phi$ term is called *Yukawa interaction*. To this expression we could add terms such as ϕ^4, $\bar{\psi} \gamma^\mu \psi \partial_\mu \phi$. In Chapter 6 we wrote the isospin invariant pion–nucleon Yukawa interaction in equations (6.14) and (6.15). At the end of this chapter we propose Problem 11.3, in which we ask the reader to write interaction Lagrangians incorporating internal symmetries such as isospin or $SU(3)$.

---

[8]A theory with $\lambda = 0$ and $g \neq 0$ is classically unstable because the Hamiltonian is unbounded from below for sufficiently large constant values of the field ϕ, no matter which sign we choose for g.

11.4 Interacting Quantum Field Theories

In this section we want to apply the quantisation rules we introduced in Chapter 10 to interacting field theories. Let us start with the simple case of a single scalar field $\phi(x)$ self-interacting through a ϕ^4 interaction. In the following sections we will extend the results to other interacting theories. The classical Lagrangian density is given by

$$\mathcal{L} = \frac{1}{2}\partial_\mu \phi(x)\partial^\mu \phi(x) - \frac{1}{2}m^2\phi^2(x) - \lambda\phi^4(x) \tag{11.15}$$

Since the interaction part has no derivatives of the field, we can still write that the conjugate momentum of the field $\phi(t, \boldsymbol{x})$ is $\pi(t, \boldsymbol{x}) = \dot{\phi}(t, \boldsymbol{x})$, so we can formally attempt to write the equal time commutation relations (10.6). But this is as far as we can go. There is no general method to find the solutions of non-linear partial differential equations and we do not know whether such commutation relations can be imposed in a consistent way. In particular, we cannot find any explicit realisation in terms of creation and annihilation operators. We faced this problem in Chapter 2, and in section 2.4.2 we developed the method of perturbation expansion according to which any quantity A is written as a formal power series

$$A = \sum_{n=0}^{\infty} \lambda^n A_n \tag{11.16}$$

where the 0 order term is computed using free field equations and the nth order contains the interaction Lagrangian $\lambda\phi^4(x)$ n times. In the rest of this chapter we shall develop precise rules for the computation of A_n for physically interesting quantities A at any order n. We call the series (11.16) "a formal power series" because the question of its convergence will not be examined. We intuitively expect this approach to be meaningful if, in some sense still to be specified, the interaction is "weak", and we expect this to be true for λ sufficiently small.

The perturbation expansion is the main computational tool in the physics of elementary particles and it will be fully developed in this book. However, it does not offer, even formally, a definition of an interacting quantum field theory and, in particular, it does not make sense for a theory in the strong coupling regime in which the coupling constant λ is large. We can obtain such a definition using Feynman's path integral approach. In formulae (9.4) and (9.6) the form of the action is not specified and it is not assumed that the resulting equations of motion are linear. So, we can generalise, albeit in a formal way, equations (9.46) or (10.25)

$$Z[J] = \frac{\int e^{i\int d^4x[\mathcal{L}+J(x)\phi(x)]}\mathcal{D}[\phi]}{\int e^{i\int d^4x\mathcal{L}}\mathcal{D}[\phi]} \tag{11.17}$$

or, equivalently,

$$\mathcal{W}_n(x_1,\ldots,x_n) = \frac{\int \phi(x_1)\ldots\phi(x_n)\,e^{i\int d^4x\mathcal{L}}\,\mathcal{D}[\phi]}{\int e^{i\int d^4x\mathcal{L}}\,\mathcal{D}[\phi]} \equiv \frac{T_n(x_1,\ldots,x_n)}{T_0} \tag{11.18}$$

These expressions are still formal because we have not defined the integration measure in the presence of the ϕ^4 term in the Lagrangian. The corresponding normalisation factor, the analogue of $(\det(m^2 - \Box_E))^{-1/2}$ in equation (10.22), is represented by the denominator, which is an unknown constant, independent of J. Despite this limitation, this formulation offers several advantages.

First, it can be approximated by truncating the four-dimensional space by a finite Euclidean lattice, the way we explained in equations (10.20) and (10.21). This way, keeping the lattice spacing a and the number of points N fixed, we can evaluate the functional integral numerically for any value of the coupling constant λ. Repeating the calculation varying a and N, we might able to study the limit $a \to 0$ and $N \to \infty$. In fact, this is the only method we have to extract physically measurable results out of a quantum field theory in the strong coupling regime. We shall present such results for the theory of strong interactions in Chapter 21.

The second merit of expressions such as (11.17) or (11.18) is that they offer a convenient framework to formulate the perturbation expansion to which we turn next.

11.5   The Perturbation Expansion

As we explained in equation (11.16), it is an expansion in powers of the coupling constant. Let us concentrate on the correlation functions $\mathcal{W}_n(x_1, \ldots, x_n)$. The numerator in equation (11.18) can be written as

$$T_n(x_1, \ldots, x_n) = \int \phi(x_1) \ldots \phi(x_n) \, \mathrm{e}^{\mathrm{i} \int \mathrm{d}^4 x \mathcal{L}} \, \mathcal{D}[\phi]$$

$$= \sum_{k=0}^{\infty} \frac{(-\mathrm{i}\lambda)^k}{k!} \int \phi(x_1) \ldots \phi(x_n) \prod_{j=1}^{k} \left(\int \mathrm{d}^4 y_j \phi^4(y_j) \right) \mathrm{e}^{\mathrm{i} \int \mathrm{d}^4 x \mathcal{L}_0} \mathcal{D}[\phi] \qquad (11.19)$$

where $\mathcal{L}_0$ is the quadratic part of the Lagrangian density (11.15), which means that every term in the expansion can be computed using the Gaussian measure. But this is precisely the expression we have computed in equations (10.25) and (10.26). We have shown that every term in this expansion is in fact a product of Feynman propagators and, in the following sections we will develop the practical tools to perform these computations.

11.5.1   The configuration space Feynman rules

Feynman invented a set of graphical rules which encode the fact that each term in the perturbation expansion of a correlation function $\mathcal{W}_n$ is expressed in terms of products of Feynman propagators. The main idea goes as follows: The building block is the propagator which, for a scalar theory, is the 2-point function $G_F(z_1 - z_2)$. It is given by (7.19). We explained in section 10.2 that it is equal to the amplitude for a particle to move from z_1 to z_2. We will represent it by a line joining these two points. Looking at the expression (11.19) we see that we will need two kinds of points: the x_i, $i = 1, 2, \ldots, n$ and the y_j, $j = 1, 2, \ldots, k$. Each one of the x-points will appear only in one propagator, so there will be a single line attached to each one of them. We will call them *line points*. On the other hand, each of the y-points will appear in four propagators (propagators starting and ending at the same point should be counted twice) so there will be four

Fig. 11.1 The zeroth order 2-point function.

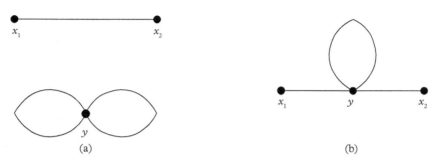

(a)   (b)

Fig. 11.2 The diagrams contributing to the 1st order 2-point function.

lines attached to it. We will call them *vertex points*, or *vertices*. Now the rules are simple: Let us consider the λ^k term in the expansion (11.19).

• In a plane mark the n line points x_i with one line segment attached to each, and the k vertices y_j with four line segments attached to each.

• Join the line segments in pairs with lines. If n is even all segments will be joined. There will be $(4k + n)/2$ lines. (As we have explained, for the ϕ^4 theory all $\mathcal{W}_{2m+1}(x_1, \dots, x_{2m+1})$ vanish identically.) We can do this in many different ways, so the result is a set of figures with different topologies. Each of these figures is called *an x-space Feynman diagram*.

• Consider any one of the diagrams with a given topology. To each line write the corresponding propagator and integrate the product of all these propagators over the $y_j, j = 1, \dots, k$. The result is the value of the diagram.

• Each diagram with a given topology has a numerical coefficient which is the product of (i) the factor $(-i\lambda)^k/k!$ of the perturbation expansion, (ii) a factor $k!$ resulting from the permutations of the k vertices, and (iii) a combinatoric factor corresponding to the number of ways the segments can be joined in pairs and lead to a diagram with this particular topology. There is no closed form which gives this factor for any diagram and the safest method is to compute it directly.

• $T_n(x_1, \dots, x_n)$ is the sum of all topologically distinct graphs.

Let us give some examples. We start with the 2-point function $\mathcal{W}_2(x_1, x_2)$. We compute separately the numerator $T_2(x_1, x_2)$ and the denominator T_0.

At zero order of perturbation theory T_2 is given by the diagram of Figure 11.1 and equals $G_F(x_1 - x_2)$. T_0 is the infinite factor which is absorbed in the definition of the integration measure.

Let us look at the first order terms. We have the two line points x_1 and x_2 and one vertex y. The diagrams for T_2 are shown in Figure 11.2. They are of two distinct topologies: one in which x_1 is joined with x_2 and the four lines of the vertex are joined together, Figure 11.2(a), and a second, shown in Figure 11.2(b), in which x_1 and x_2 are joined with the vertex y. There is one diagram for T_0, which is shown in

Fig. 11.3 The diagram contributing to T_0 at order λ.

Figure 11.3. According to our rules and putting together the zero and the first order results, we obtain

$$T_2(x_1, x_2) = G_F(x_1 - x_2) \left[1 - 3i\lambda(G_F(0)V)^2 \right]$$

$$-12\,i\lambda G_F(0) \int d^4 y G_F(x_1 - y) G_F(x_2 - y) + \mathcal{O}(\lambda^2) \qquad (11.20)$$

$$T_0 = 1 - 3i\lambda(G_F(0)V)^2 + \mathcal{O}(\lambda^2) \qquad (11.21)$$

with $V = \int d^4 y$. This gives for $\mathcal{W}_2 = T_2/T_0$,

$$\mathcal{W}_2(x_1, x_2) = G_F(x_1 - x_2) - 12\,i\lambda G_F(0) \int d^4 y G_F(x_1 - y) G_F(x_2 - y) + \mathcal{O}(\lambda^2) \quad (11.22)$$

We call diagrams with no external lines, like the one of Figure 11.3, *vacuum diagrams*. The important point is that the vacuum diagram of Figure 11.3 has cancelled between the numerator and the denominator. We can show, see Problem 11.4, that this is a general result valid for all vacuum diagrams at any order in the perturbation expansion and for any correlation function. Therefore we can add a sixth rule to our previous list:

• When computing $\mathcal{W}_n$ ignore the denominator T_0 and drop all vacuum diagrams in the expansion of the numerator.

A technical remark: In the expression (11.22) 12 is an example of a combinatoric factor. It comes from the four possibilities in joining x_1 with one of the four line segments in the vertex and the three possibilities in joining x_2 with one of the remaining three. This kind of factor will appear often, so it has become customary to change the definition of the coupling constant $\lambda \to \lambda/4!$ in order to simplify the final expression. In Problem 11.5 we ask the reader to write the expression for the 4-point correlation function $\mathcal{W}_4(x_1, \ldots x_4)$ in second order of perturbation theory corresponding to the diagrams of Figure 11.4.

A final important remark: The expressions we just obtained contain many ill defined quantities. They are of two kinds: First the propagators $G_F(x)$ at $x = 0$. As we can see in equation (7.19), they are given by a divergent integral. Second the integrals over y which are often divergent when two points coincide. We will address these problems in a later chapter.

11.5.2   The momentum space Feynman rules

The expressions we obtained for the Feynman diagrams simplify by taking the Fourier transform and going to momentum space. It is a simple exercise to translate the rules we found previously. We are interested in the Fourier transform $\widetilde{W}_n(p_1, p_2, \ldots, p_n)$ of the correlation function

$$\widetilde{W}_n(p_1, p_2, \ldots, p_n) = \int \prod_{i=0}^{n} \mathrm{d}^4 x_i \, \mathrm{e}^{\mathrm{i} p_i \cdot x_i} W_n(x_1, \ldots, x_n) \qquad (11.23)$$

Let us focus on a vertex located at the point y in x-space (see Figure 11.5). We call $G_{\mathrm{F}}(w_i - y), i = 1, \ldots, 4$ the four propagators joined to y. For the following argument it will not matter whether w_i are line points x or vertex points y. According to our previous rules we obtain

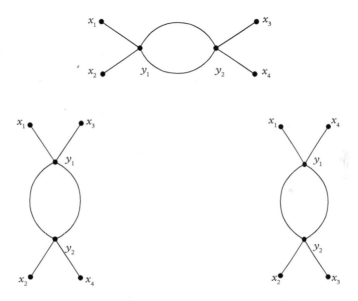

Fig. 11.4 Diagrams contributing to the 4-point function in second order of perturbation theory.

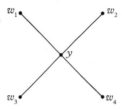

Fig. 11.5 A vertex at a point y.

$$\int d^4y \prod_{i=1}^{4} G_F(w_i - y) = \int d^4y \prod_{i=1}^{4} \frac{d^4q_i}{(2\pi)^4} \, e^{-iq_i \cdot (w_i - y)} \frac{i}{q_i^2 - m^2 + i\epsilon}$$

$$= \int (2\pi)^4 \delta^4 \left(\sum_{i=1}^{4} q_i \right) \prod_{i=1}^{4} \frac{d^4q_i}{(2\pi)^4} \, e^{-iq_i \cdot w_i} \frac{i}{q_i^2 - m^2 + i\epsilon} \quad (11.24)$$

i.e. the y integrations give rise to δ-functions implying energy and momentum conservation at every vertex. If we perform all the y integrations, we will be left with the exponentials $e^{-iq_i \cdot w_i}$, where the points w will be points x; in other words these will be the exponentials $e^{ip_i \cdot x_i}$ of equation (11.23).

We are now in a position to formulate the Feynman rules in momentum space. $\widetilde{W}_n(p_1, p_2, \ldots, p_n)$ is given by a sum of Feynman diagrams with $n = 2m$ external lines each line carrying a momentum p_i. A diagram of kth order in the perturbation expansion has k vertices. For the ϕ^4 theory each vertex has four lines attached to it. A useful concept, which will simplify the discussion, is the distinction between *connected* and *disconnected* diagrams, with the obvious meaning of the terms. Without loss of generality we can restrict ourselves to the study of the connected diagrams, since any disconnected one is made out of connected pieces. A simple counting shows that a connected diagram with $2m$ external lines and k vertices has $2k - m$ ($k \geq 1$, $m \geq 1$) internal lines.[9] Each internal line carries a momentum q. The rules of calculation of the order $\mathcal{O}(\lambda^k)$ term of $\widetilde{W}_n$ are the following:

- Draw all connected diagrams with n external lines and k vertices.
- To each external line of momentum p_i corresponds a propagator $i(p_i^2 - m^2 + i\epsilon)^{-1}$
- To each internal line of momentum q_j corresponds a propagator $i(q_j^2 - m^2 + i\epsilon)^{-1}$
- Each vertex carries the coupling constant $-i\lambda/4!$ and a δ-function of energy and momentum conservation of the form $(2\pi)^4\delta^4(k_1 + k_2 + k_3 + k_4)$, where all momenta are considered flowing into the vertex. The momenta k may be either external momenta p or internal momenta q.
- Integrate over all internal momenta with the measure $d^4q/(2\pi)^4$.
- Each diagram has the numerical factor we found in the x-space Feynman rules.
- Sum over all topologically distinct diagrams.

Since energy and momentum are conserved at every vertex, they are also conserved for the entire diagram. Indeed, for a connected diagram we can show (see Problem 11.6) that we can combine all δ-functions to factor out an overall δ-function in the form $(2\pi)^4\delta^4(\sum_{i=1}^{n} p_i)$, where again we consider all momenta as incoming to the diagram. We can use the remaining $k - 1$ δ-functions in order to perform $k - 1$ q integrations. We will be left with $k + 1 - m$ integrations for which there are no δ-functions to help. It is useful to introduce another notion from graph topology, that of a *closed loop*. It is a sequence of lines in a diagram, which starts and ends at the same vertex. A simple counting shows that a connected diagram has precisely $k + 1 - m$ loops. Therefore, we are left with one integration for every closed loop. A diagram which contains no

---

[9] There is also the trivial case of a connected diagram with $k = 0$ and $m = 1$. It is the free propagator.

Fig. 11.6 The fermion and the photon lines in QED.

closed loops is called a *tree diagram*. In Problem 11.5 we ask the reader to write the momentum space expression for the diagrams of Figure 11.4.

11.5.3 Feynman rules for quantum electrodynamics

The formalism of Feynman diagrams and Feynman rules can be easily extended to any interacting quantum field theory. We give here the result for quantum electrodynamics (QED), with the interaction Lagrangian density given by

$$\mathcal{L}_I^{\text{QED}} = -e\bar{\psi}(x)\gamma^\mu\psi(x)A_\mu(x) \tag{11.25}$$

We just list the differences with the ϕ^4 theory we studied before:
- In QED we have two kinds of fields, the Dirac field $\psi(x)$ for the electrons and the positrons, and the Maxwell field $A_\mu(x)$ for the photons. Correspondingly, we will have two kinds of lines in a Feynman diagram, the fermion lines represented by the propagator S_F given by equation (7.68), and the photon lines with propagator $G_F^{\mu\nu}$ given by equation (7.100) (see Figure 11.6). Similarly, we must distinguish two kinds of external lines in the correlation functions $\widetilde{\mathcal{W}}$, those corresponding to electrons with momenta p_i, $i = 1, ..., 2m$, and those to photons with momenta k_j, $j = 1, ..., r$.
- In QED, electrons carry electric charge and the corresponding Dirac fields are complex valued. As a result the propagators connect only $\bar{\psi}$ and ψ, as we explained in section 10.3.1. In order to keep track of this property, we mark every fermion line with an arrow, which represents the direction of charge flow. One arrow flows into and the other out of each vertex in any diagram. This encodes the conservation of electric charge.
- The correlation functions $\widetilde{\mathcal{W}}$ are matrix valued. They carry a spinor index α for each external fermion line and a vector index μ for each external photon line. Correspondingly, fermion and photon propagators are matrices $S_F^{\alpha\beta}$ and $G_F^{\mu\nu}$.
- The interaction vertex is the matrix $-ie\gamma^\mu_{\alpha\beta}$

With these rules we can compute any QED correlation function at any order of perturbation theory. In Problem 11.7 we ask the reader to prove that, as a result, the following rules should be added to the list:
- Every closed fermion loop gives a trace of the product of the matrices corresponding to the vertices and the fermion propagators taken around the loop.
- There is a minus sign for every closed fermion loop. This last property follows from the integration rules of Grassmannian variables.

11.5.4 Feynman rules for other theories

It is now straightforward to extend these rules to any other interacting theory with an interaction Lagrangian which is a polynomial in the fields and their derivatives.

In perturbation theory a correlation function will be expressed, order by order, as a sum of Feynman diagrams. For its computation we must give the rules for the propagators and the vertices.

We have already derived the Feynman propagators for scalar, spinor and vector fields. For the last we distinguish the massless case, for which a gauge fixing is required, and the massive one. Although it is straightforward to derive propagators for fields of any spin, we shall never need them in this book.

So, the missing element for other field theories is the rule for the vertices. It is easily derived by going back to the perturbation expansion.

Let us first consider a term in the interaction Lagrangian which is a monomial in the fields with no derivatives. It has the general form

$$\mathcal{L}_I = \Phi_{\{i_1\}}\Phi_{\{i_2\}}\cdots\Phi_{\{i_k\}}\Gamma^{\{\alpha_1\beta_1\}}_{\{s_1\}}\cdots\Gamma^{\{\alpha_r\beta_r\}}_{\{s_r\}} \tag{11.26}$$

where the Φ's are the various fields, the set of indices $\{i_j\}$ denote collectively the nature of the field, and they include Lorentz as well as internal symmetry indices. $\Gamma^{\{\alpha\beta\}}$ are numerical matrices with indices, which may also refer to Lorentz or internal symmetry. The lower index $\{s\}$ indicates the type of the matrix, for example a γ matrix, and the upper indices the matrix element. We have included the coupling constant in these matrices.

This interaction term generates a vertex in the Feynman diagrams, which has the following form:

• It has k lines, one for each field.

• Each line carries the set of $\{i_j\}, j = 1, \ldots, k$ indices.

• The term is proportional to the product of the matrix elements of the corresponding numerical matrices.

• At each vertex there is the factor $(2\pi)^4\delta^{(4)}(\Sigma p)$. The argument of the delta function is the sum of all momenta (assumed flowing into the vertex) and ensures energy-momentum conservation at the vertex.

In most cases the interaction Lagrangian is a Lorentz scalar, so the corresponding Lorentz indices are contracted. The same happens also for internal symmetry indices. These summations are also understood in the expression for the vertex.

Including derivatives in some of the fields does not require more work. Going back to the expression for the path integral we see that the result is very simple: in the momentum space Feynman rules, a derivative ∂_μ on the field ϕ will give a factor ip_μ, so the above set of rules is complemented by:

• A derivative $\partial_\mu\phi_{\{i_j\}}$ gives a factor ip_μ, where p is the momentum of the line corresponding to the field $\phi_{\{i_j\}}$.

• As we said for QED, there is no simple rule to compute the combinatoric factor for an arbitrary diagram. However, the cancellation of the $1/n!$ factor from the expansion of the exponential $\exp(i\int d^4x\mathcal{L}_I)$ in, for example, (11.17) holds for any theory.

• As for QED, a closed loop leads to a trace of the product of the vertex and propagator matrices. This trace refers to both the internal symmetry matrices and the γ matrices, if the loop is made out of spinor fields.

• There is an overall minus sign for every closed fermion loop.

   This completes the Feynman rules for arbitrary interaction Lagrangians, which are given as a sum of monomials in the fields and their derivatives. In Appendix A we give a list with the ones most commonly used. Using these rules we can compute any correlation function in lowest order of perturbation theory. By this we mean diagrams, which contain no closed loops, the ones we called "tree diagrams". We still have two important questions to address: First, how to choose the right theory among the infinity of polynomial Lagrangians. Second, how to go beyond the lowest order and consider complicated diagrams involving loops. As it will turn out, the two questions are not unrelated. They will be addressed briefly in a later chapter.

   A final remark: The analysis we presented in this chapter gave us a set of Feynman rules for the perturbative computation of correlation functions starting from a Lagrangian density. The derivation was purely formal and the various steps were not mathematically rigorous. We were manipulating expansions whose convergence we were in no position to control. However, the final result, the set of rules, appears to be well-defined and unambiguous. In Chapter 15 we will further establish their consistency. Here we want only to remark that we could have started from the rules. We could have *defined* the theory by postulating the set of Feynman rules and use the Lagrangian density only as a short-hand expression encoding the rules, forgetting everything about their derivation. At the end, to the extent that we will perform only computations in perturbation theory, it wouldn't make any difference.

11.6 Problems

Problem 11.1 Show that invariance under CPT implies the equality of masses between particles and anti-particles.

Problem 11.2 1. Consider a Lagrangian density describing the interaction of a real scalar field and a Dirac spinor field of the form

$$\mathcal{L}_1 = \overline{\psi}(i\partial\!\!\!/ - m)\psi + \frac{1}{2}\partial^\mu\phi\partial_\mu\phi - \frac{1}{2}M^2\phi^2 + g\overline{\psi}(1 + i\lambda\,\gamma_5)\psi\,\phi$$

with g and λ two real dimensionless coupling constants. Show that $\mathcal{L}_1$ violates the separate conservation of parity and time reversal, but conserves the product CPT.

   2. We replace the real scalar field by a complex one and we introduce a derivative coupling of the form

$$\mathcal{L}_2 = \overline{\psi}(i\partial\!\!\!/ - m)\psi + \partial^\mu\phi^*\partial_\mu\phi - M^2\phi^*\phi + \left(\frac{1}{M}\overline{\psi}\gamma^\mu(\alpha + \beta\gamma_5)\psi\,i\partial_\mu\phi + \text{h.c.}\right)$$

where α and β are complex dimensionless coupling constants and h.c. stands for "hermitian conjugate". Show that, despite the complex coupling constants, $\mathcal{L}_2$ is invariant under CP and time reversal.

Problem 11.3 The purpose of this exercise is to get used to writing interaction Lagrangian densities invariant under various symmetries. We will consider various particles and assume that they are all described by local quantum fields. In the following

the full Hamiltonian H and the free Hamiltonian H_0, and (ii) the short range nature of the interactions which, as we have noted already, implies the absence of zero mass particles. This last assumption excludes several fundamental interactions, such as the electromagnetic interaction, which are mediated by massless particles. We will show briefly in section 12.3.2 how we can get around this problem and we will present a more complete treatment in Chapter 17.

The first step towards the construction of the asymptotic theory consists in expressing the initial and final states in the amplitude of (12.2) as eigenstates of the unperturbed free Hamiltonian H_0, because they are the only ones we know how to construct explicitly. They form the Fock space we introduced in Chapter 10.[1] The physical justification for using this space in describing scattering experiments is based on the assumption that the interactions have short range, which means that they are mediated by massive particles. Let us give some orders of magnitude: Experimentally, the range of the nuclear forces has been determined to be on the order of one fermi, or 10^{-13} cm. Weak interactions have an even shorter range. By comparison, the size of an atom is of order 10^{-8} cm. In scattering experiments the smallest separation among particles in the beams we shall ever have to consider, either in the initial or in the final state, is much larger than interatomic distances in the target, so it is safe to assume that these particles are free.[2] This requirement of space separation can be translated into a similar requirement on the time interval $|t_1 - t_2|$, which separates the initial and final states. In modern particle physics experiments the typical relative speed among particles is a sizable fraction of the speed of light, so the characteristic time for a strong interaction collision is on the order of, or smaller than, 10^{-22} s. Compared to that, even the lifetime of a neutral pion π^0, which is about 10^{-16} s, can be considered as infinite.[3] This led J.A. Wheeler in 1937, and W. Heisenberg in 1943, to introduce and develop the concept of the *S-matrix*, defined formally as the double limit $t_1 \to -\infty$, $t_2 \to +\infty$, of the evolution operator $U(t_2, t_1)$

$$S = U(+\infty, -\infty) \tag{12.3}$$

This leads us to introduce the notion of the incoming Hilbert space, i.e. the space of initial or incoming states $\mathcal{H}_{\mathrm{in}}$. It consists of non-interacting states, composed of wave packets well separated and evolving independently. Symmetrically, we introduce the Fock space of final or outgoing states $\mathcal{H}_{\mathrm{out}}$. It is isomorphic to $\mathcal{H}_{\mathrm{in}}$. Finally, we assume that these two spaces are also isomorphic to the space of the physical states $\mathcal{H}$, i.e.

$$\mathcal{H}_{\mathrm{in}} \cong \mathcal{H}_{\mathrm{out}} \cong \mathcal{H} \tag{12.4}$$

---

[1]We may have to enlarge this space with the addition of possible stable, or quasi-stable bound states, which are new elementary systems generated by the interaction. There is no fully systematic way to do so and we will address this problem only when it is needed.

[2]Here again the assumption about the absence of zero mass particles is important. In a real experiment a beam of charged particles radiates energy in the form of soft photons even before the actual collision and this effect requires a special treatment, which we present in Chapter 17.

[3]The very short-lived hadronic resonances have mean lifetimes comparable to, or shorter than, the above strong interaction time. The same is true for some of the "elementary particles" we presented in Table 6.6, such as the heavy quark t and the intermediate bosons $W^\pm$ and Z. We will discuss these points later.

This relation implies the existence of a unitary operator S, which maps $\mathcal{H}_{\text{in}}$ to $\mathcal{H}_{\text{out}}$. The probability amplitude of the transition from an initial state Ψ_{in} to a final state Φ_{out} is then given by

$$(\Phi_{\text{out}}, \Psi_{\text{in}}) = (\Phi_{\text{in}}, S\Psi_{\text{in}}) \qquad (12.5)$$

The operator S, which for historical reasons is called the *S-matrix* or *Scattering matrix*, will become an important object of study all along in this book. From its formal definition $S = U(+\infty, -\infty)$ it follows that S is a unitary operator, i.e. it satisfies

$$SS^\dagger = S^\dagger S = \mathbf{1} \qquad (12.6)$$

It will be convenient to take out from S the identity operator and define the *transition matrix T* by

$$S = \mathbf{1} + \mathrm{i}T \qquad (12.7)$$

The unit operator represents the process in which there is no scattering, i.e. $\Psi_{\text{in}} = \Phi_{\text{in}}$, and T is the actual scattering matrix. The unitarity of S implies

$$-\mathrm{i}(T - T^\dagger) = TT^\dagger = T^\dagger T \qquad (12.8)$$

An interesting immediate consequence of this unitarity relation is obtained by considering the diagonal matrix element of both sides between an incoming state $\Psi_{\text{in}} = |\boldsymbol{k}_1, \boldsymbol{k}_2\rangle$, consisting of two particles with momenta $\boldsymbol{k}_1$ and $\boldsymbol{k}_2$. We obtain

$$2\operatorname{Im}\langle\boldsymbol{k}_1, \boldsymbol{k}_2|\, T\, |\boldsymbol{k}_1, \boldsymbol{k}_2\rangle = \langle\boldsymbol{k}_1, \boldsymbol{k}_2|\, T^\dagger T\, |\boldsymbol{k}_1, \boldsymbol{k}_2\rangle = \sum_n |\langle n|\, T\, |\boldsymbol{k}_1, \boldsymbol{k}_2\rangle|^2 \qquad (12.9)$$

where the sum is over a complete set of states. On the left-hand side there is the imaginary part of the forward elastic scattering amplitude of the two particles. The right-hand side is, up to a kinematical factor, the total cross section of these two particles. Therefore, this relation gives the quantum field theory version of the *optical theorem*, we discussed in potential scattering.

12.1.2 The asymptotic fields

In the previous section we introduced the *in* and *out* states, which are states of free particles. Following our previous discussion on the scattering process, we want to use them to describe states in the remote past, i.e. $t \to -\infty$ and the distant future $t \to +\infty$. For this, we will introduce quantum field operators $\phi_{\text{in}}(x)$ and $\phi_{\text{out}}(x)$, which will act in $\mathcal{H}_{\text{in}}$ and $\mathcal{H}_{\text{out}}$, respectively. We shall call them *asymptotic fields*.

We start with the field operators $\phi(x)$, which evolve with the full Hamiltonian H. According to our axioms of section 11.1 the polynomials of ϕ form a complete set of operators in the Hilbert space $\mathcal{H}$. Let us consider the state $\phi|\Omega\rangle$, obtained by the action of ϕ on the ground state of the full Hamiltonian. It is a vector in $\mathcal{H}$ and, according to our assumption (12.4), it can be expanded in the Fock space basis of either $\mathcal{H}_{\text{in}}$ or $\mathcal{H}_{\text{out}}$, i.e.

$$\phi|\Omega\rangle = C_0 |0\rangle + C_1 |1\rangle + \sum_{n\geq 2} C_n |n\rangle \qquad (12.10)$$

In momentum space the first term on the right-hand side has $p^0 = 0$, the second, being a one-particle state, has support on the $p^2 = m^2$ hyperboloid and all the other ones correspond to $p^2 \geq 4m^2$. We want to isolate the coefficient C_1 of the one-particle state $|1\rangle$. Our basic assumption is that $C_1 \neq 0$; in other words, we assume that the field operator has a non-vanishing probability to create a one-particle state out of the vacuum. With this assumption, which is clearly "natural" in perturbation theory, we will construct the asymptotic fields $\phi_{\text{in}}(x)$ and $\phi_{\text{out}}(x)$, which will satisfy free field equations of motion.

Let us choose a test function $h(p)$ in momentum space, whose support contains a neighbourhood of the mass hyperboloid $p^2 = m^2$ and does not intersect $\{p = 0\} \cup V_+^{2m}$, i.e. it does not contain the four-momentum $p^\mu = 0$ of the perturbative vacuum, and also it does not intersect the space of energy and momentum of two or more free particles $p^2 \geq 4m^2$. Moreover, we suppose that $h(p) = 1$ on the one-particle hyperboloid, i.e. for $p \in \{p \,|\, p^2 = m^2\}$. Note that this is possible only if $m^2 \neq 0$; otherwise there is no mass gap between the vacuum, the one-particle and the multi-particle states.

Starting from the field ϕ, we define the auxiliary field

$$B(x) = \phi * \tilde{h}(x) = \frac{1}{(2\pi)^4} \int e^{ip\cdot x}\, \tilde{\phi}(p) h(p)\, \mathrm{d}^4 p \tag{12.11}$$

and we check easily (see Problem 12.2) that $B(x)\,|\Omega\rangle$ is a solution of the Klein–Gordon equation: $(\Box + m^2)B(x)\,|\Omega\rangle = 0$. This property is due to the fact that $h(p)$ restricts the spectral measure[4] in the expansion (12.10) to the hyperboloid $p^2 = m^2$, therefore to the projector over the one particle state. This result can be extended to any state of the form $B(x)\,|\Psi\rangle$, which shows that $B(x)$ satisfies a free field equation of motion. With the help of $B(x)$ we construct the free field operators for *in* and *out* states. For example, the annihilation operator $a_{\text{in}}(\boldsymbol{p})$ is given by equation (7.8)

$$a_{\text{in}}(\boldsymbol{p}) = \mathrm{i} \int_{x_0 \to -\infty} \mathrm{d}^3 x \left(e^{ip\cdot x} \overleftrightarrow{\partial_0} B(x) \right) \tag{12.12}$$

Similarly, we construct $a_{\text{out}}(\boldsymbol{p})$ by taking $x^0 \to +\infty$. The ϕ_{in} and ϕ_{out} are then constructed out of the corresponding creation and annihilation operators following equation (7.7).

12.2 The Reduction Formula

Let us now return to the description of a scattering experiment. At t very large and negative we have the initial state formed by n well separated free particles with momenta q_i, $i = 1, \ldots, n$.[5] We denote it by $|\boldsymbol{q}_1, \ldots, \boldsymbol{q}_n; \text{in}\rangle$. Similarly, at t very large and positive, we have the final state $|\boldsymbol{p}_1, \ldots, \boldsymbol{p}_m; \text{out}\rangle$ of m free particles with momenta p_a, $a = 1, \ldots, m$. Since the initial and final particles are free, we have

$$p_a^2 = q_i^2 = m^2 \tag{12.13}$$

---

[4]i.e. the support in momentum space of $\phi\,|\Omega\rangle$

[5]Strictly speaking these properties are mutually inconsistent: particles which are spatially well separated cannot be described by plane waves with definite momenta which are not localisable. However, with the price of introducing a rather heavy formalism, it is straightforward to rewrite this section using wave packets.

We compute next the scalar product of the *in* and the *out* states. We obtain in steps

$$\langle \text{out}; \boldsymbol{p}_m, \ldots, \boldsymbol{p}_1 | \boldsymbol{q}_1, \ldots, \boldsymbol{q}_n; \text{in} \rangle = \langle \text{out}; \boldsymbol{p}_m, \ldots, \boldsymbol{p}_1 | a_{\text{in}}^\dagger(\boldsymbol{q}_1) | \boldsymbol{q}_2, \ldots, \boldsymbol{q}_n; \text{in} \rangle$$

$$= -\mathrm{i} \lim_{q_1^2 \to m^2} \int_{t \to -\infty} \mathrm{d}^3 x \, \langle \text{out}; \boldsymbol{p}_m, \ldots, \boldsymbol{p}_1 | \, [e^{-\mathrm{i} q_1 \cdot x} \overset{\leftrightarrow}{\partial_0} B(x)] \, | \boldsymbol{q}_2, \ldots, \boldsymbol{q}_n; \text{in} \rangle$$

$$= \mathrm{i} \lim_{q_1^2 \to m^2} \int \mathrm{d}^4 x \frac{\partial}{\partial t} \langle \text{out}; \boldsymbol{p}_m, \ldots, \boldsymbol{p}_1 | \, [e^{-\mathrm{i} q_1 \cdot x} \overset{\leftrightarrow}{\partial_0} B(x)] \, | \boldsymbol{q}_2, \ldots, \boldsymbol{q}_n; \text{in} \rangle$$

$$- \mathrm{i} \lim_{q_1^2 \to m^2} \int_{t \to +\infty} \mathrm{d}^3 x \, \langle \text{out}; \boldsymbol{p}_m, \ldots, \boldsymbol{p}_1 | \, [e^{-\mathrm{i} q_1 \cdot x} \overset{\leftrightarrow}{\partial_0} B(x)] \, | \boldsymbol{q}_2, \ldots, \boldsymbol{q}_n; \text{in} \rangle$$

$$= \mathrm{i} \lim_{q_1^2 \to m^2} \int \mathrm{d}^4 x \frac{\partial}{\partial t} \langle \text{out}; \boldsymbol{p}_m, \ldots, \boldsymbol{p}_1 | \, [e^{-\mathrm{i} q_1 \cdot x} \overset{\leftrightarrow}{\partial_0} B(x)] \, | \boldsymbol{q}_2, \ldots, \boldsymbol{q}_n; \text{in} \rangle$$

$$+ \langle \text{out}; \boldsymbol{p}_m, \ldots, \boldsymbol{p}_1 | a_{\text{out}}^\dagger(\boldsymbol{q}_1) | \boldsymbol{q}_2, \ldots, \boldsymbol{q}_n; \text{in} \rangle \tag{12.14}$$

where we have used the definition of creation and annihilation operators for *in* and *out* states of equation (12.12).[6] Let us look at the last term: it is non-zero only if $\boldsymbol{q}_1$ is equal to the momentum of one among the final particles, $\boldsymbol{p}_a$. If this is the case, this particle goes unaffected from the initial to the final state. In the terminology we introduced in section 11.5.2 this term contributes only to disconnected diagrams. The same will be true if any of the momenta of the incoming particles equals any of the momenta of the outgoing particles. Therefore, without loss of generality, we can assume that $\boldsymbol{p}_a \neq \boldsymbol{q}_i$ for all $a = 1, \ldots, m$ and $i = 1, \ldots, n$. This way we focus on the connected part of the scattering amplitude.

We focus now on the remaining term. We perform the differentiation with respect to time and take into account the fact that $B(x)$ satisfies the Klein–Gordon equation $\partial_t^2 B(x) = (\Delta - m^2) B(x)$. We obtain

$$\langle \text{out}; \boldsymbol{p}_m, \ldots, \boldsymbol{p}_1 | \boldsymbol{q}_1, \ldots, \boldsymbol{q}_n; \text{in} \rangle_c$$

$$= \mathrm{i} \lim_{q_1^2 \to m^2} (q_1^2 - m^2) \int \mathrm{d}^4 x \, e^{-\mathrm{i} q_1 \cdot x} \langle \text{out}; \boldsymbol{p}_m, \ldots, \boldsymbol{p}_1 | \, B(x) \, | \boldsymbol{q}_2, \ldots, \boldsymbol{q}_n; \text{in} \rangle \tag{12.15}$$

where the subscript c means "connected".

Several important remarks are in order about equation (12.15). First, the limit on q_1^2 shows that the right-hand side vanishes unless the matrix element has a pole at $q_1^2 = m^2$. The resulting expression is i times the residue of that pole. Second, since q_1^2 is restricted to be equal to m^2, we can replace $B(x)$ by $\phi(x)$ in the matrix element because $h(p)$ equals 1 on the mass hyperboloid. Third, and very important, the only property of the field operator $\phi(x)$ that we actually used was the fact that the coefficient C_1 in the expansion (12.10) was non-zero. It means that any operator that has the right quantum numbers to produce a one-particle state out of the vacuum, appropriately rescaled, can be used in formula (12.15).

---

[6]In these formulae we should have included a numerical multiplicative factor C_1 since we isolated only the one particle state, but we have absorbed it in a rescaling of the field.

We can continue, repeating the above steps with a second particle from either the incoming or the outgoing ones. Let us, for example, choose the particle with momentum $\boldsymbol{p}_1$ in the final state, for which we have

$$\langle \text{out}; \boldsymbol{p}_m, \ldots, \boldsymbol{p}_1 | = \langle \text{out}; \boldsymbol{p}_m, \ldots, \boldsymbol{p}_2 | \, a_{\text{out}}(\boldsymbol{p}_1) \tag{12.16}$$

We repeat the steps leading to equation (12.14) with only one difference: when we rewrite $a_{\text{out}}(\boldsymbol{p}_1)$ as $a_{\text{in}}(\boldsymbol{p}_1)$, the latter is placed on the left of the field ϕ and does not act directly on the *in* state on the right. We go around this problem by time ordering the product of the field operators and obtain

$$\langle \text{out}; \boldsymbol{p}_m, \ldots, \boldsymbol{p}_1 | \boldsymbol{q}_1, \ldots, \boldsymbol{q}_n; \text{in} \rangle_c = \mathrm{i}^2 \lim_{q_1^2 \to m^2} \lim_{p_1^2 \to m^2} (q_1^2 - m^2)(p_1^2 - m^2)$$

$$\times \int \mathrm{d}^4x \, \mathrm{d}^4y \, \mathrm{e}^{-\mathrm{i}q_1 \cdot x} \mathrm{e}^{\mathrm{i}p_1 \cdot y} \, \langle \text{out}; \boldsymbol{p}_m, \ldots, \boldsymbol{p}_2 | \, T(\phi(x)\phi(y)) \, | \boldsymbol{q}_2, \ldots, \boldsymbol{q}_n; \text{in} \rangle \tag{12.17}$$

It is now straightforward to continue this reduction (see Problem 12.3) until all particles in the *in* and *out* states are exhausted. We end up with[7]

$$\langle \text{out}; \boldsymbol{p}_m, \ldots, \boldsymbol{p}_1 | \boldsymbol{q}_1, \ldots, \boldsymbol{q}_n; \text{in} \rangle_c = \mathrm{i}^{n+m} \prod_{i=1}^{n} \prod_{a=1}^{m} \lim_{q_i^2, p_a^2 \to m^2} (q_i^2 - m^2)(p_a^2 - m^2)$$

$$\times \int \mathrm{d}^4x_i \, \mathrm{d}^4y_a \, \mathrm{e}^{-\mathrm{i}q_i \cdot x_i} \mathrm{e}^{\mathrm{i}p_a \cdot y_a} \, \langle 0 | \, T(\phi(x_i)\phi(y_a)) \, | 0 \rangle \tag{12.18}$$

This is our final formula, called *the reduction formula*. It expresses the connected part of a scattering amplitude in terms of the vacuum expectation value of the time-ordered product of field operators. But this is precisely the quantity we know how to compute in the perturbation expansion, see section 11.5.

An important property of the scattering amplitude can immediately be deduced from this expression. We derived (12.18) by considering the scattering of n initial particles with momenta $\boldsymbol{q}_1, \ldots, \boldsymbol{q}_n$ to m final particles with momenta $\boldsymbol{p}_1, \ldots, \boldsymbol{p}_m$. It is a function $\mathrm{i}\,T_{fi}(\boldsymbol{p}_1, \ldots, \boldsymbol{p}_m; \boldsymbol{q}_1, \ldots, \boldsymbol{q}_n) = \langle \text{out}; \boldsymbol{p}_m, \ldots, \boldsymbol{p}_1 | \boldsymbol{q}_1, \ldots, \boldsymbol{q}_n; \text{in} \rangle_c$ with the four-momenta satisfying the energy–momentum conservation $q_1 + q_2 + \ldots - p_1 - p_2 - \ldots = 0$. We note that the only difference between the qs and the ps is in the sign with which they appear, both in the exponentials of formula (12.18) as well as in the expression for the energy-momentum conservation. It follows that the same function S will describe the scattering amplitude $\langle \text{out}; \boldsymbol{p}_m, \ldots, \boldsymbol{p}_2 | \boldsymbol{q}_1, \ldots, \boldsymbol{q}_n, -\boldsymbol{p}_1; \text{in} \rangle_c$ of $n+1$ initial particles with momenta $\boldsymbol{q}_1, \ldots, \boldsymbol{q}_n, -\boldsymbol{p}_1$ to $m-1$ final particles with momenta $\boldsymbol{p}_2, \ldots, \boldsymbol{p}_m$. In other words, we can move particles between the initial and the final states by changing the sign of their momentum. Let us make this statement precise in an example.

Consider the $2 \to 2$ scattering of particles with four-momenta $q_1, q_2 \to p_1, p_2$. By Lorentz invariance the amplitude is a function of the scalar products we can build out of the momenta. For the process at hand, there are two independent such products.

---

[7]We hope the notation is obvious: For example, the products over i and a of $(\phi(x_i)\phi(y_a))$ mean $(\phi(x_1)\phi(x_2)\ldots\phi(x_n)\phi(y_1)\phi(y_2)\ldots\phi(y_m))$

They are usually expressed in terms of the, so-called, *Mandelstam variables* s, t and u, which are defined by

$$(q_1 + q_2)^2 = (p_1 + p_2)^2 \equiv s$$
$$(q_1 - p_1)^2 = (q_2 - p_2)^2 \equiv t$$
$$(q_1 - p_2)^2 = (q_2 - p_1)^2 \equiv u \qquad (12.19)$$

and satisfy the condition $s + t + u = 4m^2$. Note that s represents the square of the total energy in the centre-of-mass frame, while t and u are the squares of the two invariant momentum transfers. For the process $q_1, q_2 \to p_1, p_2$ the physical kinematical region is $s \geq 4m^2$, $t \leq 0$ and $u \leq 0$. Let us now interchange one initial and one final particle, for example q_1 and p_1. According to the rule we must put $q_1 \to -p_1$ and $p_1 \to -q_1$. The new process is $-p_1, q_2 \to -q_1, p_2$, and in terms of the Mandelstam variables, it is related to the initial by

$$t \leftrightarrow t , \quad s \leftrightarrow u \qquad (12.20)$$

i.e. the new process is described by the same function $\mathrm{i} T_{fi}$, but in the region $u \geq 4m^2$, $t \leq 0$ and $s \leq 0$. The two regions are related by analytic continuation. We will not give a general discussion of the analyticity properties of the scattering amplitude, but in all explicit examples we will compute, they can be verified directly. This property, namely the same analytic function describing the amplitudes of such related processes in the corresponding kinematical regions, is called *crossing symmetry*.

12.3 The Feynman Rules for the Scattering Amplitude

12.3.1 One scalar field

In writing the reduction formula (12.18) we have considered a single scalar field $\phi(x)$. We have shown in section 11.5 that the correlation functions which emerge on the right-hand side are obtained, order by order, in the perturbation expansion, by products of free Feynman propagators. In particular, we have a propagator for every external line given by $\mathrm{i}(p_a^2 - m^2 + \mathrm{i}\epsilon)^{-1}$ and $\mathrm{i}(q_i^2 - m^2 + \mathrm{i}\epsilon)^{-1}$. These propagators on the external lines of the relevant graphs are precisely cancelled by the corresponding factors in the reduction formula. Therefore, the Feynman rules for the perturbative computation of the Green functions, we derived in section 11.5, can be trivially adapted to give directly the scattering amplitudes by the following additional steps:

- Consider only connected diagrams.
- Drop the propagators of the external lines.[8]
- Put all external momenta on the mass shell.

12.3.2 Quantum electrodynamics

Extending the rules we obtained for the scalar field to the scattering amplitude of particles carrying charge and/or spin requires only simple modifications. However, going

---

[8]A diagram in which the propagators of the external lines have been omitted is called *amputated*.

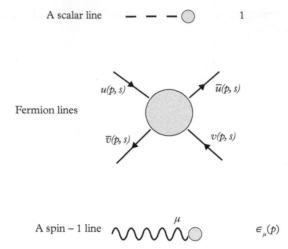

A scalar line — — — ◯   1

Fermion lines $u(p, s)$ $\bar{u}(p, s)$

$\bar{v}(p, s)$ $v(p, s)$

A spin – 1 line 〜〜〜◯ $^{\mu}$ $\epsilon_{\mu}(p)$

Fig. 12.1 The scalar, fermion and vector boson external lines. For the fermions the lines on the left are considered to be incoming and those on the right outgoing.

to quantum electrodynamics will require some extra work. The reason is that one of our basic assumptions so far has been the absence of zero mass particles. This assumption fails for QED since the photon is massless and, as a result, the electromagnetic interactions are long range. We go around this difficulty using the following artifact: in any given gauge, we introduce a fictitious mass μ^2 for the photon, thus replacing the factor $1/k^2$ in the photon propagator by $1/(k^2 - \mu^2 + i\epsilon)$. As long as the parameter μ is different from zero, everything we have said until now applies. We could re-derive the reduction formula including Dirac and (massive) Maxwell fields (see Problem 12.4), but we can easily guess the answer. As we have explained in section 11.5.3, the fermion lines in a Feynman diagram are oriented with the arrows indicating the flow of charge. Furthermore, the vacuum expectation value of the time-ordered product of the fields is a matrix with spinor indices for the external fermion lines and vector indices for the external photon lines. On the other hand, the scattering amplitude is a scalar, so these indices should be contracted with similar ones characterising the external particles. The resulting rules are summarised in Figure 12.1.

Note also that the anti-commutation property of the Dirac field leads to an extra rule, which is often added to the previous ones. It is easy to check that diagrams in which a positron line goes through the diagram from the initial to the final state have an extra minus sign because they require an odd number of transpositions of a fermion operator. See Problem 12.7.

The resulting Feynman rules depend on the unphysical parameter μ. At the end of the calculation we must take the limit $\mu \to 0$. As we shall see in explicit examples, this limit is often singular, which means that, strictly speaking, there is no S-matrix for QED. However, in Chapter 17 we shall show that this problem has a natural solution which involves a more precise definition of observable quantities in a QED scattering experiment.

A final remark: it is straightforward to extend the property of crossing symmetry that we introduced in section 12.2 to cover the more general case of scattering of particles carrying charge and/or spin. The result, see Problem 12.4, is that the symmetry also involves the change of a particle to its anti-particle, whenever we transfer it from one side of the reaction to the other.

In Appendix A we give a complete list of the Feynman rules for the theories we study in this book.

12.4   The Transition Probability

We would like next to compute the transition probability w_{fi} from an initial state $|i\rangle = |q_1, \ldots, q_n; \alpha_i\rangle$ to a final state $|f\rangle = |p_1, \ldots, p_m; \alpha_f\rangle$. The parameters α_i and α_f stand for spin and internal symmetry indices associated with the initial and final particles. In most formulae they will be suppressed in order to simplify the notation. To keep the discussion as simple as possible we will also assume that the particles are described by plane waves, although, as we have already pointed out, this description is only approximate and we should instead use wave packets.

Given the probability amplitude iT_{fi} for the process $|i\rangle = |q_1, \ldots, q_n\rangle \to |f\rangle = |p_1, \ldots, p_m\rangle$, the probability of a process leading to a final state with the particles having momenta in a set $\mathcal{V}$, defined by the ranges $(p_a, p_a + dp_a), a = 1, \ldots, m$, of the individual momenta, should be proportional to

$$dw_{fi} \sim |T_{fi}|^2 d\Omega_1 \ldots d\Omega_m \tag{12.21}$$

with, see equation (2.19),

$$d\Omega_a = \frac{1}{(2\pi)^3} \frac{d^3 p_a}{2E_a} \ , \quad E_a = \sqrt{p_a^2 + m_a^2} \ , \quad a = 1, 2, \ldots, m \tag{12.22}$$

in which we allow the particles to have in general different masses. The matrix element iT_{fi} can be computed following the rules we derived in section 12.3. Its general form is

$$T_{fi} = (2\pi)^4 \delta^{(4)} \left(\sum_{i=1}^{n} q_i - \sum_{j=1}^{m} p_j \right) \mathcal{M}_{fi} \tag{12.23}$$

The four-dimensional δ-function expresses the overall energy-momentum conservation. As in the case of quantum mechanics, the expression (12.21) is singular since it contains the square of a delta function $\delta^{(4)}(P_{\text{in}} - P_{\text{out}})$. This is the price we pay for describing the particles using plane waves, which are not normalisable. It is as if a collision can occur at any point in space and at any moment in time, which means that (12.21) has to be properly normalised. It is easy to find the corresponding normalisation factor: One of the delta functions can be replaced by the total volume of space-time VT and introduce the probability of transition per unit volume and unit time

$$dw_{fi} \to \frac{dw_{fi}}{(\prod_1^n \rho_a) VT} \sim (2\pi)^4 \delta^{(4)} \left(\sum_{i=1}^{m} p_i - \sum_{j=1}^{n} q_j \right) \frac{1}{\prod_1^n \rho_a} |\mathcal{M}_{fi}|^2 d\Omega_1 \ldots d\Omega_m \tag{12.24}$$

In writing (12.24) we have included in the normalisation factor the product $\prod_1^n \rho_a$ with ρ_a representing the densities of the initial particles. The physical reason is clear and it is the same as the one we used in potential scattering. The quantity, which has an intrinsic physical meaning, is the transition probability per unit volume and unit time *and* unit initial densities.

12.4.1   The cross section

The formalism we developed for the reduction formulae applies, in principle, to any number of initial and/or final particles. However, in describing actual scattering experiments, we will consider only the case of two incoming particles: $A + B \to C + D + \ldots$ It is for such processes that we can extend the definition of the cross section given in Chapter 4. In order to simplify the discussion, let us assume that we have only massive scalar particles. The cross section of a specific process, defined in (4.1), is the number of the corresponding events per unit time and unit target volume divided by the incident flux and by the initial densities. If the incoming beams contain one particle per unit volume, the division factor reduces to the incoming flux whose value is

$$v = |v_1 - v_2| \tag{12.25}$$

where v_1 and v_2 are the velocities of the two incoming particles, assumed to be collinear, since this is the most interesting case in actual particle physics experiments.[9] In general, the density of initial states is given by $\varrho_i = 2\mathcal{E}_i = 2\sqrt{q_i^2 + m_i^2}, i = 1, 2$ so the overall normalisation factor is $(v 2\mathcal{E}_1 2\mathcal{E}_2)^{-1}$. Let us compute the expression $\varrho_1 \varrho_2 v$ for collinear initial momenta q_1 and q_2 in the centre-of-mass frame

$$(\varrho_1 \varrho_2 v)^2 = 16\mathcal{E}_1^2 \mathcal{E}_2^2 \left| \frac{q_1}{\mathcal{E}_1} - \frac{q_2}{\mathcal{E}_2} \right|^2 = 16|q_1|^2 (\mathcal{E}_1 + \mathcal{E}_2)^2$$

$$= 16\left((\omega(q_1)\omega(q_2) + q_1^2)^2 - \omega^2(q_1)\omega^2(q_2) + q_1^2(\omega^2(q_1) + \omega^2(q_2)) - q_1^4 \right)$$

$$= 16\left((q_1 \cdot q_2)^2 - q_1^2 q_2^2 \right) \tag{12.26}$$

The last expression is invariant under Lorentz boosts in the direction of the momenta of the colliding particles and positive.

Therefore, the cross section can be written as

$$\sigma = \frac{(2\pi)^4}{\varrho_1 \varrho_2 v} \int \delta^{(4)}\left(q_1 + q_2 - \sum p_j \right) |\mathcal{M}_{fi}|^2 \mathrm{d}\Omega_1 \ldots \mathrm{d}\Omega_m \tag{12.27}$$

$$= \frac{1}{4\sqrt{(q_1 \cdot q_2)^2 - m_1^2 m_2^2}} \int (2\pi)^4 \delta^{(4)}\left(q_1 + q_2 - \sum p_j \right) |\mathcal{M}_{fi}|^2 \mathrm{d}\Omega_1 \ldots \mathrm{d}\Omega_m$$

where m_1 and m_2 are the masses of the incoming particles.

----

[9]The only particle physics collider in which the beams crossed with an appreciable angle (almost 15 degrees) was ISR. In all other cases the crossing angle is a small fraction of a degree.

12.4.2   The decay rate

Although the asymptotic theory on which the previous results are based applies, strictly speaking, only to stable particles,[10] we have argued that it can still be used provided the lifetime of the unstable particle is much longer than the typical interaction time. Under this assumption we obtain the decay rate by simply putting into equation (12.24) the number of initial particles $n = 1$. For spinless particles we obtain

$$w = \frac{(2\pi)^4}{2\mathcal{E}_q} \int \delta^{(4)}\Big(q - \sum p_j\Big) |\mathcal{M}_{fi}|^2 \mathrm{d}\Omega_1 \ldots \mathrm{d}\Omega_m \qquad (12.28)$$

which, in the rest frame of the decaying particle of mass M, becomes

$$w = \frac{1}{\tau} = \frac{(2\pi)^4}{2M} \int \delta\Big(M - \sum p_j^0\Big)\, \delta^{(3)}\Big(\sum \boldsymbol{p}_j\Big) |\mathcal{M}_{fi}|^2 \mathrm{d}\Omega_1 \ldots \mathrm{d}\Omega_m \qquad (12.29)$$

and is equal to the inverse of the mean lifetime τ of the unstable particle.

12.5   Examples

In this section we will apply the above rules to the detailed computation of two important examples of cross sections and one decay rate.

12.5.1   The electron Compton scattering

Our first example will be the cross section of the Compton process, namely the scattering of photons off electrons

$$\gamma(k_1, \epsilon_1) + e^-(p_1, \alpha_1) \rightarrow \gamma(k_2, \epsilon_2) + e^-(p_2, \alpha_2) \qquad (12.30)$$

where α_1 (α_2) and ϵ_1 (ϵ_2) denote the spin indices and the polarisation vectors of the initial (final) electron and photon, respectively. The theory is quantum electrodynamics with Lagrangian density given in equation (13.6). At lowest order in perturbation theory there are two diagrams contributing to this process shown in Figure 12.2. We start by computing the matrix element $\mathcal{M}_{fi}$ following the rules we derived in section 12.3. By convention, we write the matrices following the arrow of fermion lines in the inverse direction. We obtain [11]

$$\mathcal{M}_{fi} = -\mathrm{i}e^2 \overline{u}(\boldsymbol{p}_2, \alpha_2) \left[\frac{\not{\epsilon}_2^*(\not{p}_1 + \not{k}_1 + m)\not{\epsilon}_1}{(p_1 + k_1)^2 - m^2 + \mathrm{i}\epsilon} + \frac{\not{\epsilon}_1(\not{p}_1 - \not{k}_2 + m)\not{\epsilon}_2^*}{(p_1 - k_2)^2 - m^2 + \mathrm{i}\epsilon} \right] u(\boldsymbol{p}_1, \alpha_1)$$

$$\qquad (12.31)$$

$$= -\mathrm{i}e^2 \overline{u}(\boldsymbol{p}_2, \alpha_2) \left[\frac{p_1 \cdot \epsilon_1}{p_1 \cdot k_1}\not{\epsilon}_2^* - \frac{p_1 \cdot \epsilon_2^*}{p_1 \cdot k_2}\not{\epsilon}_1 + \frac{\not{\epsilon}_2^*\not{k}_1\not{\epsilon}_1}{2p_1 \cdot k_1} + \frac{\not{\epsilon}_1\not{k}_2\not{\epsilon}_2^*}{2p_1 \cdot k_2} \right] u(\boldsymbol{p}_1, \alpha_1)$$

where $\not{\epsilon}^* \equiv \epsilon_\mu^* \gamma^\mu$, and where, in addition, we used the fact that the external particles are on their mass shells, i.e. they satisfy $(\not{p}_i - m)u(p_i, \alpha_i) = 0$, $p_i^2 = m^2$ and $k_i^2 = 0, i = 1, 2,$

---

[10]It is obvious that an unstable particle cannot exist at $t \to \pm\infty$.

[11]To simplify the formulae we denote the Dirac spinors by $u(\boldsymbol{p}, \alpha)$ and $v(\boldsymbol{p}, \alpha)$ instead of $u^{(\alpha)}(\boldsymbol{p})$ and $v^{(\alpha)}(\boldsymbol{p})$.

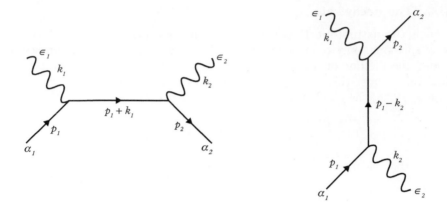

Fig. 12.2 The diagrams contributing to the electron Compton scattering in second order of perturbation theory.

which imply $(p_1+k_1)^2 - m^2 = 2p_1 \cdot k_1$ and $(p_1-k_2)^2 - m^2 = -2p_1 \cdot k_2$.[12] This expression simplifies further if we choose the initial electron rest frame, in which $p_1 = (m, 0, 0, 0)$, and the photon polarisation vectors $\epsilon_1 = (0, \boldsymbol{e}_1)$ and $\epsilon_2 = (0, \boldsymbol{e}_2)$ to have only spatial components perpendicular to $\boldsymbol{k}_1$ and $\boldsymbol{k}_2$, respectively. Then the first two terms in the square bracket vanish, and we end up with

$$\mathcal{M}_{fi} = -\mathrm{i}e^2 \overline{u}(\boldsymbol{p}_2, \alpha_2) \left(\frac{\not{\epsilon}_2^* \not{k}_1 \not{\epsilon}_1}{2m\omega_1} + \frac{\not{\epsilon}_1 \not{k}_2 \not{\epsilon}_2^*}{2m\omega_2} \right) u(\boldsymbol{p}_1, \alpha_1) \tag{12.32}$$

where $\omega_i = k_i^0 = |\boldsymbol{k}_i|$. Let us next compute $|\mathcal{M}_{fi}|^2$ in the simple case in which the initial electron is unpolarised and, furthermore, we do not measure the polarisation of the final one. This means, that we must sum over the final spin orientations and average over the initial ones. Let us define the quantity

$$\mathcal{M} = \overline{u}(\boldsymbol{p}_2, \alpha_2) \Gamma u(\boldsymbol{p}_1, \alpha_1)$$

with Γ any 4×4 matrix. Using the formula we obtained in Problem 7.5, we have

$$\sum_{\alpha_1, \alpha_2} |\mathcal{M}|^2 = \sum_{\alpha_1, \alpha_2} [\overline{u}(\boldsymbol{p}_2, \alpha_2) \Gamma u(\boldsymbol{p}_1, \alpha_1)][\overline{u}(\boldsymbol{p}_1, \alpha_1) \gamma^0 \Gamma^\dagger \gamma^0 u(\boldsymbol{p}_2, \alpha_2)]$$

$$= \mathrm{Tr} \left[(\not{p}_2 + m) \Gamma (\not{p}_1 + m) \gamma^0 \Gamma^\dagger \gamma^0 \right] \tag{12.33}$$

which, applied to the expression (12.32), gives

$$\frac{1}{2} \sum_{\alpha_1, \alpha_2} |\mathcal{M}_{fi}|^2 = \left[\frac{\omega_1^2 + \omega_2^2}{\omega_1 \omega_2} + 4(\epsilon_1 \cdot \epsilon_2)^2 - 2 \right] \tag{12.34}$$

Using energy-momentum conservation we can express ω_2, the energy of the outgoing photon, in terms of ω_1, the energy of the incident photon,

---

[12]It is a straightforward exercise to verify that the amplitude (12.31) is indeed gauge invariant, i.e. invariant under the substitution e.g. $\epsilon_2^\mu \to \epsilon_2^\mu + \lambda k_2^\mu$ for arbitrary constant λ.

and the laboratory frame scattering angle. A simple way to do it is to write

$$(k_1 - k_2)^2 = -2k_1 \cdot k_2 = -2\omega_1\omega_2(1 - \cos\theta) = (p_1 - p_2)^2 = 2m^2 - 2mE_2 = 2m(\omega_2 - \omega_1) \tag{12.35}$$

which implies, in particular

$$\omega_2 = \frac{\omega_1}{1 + (\frac{\omega_1}{m})(1 - \cos\theta)} \tag{12.36}$$

from which we can obtain the photon frequency shift as a function of the scattering angle.

The differential cross section is obtained from the general formula (12.27). For a two-particle final state the phase space integration is trivial because of the δ-function and we obtain

$$\int \frac{d^3k_2}{(2\pi)^3 2\omega_2} \frac{d^3p_2}{(2\pi)^3 2E_2} (2\pi)^4 \delta^{(4)}(p_2 + k_2 - p_1 - k_1) = \int d\Omega \frac{1}{(4\pi)^2} \frac{\omega_2^2}{m\omega_1} \tag{12.37}$$

where $d\Omega$ denotes the solid angle element in the direction of the scattered photon in the laboratory reference frame.

The kinematic factor in the denominator of (12.27) is $4\sqrt{(k_1 \cdot p_1)^2} = 4m\omega_1$.

Upon substitution into (12.27) we finally obtain

$$\frac{d\sigma}{d\Omega} = \frac{\alpha^2}{4m^2} \frac{\omega_2^2}{\omega_1^2} \left[\frac{\omega_1^2 + \omega_2^2}{\omega_1\omega_2} + 4(\epsilon_1 \cdot \epsilon_2)^2 - 2 \right] \tag{12.38}$$

where $\alpha = e^2/4\pi$ is the dimensionless fine structure constant. The differential cross section depends through ω_2 (12.36) on the photon scattering angle, and through $\epsilon_1 \cdot \epsilon_2 = -e_1 \cdot e_2 = -\cos\Theta$ on the angle Θ between the initial and final polarisation three-vectors. This last formula is called the *Klein–Nishina formula*.

In the low energy limit $\omega_1/m \to 0$, $\omega_2 \to \omega_1$ and we obtain

$$\frac{d\sigma}{d\Omega} = \frac{\alpha^2}{m^2}(\epsilon_1 \cdot \epsilon_2)^2 = \frac{\alpha^2}{m^2}(e_1 \cdot e_2)^2 \tag{12.39}$$

This is the Thomson formula, known from classical electrodynamics. If the incoming beam of photons is not polarised, we must average over the initial polarisations, that is to say over the vectors $\mathbf{e}_1$ orthogonal to $\mathbf{k}_1$,[13] and we find

$$\frac{d\sigma}{d\Omega} = \frac{\alpha^2}{2m^2}\left(1 - (\hat{\mathbf{k}}_1 \cdot \mathbf{e}_2)^2\right) \tag{12.40}$$

From this we learn that scattering of unpolarised radiation by free electrons can produce linearly polarised light. Let us look at ninety-degree scattering ($\theta = \pi/2$). In

---

[13]To compute the average over the initial polarisations we first write $\mathbf{e}_1(\alpha) = \mathbf{n}\cos\alpha + \mathbf{n} \times \hat{\mathbf{k}}_1 \sin\alpha$, where $\mathbf{n}$ is any unit vector perpendicular to $\mathbf{k}_1$. A convenient choice is often to take $\mathbf{n}$ perpendicular to the scattering plane, i.e. $\mathbf{n} = \hat{\mathbf{k}}_1 \times \hat{\mathbf{k}}_2/\sin\theta$. Then, the average of the product $e_{1i}e_{1j}$ is $\overline{e_{1i}e_{1j}} = (1/2\pi)\int_0^{2\pi} d\alpha\, e_{1i}e_{1j} = (\delta_{ij} - \hat{k}_{1i}\hat{k}_{1j})/2$.

this case, the two independent polarisations of the final photon can be chosen to be $e_2 = n = \hat{k}_1 \times \hat{k}_2$ and $e_2' = n \times \hat{k}_2 = -\hat{k}_1$. Then, according to (12.40) the differential cross section for the polarisation e_2' vanishes. The scattered photons at $\theta = \pi/2$ are linearly polarised, having polarisation $\hat{k}_1 \times \hat{k}_2$, perpendicular to the scattering plane. This property of Compton scattering may have important applications in early cosmology by inducing polarisation to the microwave background radiation.

Finally, if we do not observe the polarisation of the final photons, the corresponding differential cross section is the sum of (12.40) over the final polarisations, i.e. twice the average over the e_2, which is perpendicular to k_2. We obtain

$$\left(\frac{d\sigma}{d\Omega}\right)_{\text{unpolarised}} = 2 \times \frac{\alpha^2}{2m^2}\left(1 - \frac{1}{2}(\delta_{ij} - \hat{k}_{2i}\hat{k}_{2j})\hat{k}_{1i}\hat{k}_{1j}\right) = \frac{\alpha^2}{2m^2}(1 + \cos^2\theta) \quad (12.41)$$

and a total cross section

$$\sigma_{\text{total}} = \frac{8\pi\alpha^2}{3m^2} \quad (12.42)$$

a well known formula of classical electrodynamics.

12.5.2 The electron–positron annihilation into a muon pair

Our second example will be the cross section of the process

$$e^+ + e^- \rightarrow \mu^+ + \mu^- \quad (12.43)$$

This is a very important process for various reasons: First, electron–positron colliders have proven to be an extremely powerful tool to probe the structure of matter at very short distances. The reason is that the initial state, an electron and a positron, has the quantum numbers of the vacuum and, as such, it does not privilege any particular final state. Second, the amplitude for the reaction (12.43) can be computed in quantum electrodynamics, but also, it can be accurately measured experimentally. Therefore, it provides a good test of the fundamental theory. Third, because of these properties, it can be used to calibrate the accelerator and to provide a measure with which other processes can be compared.

As we know, the muons have the same spin and electric charge with the electrons, but their mass is about 207 times bigger. They are unstable with a lifetime of 2.2×10^{-6} s. Their electromagnetic interactions are the same as those of the electrons, so the interaction Lagrangian density relevant to the process (12.43) is given by

$$\mathcal{L}_I = -e\left[\overline{\psi}(x)\gamma_\mu\psi(x) + \overline{\Psi}(x)\gamma_\mu\Psi(x)\right]A^\mu(x) \quad (12.44)$$

where ψ and Ψ are the Dirac quantum fields for the electron and the muon respectively, A^μ is the electromagnetic four-potential and e the electric charge of the electron and the muon. There is only one diagram contributing to this process, the one of Figure 12.3. The (p, s) and (p', s') are the four-momenta and spin indices of the electron and the positron, respectively. Similarly, (q, r) and (q', r') are the momenta and the spins of the final muon and antimuon. Energy-momentum conservation implies $p + p' = k = q + q'$, where k is the momentum of the virtual photon.

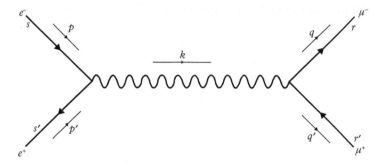

Fig. 12.3 The Feynman diagram for the reaction $e^- + e^+ \rightarrow \mu^- + \mu^+$. The arrows on the fermion lines indicate the charge flow, while those next to the lines indicate the momentum flow.

Using the Feynman rules we developed previously, we find for the amplitude for this reaction

$$\mathcal{M}_{fi} = \mathrm{i}e^2 \left[\bar{v}(p',s')\gamma_\mu u(p,s)\right] \frac{\eta^{\mu\nu}}{k^2} \left[\overline{U}(q,r)\gamma_\nu V(q',r')\right] \tag{12.45}$$

where u and v are the spinor wave functions for the initial electron and positron and U and V those of the final μ^- and μ^+.[14]

In most cases the electron and positron beams are unpolarised and we do not measure the polarisation of the final muons, (see however Problem 12.5). So, in computing the square of the amplitude, we sum over final polarisations and average over the initial ones, so that, according to (12.27), the cross section of interest in this case is

$$\mathrm{d}\sigma = \frac{1}{4\sqrt{(p\cdot p')^2 - m^4}}(2\pi)^4\delta^{(4)}\left(p + p' - q - q'\right)\frac{1}{4}\sum_{\mathrm{spins}}|\mathcal{M}_{fi}|^2\,\mathrm{d}\Omega_q\,\mathrm{d}\Omega_{q'} \tag{12.46}$$

where, suppressing the momentum and spin arguments of the spinors, the quantity $|\mathcal{M}_{fi}|^2$ is given by

$$|\mathcal{M}_{fi}|^2 = (e^4/(k^2)^2)(\bar{v}\gamma_\mu u)\,(\overline{U}\gamma^\mu V)(\bar{v}\gamma_\nu u)^\dagger\,(\overline{U}\gamma^\nu V)^\dagger$$

The summation over the spin indices simplifies this expression, because the spinors satisfy the completeness relations we proved in Problem 7.5. We thus obtain

$$\frac{1}{4}\sum_{\mathrm{spins}}|\mathcal{M}_{fi}|^2 = \frac{e^4}{4(k^2)^2}\mathrm{Tr}[(\not{p}' - m)\gamma_\mu(\not{p} + m)\gamma_\nu]\,\mathrm{Tr}[(\not{q} + M)\gamma^\mu(\not{q}' - M)\gamma^\nu] \tag{12.47}$$

where m is the mass of the electron and M that of the muon. The traces of the γ matrices can be easily computed using the identities (7.48) which show that the trace

[14]Following our discussion of section 12.3.2, we should replace the $1/k^2$ factor of the photon propagator by $1/(k^2 - \mu^2 + \mathrm{i}\epsilon)$, but it won't be necessary in this calculation.

of the product of an odd number of γ matrices vanishes and that of four γ matrices is given by

$$\text{Tr}\,\gamma_\mu\gamma_\nu\gamma_\rho\gamma_\sigma = 4\left(\eta_{\mu\nu}\eta_{\rho\sigma} - \eta_{\mu\rho}\eta_{\nu\sigma} + \eta_{\mu\sigma}\eta_{\nu\rho}\right) \tag{12.48}$$

Using all these formulae we obtain

$$\frac{1}{4}\sum_{\text{spins}}|\mathcal{M}_{fi}|^2 = \frac{8e^4}{(k^2)^2}\left(p'\cdot q\,p\cdot q' + p\cdot q\,p'\cdot q' + M^2 p\cdot p' + m^2 q\cdot q' + 2M^2 m^2\right)$$

In terms of the Mandelstam variables for the process at hand, taking also into account the conservation of energy and momentum $p+p' = q+q'$, we have $2p\cdot p' = s-2m^2$, $2q\cdot q' = s - 2M^2$, $2p\cdot q = 2p'\cdot q' = M^2 + m^2 - t$, $2p\cdot q' = 2p'\cdot q = M^2 + m^2 - u$, and upon substitution in the above Lorentz invariant expression, we obtain

$$\frac{1}{4}\sum_{\text{spins}}|\mathcal{M}_{fi}|^2 = e^4\left(\frac{s + 4(M^2 + m^2)}{s} + \frac{(t-u)^2}{s^2}\right) \tag{12.49}$$

To complete the calculation of the cross section (12.46) we must perform the phase space integral using equation (12.37). Putting all these factors together we obtain for the differential cross section in the centre-of-mass frame

$$\left(\frac{d\sigma}{d\Omega}\right)_{\text{cm}} = \frac{\alpha^2}{4s}\sqrt{\frac{s-4M^2}{s-4m^2}}\left[\frac{s+4(M^2+m^2)}{s} + \frac{(s-4m^2)(s-4M^2)}{s^2}\cos^2\theta\right] \tag{12.50}$$

where we have introduced the scattering angle θ in the centre-of-mass frame, i.e. the angle formed by the momenta $\boldsymbol{p}$ and $\boldsymbol{q}$ of the initial e^- and the final μ^-. It is given, in terms of Lorentz invariant quantities, by

$$\cos\theta = \frac{t-u}{\sqrt{(s-4m^2)(s-4M^2)}} \tag{12.51}$$

Following the same method and similar steps, we can compute the cross section for any elementary reaction in quantum electrodynamics. In the list of problems at the end of this chapter we propose the calculation of two other important processes, the *Bhabha scattering*, i.e., the electron–positron elastic scattering $e^- + e^+ \to e^- + e^+$, named after the Indian physicist H.J. Bhabha, and the *Møller scattering*, i.e. the electron–electron elastic scattering $e^- + e^- \to e^- + e^-$, named after the Danish physicist C. Møller. The first receives contributions from two diagrams, the annihilation diagram, which is the analogue of the one shown in Figure 12.3, and the one-photon exchange diagram in which a photon line is exchanged between the electron and the positron lines. The second has no annihilation diagram, but it has two exchange diagrams, in which the two electrons in the final state are interchanged. The resulting cross sections, which the reader is asked to verify, are given by

$$\left.\frac{d\sigma}{d\Omega}\right|_{\text{Bhabha}} = \frac{\alpha^2}{2s}\left[u^2\left(\frac{1}{s} + \frac{1}{t}\right)^2 + \left(\frac{t}{s}\right)^2 + \left(\frac{s}{t}\right)^2\right] \tag{12.52}$$

$$\frac{d\sigma}{d\Omega}\bigg|_{\text{Møller}} = \frac{\alpha^2}{2s}\left[s^2\left(\frac{1}{u}+\frac{1}{t}\right)^2 + \left(\frac{t}{u}\right)^2 + \left(\frac{u}{t}\right)^2\right] \tag{12.53}$$

where we have assumed $s \gg m^2$ and neglected the electron mass compared to the centre-of-mass energy of the process. Both expressions become singular when $t = 0$ and (12.53) also when $u = 0$. These singularities are due to the photon propagator in the exchange diagrams because of the zero photon mass. Had we used the fictitious mass propagator $1/(k^2 - \mu^2 + i\epsilon)$, they would have been shifted to the unphysical region $t = \mu^2$. They are what we have called earlier *infrared singularities* and they will be briefly discussed in Chapter 17.

The expressions (12.52) and (12.53) provide an example of the property we called in section 12.2 "crossing symmetry". The two processes $e^- + e^- \to e^- + e^-$ and $e^- + e^+ \to e^- + e^+$ are related by interchanging an initial electron with a final positron, which amounts to changing $s \leftrightarrow u$, while $t \leftrightarrow t$. And indeed, putting aside the common kinematical factor $1/s$, the two expressions (12.52) and (12.53) are related with each other by this substitution.

12.5.3 The charged pion decay rate

As a third example we study a decay process and we choose that of a charged pion. If we restrict ourselves to two-body final states, the two possible processes are

$$\pi^+ \to e^+ + \nu_e \quad \text{and} \quad \pi^+ \to \mu^+ + \nu_\mu \tag{12.54}$$

The presence of the neutrinos shows that they are due to the weak interactions. Experimentally, both modes have been measured and the ratio of the corresponding partial decay rates is found to be

$$R = \frac{\Gamma(\pi^+ \to e^+ + \nu_e)}{\Gamma(\pi^+ \to \mu^+ + \nu_\mu)} \sim 10^{-4} \tag{12.55}$$

It is this number we would like to reproduce. Note that phase space favours the electron mode because of the small mass of the electron, so this counterintuitive result must be of dynamical origin. We want to find the Hamiltonian responsible for these decays. Let us denote by $\Phi_\pi(x)$ the pion field, $\Psi_l(x)$ the fields of the charged leptons, l stands for e or μ, and $\psi_{\nu_l}(x)$ the fields of the corresponding neutrinos. We will try the two simplest forms for the interaction Hamiltonian: a scalar coupling $\mathcal{H}_I^{(S)}$ and a vector coupling $\mathcal{H}_I^{(V)}$ given, respectively, by

$$\mathcal{H}_I^{(S)} = g_\pi \Phi_\pi \overline{\Psi}_l(1+\gamma^5)\psi_{\nu_l} + \text{h.c.}, \quad \mathcal{H}_I^{(V)} = \frac{if_\pi}{m_\pi}(\partial^\mu \Phi_\pi)\overline{\Psi}_l\gamma_\mu(1+\gamma^5)\psi_{\nu_l} + \text{h.c.} \tag{12.56}$$

where h.c. stands for "hermitian conjugate". The g_π and f_π are two coupling constants. In writing these Hamiltonians we have made the following assumptions: (i) parity conservation is violated and the neutrinos are only left-handed, (ii) we have an electron–muon universality (the same form of the interaction Hamiltonian with the same coupling constants for electrons and muons). The decay diagram is shown in

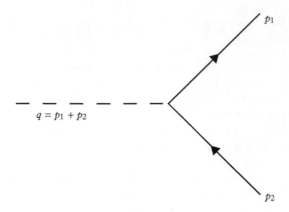

Fig. 12.4 The pion decay diagram.

Figure 12.4. The decay amplitudes for the scalar and vector coupling, respectively, are given by

$$\mathcal{M}^{(S)} = -ig_\pi \bar{u}_{\nu_l}(\boldsymbol{p}_1, \alpha_1)(1 + \gamma^5)v_l(\boldsymbol{p}_2, \alpha_2) \tag{12.57}$$

$$\mathcal{M}^{(V)} = \frac{f_\pi}{m_\pi} \bar{u}_{\nu_l}(\boldsymbol{p}_1, \alpha_1)(1 + \gamma^5)(\not{p}_1 + \not{p}_2)v_l(\boldsymbol{p}_2, \alpha_2) \tag{12.58}$$

$$= -m_l \frac{f_\pi}{m_\pi} \bar{u}_{\nu_l}(\boldsymbol{p}_1, \alpha_1)(1 + \gamma^5)v_l(\boldsymbol{p}_2, \alpha_2)$$

where we have used the Dirac equation and we have set the neutrino mass equal to zero. In both cases $|\mathcal{M}|^2$, summed over the final spin states, is proportional to the trace

$$\sum_{\text{spins}} |\mathcal{M}|^2 \sim \text{Tr}\left[\not{p}_2(\not{p}_1 - m_l)\right] = 4p_1 \cdot p_2 = 2(m_\pi^2 - m_l^2) \tag{12.59}$$

It is independent of the final momenta, so the phase space factor is the one we compute in Problem 12.10. The final result for the ratio (12.55) is

$$R^{(S)} = \frac{(m_\pi^2 - m_e^2)^2}{(m_\pi^2 - m_\mu^2)^2} \quad , \quad R^{(V)} = \frac{m_e^2}{m_\mu^2} \frac{(m_\pi^2 - m_e^2)^2}{(m_\pi^2 - m_\mu^2)^2} \tag{12.60}$$

The scalar coupling gives the naïvely expected result, namely that the electron mode is favoured by a factor roughly equal to 5.5 and it is ruled out by the measurement. On the contrary, the vector coupling gives an extra factor $(m_e/m_\mu)^2 \simeq 2.32 \times 10^{-5}$, which leads to a final value 1.3×10^{-4} for the ratio, in very good agreement with the experiment. Historically, this decay gave an excellent confirmation of the vectorial nature of the weak interaction Hamiltonian. The identification of the electron mode with the correct ratio was the first great success of CERN, where it was found in 1958.

12.6 Problems

Problem 12.1 This is a continuation of Problem 11.1. Show that invariance under CPT implies:

1. $\mathcal{M}(A \to B) = \mathcal{M}(\bar{B}' \to \bar{A}')$, where $\mathcal{M}(A \to B)$ is the transition amplitude for the process $A \to B$, "bar" denotes replacement of particles with antiparticles and vice versa, and prime denotes reversal of spin directions.

2. The equality of the lifetimes between unstable particles and anti-particles.

Problem 12.2 Show that the field $B(x)$ defined in equation (12.11) satisfies the free Klein–Gordon equation.

Problem 12.3 Prove the general reduction formula of equation (12.18).

Problem 12.4 Derive the reduction formula for quantum electrodynamics and establish the properties of the scattering amplitude under crossing symmetry.

Problem 12.5 For some physical questions the use of polarised electron and positron beams in $e^+ - e^-$ colliders allows to extract information, which is washed out when we average over the initial spins. Repeat the calculation of the cross section for the process $e^- + e^+ \to \mu^- + \mu^+$ we presented above and find the dependence on the spins of the initial particles.

Problem 12.6 The spin-$\frac{1}{2}$ fermions ψ and Ψ, with masses m_ψ and m_Ψ, respectively, interact with the scalar ϕ with mass M via the Yukawa interaction $\mathcal{L} = g(\bar{\psi}\psi + \overline{\Psi}\Psi)\phi$.

1. Write the scattering amplitude for the processes $\Psi + \psi \to \Psi + \psi$ and $\Psi + \bar{\psi} \to \Psi + \bar{\psi}$.

2. Consider the non-relativistic limit of these amplitudes and show that up to an inessential factor, having to do with the different normalisation of the states, it is identical with the amplitude obtained in Problem 4.5 for scattering off the Yukawa potential U_Y. Conclude that the interaction of ψ and of $\bar{\psi}$ with Ψ is attractive.

Problem 12.7 Compute the differential cross section for the elastic electron–positron scattering $e^- + e^+ \to e^- + e^+$ in the tree approximation. Show that at the limit when the mass of the electron is negligible compared to the centre-of-mass energy, we obtain the formula (12.52).

Problem 12.8 Compute the differential cross section for the elastic electron–electron scattering $e^- + e^- \to e^- + e^-$ in the tree approximation. Show that at the limit of negligible electron mass we obtain the formula (12.53).

Problem 12.9 Repeat the calculations of the processes $e^- + e^+ \to \mu^- + \mu^+$, $e^- + e^+ \to e^- + e^+$ and $e^- + e^- \to e^- + e^-$ in the tree approximation using the photon propagator given in equation (7.100) which depends on the gauge parameter λ. Prove that the scattering amplitudes are λ-independent.

Problem 12.10 Phase space calculations. If in the scattering formula (12.27) we take the matrix element $\mathcal{M}_{fi}$ to be a constant, independent of the momenta p_i, the calculation of the cross section is reduced to the integral over the final on-shell momenta constrained by the δ-function of the overall energy-momentum conservation. For n final particles we call the resulting factor the *n-particle phase space* and we denote it by R_n. We define:

$$\mathrm{d}R_n(P; p_1, \ldots, p_n) = \delta^4\left(P - \sum_i p_i\right) \prod_{i=1}^{n} \mathrm{d}\Omega_i$$

1. Compute R_2 and derive formulae (12.60).
2. Show that R_n satisfies the recursion formula:

$$\mathrm{d}R_n(P; p_1, \ldots, p_n) = \mathrm{d}R_j(q; p_1, \ldots, p_j)\, \mathrm{d}R_{n-j+1}(P; q, p_{j+1}, \ldots, p_n)(2\pi)^3 \mathrm{d}q^2$$

where q^2 is the square of the total energy in the j-particle centre-of-mass system: $q^2 = (\sum_{i=1}^{j} p_i^0)^2 - (\sum_{i=1}^{j} \boldsymbol{p}_i)^2$.

Problem 12.11 In this book we shall often study decays producing three particles in the final state. Examples are the neutron β-decay $n \to p + e^- + \overline{\nu}$, the muon decay $\mu^+ \to e^+ + \nu_e + \overline{\nu}_\mu$, various kaon decays etc. Experimentally we often measure the energy spectrum or the angular distribution of one of the final particles, which means that we must integrate over the phase space of the other two. The purpose of this exercise is to learn how to perform such integrals.

We consider the decay $A(P_i) \to B(P_f) + c_1(k_1) + c_2(k_2)$ with the energy-momentum conservation $P_i = P_f + k_1 + k_2$. We want to integrate over the phase space of c_1 and c_2, which means that we must compute integrals of the form

$$I_F = \int F \frac{\mathrm{d}^3 k_1}{E_1} \frac{\mathrm{d}^3 k_2}{E_2} \delta^4(P_i - P_f - k_1 - k_2)$$

where F is a Lorentz invariant function describing the decay process. Let $Q = k_1 - k_2$. We want to compute I_F for : (1) $F_1 = 1$, (2) $F_2 = Q^2$, (3) $F_3 = P_1 \cdot Q$ and (4) $F_4 = (P_1 \cdot Q)(P_2 \cdot Q)$ where P_1 and P_2 are four-vectors which do not depend on Q.

13

Gauge Interactions

In the previous chapter, we wrote several interaction terms in quantum field theory with the criteria of Lorentz invariance and simplicity. In this chapter, we want to introduce a geometric method to construct a new class of interactions, which will turn out to be very important in the study of the fundamental forces among elementary particles. It is the method of gauge invariance.

13.1   The Abelian Case

In section 8.3.3 we derived the coupling of an electron to the electromagnetic field by imposing the invariance of the Dirac equation under local or *gauge* transformations of the wave function. Here we want to put this result in a more general context.

We start with the free Dirac Lagrangian (7.60)

$$\mathcal{L} = \overline{\psi}(x)(i\slashed{\partial} - m)\psi(x) \tag{13.1}$$

It is invariant under the $U(1)$ group of phase transformations acting on the fields $\psi(x)$, namely

$$\psi(x) \longrightarrow \psi'(x) = e^{ie\theta}\psi(x) \tag{13.2}$$

where, for future convenience, we introduced an arbitrary constant e multiplying the parameter θ of the transformation. As we have seen, this invariance leads to the conservation of the Dirac current $j_\mu = \overline{\psi}(x)\gamma_\mu\psi(x)$. It is a *global* invariance, in the sense that the parameter θ in (13.2) is independent of the space-time point x.

Now, we shall ask the following question: Is it possible to extend this invariance to a *local* one, namely one in which the constant θ is replaced by an arbitrary differentiable function of x, $\theta(x)$?

There may be various, essentially aesthetic, reasons for which we may wish to do that. In physical terms, we could argue that the invariance under the global phase transformations means that the phase of the field is not a measurable quantity. Therefore, it is natural to look for a formalism, which makes it possible to choose this phase locally. It is obvious that the term $\overline{\psi}(x)\psi(x)$ is invariant under such local transformations. More generally, every term which is polynomial in the fields with no derivatives will be invariant under local transformations if it is invariant under the corresponding global ones. Problems will arise only with terms containing derivatives of the fields, as we explained in section 2.4. Indeed, a derivative couples the fields at neighbouring points, $\overline{\psi}(x)$ and $\psi(x+\epsilon)$. If the phase is defined locally, the first will be multiplied by $e^{-ie\theta(x)}$ and the second by $e^{ie\theta(x+\epsilon)}$. In order to restore invariance, we need a means

Elementary Particle Physics. John Iliopoulos and Theodore N. Tomaras, Oxford University Press.
© John Iliopoulos and Theodore N. Tomaras (2021). DOI: 10.1093/oso/9780192844200.003.0013

to connect the two points. This requirement can be made more precise using a mathematical language. The corresponding branch of mathematics is called *differential geometry* and studies the geometrical properties of differentiable manifolds. For the case at hand we can consider the manifold of the field configurations $\psi(x)$ defined on the four-dimensional Minkowski space. We can prove that the global transformations of (13.2) imply only a trivial structure on this manifold with vanishing curvature. A much richer geometry can be obtained by allowing the group action to depend on the point x, in other words by allowing local, or *gauge*, transformations.

Whichever one's motivations may be, physical or mathematical, it is clear that (13.1) is not invariant under (13.2) with θ replaced by $\theta(x)$ because the derivative gives rise to a term proportional to $\partial_\mu \theta(x)$. In order to restore invariance, we must modify (13.1), in which case it will no longer describe a free Dirac field; invariance under gauge transformations leads to the introduction of interactions. Both physicists and mathematicians know the answer to the particular case of (13.1) and we used it extensively in the first chapters of this book. To achieve our goal, we had to introduce a new field $A_\mu(x)$, given at this stage the generic name "gauge field", and replace the derivative operator ∂_μ by a "covariant derivative" D_μ, i.e.

$$\partial_\mu \;\longrightarrow\; D_\mu = \partial_\mu + ieA_\mu \tag{13.3}$$

D_μ is called "covariant" because it satisfies

$$D_\mu \left[e^{ie\theta(x)} \psi(x) \right] = e^{ie\theta(x)} D_\mu \psi(x) \tag{13.4}$$

valid if, at the same time, the gauge field $A_\mu(x)$ undergoes the transformation

$$A_\mu(x) \;\longrightarrow\; A'_\mu(x) = A_\mu(x) - \partial_\mu \theta(x) \tag{13.5}$$

Under the substitution (13.3), the Dirac Lagrangian density becomes

$$\mathcal{L} = \overline{\psi}(x)(i\slashed{D} - m)\psi(x) = \overline{\psi}(x)(i\slashed{\partial} - e\slashed{A} - m)\psi(x) \tag{13.6}$$

and it is invariant under the gauge transformations (13.2) and (13.5). As we have seen, upon identification of $A_\mu(x)$ with the four-vector potential of electromagnetism, the Lagrangian density (13.6) describes the interaction of a Dirac spinor of mass parameter m and electric charge e with an external electromagnetic field! We can complete the picture by including the degrees of freedom of the electromagnetic field itself and add to (13.6) the corresponding Lagrangian density. Again, gauge invariance determines its form uniquely and we are led to (2.7). A simple rule to derive the field strength tensor $F_{\mu\nu}$ is to notice that it is given by the commutator of two covariant derivatives, i.e.

$$[D_\mu, D_\nu] = ieF_{\mu\nu} \tag{13.7}$$

Let us summarise: We started with a field theory invariant under a group $U(1)$ of global phase transformations. The extension to a local invariance can be interpreted as a $U(1)$ symmetry at each point x. In a qualitative sense we can say that gauge invariance induces an invariance under $U(1)^\infty$. This extension, a purely geometrical requirement, implies the introduction of new interactions. The surprising result is that these "geometrical" interactions describe the well-known electromagnetic forces.

The above construction of the electromagnetic interaction can be generalised to any number of charged fields. Consider, for example, N Dirac fields ψ_a, $a = 1, \ldots, N$ with mass parameters m_a, and electric charges q_a.[1] The free Dirac theory of this system is described by

$$\mathcal{L}_0 = \sum_a \overline{\psi}_a (i\slashed{\partial} - m_a)\psi_a \tag{13.8}$$

We choose to gauge the global symmetry

$$\psi_a(x) \longrightarrow \psi'_a(x) = e^{iq_a\theta}\,\psi_a(x) \tag{13.9}$$

For that, we have to introduce a four-vector field $A_\mu(x)$ and modify (13.8) to

$$\mathcal{L} = \sum_a \overline{\psi}_a (i\slashed{\partial}_a - q_a\slashed{A} - m_a)\psi_a - \frac{1}{4}F_{\mu\nu}F^{\mu\nu} \tag{13.10}$$

including also the free Maxwell term for A_μ.

$\mathcal{L}$ is invariant under the $U(1)$ gauge transformation

$$\psi_a(x) \to \psi'_a(x) = e^{iq_a\theta(x)}\,\psi_a(x)\,, \quad A_\mu(x) \to A'_\mu(x) = A_\mu(x) - \partial_\mu\theta(x) \tag{13.11}$$

and correctly describes the interaction of the charged fields with the electromagnetic field.

A few remarks before closing this section:

• Gauge invariance is already present in the classical Maxwell theory, but its interpretation is physically more intuitive in the framework of quantum mechanics. Indeed, it is the invariance under phase transformations of the wave function, or of the quantum field, which naturally extends to a local invariance.

• In studying the vector field we noted that a term $A_\mu A^\mu$ could be interpreted as a mass term for the corresponding spin-1 particle. Such a term is not invariant under the transformation (13.5) and this seems to imply that local invariance requires the gauge particle to be massless. We shall come back to this very important point in Chapter 14.

• The same reasoning can be applied to space-time transformations. Let us consider the example of translations. Classical mechanics is assumed to be invariant under global space-time translations $x_\mu \to x_\mu + a_\mu$, where a_μ is a constant four-vector. We can again attempt to extend this invariance to local translations, with a_μ arbitrary functions of the space-time point x. This can be done in a mathematically consistent way by extending the invariance to the group of diffeomorphisms, which can be defined on any manifold, and gives the correct mathematical definition of the "equivalence principle". Again, we can show that this extension implies the introduction of new forces. The resulting gauge invariant theory turns out to be classical general relativity. Thus, gravitational interactions too have a geometric origin. In fact, the mathematical

---

[1]A very well established experimental fact is that all particles in the world have electric charges which are integer multiples of an "elementary" charge. This phenomenon is called *charge quantisation*. and does not have a natural explanation in the context of the electromagnetic theory we present here.

formulation of Einstein's original motivation to extend the principle of equivalence to accelerated frames, is precisely the requirement of gauge invariance. Historically, many mathematical techniques, which are used in today's gauge theories, were developed in the framework of general relativity.

13.2   Non-Abelian Gauge Invariance and Yang–Mills Theories

The extension of the formalism of gauge theories to non-Abelian groups has not been a trivial task and it was first discovered by trial and error.

Here, we shall restrict ourselves to internal symmetries, which are simpler to analyse and they are the ones we will use to describe all fundamental forces with the exception of gravitation.

Let us consider a classical field theory given by a Lagrangian density $\mathcal{L}$. It depends on a set of r fields $\psi_i(x)$, $i = 1, ..., r$ and their first space-time derivatives. The Lorentz transformation properties of these fields will play no role in this section. We assume that the fields ψ_i transform linearly according to an r-dimensional representation, not necessarily irreducible, of a compact, simple, Lie group G, i.e.

$$\Psi(x) \to \Psi'(x) = U(\omega)\Psi(x), \quad \omega \in G; \quad \Psi(x) = \begin{pmatrix} \psi_1 \\ \psi_2 \\ \vdots \\ \psi_r \end{pmatrix}, \tag{13.12}$$

where $U(\omega)$ is the $r \times r$ matrix representing $\omega \in G$. As we know, if m is the dimension of the Lie algebra of G, $U(\omega)$ can be written in the form

$$U(\omega) = e^{i\Theta}, \quad \Theta = \sum_{a=1}^{m} \theta^a T^a \tag{13.13}$$

where θ^a, $a = 1, ..., m$ are a set of m constant real parameters parametrising the elements $\omega \in G$, and the T^as are m $r \times r$ Hermitean matrices representing the m generators of the Lie algebra of G. They satisfy the Lie algebra

$$[T^a, T^b] = \mathrm{i} f^{abc} T^c \tag{13.14}$$

The fully antisymmetric real numbers f^{abc} are the structure constants of G and a summation over repeated indices is understood. The normalisation of the structure constants is usually fixed by requiring that, in the fundamental representation, the corresponding matrices t^a of the generators are normalised to satisfy

$$\mathrm{Tr}\left(t^a t^b\right) = \frac{1}{2}\delta^{ab} \tag{13.15}$$

The Lagrangian density $\mathcal{L}(\Psi, \partial\Psi)$ is assumed to be invariant under the global transformations (13.12), (13.13). As was done for the Abelian case, we wish to find a new $\mathcal{L}$, invariant under the corresponding gauge transformations in which the θ^a's of (13.13) are arbitrary functions of x. In the same qualitative sense, we look for a theory invariant under G^∞. This problem has a long history but, in the form we present it here, it was first solved by trial and error for the case of $SU(2)$ by C.N. Yang and R.L. Mills in 1954. They gave the underlying physical motivation and these theories are called since "Yang–Mills theories". The steps are direct generalisations of the ones followed in the Abelian case. Under an infinitesimal transformation of (13.12), (13.13) we have

$$\delta\Psi = iT^a \delta\theta^a \Psi \tag{13.16}$$

Once we allow the parameters $\delta\theta^a$ to be space-time dependent, this transformation is no longer a symmetry of $\mathcal{L}$ because of the last term in

$$\delta(\partial_\mu \Psi) = iT^a \delta\theta^a \partial_\mu \Psi + iT^a (\partial_\mu \delta\theta^a)\Psi \tag{13.17}$$

From our experience with the Abelian case, we expect to be able to restore the symmetry if we replace in $\mathcal{L}$ the ordinary space-time derivatives $\partial_\mu \Psi$ by the covariant derivatives $\mathcal{D}_\mu \Psi$, provided they transform according to

$$\delta(\mathcal{D}_\mu \Psi) = iT^a \delta\theta^a \mathcal{D}_\mu \Psi \tag{13.18}$$

Then, the new Lagrangian $\mathcal{L}(\Psi, \mathcal{D}_\mu \Psi)$ will clearly be invariant under these gauge transformations.

We can now guess how to construct $\mathcal{D}_\mu \Psi$. We introduce m real vector fields A_μ^a, one for each parameter $\delta\theta^a$ and define

$$\mathcal{D}_\mu \Psi = \partial_\mu \Psi + igT^a A_\mu^a \Psi \tag{13.19}$$

with g an arbitrary constant. To satisfy the requirement (13.18) we postulate that together with the transformation (13.16) of Ψ, the fields A_μ^a also transform according to

$$\delta A_\mu^a = f^{abc} A_\mu^b \delta\theta^c - \frac{1}{g}\partial_\mu \delta\theta^a \tag{13.20}$$

The last term in (13.20) is the obvious generalisation of the corresponding derivative term of the Abelian case, while the first corresponds to the statement that the fields A_μ^a transform according to the adjoint representation of the group G.

The construction of the field strength, the analogue of $F_{\mu\nu}$ of electromagnetism, and of the analogue of the term $F_{\mu\nu}^2$ term in the Lagrangian, is a little less straightforward. As in electromagnetism, it is instructive to compute the commutator of two covariant derivatives. We find

$$[\mathcal{D}_\mu, \mathcal{D}_\nu]\Psi = igT^a F_{\mu\nu}^a \Psi \tag{13.21}$$

where

$$F_{\mu\nu}^a = \partial_\mu A_\nu^a - \partial_\nu A_\mu^a + gf^{abc} A_\mu^b A_\nu^c \tag{13.22}$$

Computing the variation of this relation under the gauge transformation (13.20) we conclude that the $F_{\mu\nu}^a$, contrary to the field strength of electromagnetism, are not invariant. Instead they vary according to

$$\delta F^a_{\mu\nu} = f^{abc} F^b_{\mu\nu} \delta\theta^c(x) \tag{13.23}$$

i.e. they belong to the adjoint representation of G. Nevertheless, using the antisymmetry of the structure constants we see that the obvious candidate $(F^a_{\mu\nu})^2$ for the gauge field "kinetic" term is indeed gauge invariant, since

$$\delta(F^a_{\mu\nu} F^{a\mu\nu}) = 2F^{a\mu\nu} f^{abc} F^b_{\mu\nu} \delta\theta^c = 0$$

We conclude that the full gauge invariant Lagrangian density is

$$\mathcal{L}_{\text{inv}} = -\frac{1}{4} F^a_{\mu\nu} F^{a\mu\nu} + \mathcal{L}(\Psi, \mathcal{D}\Psi) \tag{13.24}$$

Finally, defining

$$\mathcal{A}_\mu = A^a_\mu t^a \quad \text{and} \quad \mathcal{F}_{\mu\nu} = F^a_{\mu\nu} t^a \tag{13.25}$$

we find

$$\mathcal{F}_{\mu\nu} = \partial_\mu \mathcal{A}_\nu - \partial_\nu \mathcal{A}_\mu - i g [\mathcal{A}_\mu, \mathcal{A}_\nu] \tag{13.26}$$

and we can rewrite (13.24) in the form

$$\mathcal{L}_{\text{inv}} = -\frac{1}{2} \text{Tr} \mathcal{F}_{\mu\nu} \mathcal{F}^{\mu\nu} + \mathcal{L}(\Psi, \mathcal{D}\Psi) \tag{13.27}$$

This completes the construction of the gauge invariant Lagrangian. A few remarks are in order here:

• All the steps described above may seem arbitrary but, in fact, they are uniquely determined by the rules of differential geometry plus some obvious requirements such as absence of higher order derivatives. In particular, the field strength tensor $\mathcal{F}_{\mu\nu}(x)$ has a simple geometrical interpretation: it is the curvature in the manifold formed by the field configurations over the four-dimensional Minkowski space.

• As was the case with the electromagnetic field, the Lagrangian (13.24) does not contain terms proportional to $A^a_\mu A^{a\mu}$. This means that the gauge fields describe massless particles.

• Since $F^a_{\mu\nu}$ is not linear in the fields A^b_μ, the term $(F^a_{\mu\nu})^2$ in (13.24), besides the usual kinetic term, which is bilinear in the fields, contains cubic and quartic terms with a coupling constant g. In other words, the non-Abelian gauge fields are self-coupled, while the Abelian (photon) field is not. A Yang–Mills theory, containing only gauge fields, is still an interacting quantum field theory, while a theory with the electromagnetic field alone is a trivial free theory.

• The same coupling constant g appears in the covariant derivative of the fields Ψ in (13.19). This simple consequence of gauge invariance has an important physical application: if we add another field Ψ', its coupling strength with the gauge fields will still be given by the same constant g. Contrary to the Abelian case studied before, if electromagnetism is part of a non-Abelian simple group, gauge invariance implies electric charge quantisation.[2]

---

[2]This remark connecting electric charge quantisation and non-Abelian gauge theories was first made by M. Gell-Mann and S.L. Glashow in 1961.

- The above analysis can be extended in a straightforward way to the case where the group G is the product of simple groups $G = G_1 \times ... \times G_n$. The only difference is that one should introduce n coupling constants $g_1, ..., g_n$, one for each simple factor. Charge quantisation is still true inside each subgroup, but electric charges belonging to different factors are no more related.
- The situation changes if one considers non semi-simple groups, where one or more of the factors G_a is Abelian. In this case the associated coupling constants can be chosen differently for each field and the corresponding Abelian charges are not quantised.

13.2.1 Quantisation of Yang–Mills theories

The quantisation of Yang–Mills theories is particularly simple if we adopt the path integral method. The steps which led us to equation (10.61) in section 10.4.2 do not depend on the Abelian, or non-Abelian, character of the gauge group, so we obtain again

$$D = \int \mathcal{D}[\Omega]\mathcal{D}[A]e^{-S_{\mathrm{inv}}[A]-\frac{1}{2}\int \mathrm{d}^4 x G^2[A]} \det\left(\frac{\delta G}{\delta \Omega}\right) \tag{13.28}$$

where D is the denominator of the generating functional in (10.59).

Up to this point the result looks the same as the one we obtained in the Abelian case. We can again factor out the volume of the gauge group, and this factor cancels between the numerator and the denominator in equation (10.59). However, the remaining determinant is not so easy to handle. Let us choose again $G = \partial^\mu A_\mu = t^a \partial^\mu A_\mu^a \equiv t^a G^a$ as the gauge fixing function. Using the transformation properties of the field given in equation (13.20), we obtain for the Jacobian

$$\det\frac{\delta G}{\delta \Omega} = \det\partial^\mu \mathcal{D}_\mu = -\frac{1}{g}\det\left(\Box\delta^{ab} + gf^{abc}\partial^\mu A_\mu^c\right) \tag{13.29}$$

The difference with the Abelian case is crucial: instead of $\det\Box$, now the determinant depends on the gauge field A_μ^a and cannot be taken out of the functional integral.

Here the integration rules over Grassmann variables become important. According to equation (10.45), if $\mathcal{R}$ is a differential operator, we have

$$\int \mathcal{D}[\overline{\chi}]\mathcal{D}[\chi]e^{-\int \mathrm{d}^4 x\, \overline{\chi}(x)\mathcal{R}\chi(x)} = \det \mathcal{R} \tag{13.30}$$

where $\chi(x)$ and $\overline{\chi}(x)$ are two fields taking values in a Grassmann algebra. In spite of the notation, they should not be viewed as being necessarily complex conjugates of each other. Using this formula, we can rewrite (13.28) as

$$D = \int \mathcal{D}[\Omega]\mathcal{D}[\overline{\chi}]\mathcal{D}[\chi]\mathcal{D}[A]\, e^{-S_{\mathrm{inv}}[A]-\int \mathrm{d}^4 x\left(\frac{1}{2}G^2[A]+\overline{\chi}^a(\Box\delta^{ab}+gf^{abc}\partial^\mu A_\mu^c)\chi^b\right)} \tag{13.31}$$

This is the non-Abelian generalisation of equation (10.62). It says that quantising the theory in a covariant gauge amounts to replacing the gauge invariant action by an effective, gauge fixed one. In the Abelian case this was given by $S_{\mathrm{eff}} = S_{\mathrm{inv}} + \frac{1}{2}\int \mathrm{d}^4 x G^2$.

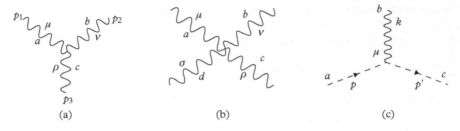

Fig. 13.1 The three vertices of the Yang–Mills theory.

In the non-Abelian case we must add to it the term $\int \mathrm{d}^4 x \overline{\chi}^a (\square \delta^{ab} + g f^{abc} \partial^\mu A_\mu^c) \chi^b$ which involves two new fields, $\chi(x)$ and $\overline{\chi}(x)$. This term seems to have several strange properties: First, it is not Hermitian. Second, the fields $\chi(x)$ and $\overline{\chi}(x)$ are Grassmannian fields, which implies that they are quantised according to the rules of Fermi–Dirac statistics, yet they have no Lorentz indices, which means that they are Lorentz scalars. This is consistent with the fact that their equations of motion contain second order derivatives. We have learned in Chapter 11 that this is not possible for a field theory which satisfies the usual axioms and, in particular, that of positive energy. We conclude that $\chi(x)$ and $\overline{\chi}(x)$ do not correspond to physical degrees of freedom. In the physicists jargon they are called *ghosts*. This is not really a new result. Even in the Abelian case we saw that Lorentz covariance forces us to quantise the theory in a Hilbert space containing unphysical degrees of freedom, the longitudinal and scalar photons. For the non-Abelian case we just saw that we must consider an even larger unphysical space including the $\chi(x)$ and $\overline{\chi}(x)$ ghosts. Like the longitudinal and scalar photons we should not include them in the initial or final states when computing a scattering amplitude, but they will appear in closed loops.

The necessity of such ghosts was first discovered by Feynman, but the formal method we just described is due to L.D. Faddeev and V.N. Popov, so they are called *Faddeev–Popov ghosts*.

13.2.2   Feynman rules for Yang–Mills theories

These theories have two interaction terms among gauge fields:

(i) The cubic $\mathcal{L}_I^{(1)} = -\frac{1}{2} g f_{abc} (\partial_\mu A_\nu^a - \partial_\nu A_\mu^a) A^{b\mu} A^{c\nu}$ where f_{abc} are the structure constants of the Lie algebra of the gauge group G. The vertex is represented in Figure 13.1 (a) and is given by

$$g f_{abc} [\eta_{\mu\nu}(p_1 - p_2)_\rho + \eta_{\nu\rho}(p_2 - p_3)_\mu + \eta_{\rho\mu}(p_3 - p_1)_\nu](2\pi)^4 \delta^{(4)}(\Sigma p) \qquad (13.32)$$

(ii) The quartic $\mathcal{L}_I^{(2)} = -\frac{1}{4} g^2 f_{abc} f_{ab'c'} A_\mu^b A_\nu^c A^{b'\mu} A^{c'\nu}$. The vertex is represented in Figure 13.1 (b) and is given by

$$-\mathrm{i} g^2 [f_{eab} f_{ecd}(\eta_{\mu\rho}\eta_{\nu\sigma} - \eta_{\mu\sigma}\eta_{\nu\rho}) + f_{eac} f_{edb}(\eta_{\mu\sigma}\eta_{\rho\nu} - \eta_{\mu\nu}\eta_{\rho\sigma})$$

$$+ f_{ead} f_{ebc}(\eta_{\mu\nu}\eta_{\rho\sigma} - \eta_{\mu\rho}\eta_{\nu\sigma})](2\pi)^4 \delta^{(4)}(\Sigma p) \qquad (13.33)$$

In addition, as we just saw, the quantisation introduced two new fields, which should be included in the Feynman rules. We have a $\chi - \overline{\chi}$ propagator given by

$$S^{ab}(q) = \langle \chi^a(q)\overline{\chi}^b(-q)\rangle = -\frac{i\delta^{ab}}{q^2 + i\epsilon} \tag{13.34}$$

Note that there are no $\langle \chi\chi\rangle$ or $\langle \overline{\chi}\,\overline{\chi}\rangle$ propagators. There is also a $\overline{\chi} - \chi - A$ vertex. After partial integration the relevant interaction term becomes

$$\mathcal{L}_{\text{YMgh}} = g f_{abc}(\partial^\mu \overline{\chi}^a) A_\mu^b \chi^c \tag{13.35}$$

Notice that the derivative operator ∂^μ applies to $\overline{\chi}$ but not to χ another indication that these fields should not be considered as complex conjugates to each other. The vertex is represented in Figure 13.1 (c) and it is given by

$$+ g p^\mu f_{abc}(2\pi)^4 \delta^4(k + p - p') \tag{13.36}$$

A further rule should be added to the above ones stemming from the fact that $\overline{\chi}$ and χ are anti-commuting ghost fields. It follows that there is a minus sign for every closed ghost loop.

13.3 Gauge Theories on a Space-Time Lattice

In the previous sections we derived the form of the gauge interaction for Abelian and non-Abelian groups. We followed a trial and error method, which gave us the result but obscured its geometric origin. In our formalism the gauge field $A_\mu(x)$ is a field like any other. Here we want to show that this interpretation is misleading. The correct one has its basis in differential geometry, which we will describe here in a simplified discrete version and consider field theory on a four-dimensional Euclidean space-time lattice.

Let us consider, for simplicity, a lattice with hypercubic symmetry. The space-time point x_μ is replaced by

$$x_\mu \to n_\mu a \tag{13.37}$$

where a is a constant length (the lattice spacing) and n_μ is a four-dimensional vector with components $n_\mu = (n_0, n_1, n_2, n_3)$ which take integer values $0 \le n_\mu \le N_\mu$. $N_\mu + 1$ is the number of lattice points in the direction μ. The total number of points, i.e. the volume of the system is given by $V \sim \prod_{\mu=0}^3 N_\mu$. The presence of a introduces an ultraviolet or short distance cut-off because all momenta are bounded from above by $2\pi/a$. The presence of N_μ introduces an infrared or large distance cut-off because the momenta are also bounded from below by $2\pi/Na$, where N is the maximum of N_μ. The infinite volume continuum space is recovered at the appropriate double limit $a \to 0$ and $N_\mu \to \infty$.

The dictionary between quantities defined in the continuum and the corresponding ones on the lattice is easy to establish (we take the lattice spacing a equal to one):

1. A field $\quad \Psi(x) \quad \Rightarrow \quad \Psi_n$
2. A local term such as $\quad \overline{\Psi}(x)\Psi(x) \quad \Rightarrow \quad \overline{\Psi}_n\Psi_n$
3. A derivative $\quad \partial_\mu \Psi(x) \quad \Rightarrow \quad \Psi_{n+\mu} - \Psi_n$

where $n + \mu$ is the nearest neighbour of n in the direction μ.

4. The kinetic energy term[3] $\overline{\Psi}(x)\partial_\mu \Psi(x) \;\Rightarrow\; \overline{\Psi}_n \Psi_{n+\mu} - \overline{\Psi}_n \Psi_n$

5. A gauge transformation $\Psi'(x) = \mathrm{e}^{\mathrm{i}\Theta(x)}\Psi(x) \;\Rightarrow\; \Psi'_n = \mathrm{e}^{\mathrm{i}\Theta_n}\Psi_n$

All local terms of the form $\overline{\Psi}_n \Psi_n$ remain invariant but the part of the kinetic energy, which couples fields at neighbouring points, does not.

6. Indeed, the gauge transformation of the term $\overline{\Psi}_n \Psi_{n+\mu}$ is

$$\overline{\Psi}_n \Psi_{n+\mu} \to \overline{\Psi}'_n \Psi'_{n+\mu} = \overline{\Psi}_n \mathrm{e}^{-\mathrm{i}\Theta_n} \mathrm{e}^{\mathrm{i}\Theta_{n+\mu}} \Psi_{n+\mu}$$

which shows that we recover the problem we had with the derivative operator in the continuum. In order to restore invariance, we must introduce a new field, which carries both indices n and $n + \mu$. We denote it by $U_{n,n+\mu}$ and we shall impose on it the constraint $U_{n,n+\mu} = U_{n+\mu,n}^{-1} = U_{n+\mu,n}^\dagger$. Under a gauge transformation, U is defined to transform as

$$U_{n,n+\mu} \;\longrightarrow\; U'_{n,n+\mu} = \mathrm{e}^{\mathrm{i}\Theta_n} U_{n,n+\mu} \mathrm{e}^{-\mathrm{i}\Theta_{n+\mu}} \tag{13.38}$$

Then, with the help of this gauge field we can modify the kinetic energy term and replace it with

$$\overline{\Psi}_n \Psi_{n+\mu} \;\longrightarrow\; \overline{\Psi}_n \, U_{n,n+\mu} \, \Psi_{n+\mu} \tag{13.39}$$

which is invariant under gauge transformations and, upon summation over n and μ, leads to a Hermitean term in the action.

U is an element of the gauge group. It corresponds to the transformation (13.12) we considered in the continuum. If the group G is a Lie group, we can write the matrices U as

$$U_{n,n+\mu} = \exp\left(\mathrm{i}ga \sum_{b=1}^{m} A_{n,n+\mu}^b T^b \right) \tag{13.40}$$

where, in order to make contact with the expressions we used in the continuum, we introduced a constant g, which will become the coupling constant, and the lattice spacing a to give A the dimensions of mass. Note that U takes values in the group G, while $A_{n,n+\mu}^b$ parameterise the elements of the corresponding Lie algebra. For the continuum limit it is convenient to introduce the following notation: $U_{n,n+\mu} \to U_\mu(n)$ and similarly for A. In Problem 13.3 we ask the reader to verify that A has the right transformation properties for a gauge field we found in equation (13.20). Note that, contrary to the field Ψ, U does not live on a single lattice point, but it has two indices, n and $n + \mu$; in other words it lives on the oriented link joining these two neighbouring points. We see that the mathematicians are right when they do not call the gauge field "a field" but "a connection".

In order to finish the story we want to obtain an expression for the kinetic energy of the gauge field, the analogue of $\mathrm{Tr}\mathcal{F}_{\mu\nu}(x)\mathcal{F}^{\mu\nu}(x)$, on the lattice. As in the continuum, the guiding principle is gauge invariance. Let us consider two points n and m on the

---

[3]We write here the expression for spinor fields which contain only first order derivatives in the kinetic energy. The extension to scalar fields with second order derivatives is straightforward.

lattice. We shall call a path $p_{n,m}$ on the lattice a sequence of oriented links which join continuously these two points. Consider next the product of the gauge fields U along all the links of the path $p_{n,m}$

$$P^{(p)}(n,m) = \prod_p U_{n,n+\mu} \cdots U_{m-\nu,m} \tag{13.41}$$

Using the transformation rule (13.38), we see that $P^{(p)}(n,m)$ transforms as

$$P^{(p)}(n,m) \rightarrow e^{i\Theta_n} P^{(p)}(n,m) e^{-i\Theta_m} \tag{13.42}$$

It follows that if we consider a closed path $c = p_{n,n}$ the quantity $\text{Tr}P^{(c)}$ is gauge invariant. The simplest closed path for a hypercubic lattice has four links and it is called *plaquette*. We introduce the notation

$$\mathcal{P}_{\mu\nu}(n) = U_\mu(n)U_\nu(n+\mu)U_\mu^\dagger(n+\nu)U_\nu^\dagger(n) \tag{13.43}$$

It is now straightforward to show (see Problem 13.3), that the action

$$S_g = \frac{1}{g^2} \sum_{n,\mu<\nu} [1 - \text{Re}(\text{Tr}\mathcal{P}_{\mu\nu}(n))] \tag{13.44}$$

is gauge invariant and tends to the space-time integral of $\text{Tr}(\mathcal{F}^2)$ when a goes to zero. The condition $\mu < \nu$ is set to avoid double counting of the plaquettes. Equation (13.44) is the action introduced by K. Wilson to describe the Yang–Mills theory on the lattice and it is called *the Wilson action*.

In mathematics $\mathcal{F}$ is called *the curvature* and we see that the terminology is not fortuitous. As we learn in elementary differential geometry, parallel transport of a vector along a closed path in a given manifold, provides information about its curvature. We see that, in the lattice example, $\mathcal{F}$ denotes the curvature in the field space.

13.4 Brief Historical Notes

Although many versions of the history of gauge theories exist already in the recent literature[4], the subject is not always covered in textbooks students usually read. We present in this section a very brief historical note.

The vector potential was introduced in classical electrodynamics during the first half of the nineteenth century, either implicitly or explicitly, by several authors independently. It appears in some manuscript notes by Carl Friedrich Gauss as early as 1835 and it was fully written by Gustav Kirchhoff in 1857, following some earlier work by Franz Neumann. It was soon noticed that it carried redundant variables and several "gauge conditions" were used. The condition, which in modern notation is written as $\partial_\mu A^\mu = 0$, was proposed by the Danish mathematical physicist Ludvig Valentin Lorenz

---

[4]See, among others, O. Darrigol, *Electrodynamics from Ampère to Einstein*, Oxford University Press, 2000 ; L. O'Raifeartaigh, *The Dawning of gauge theory*, Princeton University Press, 1997 ; L. O'Raifeartaigh and N. Straumann, Rev. Mod. Phys. **72**, 1 (2000) ; J.D. Jackson and L.B. Okun, Rev. Mod. Phys. **73**, 663 (2001)

in 1867. Incidentally, most physics books misspell Lorenz's name as *Lorentz*, thus erroneously attributing the condition to the famous Dutch H.A. Lorentz, of the Lorentz transformations.[5] However, for internal symmetries, the concept of gauge invariance, as we know it today, belongs to quantum mechanics. The first person who realised that the invariance under local transformations of the phase of the wave function in the Schrödinger theory implies the introduction of an electromagnetic field was Vladimir Aleksandrovich Fock in 1926,[6] just after Schrödinger wrote his equation. Naturally, one would expect non-Abelian gauge theories to be constructed following the same principle immediately after isospin symmetry was established in the thirties, following Heisenberg's original article of 1932. This is the method we followed in section 13.2. But in this case history took a totally unexpected route.

The development of the general theory of relativity offered a new paradigm for a gauge theory. The fact that it can be written as the theory invariant under local translations was certainly known to Hilbert,[7] hence the name of *Einstein–Hilbert action*. For the next decades it became the starting point for all studies on theories invariant under local transformations, such as the electromagnetic and the gravitational ones which were the only fundamental interactions known at that time. It was, therefore, tempting to look for a unified theory, namely one in which both interactions follow from the same gauge principle. Today we know the attempt by Theodor Kaluza, completed by Oscar Benjamin Klein,[8] which is often used in supergravity and superstring theories. These authors consider a theory of general relativity formulated in a five dimensional space-time (1+4). They remark that if the fifth dimension is compact the components $g_{4\mu}$ of the metric tensor along this dimension may look to a four dimensional observer as those of an electromagnetic vector potential. What is less known is that the idea was introduced earlier by the Finnish Gunnar Nordström[9] who had constructed a scalar theory of gravitation. In 1914 he wrote a five-dimensional theory of electromagnetism and showed that, if one assumes that the fields are independent of the fifth coordinate, the assumption made later by Kaluza, the electromagnetic vector potential splits into a four dimensional vector and a four dimensional scalar, the latter being identified to his scalar field of gravitation, in some sense the mirror theory of Kaluza and Klein. An important contribution from this period is due to Hermann Klaus Hugo Weyl.[10] He is more known for his 1918 unsuccessful attempt to enlarge diffeomorphisms to local scale transformations, but, in fact, a byproduct of this work was a different form of unification between electromagnetism and gravitation. In his 1929 paper, which contains the gauge theory for the Dirac electron we saw in section 13.1, he introduced many concepts which have become classic, such as the Weyl two-component spinors and the vierbein and spin-connection formalism. Although the theory is no more scale invariant, he still used the term *gauge invariance*, a term which has survived ever since.

---

[5]In French: On ne prête qu'aux riches.

[6]V. Fock, Z. Phys **39**, 226 (1926), *Translation:* Physics-Uspekhi **53**, 839 (2010)

[7]D. Hilbert, Gött. Nachr. (1915) p. 395

[8]Th. Kaluza, K. Preuss. Akad. Wiss. (1921), p. 966 ; O. Klein, Z. Phys. **37**, 895 (1926)

[9]G. Nordström, Phys. Z. **15**, 504 (1914)

[10]H. Weyl, Deutsch Akad. Wiss. Berlin (1918), p. 465 ; Z. Phys. **56**, 330 (1929)

In particle physics we put the birth of non-Abelian gauge theories in 1954, with the fundamental paper of Chen Ning Yang and Robert Laurence Mills.[11] It is the paper which introduced the $SU(2)$ gauge theory and, although it took some years before interesting physical theories could be built, it is since that date that non-Abelian gauge theories became part of high energy physics. It is not surprising that they were immediately named *Yang-Mills theories*. The impact of this work on high energy physics has often been emphasised, but here we want to mention some earlier and little known attempts which, according to present views, have followed a quite strange route.

The first is due to Oscar Klein. In an obscure conference in 1938 he presented a paper with the title: *On the theory of charged fields*[12] in which he attempts to construct an $SU(2)$ gauge theory for the nuclear forces. This paper is amazing in many ways. First, of course, because it was done in 1938. He starts from the discovery of the muon, misinterpreted as the Yukawa meson, in the old Yukawa theory in which the mesons were assumed to be vector particles. This provides the physical motivation. The aim is to write an $SU(2)$ gauge theory unifying electromagnetism and nuclear forces. Second, and even more amazing, because he follows an incredibly circuitous road: He considers general relativity in a five dimensional space, he compactifies *à la* Kaluza-Klein, but he takes the $g_{4\mu}$ components of the metric tensor to be 2×2 matrices. He wants to describe the $SU(2)$ gauge fields but the matrices he is using, although they depend on three fields, are not traceless. In spite of this problem he finds the correct expression for the field strength tensor of $SU(2)$. In fact, answering an objection by Møeller, he added a fourth vector field, thus promoting his theory to $SU(2) \times U(1)$. He added mass terms by hand and it is not clear whether he worried about the resulting breaking of gauge invariance. It is not known whether this paper has inspired anybody else's work and Klein himself mentioned it only once in a 1955 Conference in Berne.[13]

The second work in the same spirit is due to Wolfgang Pauli[14] who in 1953, in a letter to Abraham Pais, as well as in a series of seminars, developed precisely this approach: the construction of the $SU(2)$ gauge theory as the flat space limit of a compactified higher dimensional theory of general relativity. He was closer to the approach followed today because he considered a six dimensional theory with the compact space forming an S_2. He never published this work and we do not know whether he was aware of Klein's 1938 paper. He had realised that a mass term for the gauge bosons breaks the invariance and he had an animated argument during a seminar by Yang in the Institute for Advanced Studies in Princeton in 1954.[15] What is certainly surprising is that both Klein and Pauli, fifteen years apart one from the other, decided to construct the $SU(2)$ gauge theory for strong interactions and both

[11]C.N. Yang and R.L. Mills, Phys. Rev. **96**, 191 (1954) ; It seems that similar results were also obtained by R. Shaw in his thesis.

[12]O. Klein, in *Les Nouvelles Théories de la Physique*, Paris 1939, p. 81. Report in a Conference organised by the "Institut International de Coopération Intellectuelle", Warsaw 1938

[13]O. Klein, Helv. Phys. Acta Suppl. **IV**, 58 (1956)

[14]W. Pauli, Unpublished. It is summarised in a letter to A. Pais dated July 22-25 1953. A. Pais, *Inward Bound*, Oxford Univ. Press, 1986, p. 584

[15]C.N. Yang, *Selected Papers 1945-1980 with Commentary* Published by Freeman, San Francisco, p. 525

choose to follow this totally counter-intuitive method. It seems that the fascination which general relativity had exerted on this generation of physicists was such that, for many years, local transformations could not be conceived independently of general coordinate transformations. Yang and Mills were the first to understand that the gauge theory of an internal symmetry takes place in a fixed background space, which can be chosen to be flat, in which case general relativity plays no role.

With the work of Yang and Mills gauge theories entered particle physics. Although the initial motivation was a theory of the strong interactions, the first semi-realistic models aimed at describing the weak and electromagnetic interactions. We shall present later on in this book the present state of what is known as the *Standard Model of Elementary Particle Physics*. To complete the story, we mention only a paper by S.L. Glashow and M. Gell-Mann[16], which is often left out from the history articles. In this paper the authors extend the Yang-Mills construction, which was originally done for $SU(2)$, to arbitrary Lie algebras. The well-known result of associating a coupling constant to every simple factor in the algebra, result which we presented in section 13.2, appeared for the first time there. We can even find the seed for what we will call later *a grand unified theory*. In a footnote they write:

"The remarkable universality of the electric charge would be better understood were the photon not merely a singlet, but a member of a family of vector mesons comprising a simple partially gauge invariant theory."

13.5   Problems

Problem 13.1 Consider the Schrödinger equation for a free electron. Both the equation and the normalisation condition are invariant under a group of global $U(1)$ transformations $\Psi(t, \boldsymbol{x}) \to \mathrm{e}^{\mathrm{i}\theta}\Psi(t, \boldsymbol{x})$ with constant θ.

1. Compute the corresponding conserved current.

2. Extend the symmetry into gauge $U(1)$ transformations, i.e. to an invariance under $\Psi(t, \boldsymbol{x}) \to \mathrm{e}^{\mathrm{i}\theta(t,\boldsymbol{x})}\Psi(t, \boldsymbol{x})$ with $\theta(t, \boldsymbol{x})$ an arbitrary function of t and $\boldsymbol{x}$. Show that it describes the motion of a non-relativistic electron in an external electromagnetic field.

This was first noticed by V. Fock in 1926, immediately after Schrödinger published his equation.

Problem 13.2 Consider a triplet of Dirac spinor fields $q_i(x)$ $i = 1, 2, 3$ belonging to the fundamental representation of $SU(3)$. Write the most general Lagrangian density, polynomial in the fields and their derivatives, having terms with dimensions less than or equal to four, which is invariant under gauge $SU(3)$ transformations.

Problem 13.3 Show that at the limit when the lattice spacing a goes to zero:

1. The quantity $A^b_{n,n+\mu}$ introduced in equation (13.40) has the right transformation properties to describe the gauge potential $A^b_\mu(x)$.

2. The action given in equation (13.44) becomes the Yang–Mills action.

---

[16]S.L. Glashow and M. Gell-Mann, Ann. of Phys. **15**, 437 (1961) ; see also: R. Utiyama, Phys. Rev. **101**, 1597 (1956)

14
Spontaneously Broken Symmetries

14.1 Introduction

To symmetric problems we are used to look for symmetric solutions. To cite Pierre Curie, "when certain causes produce certain effects, the symmetry elements of the causes must be present in the produced effects", in other words, if the equations of motion are invariant under a group of transformations, the solutions will exhibit the consequences of this invariance. A simple example: Consider a spherically symmetric electric charge distribution of radius R and let us compute the electrostatic field E at a point A outside the sphere. Spherical symmetry has two immediate consequences: (i) the field points in the radial direction with the origin at the centre of the charged sphere, and (ii) it has the same modulus at every point of a spherical surface passing through A. Under these conditions, application of Gauss's theorem gives the answer immediately.

Stated this way the theorem sounds almost trivial, but a careful study shows that it contains several subtleties. One of them is that, in practice, we often need a much stronger version. The symmetry of a real physical problem is rarely exact and we often neglect small symmetry breaking effects, implicitly assuming that they will produce only small deviations from the symmetric solution. In our previous example, we assume that an almost spherically symmetric charge distribution will produce an almost symmetric electric field. But this is far from being always true, because it implies an assumption not only on the existence of a symmetric solution, but also on its stability against small perturbations. In this chapter we will study in some simple examples the ways a symmetry of the dynamical equations of a physical system is realised in the solutions.

Already at the classical level, an infinite system subject to the variation of one or more of its parameters may exhibit the phenomenon of phase transitions. This is often accompanied by a change in the symmetry of the state of equilibrium. For example, in the liquid–solid phase transition, the translational invariance of the liquid phase is reduced to the discrete subgroup which leaves invariant the lattice of the solid. A field theory describes a system with an infinite number of degrees of freedom, so we expect to encounter here also the phenomenon of phase transitions. We will study this phenomenon for field theories having both global and local symmetries. We will see that, in many cases, we encounter at least two phases:

In one of them, whichever symmetry is present is manifest in the spectrum of the theory, whose excitations appear to form irreducible representations of the symmetry group. This is called *the symmetric phase*, or *the Wigner phase*. We have already seen

Elementary Particle Physics. John Iliopoulos and Theodore N. Tomaras, Oxford University Press.
© John Iliopoulos and Theodore N. Tomaras (2021). DOI: 10.1093/oso/9780192844200.003.0014

the examples of isospin, or $SU(3)$, invariance in hadronic physics where the observed hadrons form indeed irreducible representations with nearly degenerate masses. For a gauge theory this implies that the vector gauge bosons are massless and belong to the adjoint representation of the group.[1] In the other phase, part of the symmetry does not manifest itself as a classification symmetry of the spectrum. The states do not form irreducible representations of the symmetry group. It is called *spontaneously broken*, or *Nambu–Goldstone phase* and, for gauge theories, we shall show presently that the corresponding gauge bosons become massive. Such a gauge theory, based on the group $U(2)$ spontaneously broken to $U(1)$, describes in a unified way, the electromagnetic and weak interactions. In this chapter we will study these phenomena of phase transitions in a few simple examples.

14.2   Global Symmetries

14.2.1   Simple examples

In Problem 14.1 we propose a simple example, that of the bent rod. It is a well-defined classical system, whose equations are well known, and which exhibits the phenomenon of phase transition in a simple and controlled way.

Analogous considerations apply also to quantum physics. We assume that a symmetry transformation is implemented in the Hilbert space by an operator Q, which commutes with the Hamiltonian

$$[H, Q] = 0 \tag{14.1}$$

As a result, if $|\Psi\rangle$ is an eigenstate of the Hamiltonian, so is $Q|\Psi\rangle$ with the same eigenvalue, unless $Q|\Psi\rangle = 0$. So, if the transformations can be represented by well-defined operators acting on the Hilbert space, we obtain the result that the states form irreducible representations of the corresponding algebra.

The Heisenberg ferromagnet is a good example in which phenomena of this kind can be studied numerically (and in simple cases even analytically). It is known that several materials exhibit a phase transition and, at low temperature, they develop a spontaneous magnetisation. For their description Heisenberg proposed a simple model, which captures the essential physics of the phenomenon. Let us consider a cubic lattice, whose sites are occupied by atoms, for which we make the following simplifying assumptions: (i) We neglect all atomic degrees of freedom and keep only a spin variable $\boldsymbol{S}_a$, $a = 1, 2, \ldots, N$, at each lattice point a. (ii) The spin variables are subject to short range forces and we can keep only nearest neighbour interactions. (iii) The interaction is assumed to be rotationally invariant. We are thus led to an effective interaction Hamiltonian of the form

---

[1]In Chapter 21 we will argue that non-Abelian gauge theories with semi-simple groups exhibit a remarkable phenomenon in this phase: all physical states are singlets of the group. All non-singlet states, such as those corresponding to the gauge particles, are supposed to be *confined*, in the sense that they do not appear as physically realisable asymptotic states. Although there are several arguments in favour of confinement, a rigorous proof is still missing and this constitutes one of the central unsolved problems in quantum field theory today. Such a gauge theory, based on the group $SU(3)$ in this *confinement phase* is the theory of strong interactions and it is called *quantum chromodynamics* (QCD)

$$H = -\frac{1}{2}J \sum_{a,b} \boldsymbol{S}_a \cdot \boldsymbol{S}_b \qquad (14.2)$$

where the sum extends over all pairs of nearest neighbouring points (in d dimensions each point has $2d$ nearest neighbours) and J is a positive coupling constant. For a three-dimensional lattice the system can be studied numerically and is known to exhibit the phenomenon of spontaneous breaking of the rotational $O(3)$ symmetry.

Specifically the main symmetry features of the Heisenberg ferromagnet can be described as follows:

1. The system has two phases: (i) The symmetric phase with the full $O(3)$ symmetry. The spins undergo rapid random oscillations with vanishing average value and there is no privileged direction in space. Spins belonging to distant lattice points are not correlated. (ii) The spontaneously broken phase, in which the spins are on the average oriented parallel to each other and a macroscopic magnetisation $\boldsymbol{M}$ appears. The symmetry is broken from the full $O(3)$ down to $O(2)$, the subgroup of rotations around the axis defined by $\boldsymbol{M}$. We have long range correlations which imply the existence of massless excitations in the energy spectrum.

2. There exists an external control parameter, to wit the temperature. There is a critical value T_{cr}, called *the Curie temperature*, which determines in which phase the system will settle. At any temperature there are two competing mechanisms: The short range interaction which favours parallel spin configurations in order to minimise the energy (14.2). Its strength grows with the number of nearest neighbours, i.e. with d. Thermal fluctuations on the other hand tend to destroy any long range order. For most values of d (except $d = 1$ and $d = \infty$), there is a phase transition at a critical temperature T_{cr}. At $T > T_{\mathrm{cr}}$ thermal fluctuations win and the system is in a disordered, or symmetric, phase. At $T < T_{\mathrm{cr}}$ the interaction wins and the system settles in an ordered phase with spontaneous symmetry breaking.

3. In the terminology of phase transitions there exists an order parameter, the magnetisation $\boldsymbol{M}$. It is a d-dimensional vector, so the manifold of ground states will span the surface of a $d-1$-dimensional sphere. The initial rotational symmetry is not completely broken. A subgroup remains unbroken even in the low temperature phase.

An interesting feature of this model is that we can study its physics starting from a system with a finite number N of spins and then take the limit $N \to \infty$. In three dimensions this can only be done numerically, but some properties can be understood qualitatively. Let us call J_i, $i = 1, 2, 3$ the three generators of $O(3)$. They all commute with the Hamiltonian (14.2). For simplicity we shall choose the spins $\boldsymbol{S}_a$ to be equal to $\frac{1}{2}$. At every lattice site we have a two-dimensional space of states with the basis vectors

$$|+\rangle = \begin{pmatrix} 1 \\ 0 \end{pmatrix}, \qquad |-\rangle = \begin{pmatrix} 0 \\ 1 \end{pmatrix} \qquad (14.3)$$

describing the states with values $\pm\frac{1}{2}$ of the z-component of the corresponding spin.

At zero temperature the ground state consists of all spins parallel giving a non-vanishing magnetisation $\boldsymbol{M}$. It is infinitely degenerate because we can rotate $\boldsymbol{M}$ to point in any direction. Starting from (14.3), the rotated spin state at the a-th site will be

$$|a, \pm\rangle_{\boldsymbol{\theta}} \, = \, \mathrm{e}^{\mathrm{i}\boldsymbol{J} \cdot \boldsymbol{\theta}} \, |a, \pm\rangle \tag{14.4}$$

For spin$-\frac{1}{2}$ we have $\boldsymbol{J} = \boldsymbol{\sigma}/2$. We can compute the overlap between two such states as the matrix element of the rotation operator between them

$$_{\boldsymbol{\theta_2}}\langle a, +|a, +\rangle_{\boldsymbol{\theta_1}} \, = \, _{\boldsymbol{\theta_2}}\langle a, +| \, \mathrm{e}^{\mathrm{i}\boldsymbol{J} \cdot \boldsymbol{\theta_{12}}} \, |a, +\rangle_{\boldsymbol{\theta_2}} = \cos \frac{\theta_{12}}{2} \tag{14.5}$$

where θ_{12} is the angle between the directions defined by $\boldsymbol{\theta_1}$ and $\boldsymbol{\theta_2}$. The overlap of these states at each site is smaller than one and, consequently, the overlap of the rotated ground states will be $(\cos(\theta_{12}/2))^N$, which vanishes in the limit $N \to \infty$. Thus, for a system with an infinite number of degrees of freedom, the different ground states are orthogonal and so are the Hilbert spaces, which we can build above each one of them. Physically this means that, although it costs no energy to turn a finite number of spins, it is impossible to turn simultaneously an infinite number of them. This fact is crucial for a system to undergo a phase transition. There are no phase transitions in systems with a finite number of degrees of freedom.

14.2.2   A field theory model

Let $\phi(x)$ be a complex scalar field, whose dynamics is described by the Lagrangian density

$$\mathcal{L}_1 = \partial_\mu \phi^* \, \partial^\mu \phi - M^2 \, \phi^* \phi - \lambda(\phi^* \phi)^2 \tag{14.6}$$

where $\mathcal{L}_1$ is a classical Lagrangian density and $\phi(x)$ is a classical field. Equation (14.6) is invariant under the group $U(1)$ of global transformations

$$\phi(x) \; \longrightarrow \; \phi'(x) = \mathrm{e}^{\mathrm{i}\theta} \, \phi(x) \tag{14.7}$$

To this invariance corresponds the conserved current

$$j_\mu = -\mathrm{i}(\phi^* \partial_\mu \phi - \phi \, \partial_\mu \phi^*)$$

which satisfies $\partial_\mu \, j^\mu = 0$, as can be verified using the equations of motion.

We are interested in the classical field configuration which minimises the energy of the system. The energy density of the system is the Hamiltonian density given by

$$\mathcal{H}_1 = \partial_0 \phi^* \, \partial_0 \phi + \partial_i \phi^* \, \partial_i \phi + V(\phi) \tag{14.8}$$

with

$$V(\phi) = M^2 \phi^* \phi + \lambda(\phi^* \phi)^2 \tag{14.9}$$

Given that the first two terms of $\mathcal{H}_1$ are positive definite, the ground state of the system corresponds to $\phi = $ constant $=$ global minimum of $V(\phi)$. V has a global minimum only if $\lambda > 0$. In this case the position of the minimum depends on the sign of M^2. (Notice that we are still studying a classical field theory and M^2 is just a parameter. One should not be misled by the notation into thinking that M is a "mass" and that M^2 is necessarily positive.) For $M^2 > 0$ the minimum is at $\phi = 0$ (symmetric

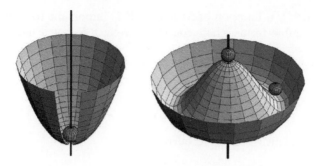

Fig. 14.1 The function $V(\phi)$ for $M^2 > 0$ (left) and $M^2 < 0$ (right). The same figure gives the energy of the bent rod of Problem 14.1 as a function of the order parameter for $F < F_{\text{cr}}$ (left) and $F > F_{\text{cr}}$ (right). *(Source: Dr A. Hoecker, CERN).*

solution, shown on the left in Figure 14.1), while for $M^2 < 0$ there is a whole circle of minima in the complex ϕ-plane with radius

$$v = (-M^2/2\lambda)^{1/2}$$

(Figure 14.1, right). Any point ϕ_0 on the circle exhibits spontaneous breaking of (14.7), since $e^{i\theta}\phi_0$ is degenerate with ϕ_0 but differs from it.

We conclude that

- the critical point for the parameter M^2 is $M^2 = 0$. The value $M^2 = 0$ corresponds to $(T = T_{\text{cr}})$ of the ferromagnet.
- For $M^2 > 0$ the symmetric solution is stable.
- For $M^2 < 0$ spontaneous symmetry breaking occurs.

Let us choose $M^2 < 0$. Our next task is to study the physics of fluctuations of $\phi(x)$ around its ground state. It is clear that the physics will not depend on the particular point on the circle of minima of $V(\phi)$ we will choose for the ground state, since they are all equivalent, obtained from any given one by an appropriate symmetry transformation (14.7). Let us, for convenience, choose the point on the real axis in the ϕ plane, and write

$$\phi(x) = v + \frac{1}{\sqrt{2}}\left(\psi(x) + i\chi(x)\right) \tag{14.10}$$

Upon substitution of (14.10) in (14.6) we find, up to the inessential additive constant λv^4,

$$\mathcal{L}_1(\phi) \;\rightarrow\; \mathcal{L}_2(\psi, \chi) = \frac{1}{2}\left(\partial_\mu \psi\right)^2 + \frac{1}{2}\left(\partial_\mu \chi\right)^2 - \frac{1}{2}\left(4\lambda v^2\right)\psi^2$$
$$-\sqrt{2}\lambda v\psi\left(\psi^2 + \chi^2\right) - \frac{\lambda}{4}\left(\psi^2 + \chi^2\right)^2 \tag{14.11}$$

Notice that $\mathcal{L}_2$ does not contain any term proportional to χ^2, which is expected since V is locally flat in the χ direction. A second remark concerns the arbitrary parameters of the theory. $\mathcal{L}_1$ contains two such parameters, a "mass" M and a dimensionless coupling constant λ. In $\mathcal{L}_2$ we have again the coupling constant λ and a new mass parameter v, which is a function of M and λ. It is important to notice that, although $\mathcal{L}_2$ also contains cubic terms, their coupling strength is not a new parameter, but is proportional to λv.

The term *spontaneously broken symmetry* is slightly misleading because the invariance is not broken. $\mathcal{L}_2$ is still invariant under the transformations with infinitesimal parameter θ

$$\delta\psi = -\theta\chi, \quad \delta\chi = \theta\psi + \sqrt{2}\,\theta\,v \tag{14.12}$$

to which corresponds the conserved current

$$j_\mu = \psi\partial_\mu\chi - \chi\partial_\mu\psi + \sqrt{2}\,v\,\partial_\mu\chi \tag{14.13}$$

The last term, which is linear in the derivative of χ, is characteristic of the phenomenon of spontaneous symmetry breaking.

It should be emphasised here that $\mathcal{L}_1$ and $\mathcal{L}_2$ are completely equivalent Lagrangians. They both describe the dynamics of the same physical system, and a change of variables, such as (14.10), cannot change the physics. However, this equivalence can be verified only by solving the theory exactly. In this case we shall find the same solution using either of them. However, we usually do not have exact solutions and we apply perturbation theory, which is an approximation scheme. Then the equivalence is no longer guaranteed and, in fact, perturbation theory has much better chances to give sensible results using one language rather than the other. In particular, if we use $\mathcal{L}_1$ as a quantum field theory and we decide to apply perturbation theory taking, as the unperturbed part, the quadratic terms of $\mathcal{L}_1$, we immediately see that we shall get nonsense. The spectrum of the unperturbed Hamiltonian would consist of particles with negative mass squared, and no perturbation corrections, at any finite order, could change that. This is essentially due to the fact that, in doing so, we are trying to calculate the quantum fluctuations around an unstable solution, and perturbation theory is just not designed for that. On the contrary, we see that the quadratic part of $\mathcal{L}_2$ gives a reasonable spectrum; thus we hope that perturbation theory will also give reasonable results. Therefore, we conclude that our physical system, considered now as a quantum system, consists of two interacting scalar particles, one with mass $m_\psi^2 = 4\lambda v^2$ and the other with $m_\chi = 0$. We believe that this is the spectrum we would find, also starting from $\mathcal{L}_1$, if we could solve the dynamics exactly. The remark we made on the Heisenberg ferromagnet applies here as well. We can choose any point on the circle of minima as our vacuum state, but the Hilbert spaces we build above each one of them are orthogonal to each other and there is no transition between states belonging to different spaces.

The appearance of a zero-mass particle in the quantum version of the model is an example of a general theorem due to J. Goldstone: To every generator of a spontaneously broken symmetry there corresponds a massless particle, called the Goldstone particle. This theorem is just the translation, into quantum field theory language, of the statement about the degeneracy of the ground state. The ground state of a system described by a quantum field theory is the vacuum state, and you need massless excitations in the spectrum of states in order to allow for the degeneracy of the vacuum. We shall give a more general proof of Goldstone's theorem in the next section.

14.2.3   Goldstone theorem

In section 14.2.2 we saw an example of Goldstone's theorem, which states that to every generator of a spontaneously broken symmetry there corresponds a massless particle. We want to give here a general proof, which will not depend on a particular field theory model and which will illustrate the underlying assumptions for its validity.[2]

Let us consider a theory with the following properties:

1. It is Lorentz invariant.
2. It is defined in a Hilbert space with only positive definite norm states.
3. There exists a minimum energy state, the "ground state"[3] $|\Omega\rangle$, although we make no assumption about its uniqueness.
4. The theory is invariant under a Lie group G of transformations, and this invariance implies the existence of a set of conserved currents $j_\mu^a(x)$, $a = 1, 2, \ldots, N$, the dimension of the Lie algebra of G. The current conservation implies the time-independence of the corresponding charges[4]

$$\partial^\mu j_\mu^a(x) = 0 \quad \Longrightarrow \quad Q^a \equiv \int \mathrm{d}^3x\, j_0^a(x)\,, \quad \dot{Q}^a = 0 \qquad (14.14)$$

We assume that the symmetry is not trivial; in other words, there exists at least one operator $A(x)$, which is not a singlet of G, i.e.

$$\delta A(x) \equiv [Q^a, A(x)] \neq 0 \qquad (14.15)$$

The assumptions (1)–(4) are the ones we usually make in local quantum field theory. Now comes the one about spontaneous symmetry breaking:

5. We assume that, for some Q^a, the expectation value of δA in the ground state, which we call "vacuum expectation value" is non-zero.

$$\langle \Omega | \, \delta A \, | \Omega \rangle = \langle \Omega | \, [Q^a, A] \, | \Omega \rangle \neq 0 \qquad (14.16)$$

Notice that this implies that the vacuum state is not invariant under G and, therefore, it is not annihilated by all the generators Q^a. This is the formal version of the degeneracy of the vacuum that we found in the previous examples. We saw there that applying a symmetry transformation to a minimum energy field configuration (a "vacuum" state), we obtained another field configuration with the same energy.

---

[2]This argument is due to W. Gilbert, *Phys. Rev. Lett.* **12**, 713 (1964), who presented it as a proof against the possibility of spontaneous symmetry breaking without massless particles. As we shall see, the argument fails if the spontaneously broken symmetry is gauged.

[3]We often call the ground state the "vacuum state", although the terminology is slightly misleading because the definition of the latter is connected to a particular set of creation and annihilation operators.

[4]At this point our proof is only heuristic. The existence of the integrals in (14.14) assumes that all fields vanish sufficiently rapidly at infinity so that surface terms can be dropped. However, in the presence of massless particles this is not always true. See, for example, the Coulomb scattering amplitude in quantum mechanics. Note also that it is precisely the existence of such massless particles that we want to prove. A more rigorous way would be to define the charges as integrals inside a given volume V, $Q_V^a \equiv \int_V \mathrm{d}^3x\, j_0^a(x)$ and study carefully the limit $V \to \infty$. The result is that the relations (14.14) are not valid in a strong sense, i.e. as operator equations, although they can still be used under suitable conditions in a weak sense as matrix elements between well localised states.

Using these assumptions, we shall prove that the spectrum of states of such a theory contains a massless particle.

Let us consider the Fourier transform of the vacuum expectation value of the commutator between the current $j_\mu^a(x)$ and the operator A

$$G_\mu(k) \equiv \int \mathrm{d}^4 x\, \mathrm{e}^{\mathrm{i}k\cdot x}\, \langle \Omega|\, [j_\mu^a(x), A(0)]\, |\Omega\rangle \qquad (14.17)$$

$G_\mu(k)$ is a vector function of the four vector k and, by assumptions (1) and (5), it is of the general form

$$G_\mu(k) = k_\mu f(k^2), \quad f(k^2) \neq 0 \qquad (14.18)$$

We now compute

$$k^\mu G_\mu(k) = k^2 f(k^2) = -\mathrm{i} \int \mathrm{d}^4 x\, (\partial^\mu \mathrm{e}^{\mathrm{i}k\cdot x})\, \langle \Omega|\, [j_\mu^a(x), A(0)]\, |\Omega\rangle = 0 \qquad (14.19)$$

where we used partial integration (see, however, the footnote on the behaviour at infinity) and the conservation of the current. The only solution of the two equations (14.18) and (14.19) is

$$f(k^2) = C\, \delta(k^2) \qquad (14.20)$$

with C some constant. In other words, the correlation function $G_\mu(k)$ has a delta-function singularity at $k^2 = 0$. Notice that $-\mathrm{i}\pi\delta(k^2) = \lim_{\epsilon\to 0} \mathrm{Im}(k^2 + \mathrm{i}\epsilon)^{-1}$, which is the propagator of a massless particle. To finish the proof we must show that this propagator corresponds to a massless physical state and it is not cancelled by some other massless state. This is guaranteed by assumption (2). Since all states have positive norm they all contribute with the same sign and no cancellations can occur.

A few final remarks:

(i) In physics we often deal with symmetries which are only approximate. This means that we can split the total Hamiltonian H into a sum of a symmetric part and a symmetry breaking part: $H = H_{\mathrm{sym}} + H_{\mathrm{br}}$. We say that the symmetry is approximately valid if the effects of H_{br} are small compared to those of H_{sym}. If such an approximate symmetry is spontaneously broken we expect the corresponding Goldstone particle to be massive but with a very small mass. We will call it the *pseudo-Goldstone particle*.

(ii) Goldstone's theorem tells us that to every generator of a spontaneously broken symmetry corresponds a massless particle. This Goldstone (or pseudo-Goldstone) particle has the same quantum numbers as the divergence of the corresponding symmetry current. We can obtain this result by looking at the intermediate states, which contribute to the matrix element of equation (14.19), but we will present a more general discussion in Chapter 18. In perturbation theory we can verify that the existence of this massless particle leads to infrared divergences, which explain why the formal definition of the charge operator as the space integral of the zero component of the current is not correct. However, it can be used to answer the opposite question: If we observe a massless, or nearly massless, particle in nature, how could we know whether it is the

Goldstone, or pseudo-Goldstone, particle of some spontaneously broken symmetry? The answer is to look for the existence of a conserved, or nearly conserved, symmetry current with the right quantum numbers. We shall give a more operational version of this property in Chapter 18. In Problem 14.2 we study a field theory model, the so-called σ-model, which exemplifies all these concepts.

Although the proof of the theorem we presented here applies only to relativistic field theories, an analogous phenomenon is known to occur in non-relativistic systems, such as the isotropic Heisenberg ferromagnet (14.2). The spontaneous breaking of rotational symmetry with magnetisation M, results in the appearance of excitations in the form of spin waves whose dispersion relation for small k is $\mathcal{E}(k) = Ck^2$. They are gapless, which is the non-relativistic counterpart of a massless particle. However, their number and their properties are model dependent.

14.3   Gauge Symmetries

In the previous chapter we introduced the concept of a gauge symmetry. We saw that it follows from a fundamental geometric principle, but it seems to imply the existence of massless vector particles. Since, apart from the photon, such particles are not present in physics, we could naïvely conclude that gauge theories and, in particular, non-Abelian gauge theories, cannot describe the interactions among elementary particles. In this chapter we studied the phenomenon of spontaneous symmetry breaking. We saw that it is also associated with the existence of massless particles, the Goldstone particles and again this requirement of massless particles seems to severely limit their applicability to particle physics. In this section we will study the consequences of spontaneous breaking of a symmetry which is gauged. We shall find a very surprising result. When combined together the two problems solve each other. It is this miracle that we want to present here. We start with the Abelian case.

14.3.1   The Abelian model

We look at the model of section 14.2.2, with the $U(1)$ symmetry (14.7) promoted to a local symmetry $\theta \to \theta(x)$. As we have explained already, this implies the introduction of a massless vector field, which we can call the "photon", and the interactions are obtained by replacing the derivative operator ∂_μ with the covariant derivative D_μ. Adding, finally, the photon kinetic energy term we end up with

$$\mathcal{L}_1 = -\frac{1}{4}F_{\mu\nu}^2 + |(\partial_\mu + \mathrm{i}eA_\mu)\phi|^2 - M^2\phi^*\phi - \lambda(\phi^*\phi)^2 \qquad (14.21)$$

$\mathcal{L}_1$ is invariant under the gauge transformation

$$\phi(x) \to \phi'(x) = \mathrm{e}^{\mathrm{i}e\theta(x)}\phi(x), \quad A_\mu(x) \to A'_\mu(x) = A_\mu(x) - \partial_\mu\theta(x) \qquad (14.22)$$

The same analysis as before shows that for $\lambda > 0$ and $M^2 < 0$ there is spontaneous breaking of the $U(1)$ symmetry. Substitution of (14.10) into (14.21) leads to

$$\mathcal{L}_1 \;\to\; \mathcal{L}_2 = -\frac{1}{4}F_{\mu\nu}^2 + \frac{2\,e^2 v^2}{2}A_\mu^2 + \sqrt{2}\,evA_\mu\partial^\mu\chi + \frac{1}{2}(\partial_\mu\chi)^2$$

$$+ \frac{1}{2}(\partial_\mu\psi)^2 - \frac{1}{2}(4\lambda v^2)\psi^2 + \qquad\qquad (14.23)$$

$$+ eA^\mu(\psi\partial_\mu\chi - \chi\partial_\mu\psi) + \sqrt{2}e^2 v A_\mu^2\,\psi + \frac{1}{2}e^2 A_\mu^2(\psi^2 + \chi^2)$$

$$- \sqrt{2}\lambda v\psi(\psi^2 + \chi^2) - \frac{\lambda}{4}(\psi^2 + \chi^2)^2$$

The surprising term is the second one, which is proportional to A_μ^2. It looks as though the photon has become massive. Notice that (14.23) is still gauge invariant since it is equivalent to (14.21). The gauge transformation is now obtained by replacing (14.10) in (14.22), to become

$$\psi'(x) = \Big[\psi(x) + \sqrt{2}v\Big]\cos(e\theta(x)) - \chi(x)\sin(e\theta(x)) - \sqrt{2}v$$

$$\chi'(x) = \Big[\psi(x) + \sqrt{2}v\Big]\sin(e\theta(x)) + \chi(x)\cos(e\theta(x)) \qquad (14.24)$$

$$A'_\mu(x) = A_\mu(x) - \partial_\mu\theta(x)$$

This means that our previous conclusion, that gauge invariance forbids the presence of an A_μ^2 term, was simply wrong. Such a term can be present, but the gauge transformation is slightly more complicated; it must be accompanied by a constant shift of the scalar field.

The Lagrangian (14.23) as a quantum field theory seems to describe the interaction of a massive vector particle (A_μ) and two scalars, one massive (ψ) and one massless (χ). However, we can see immediately that something is wrong with this counting. A warning is already present in the quadratic non-diagonal term $A_\mu\partial^\mu\chi$. Indeed, the perturbative particle spectrum can be read from the Lagrangian only after we have diagonalised the quadratic part. A more direct way to understand the problem is to count the apparent degrees of freedom in the two Lagrangians.

Lagrangian (14.21):
(i) One massless vector field: 2 degrees of freedom
(ii) One complex scalar field: 2 degrees of freedom
Total: 4 degrees of freedom

Lagrangian (14.23):
(i) One massive vector field: 3 degrees of freedom
(ii) Two real scalar fields: 2 degrees of freedom
Total: 5 degrees of freedom

Since physical degrees of freedom cannot be created by a simple change of variables, we conclude that the Lagrangian (14.23) should contain fields which do not describe physical particles. This is indeed the case, and we can exhibit a transformation, which makes the unphysical fields disappear. Instead of parametrising the complex field ϕ by its real and imaginary parts, let us choose its modulus and its phase. The choice is dictated by the fact that it is a change of phase that describes the motion along the circle of the minima of the potential $V(\phi)$. We thus write

$$\phi(x) = \left[v + \frac{1}{\sqrt{2}} \rho(x) \right] e^{i\zeta(x)/v} \ , \quad A_\mu(x) = B_\mu(x) - \frac{1}{v} \partial_\mu \zeta(x) \tag{14.25}$$

In this notation, the gauge transformation (14.22) or (14.25) is simply a translation of the field ζ: $\zeta(x) \to \zeta(x) + ev\theta(x)$. Replacing (14.25) in (14.21) we obtain

$$\mathcal{L}_1 \ \to \ \mathcal{L}_3 = -\frac{1}{4} B_{\mu\nu}^2 + \frac{e^2 v^2}{2} B_\mu^2 + \frac{1}{2} (\partial_\mu \rho)^2 - \frac{1}{2} (4\lambda v^2) \rho^2$$
$$- \frac{\lambda}{4} \rho^4 + \frac{1}{2} e^2 B_\mu^2 (2\sqrt{2} v\rho + \rho^2) - \sqrt{2} \lambda v \rho^3 \tag{14.26}$$

with

$$B_{\mu\nu} = \partial_\mu B_\nu - \partial_\nu B_\mu$$

The field $\zeta(x)$ has disappeared. The Lagrangian density $\mathcal{L}_3$ of equation (14.26) describes two massive particles, a vector (B_μ) and a scalar (ρ). It exhibits no gauge invariance, since it does not depend on the field $\zeta(x)$.

We see that we obtained three different Lagrangians describing the same physical system. $\mathcal{L}_1$ is invariant under the usual gauge transformation, but it contains a negative square mass and, therefore, it is unsuitable for quantisation. $\mathcal{L}_2$ is still gauge invariant, but the transformation law (14.25) is more complicated. It can be quantised in a space containing unphysical degrees of freedom. This, by itself, is not a great obstacle and it occurs frequently. We have seen that ordinary quantum electrodynamics is usually quantised in a space involving unphysical (longitudinal and scalar) photons. In fact, it is $\mathcal{L}_2$, in a suitable gauge, which is often used for practical calculations. Finally, $\mathcal{L}_3$ is no longer invariant under any kind of gauge transformation, but it exhibits clearly the particle spectrum of the theory. It contains only physical particles and they are all massive. This is the miracle that was announced earlier. Although we start from a gauge theory, the final spectrum contains massive particles only.

Actually, $\mathcal{L}_3$ can be obtained from $\mathcal{L}_2$ by an appropriate choice of gauge. Indeed, following the method we introduced in section 7.4, we can add to the Lagrangian $\mathcal{L}_2$ the term $-G^2/2\alpha$ with $G = \partial_\mu A^\mu$. The quadratic part of the $A - \chi$ system now becomes

$$\mathcal{L}_{A-\chi} = -\frac{1}{4} F_{\mu\nu}^2 - \frac{1}{2\alpha} (\partial_\mu A^\mu)^2 + \frac{2e^2 v^2}{2} A_\mu^2 + \sqrt{2} ev A_\mu \partial^\mu \chi + \frac{1}{2} (\partial_\mu \chi)^2 \tag{14.27}$$

The propagators are well defined, but there is still a non-diagonal $A - \chi$ term. A more convenient gauge choice is given by $G = \partial_\mu A^\mu - \sqrt{2} ev\alpha\chi$. In this case $\mathcal{L}_{A-\chi}$ takes the form

$$\mathcal{L}_{A-\chi}^Q = -\frac{1}{4} F_{\mu\nu}^2 - \frac{1}{2\alpha} (\partial_\mu A^\mu)^2 + \frac{2e^2 v^2}{2} A_\mu^2 + \frac{1}{2} (\partial_\mu \chi)^2 - \frac{\alpha 2 e^2 v^2}{2} \chi^2 \tag{14.28}$$

The A_μ and χ propagators decouple, and in momentum space they are

$$G_{\mu\nu}(k) = \frac{-i}{k^2 - 2e^2 v^2} \left[\eta_{\mu\nu} - (1-\alpha) \frac{k_\mu k_\nu}{k^2 - 2\alpha e^2 v^2} \right]$$

and

$$G(k) = \frac{i}{k^2 - 2\alpha e^2 v^2} \tag{14.29}$$

respectively. We note that we find the same singularity in both the χ-propagator and the $k_\mu k_\nu$ part of the A-propagator. We can check, see problem 14.6, that these singularities cancel in all physical amplitudes, as they should, since physical amplitudes must be independent of the gauge parameter α. Some special choices of α are: (i) $\alpha = 0$; both χ and the $k_\mu k_\nu$ part of A correspond to massless propagators, (ii) $\alpha = 1$; all poles in the propagators correspond to mass $\sqrt{2}ev$, (iii) $\alpha \to \infty$; $G_{\mu\nu}$ becomes the propagator of a massive vector field with mass $\sqrt{2}ev$, while χ becomes infinitely heavy and does not contribute to any Green function of the theory, because its propagator vanishes. We say that the field χ *decouples.* The Lagrangian becomes $\mathcal{L}_3$. This choice is often referred to as *the unitary gauge.* Only physical degrees of freedom propagate.

The conclusion of this section can now be stated as follows:

In a theory with spontaneous symmetry breaking in the presence of gauge interactions, the gauge vector bosons acquire a mass and the would-be massless Goldstone bosons decouple and disappear. Their degrees of freedom are used in order to make possible the transition from massless to massive vector bosons. This phenomenon has a complicated history. It was implicit in the first phenomenological description of superconductivity by F. London and H. London as well as in the L.D. Landau and V.L. Ginzburg theory of 1950. In the framework of the BCS theory it was studied by Ph. Anderson in 1962. J. Schwinger was the first to understand the physical principles in particle physics. In four-dimensional field theory it was first introduced by R. Brout and F. Englert as well as by P. Higgs in 1964. It was further studied by G.S. Guralnik, C.R. Hagen and T.W.B. Kibble. It is commonly known as the "Higgs" or the "BEH mechanism".

In a qualitative sense we can understand the phenomenon by looking at the correlation length in the two phases. In the symmetric phase the presence of a gauge interaction implies the existence of a zero mass gauge boson which induces long range correlations. In the other phase the spontaneous breaking of the global symmetry would imply the existence of a massless Goldstone mode, thus creating new long range correlations. The BEH mechanism tells us that in the final spectrum of the theory there are no zero mass excitations, which can be understood by saying that the two long range correlations, the one due to gauge interactions and the one due to the Goldstone mode, cancel each other. The commonly used expression "spontaneous breaking of a gauge symmetry" is, in fact, quite misleading.

A final remark: The BEH phenomenon seems to violate Goldstone's theorem: there is a spontaneous symmetry breaking and no massless Goldstone particle. The reason for the apparent paradox is that, in quantising a gauge theory, we do not respect the assumptions we stated in section 14.2.3. In particular, assumptions (1) (explicit Lorentz invariance) and (2) (positivity in Hilbert space), cannot be enforced simultaneously. As we saw in the example of quantum electrodynamics, in the covariant gauge we had to introduce unphysical degrees of freedom, such as scalar photons, which come with negative metric. If we want to have only physical, positive metric degrees of freedom, we must choose a non-covariant gauge, such as the Coulomb gauge.

14.3.2   The non-Abelian case

The extension to the non-Abelian case is straightforward. Let us consider a gauge group G with n generators and, thus, n massless gauge bosons. The claim is that, if part of this symmetry is broken spontaneously, leaving a subgroup H with m generators unbroken, then the m gauge bosons associated to H will remain massless, while the remaining $n - m$ will acquire non-zero masses. For this picture to be realised, we need $n - m$ scalar fields with the quantum numbers of the broken generators, which will be transformed to the zero helicity states of the massive vector bosons. Furthermore, as in the previous example, we shall see that we need at least one more scalar particle, which remains physical.

Let us prove the above claims. We introduce a multiplet of scalar fields Φ_i, which transform according to some N-dimensional representation of G, not necessarily irreducible. According to the rules we explained in the previous chapter, the Lagrangian of the system is given by

$$\mathcal{L} = -\frac{1}{2}\mathrm{Tr}(F_{\mu\nu}F^{\mu\nu}) + (D_\mu\Phi)^\dagger D^\mu\Phi - V(\Phi) \tag{14.30}$$

In component notation, the covariant derivative is, as usual, $(D_\mu\Phi)_i = \partial_\mu\Phi_i - ig^{(a)}T_{ij}^a A_\mu^a \Phi_j$, where we have allowed for the possibility of having arbitrary coupling constants $g^{(a)}$ for the various generators of G, because we do not assume that G is simple or semi-simple. $V(\Phi)$ is a polynomial in Φ, invariant under G and of degree equal to four. As before, we assume that we can choose the parameters in V, so that its minimum is not at $\Phi = 0$ but rather at $\Phi = v$, where v is a constant vector in the representation space of Φ, but v is not unique. The n generators of G can be separated into two classes: m generators which annihilate v and form the Lie algebra of the unbroken subgroup H, and $n - m$ generators, represented in the representation of Φ by matrices T^a, such that (i) $T^a v \neq 0$ and (ii) all vectors $T^a v$ are independent and can be chosen orthogonal. Any vector in the orbit of v, i.e. of the form $e^{iw^a T^a}v$, is an equivalent minimum of the potential. We shift the scalar fields Φ by $\Phi \rightarrow \Phi + v$. It is convenient to decompose Φ into components along the orbit of v and orthogonal to it, the analogue of the χ and ψ fields of the previous section. We write

$$\Phi = \frac{i}{\sqrt{2}} \sum_{a=1}^{n-m} \frac{\chi^a T^a v}{|T^a v|} + \frac{1}{\sqrt{2}} \sum_{b=1}^{N-n+m} \psi^b u^b + v \tag{14.31}$$

where the vectors u^b form an orthonormal basis in the space orthogonal to all $T^a v$. The corresponding generators span the coset space G/H. As before, we will show that the fields χ^a will be absorbed in accordance with the BEH mechanism and the fields ψ^b will remain physical. Note that the set of vectors u^b contains at least one element since, for all a, we have

$$v \cdot T^a v = 0 \tag{14.32}$$

This shows that the dimension N of the representation of Φ must be larger than $n-m$ and, therefore, there will remain at least one physical scalar field.

Next, we substitute (14.31) into (14.30) and obtain

$$\mathcal{L} = -\frac{1}{2}\text{Tr}(F_{\mu\nu}F^{\mu\nu}) + \frac{1}{2}\sum_{a=1}^{n-m}(\partial_\mu\chi^a)^2 + \frac{1}{2}\sum_{b=1}^{N-n+m}(\partial_\mu\psi^b)^2$$

$$+ \sum_{a=1}^{n-m} g^{(a)2}|T^a v|^2 A_\mu^a A^{\mu a} - \sqrt{2}\sum_{a=1}^{n-m} g^{(a)}|T^a v|\,\partial^\mu\chi^a A_\mu^a$$

$$-V(\Phi) + \dots \tag{14.33}$$

where the dots stand for coupling terms between the scalars and the gauge fields. In writing (14.33) we took into account that $T^b v = 0$ for $b > n - m$ and that the vectors $T^a v$ are orthogonal.

The analysis we presented in section 14.2.2 gives

$$\frac{\partial^2 V}{\partial\Phi_k\partial\Phi_l}\Big|_{\Phi=v}(T^a v)_l = 0 \tag{14.34}$$

which shows that the χ-fields would correspond to the Goldstone modes. As a result, the only mass terms which appear in V in equation (14.33) are of the form $\psi^b M^{bb'}\psi^{b'}$ and do not involve the χ-fields.

As far as the quadratic terms in the fields are concerned, the Lagrangian(14.33) is the sum of terms of the form found in the Abelian case. All gauge bosons which do not correspond to H generators acquire masses equal to $m_a = \sqrt{2}\,g^{(a)}|T^a v|$, with the corresponding zero-helicity states provided by the would-be Goldstone fields χ. The gauge bosons associated to the H-generators remain massless. Finally, the ψ^bs represent the remaining physical BEH fields.

14.4 Problems

Problem 14.1 *The bent rod.* A simple classical example of a system exhibiting the phenomenon of spontaneous symmetry breaking is provided by a bent rod shown in Figure 14.2. An ideal cylindrical rod of length l and radius R is loaded with a load F along the z-axis. We are interested in the equilibrium configuration of the rod. We call z the distance along the z-axis from the lower end O of the rod hinged at the origin of the coordinate system, and $X(z)$ and $Y(z)$ the deviations, along the x and y directions respectively, of the axis of the rod at the point z from the symmetric position. The general equations of elasticity are non-linear but, for small deviations, we can obtain a simple linearised approximation which exhibits the existence of the buckling transition. For the rod of Figure 14.2 which is hinged at both ends, the equations take the form

$$IE\frac{\mathrm{d}^2 X}{\mathrm{d}z^2} + FX = 0\ ,\quad IE\frac{\mathrm{d}^2 Y}{\mathrm{d}z^2} + FY = 0 \tag{14.35}$$

with the boundary conditions $X(0) = 0 = X(l)$, $Y(0) = 0 = Y(l)$. $I = \pi R^4/4$ is the moment of inertia of the rod with respect to any axis parallel to the x-y plane and passing through the centre of mass of the rod and E is the Young modulus characterising the material of the rod.

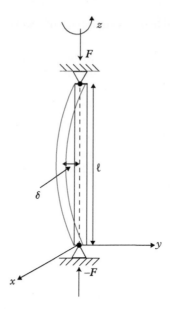

Fig. 14.2 The rod bends under the action of the load $\boldsymbol{F}$.

Show that for F larger than a critical value F_{cr} we obtain asymmetric solutions X or Y different from zero, resulting in a non-zero value of the buckling parameter δ shown in the figure. Compute the value of F_{cr}. In the terminology of section 14.2.1 identify the external control parameter and the order parameter.

Problem 14.2 *The linear σ-model.*

Consider four real spin-zero fields $\phi^i(x)$, $i = 1, .., 4$.

1. Write the $O(4)$ invariant Lagrangian density with a mass parameter M^2 and a dimensionless coupling constant λ. Compute the Hamiltonian density $\mathcal{H}$.

2. Show that for $M^2 < 0$, we have the phenomenon of spontaneous symmetry breaking with the set of minima of the potential forming the surface of a three-dimensional hypersphere. Choose a point in the ϕ^4 direction and break the symmetry from $O(4)$ to $O(3)$. Identify the corresponding Goldstone bosons.

3. Since $O(4)$ is locally isomorphic to $SU(2) \times SU(2)$ (see Problem 5.6) write the set of scalar fields as a two-by-two matrix Φ

$$\Phi(x) = \frac{i\boldsymbol{\pi}(x) \cdot \boldsymbol{\tau} + \sigma(x)\mathbf{1}}{\sqrt{2}}$$

where $\boldsymbol{\tau}$ and $\mathbf{1}$ are the three Pauli matrices and the two-by-two unit matrix, respectively, and we have renamed the fields $\phi^{1,2,3}(x) = i\pi^{1,2,3}(x)$ and $\phi^4(x) = \sigma(x)$. Write the $O(4)$ transformations acting on Φ and rewrite accordingly the Lagrangian you found in question 1.

4. We add a doublet of Dirac fermion fields $\Psi(x) = \begin{pmatrix} p(x) \\ n(x) \end{pmatrix}$ which we split in the left- and right-hand components $\Psi_{L,R}(x) = (\frac{1}{2})(1 \pm \gamma_5)\Psi(x)$. Under $SU(2) \times SU(2)$ Ψ_L

and Ψ_R are assumed to transform as members of the $(\frac{1}{2}, 0)$ and $(0, \frac{1}{2})$ representation, respectively.

Write the complete $SU(2) \times SU(2)$ invariant Lagrangian containing terms with dimensions less than or equal to four. Show that the fermions are massless. Show that σ is a scalar and the three pions are pseudoscalars.

5. For $M^2 < 0$ shift the σ-field $\sigma \to \sigma + v$ and study the resulting theory. Show that it exhibits the phenomenon of spontaneous symmetry breaking $SU(2) \times SU(2) \to SU(2)_V$. Show that this $SU(2)_V$ which is left intact is the diagonal subgroup of the initial $SU(2) \times SU(2)$ with pure vector currents. Find the transformation properties under $SU(2)_V$ of σ and the three pions. Find the mass spectrum of the fermions.

6. Show that the original Lagrangian has a further $U(1)$ invariance and interpret the corresponding conserved quantum number.

A first version of this model was introduced in 1960 by M. Gell-Mann and M. Lévy.

Problem 14.3 *The Georgi–Glashow model.* We consider an $O(3)$ Yang–Mills theory.

1. Write the Yang–Mills Lagrangian.

2. We want to induce a spontaneous symmetry breaking $O(3) \to U(1)$. Find the minimum number of Brout–Englert–Higgs fields.

3. Write the corresponding Lagrangian having only terms with dimension smaller than or equal to 4.

4. After spontaneous symmetry breaking find the mass spectrum of the physical degrees of freedom.

Problem 14.4 *Breaking patterns of $SU(3)$ with a scalar octet.*

Consider an $SU(3)$ Yang–Mills theory coupled to a multiplet of scalar fields $\Phi(x)$ belonging to the adjoint eight-dimensional representation. Write $\Phi(x)$ as a 3×3 Hermitian traceless matrix.

1. Write the resulting most general Lagrangian density, invariant under gauge $SU(3)$ transformations, having terms with dimensions less than or equal to four.

2. Find the possible symmetry breaking patterns of $SU(3)$, which can be realised with this one scalar adjoint and determine the range of parameters for which it is obtained.

3. Determine the spectrum of the resulting physical degrees of freedom in each case.

Problem 14.5 *Breaking patterns of $SU(5)$.* Consider an $SU(5)$ Yang–Mills theory coupled to a multiplet of scalar fields $\Phi(x)$ belonging to the adjoint 24-dimensional representation. Write $\Phi(x)$ as a 5×5 traceless matrix.

1. Write the resulting most general Lagrangian density, invariant under gauge $SU(5)$ transformations, having terms with dimensions less than or equal to four.

2. Find the range of parameters which induce spontaneous symmetry breaking: (i) $SU(5) \to SU(4) \times U(1)$, (ii) $SU(5) \to SU(3) \times SU(2) \times U(1)$.

3. Find the spectrum of the physical degrees of freedom in each case.

Problem 14.6 We consider a field theory model with the following fields: (i) A Dirac spinor $\Psi(x)$ which we split in its left- and right-handed parts: $L = \frac{1}{2}(1 + \gamma_5)\Psi$,

$R = \frac{1}{2}(1 - \gamma_5)\Psi$. (ii) A complex scalar field $\phi(x)$. (iii) A real massless vector field $A_\mu(x)$. We assume that the theory is invariant under the Abelian group of gauge transformations given by

$$L \to e^{ie\theta(x)}L, \quad R \to e^{-ie\theta(x)}R, \quad \phi \to e^{2ie\theta(x)}\phi, \quad A_\mu \to A_\mu + \partial_\mu\theta(x)$$

1. Write the most general gauge invariant Lagrangian density with terms of dimension smaller than or equal to four. Show that the fermion field is massless and the gauge field couples to the axial fermionic current.

2. Choose the mass-square of the scalar field negative and study the model in the phase with spontaneous symmetry breaking. Find the mass spectrum in this phase.

3. In the gauge given by the function $G = \partial^\mu A_\mu - ev\alpha\chi$ which we studied in section 14.3.1 compute the boson propagators and verify the equations (14.28) and (14.29).

4. Compute the amplitude for the elastic scattering of two Dirac particles in the one particle exchange approximation and show that the gauge-dependent poles cancel.

Problem 14.7 *The T.D. Lee model of spontaneous CP breaking.*

We consider a model with two complex scalar fields $\phi_1(x)$ and $\phi_2(x)$ which generalises the model of section 14.2.2.

1. Write the most general, Hermitian Lagrangian density, which is polynomial in the fields ϕ_1 and ϕ_2 and their first derivatives, with terms of dimension smaller than or equal to 4, and which is left invariant under CP as well as the global $U(1)$ transformations $\phi_i \to e^{i\theta}\phi_i$, $i = 1, 2$.

2. Show that there is a range of the parameters of the theory in which the minimum of the potential $V(\phi_1, \phi_2)$ occurs at $\phi_1 = v_1$ and $\phi_2 = v_2 e^{i\eta}$, with real v_1 and v_2 and $\eta \neq 0$.

3. Show that the resulting Lagrangian written in terms of the shifted fields breaks spontaneously the $U(1)$ symmetry *and CP*.

15

The Principles of Renormalisation

In this chapter we want to summarise the basic principles of perturbative renormalisation theory. Since renormalisation has a well-deserved reputation of complexity, this will be done by omitting many technical details. Our purpose is to explain the physical principles underlying this beautiful theory. We want to show that it offers the only known mathematically consistent way to define the perturbation expansion in a quantum field theory.

15.1   The Need for Renormalisation

Everybody who has attempted to compute a 1-loop Feynman diagram knows that you often encounter divergent expressions. For example, in the ϕ^4 theory we found the diagrams of Figure 11.4 involving the integral

$$I = \int \frac{\mathrm{d}^4 k}{(k^2 - m^2 + \mathrm{i}\epsilon)[(k-p)^2 - m^2 + \mathrm{i}\epsilon]} \qquad (15.1)$$

which diverges logarithmically at large k. Similar divergences can be found in any theory, such as QED. They have no place in a well defined mathematical theory, so if we meet them it means that we have made somewhere a mathematical mistake. Where is it? Let us first notice that the divergence in (15.1) occurs at large values of the internal momentum, which, by Fourier transform, implies short distances. Did we make a mistake at short distances? Yes we did! We wrote for the Lagrangian density of a scalar field $\phi(x)$ the expression

$$\mathcal{L} = \frac{1}{2} \partial_\mu \phi(x)\, \partial^\mu \phi(x) - \frac{1}{2} m^2 \phi^2(x) - \frac{\lambda}{4!} \phi^4(x) \qquad (15.2)$$

while, at the same time, we imposed the canonical commutation relations for the quantum field

$$\left[\phi(t, \boldsymbol{x}), \dot{\phi}(t, \boldsymbol{y}) \right] = \mathrm{i}\hbar \delta^3(\boldsymbol{x} - \boldsymbol{y}). \qquad (15.3)$$

We know that the Dirac δ-function is not really a "function" but a special form of what we call "a distribution". Many properties of well-behaved functions do not apply to it. In particular, multiplication is not always a well defined operation and $(\delta(x))^2$ is meaningless. The presence of the δ-function on the right-hand side of (15.3) implies that the field $\phi(x)$ is not an operator valued function but an operator valued distribution, so the product $(\phi(x))^2$ is ill defined. Yet, it is precisely expressions of this kind that we wrote in every single term of our Lagrangian (15.2). Since our initial Lagrangian is not

Elementary Particle Physics. John Iliopoulos and Theodore N. Tomaras, Oxford University Press.
© John Iliopoulos and Theodore N. Tomaras (2021). DOI: 10.1093/oso/9780192844200.003.0015

well defined, it is not surprising that our calculations yield divergent results. We have already encountered this problem, when we tried to evaluate the vacuum expectation value of the free field Hamiltonian.

Now that we have identified the origin of the problem, we can figure out ways to solve it. A conceptually simple one would be to replace the field products in (15.2) by splitting the points at which the distributions are defined,

$$\phi(x)\phi(x) \to \lim_{a \to 0} \phi\left(x + \frac{a}{2}\right)\phi\left(x - \frac{a}{2}\right) \qquad (15.4)$$

This expression is perfectly well defined for all values of the parameter a, except $a = 0$. In terms of distributions this means that the product is defined up to an arbitrary distribution $\mathcal{F}(a)$, which has support only at $a = 0$. Such a distribution is a superposition of the δ-function and its derivatives, i.e.

$$\mathcal{F}(a) = \sum_i C_i \delta^{(i)}(a) \qquad (15.5)$$

with the C_i's arbitrary real constants. The moral of the story is that the divergent expressions which appear in the perturbation expansion of a local quantum field theory are due to an incorrect application of the quantisation rules. As we see in equation (15.5), these rules imply that every term in the Lagrangian contains, in fact, a set of arbitrary constants, which must be determined by experiment. In this chapter we will show that when these constants are assigned prescribed values, every term in the perturbation expansion of any correlation function is well defined and calculable. Renormalisation is the mathematical procedure which exhibits this property explicitly. A final remark: how many parameters are needed in order to define a given field theory? The answer involves the distinction between *renormalisable* and *non-renormalisable* theories. For the first a finite number suffices. For the second we need an infinite number, which means that non-renormalisable theories have no predictive power. We will explain this difference shortly.

15.2   The Theory of Renormalisation

In this section we want to give some more information concerning the renormalisation prescription. The process we outlined above was formulated in x-space. It is intuitively easier to understand, but not very convenient for practical calculations, which are usually performed in momentum space. The connection is by Fourier transform. The derivatives of the δ-function in (15.5) become polynomials in the external momenta.

The renormalisation programme follows three steps:

- *The power counting*, which determines how many constants C_i we will need for a given field theory.
- *The regularisation*, which is a prescription to make every Feynman diagram finite at the price of introducing a new parameter in the theory, the analogue of the point-splitting parameter a we used in equation (15.4).
- *The renormalisation*, which is the mathematical procedure to remove the regularisation parameter and determine the values of the constants C_i

15.2.1 The power counting

As the term indicates, it is the counting which determines whether a given diagram is divergent or not. We shall need to introduce some terminology: We know already the notions of *disconnected*, *connected* and *amputated* diagrams. A connected amputated diagram is *one-particle irreducible* (1PI) if it cannot be separated in two disconnected pieces by cutting a single internal line. In Problem 15.1 we present a formal way to relate the sets of connected and 1PI diagrams. It is done through a Legendre transformation, similar to the one that relates the Hamiltonian and the Lagrangian in classical field theory. We show that the 2-point 1PI function $\Gamma^{(2)}(p^2)$ and the corresponding connected 2-point function $G_c^{(2)}(p^2)$ satisfy the relation

$$G_c^{(2)}(p^2)\Gamma^{(2)}(p^2) = -1 \tag{15.6}$$

A general connected diagram is constructed by joining together 1PI pieces, see Figure 15.1. It is obvious that a connected diagram is divergent if and only if at least one of its 1PI pieces is divergent, because the momenta of the internal connecting lines are fixed by energy-momentum conservation in terms of the external momenta and bring no new integrations.

This brings us to the power counting argument. A single-loop integral will be divergent in the ultraviolet if and only if the polynomial in the numerator is of degree equal or higher in the loop momentum than the denominator. For multi-loop diagrams this may not be the case, since the divergence may be entirely due to a particular sub-diagram; however, in the spirit of perturbation theory, the divergent sub-diagram must be treated first. We thus arrive at the notion of *superficial degree of divergence D* of a given 1PI diagram, defined as the difference between the degree of integration momenta of the numerator minus that of the denominator. The diagram will be called *primitively divergent* if $D \geq 0$. Let us compute, as an example, D for the diagrams of the scalar field theory (15.2), in the generalisation in which we replace the interaction term ϕ^4 by ϕ^m with m integer, $m \geq 3$. Let us consider a 1PI diagram with n vertices, I internal and E external lines. Every internal line contributes four powers of k in the numerator through the $\mathrm{d}^4 k$ integration measure and two powers in the denominator through the propagator. Every vertex brings a δ^4-function of the energy-momentum conservation. All but one of them can be used to eliminate one integration each, the last reflecting the overall conservation, which involves only external momenta. Therefore, we obtain

$$D = 2I - 4n + 4 \tag{15.7}$$

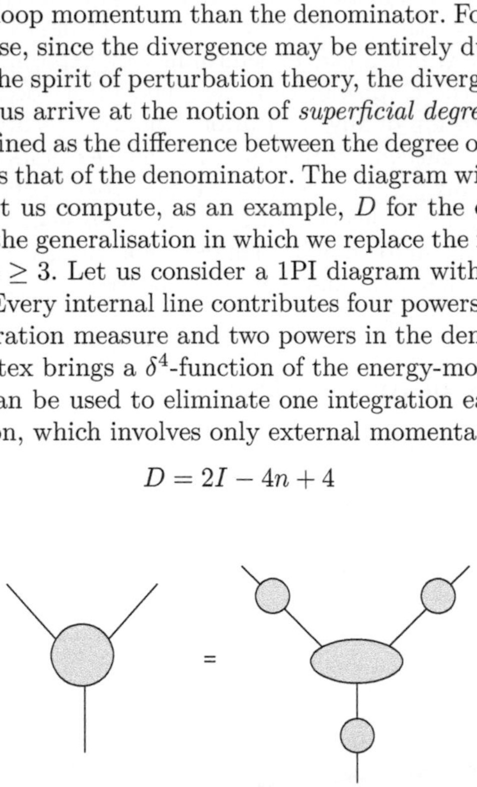

Fig. 15.1 The 1P-I decomposition of the 3-point function.

This expression can be made more transparent by expressing I in terms of E and m. A simple counting gives $2I + E = mn$ and (15.7) becomes

$$D = (m - 4)n - E + 4 \tag{15.8}$$

This is the main result. Although it is shown here as a plausibility argument, it is in fact a rigorous result. We see that $m = 4$ is a critical value and we can distinguish three cases:

(1) $m = 3$, $D = 4 - n - E$. D is a decreasing function of n, the order of the perturbation expansion. Only a limited number of diagrams are primitively divergent. Above a certain order they are all convergent. For reasons that will be clear soon, we shall call such theories *super-renormalisable*.

(2) $m = 4$, $D = 4 - E$. D is independent of the order of the perturbation expansion. If a Green function is divergent at some order, it will be divergent at all orders. For the ϕ^4 theory we see that the primitively divergent diagrams are those with $E = 2$, which have $D = 2$ and are quadratically divergent and those with $E = 4$, which have $D = 0$ and are logarithmically divergent. (Notice that, for this theory, all Green functions with odd E vanish identically because of the symmetry $\phi \to -\phi$). We shall call such theories *renormalisable*.

(3) $m > 4$, D is an increasing function of n. Every Green function, irrespectively of the number of external lines, will be divergent above some order of perturbation. We call such theories *non-renormalisable*.

This power counting analysis can be repeated for any quantum field theory. As a second example, we can look at quantum electrodynamics. We should now distinguish between photon and electron lines, which we shall denote by I_γ, I_e, E_γ and E_e for internal and external lines respectively. Counting the lines in a diagram gives the relations $2n = 2I_e + E_e$ and $n = 2I_\gamma + E_\gamma$, where n is again the order of perturbation expansion, i.e. the number of vertices. On the other hand the fermion propagator behaves like k^{-1} at large momenta. As a result, we obtain for the superficial degree of divergence of a 1PI diagram

$$D = 2I_\gamma + 3I_e - 4n + 4 = 4 - E_\gamma - \frac{3}{2}E_e \tag{15.9}$$

We see that D is independent of the order of the perturbation expansion and therefore the theory of quantum electrodynamics is renormalisable.

We leave it as an exercise to the reader to establish the renormalisation properties of other field theories. In four space-time dimensions the result is the following:

(1) The only super-renormalisable theories are interactions of the general form $\lambda_{ijk}\phi_i\phi_j\phi_k$, where ϕ_i are scalar fields and λ_{ijk} are coupling constants. We will denote them generically as ϕ^3.

(2) There exist five renormalisable interactions, namely
- ϕ^4
- Scalar, or pseudo-scalar Yukawa interactions of the general form $\bar{\psi}\psi\phi$ and $\bar{\psi}\gamma^5\psi\phi$
- Spinor electrodynamics $\bar{\psi}\gamma_\mu\psi A^\mu$
- "Scalar electrodynamics". It contains the two terms $\left[\phi^\dagger\partial_\mu\phi - (\partial_\mu\phi^\dagger)\phi\right]A^\mu$ and $A^\mu A_\mu\phi^\dagger\phi$.
- Yang–Mills theory $\mathrm{Tr}\mathcal{F}_{\mu\nu}\mathcal{F}^{\mu\nu}$.

(3) All other theories are non-renormalisable.

These results require some explanations: First, the notation is symbolic. For example, by ϕ^4 we mean any interaction among four scalar fields with non-derivative couplings. Similarly, by $\bar{\psi}\psi\phi$ we mean any term of the form $\bar{\psi}_i\Gamma_a^{ij}\psi_j\phi^a$, $i,j = 1, \ldots, N$ and $a = 1, \ldots, m$ with Γ a set of m numerical $N \times N$ matrices. Second, a sum of renormalisable interactions remains renormalisable, $g_1\bar{\psi}\psi\phi + g_2\phi^4$, with g_1 and g_2 two arbitrary coupling constants, is a renormalisable theory. Third, in the Yang–Mills theory we can include the gauge invariant couplings of multiplets of spin-0, or spin $-\frac{1}{2}$ fields.

A most *remarkable fact* is that, as we shall see later, nature uses *all* renormalisable interactions, and only those, to describe the strong, weak and electromagnetic interactions among elementary particles.

Before closing this section we want to make a remark, which is based on ordinary dimensional analysis. In four-dimensional field theories a boson field has dimensions of mass,[1] and a fermion field of mass to the power 3/2. Since all terms in a Lagrangian density must have dimensions equal to four, we conclude that the coupling constants of super-renormalisable interactions have dimensions of mass, those of renormalisable interactions are dimensionless, and those of non-renormalisable interactions have dimensions of negative powers of mass. This argument of dimensional analysis is used occasionally as a criterion for renormalisability. However, it offers an indication but it is not always correct. The most important counter-example is an interaction containing massive vector fields whose propagator has terms of the form $p_\mu p_\nu / m^2$, as we have established in equation (7.103). As a result such theories, although they may have dimensionless coupling constants, are in fact non-renormalisable. This can be checked explicitly by using the power counting argument, see Problem 15.2. The important point is that the propagator of a massive vector field behaves at large momenta as k^0 and not as k^{-2}.

15.2.2   Regularisation

The point-splitting we presented in (15.4) is an example of a procedure we shall call *regularisation*. It consists of introducing an extra parameter in the theory (in the point-splitting method in (15.4) this was the distance a), to which we do not necessarily associate a physical meaning, with the following properties: (i) the initial theory is recovered for a particular value of the parameter, in the example (15.4) this value is $a = 0$, (ii) the theory is finite for all values of the parameter in a region around the "physical" one $a = 0$, (iii) at the "physical" value we recover the divergences of the initial theory. We shall call this parameter *cut-off*.

If our purpose is to perform computations of Feynman diagrams, we may choose any cut-off procedure that renders these diagrams finite. There exists a plethora of such methods and there is no need to give a complete list. A direct method would be to cut off all integrations over loop momenta at some scale Λ. The initial theory is recovered in the limit $\Lambda \to \infty$. For practical calculations it is clear that we must choose a cut-off procedure that renders these computations as simple as possible. By trial and error, the simplest regularisation scheme turned out to be a quite counter-intuitive one.

---

[1]Remember, we are using units such that the speed of light c and Planck's constant h are dimensionless

We start by illustrating it in the simple example of the divergent integral (15.1). The integrand, considered as a function of the external momentum p, is analytic around $p = 0$, so we can expand it in a Taylor series. It is clear that only the first term in the expansion leads to a divergent integral, therefore, we can simplify the discussion by considering the value of I at $p = 0$. We get

$$I = \int \frac{\mathrm{d}^4 k}{(2\pi)^4} \frac{1}{(k^2 - m^2 + \mathrm{i}\epsilon)^2} \tag{15.10}$$

It will be convenient to perform a Wick rotation and consider the integral computed in Euclidean space, following the discussion we had in section 9.2.2. The integral takes the form

$$I = \mathrm{i} \int \frac{\mathrm{d}^4 k_E}{(2\pi)^4} \frac{1}{(k_E^2 + m^2)^2} \tag{15.11}$$

where $k_0 \equiv \mathrm{i}k_4$, $k_E^2 = k_4^2 + \mathbf{k}^2 = -k^2$ and $\mathrm{d}^4 k_E = \mathrm{d}k_4 \, \mathrm{d}^3\mathbf{k}$. Ignoring for the moment the divergence, we notice that the integrand depends only on k_E^2, so we choose spherical coordinates and write $\mathrm{d}^4 k_E = k_E^3 \, \mathrm{d}k_E \, \mathrm{d}\Omega_3$, where $\mathrm{d}\Omega_3$ is the surface element on the three-dimensional unit sphere. We further notice that I would have been convergent if we were working in a space-time of three, two or one, dimensions. The crucial observation is that in all these three cases we can write the result in a compact form as

$$\int \frac{\mathrm{d}^d k_E}{(2\pi)^d} \frac{1}{(k_E^2 + m^2)^2} = \frac{1}{(4\pi)^{d/2}} \frac{\Gamma(2 - d/2)}{(m^2)^{2 - d/2}} \,, \quad d = 1, 2, 3, \tag{15.12}$$

where $\Gamma(z)$ is the well-known special function which generalises for complex z the concept of the factorial. The values of interest in (15.12) are obtained using the formulae

$$\Gamma(n) = (n - 1)! \,, \quad \Gamma(n + 1/2) = \frac{\pi^{1/2}}{2^n}(2n - 1)!! \,; \quad n = 1, 2, \ldots \tag{15.13}$$

And now comes the big step. Nothing on the right-hand side of (15.12) forces us to consider this expression only for $d = 1$, 2 or 3. In fact, $\Gamma(z)$ is a meromorphic function in the entire complex plane with poles at the non-positive integers $n \leq 0$. For the integral (15.12), using the identity $n\Gamma(n) = \Gamma(n + 1)$, we see that when $d \to 4$ the Γ function behaves as $\Gamma(2 - d/2) \simeq 2/(4 - d)$. So, we can argue that, at least for this integral, we have introduced a regularisation, i.e. a new parameter, namely $\epsilon = 4 - d$, such that the expression is well defined for all values of $\epsilon \neq 0$ in a region around $\epsilon = 0$ and diverges when $\epsilon \to 0$.

Before showing how to generalise this approach to all other integrals we may encounter in the calculation of Feynman diagrams, let us try to make the logic clear by emphasising what this regularisation does not claim to be: First, it does not claim to be the result one would have obtained by quantising the theory in a complex number of dimensions. In fact, we do not know how to do that consistently. In this sense, dimensional regularisation does not offer a non-perturbative definition of the field theory. The prescription applies directly to the integrals obtained order by order in the perturbation expansion. Second, it cannot even be viewed as the analytic continuation to the complex d plane of the results we obtain in performing the integral for $d = 1, 2, 3$. Indeed, the knowledge of the values of a function on a finite number of points on the real axis does not allow for a unique analytic continuation. Instead, the claim is that (15.12), appropriately generalised, offers an unambiguous prescription to obtain a well defined answer for any Feynman diagram as long as ϵ stays away from zero.

The observation which allows for such a generalisation is that the Feynman rules always yield a special class of integrals. In all theories, whether renormalisable or not, after the integration momenta are Wick rotated, they are of the form

$$I(p_1, p_2, \ldots, p_n) = \int \prod_i \left(\frac{\mathrm{d}^d k_i}{(2\pi)^d} \right) \frac{N(k_1, k_2, \ldots)}{D(k_1, k_2, \ldots)} \prod_r \left((2\pi)^d \delta^d(k, p) \right) \tag{15.14}$$

where the ks and the ps are the momenta of the internal and external lines respectively, the product over i runs over all internal lines, that over r runs over all vertices and the δ functions denote the energy and momentum conservation at each vertex. D is equal to the product of the denominators of the propagators

$$D(k_1, k_2, \ldots) = \prod_i (k_i^2 + m_i^2) \tag{15.15}$$

with m_i the mass of the i-th internal line, while N is a polynomial in k with terms of the form

$$N(k_1, k_2, \ldots) \sim k_1^{\mu_1} k_1^{\mu_2} \ldots k_2^{\nu_1} k_2^{\nu_2} \ldots \tag{15.16}$$

N is equal to one in theories with only scalar fields and non-derivative couplings, such as ϕ^4. It may acquire factors of the form shown in equation (15.16) through derivative couplings, the $k^\mu k^\nu$ parts of the propagators of higher spin bosonic fields and the $\not{k}$ term of fermion propagators. All scalar products and integration measures are meant with respect to the d-dimensional Euclidean metric $\delta_{\mu\nu}$, which satisfies

$$\delta_\mu^\mu = \mathrm{Tr}\, \mathbf{1} = d \tag{15.17}$$

The dimensional regularisation consists of giving a precise expression for $I(p_1, p_2, \ldots, p_n)$ as a function of d, which coincides with the usual value whenever the latter exists and is well defined for every value of d in the complex d plane, with the exception of those positive integer values for which the original integral is divergent.

At one loop the integral (15.14) reduces to a factor $(2\pi)^d \delta^d(p_1 + p_2 + \cdots + p_n)$ times the integral

$$\tilde{I}(p_1, p_2, \ldots, p_n) = \int \frac{\mathrm{d}^d k}{(2\pi)^d} \frac{N(k; p_1, p_2, \ldots)}{D(k; p_1, p_2, \ldots)} \tag{15.18}$$

with k being the loop momentum. The denominator D takes the form

$$D(k; p_1, p_2, \ldots) = \prod_i \left[(k - \Sigma_{(i)} p)^2 + m_i^2 \right] \tag{15.19}$$

where $\Sigma_{(i)}p$ denotes the combination of external momenta, flowing in the i-th internal line. This product of propagators can be cast in a more convenient form by using the formula, first introduced by Feynman,

$$\frac{1}{P_1 P_2 \ldots P_l} = (l-1)! \int_0^1 \frac{\mathrm{d}z_1 \mathrm{d}z_2 \ldots \mathrm{d}z_l \, \delta(1 - \Sigma_i z_i)}{[z_1 P_1 + z_2 P_2 + \ldots + z_l P_l]^l} \tag{15.20}$$

With the help of (15.20) and an appropriate change of variables, all 1-loop diagrams lead to linear combinations of integrals of the general form

$$\hat{I}(p_1, p_2, \ldots, p_n) = \int \frac{\mathrm{d}^d k}{(2\pi)^d} \frac{k_{\mu_1} k_{\mu_2} \ldots k_{\mu_r}}{[k^2 + F^2(p, m, z)]^l} \tag{15.21}$$

with, generically, p-dependent coefficients. The Fs are scalar functions of the external momenta, the masses and the Feynman parameters z_i and have dimensions of mass. $I(p_1, p_2, \ldots, p_n)$ is obtained from the set of relevant $\hat{I}(p_1, p_2, \ldots, p_n)$ after integration with respect to the Feynman parameters z_i of (15.20). For odd values of r, $\hat{I}$ vanishes by symmetric integration. For even r, it can be easily computed in spherical coordinates. Two simple cases are

$$\int \frac{\mathrm{d}^d k}{(2\pi)^d} \frac{1}{[k^2 + F^2(p, m, z)]^l} = \frac{1}{(4\pi)^{d/2}} \frac{\Gamma(l - d/2)}{\Gamma(l)} \left[F^2\right]^{d/2 - l} \tag{15.22}$$

$$\int \frac{\mathrm{d}^d k}{(2\pi)^d} \frac{k_\mu k_\nu}{[k^2 + F^2(p, m, z)]^l} = \frac{1}{(4\pi)^{d/2}} \frac{\delta_{\mu\nu}}{2} \frac{\Gamma(l - 1 - d/2)}{\Gamma(l)} \left[F^2\right]^{d/2 + 1 - l} \tag{15.23}$$

At the end we are interested in the limit $d \to 4$. For that we need the behaviour of $\Gamma(z)$ near its poles. This is obtained by the use of the following two properties of the Γ-function:

$$\Gamma(z+1) = z\,\Gamma(z) \quad \text{and} \quad \Gamma(z) \simeq \frac{1}{z}$$

As a consequence, the integral (15.22) diverges for $l \leq 2$ and the one of (15.23) for $l \leq 3$. For $l = 2$ and $d = 4$ (15.22) is logarithmically divergent. Our regularised expression is regular for $\mathrm{Re}\,d < 4$ and has a simple pole $\sim 1/(4 - d)$. For $l = 1$ it is quadratically divergent, but our expression still has a simple pole at $d = 4$. The difference is that now the first pole for $d > 0$ is at $d = 2$. We arrive at the same conclusions looking at the integral (15.23). By dimensionally regularising a 1-loop integral corresponding to a Feynman diagram which, by power counting, diverges as Λ^{2n}, we obtain a meromorphic function of d with simple poles starting at $d = 4 - 2n$. By convention, $n = 0$ corresponds to a logarithmic divergence.

15.2.3 Renormalisation

In this section we want to address the physical question, namely under what circumstances a meaningful four-dimensional theory can be obtained from the regularised ϵ-dependent expressions. As one could have anticipated, the answer will turn out to

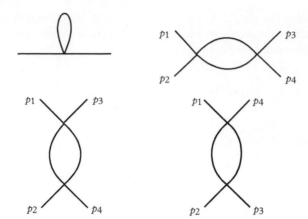

Fig. 15.2 The 1-loop primitively divergent diagrams of the ϕ^4 theory. We denote the first by $\Gamma_1^{(2)}$ and the sum of the other three by $\Gamma_1^{(4)}$.

be that this is only possible for the renormalisable and the super-renormalisable theories we introduced before. The procedure of actually obtaining such a meaningful and well defined quantum field theory is called *renormalisation*. In this section we will present it for some simple examples.

15.2.3.1 Renormalisation of the ϕ^4 theory. Let us start with the simplest four-dimensional renormalisable theory (15.2). In $d = 4$ the field ϕ has the dimensions of mass and the coupling constant λ is dimensionless. Since we intend to use dimensional regularisation, we introduce a mass parameter μ and write the coefficient of the interaction term $\lambda \to \mu^\epsilon \lambda$, so that the coupling constant λ remains dimensionless at all values of ϵ. We shall present the renormalisation programme for this theory at the lowest non-trivial order that deals with all diagrams with at most one loop.

The power counting argument presented previously shows that, at one loop, the only divergent 1PI diagrams are the ones with two or four external lines, shown in Figure 15.2.

The 2-point diagram is quadratically divergent and the four-point one logarithmically.[2] We choose to work entirely with dimensional regularisation and we obtain for these diagrams in Minkowski space-time, using (15.22) in the limit $d = 4 - \epsilon \to 4$

$$\Gamma_1^{(2)} = -\mathrm{i}\frac{\lambda\mu^\epsilon}{2} \int \frac{\mathrm{d}^d k}{(2\pi)^d} \frac{\mathrm{i}}{k^2 - m^2 + \mathrm{i}\varepsilon} = \frac{\mathrm{i}\lambda m^2}{16\pi^2}\frac{1}{\epsilon} + \text{finite terms} \tag{15.24}$$

and

---

[2]We could prevent the appearance of the first diagram by "normal ordering" the ϕ^4 term in the interaction Lagrangian, but, for pedagogical purposes, we prefer not to do so. Normal ordering is just a particular prescription to avoid certain divergences, but it is not always the most convenient one. First it is not general; for example, it will not prevent the appearance of divergence in the 2-point function at higher orders and, second, its use may complicate the discussion of possible gauge symmetries of the theory.

$$\Gamma_1^{(4)}(p_1,\ldots,p_4) = \frac{1}{2}\lambda^2\mu^{2\epsilon}\int\frac{\mathrm{d}^d k}{(2\pi)^d}\frac{1}{(k^2-m^2+\mathrm{i}\varepsilon)\left[(k-P)^2-m^2+\mathrm{i}\varepsilon\right]} + \text{crossed}$$

$$= \frac{1}{2}\lambda^2\mu^{2\epsilon}\int_0^1\mathrm{d}z\int\frac{\mathrm{d}^d K}{(2\pi)^d}\frac{1}{\left[K^2-m^2+P^2 z(1-z)+\mathrm{i}\varepsilon\right]^2} + \text{crossed}$$

$$= \mathrm{i}\frac{1}{2}\lambda^2\mu^{2\epsilon}\int_0^1\mathrm{d}z\int\frac{\mathrm{d}^d K_E}{(2\pi)^d}\frac{1}{\left[K_E^2+m^2-P^2 z(1-z)\right]^2} + \text{crossed}$$

$$= \frac{3\mathrm{i}\lambda^2}{16\pi^2}\frac{1}{\epsilon} + \text{finite terms}$$

$$(15.25)$$

where $P = p_1 + p_2$ and "crossed" stands for the contribution of the two crossed diagrams in Figure 15.2. With "finite terms" we mean the contributions which are regular for $d = 4$. In the second equation we defined $K = k - P(1-z)$, while in the following step we performed a Wick rotation to Euclidean K-space.

A few remarks are in order here:

(1) The divergent contributions are constants, independent of the external momenta. We shall see shortly, in the example of quantum electrodynamics, that this is a particular feature of the ϕ^4 theory. In fact, even for ϕ^4, it is no more true when higher loops are considered. However, we can prove the following general property: All divergent terms are proportional to monomials in the external momenta. We have already mentioned this result. For 1-loop diagrams the proof is straightforward. We start from the general expression (15.21) and notice that we can expand the integrand in powers of the external momenta p taken around some fixed point. The coefficients of this expansion are of the form $(k^2 + F^2(p))^{-n}$ with increasing $n \geq l$ so, after a finite number of terms, the integral becomes convergent. It takes some more work to generalise the proof to multi-loop diagrams, but it can be done.

(2) The dependence of the divergent terms on m^2 could be guessed by dimensional analysis. This is one of the attractive features of dimensional regularisation.

(3) The finite terms in (15.25) depend on the parameter μ. The Laurent expansion in ϵ leads to terms of the form $\ln\left([m^2 - P^2 z(1-z)]/\mu^2\right)$.

The particular form of the divergent terms suggests the prescription to remove them. Let us start with the 2-point function. In the loop expansion we write

$$\Gamma^{(2)}(p^2) = \sum_{l=0}^{\infty}\Gamma_l^{(2)}(p^2) = \Gamma_0^{(2)}(p^2) + \Gamma_1^{(2)}(p^2) + \cdots \qquad (15.26)$$

where the index l denotes the contribution of the diagrams with l loops. In the tree approximation, using the relation (15.6), we get

$$\Gamma_0^{(2)}(p^2) = \mathrm{i}(p^2 - m^2) \qquad (15.27)$$

The 1-loop diagram adds the term given by (15.24). Since it is a constant, it can be interpreted as a correction to the value of the mass in (15.27). Therefore, we can introduce a *renormalised* mass m_R^2, which is a function of m, λ and ϵ. Of course, this function can only be computed as a formal power series in λ. Up to and including 1-loop diagrams we write

$$m_R^2(m, \lambda, \epsilon) = m^2 \left(1 - \frac{\lambda}{16\pi^2} \frac{1}{\epsilon} \right) + O(\lambda^2) \tag{15.28}$$

A formal power series, whose zeroth order term is non-vanishing, is invertible in terms of another formal power series. That is, we can write m as a function of m_R, λ and ϵ, namely

$$m^2(m_R, \lambda, \epsilon) = m_R^2 \left(1 + \frac{\lambda}{16\pi^2} \frac{1}{\epsilon} \right) + O(\lambda^2) \equiv m_R^2 Z_m \tag{15.29}$$

where we have defined the function $Z_m(\lambda, \epsilon)$ as a formal power series in λ with ϵ-dependent coefficients.

The parameter m is often called *bare* mass. Replacing in the Lagrangian (15.2) the bare mass m with the help of (15.29), results in: (i) changing in the Feynman rules m to m_R and (ii) introducing a new term in $\mathcal{L}$ of the form

$$\delta\mathcal{L}_m = -m_R^2 \frac{\lambda}{32\pi^2} \frac{1}{\epsilon} \phi^2(x) \tag{15.30}$$

Since $\delta\mathcal{L}_m$ is proportional to the coupling constant λ, it can be viewed as a new vertex in the perturbation expansion, which, to first order, gives the diagram of Figure 15.3. In this case the complete 1PI two-point function to first order in λ is given by

$$\begin{aligned}
\Gamma^{(2)}(p^2) &= \mathrm{i}(p^2 - m_R^2) - \frac{\mathrm{i}\lambda m_R^2}{16\pi^2} \frac{1}{\epsilon} + \frac{\mathrm{i}\lambda m_R^2}{16\pi^2} \frac{1}{\epsilon} + O(\lambda^2) \\
&= \mathrm{i}(p^2 - m_R^2) + O(\lambda^2)
\end{aligned} \tag{15.31}$$

which, using again the relation (15.6), implies that, to this order the ϕ-propagator becomes just

$$G^{(2)}(p) = \frac{\mathrm{i}}{p^2 - m_R^2} \tag{15.32}$$

It follows that, if we keep fixed m_R and λ instead of m and λ, we can take the limit $d \to 4$ and encounter no divergences in the 2-point function up to and including 1-loop diagrams.

Now that we have understood the principle, it is straightforward to apply it to the four-point function as well. In the same spirit we write

$$\Gamma^{(4)}(p_1, \ldots, p_4) = \sum_{l=0}^{\infty} \Gamma_l^{(4)}(p_1, \ldots, p_4) = \Gamma_0^{(4)}(p_1, \ldots, p_4) + \Gamma_1^{(4)}(p_1, \ldots, p_4) + \cdots \tag{15.33}$$

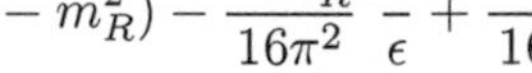

Fig. 15.3 The new interaction vertex resulting from $\delta\mathcal{L}_m$ of equation (15.30). The corresponding Feynman rule is $-\mathrm{i}\lambda m_R^2/(16\pi^2\epsilon)$.

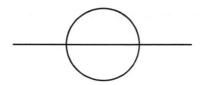

Fig. 15.4 A 2-loop diagram for the 2-point function with a divergent term proportional to p^2.

In the tree approximation $\Gamma_0^{(4)}(p_1,\ldots,p_4) = -\mathrm{i}\lambda$. Including the 1-loop diagrams we obtain

$$\Gamma^{(4)}(p_1,..,p_4) = -\mathrm{i}\lambda\left(1 - \frac{3\lambda}{16\pi^2}\frac{1}{\epsilon} + \text{finite terms}\right) + O(\lambda^3) \tag{15.34}$$

We change from the *bare* coupling constant λ to the *renormalised* one λ_R by writing

$$\lambda_R(\lambda,\epsilon) = \lambda\left(1 - \frac{3\lambda}{16\pi^2}\frac{1}{\epsilon} + O(\lambda^2)\right) \tag{15.35}$$

or, equivalently,

$$\lambda(\lambda_R,\epsilon) = \lambda_R\left(1 + \frac{3\lambda_R}{16\pi^2}\frac{1}{\epsilon} + O(\lambda_R^2)\right) \equiv \lambda_R Z_\lambda \tag{15.36}$$

Again, replacing λ with λ_R in $\mathcal{L}$ produces a new four-point vertex, described by the additional term

$$\delta\mathcal{L}_{\text{int}} = -\frac{3\lambda_R^2}{16\pi^2}\frac{1}{\epsilon}\frac{1}{4!}\phi^4 \tag{15.37}$$

in the Lagrangian, which cancels the divergent part of the 1-loop diagrams of Figure 15.2. Let us also note that we can replace λ with λ_R in (15.29) since the difference will appear only at higher order.

Until now we have succeeded in building a new, *renormalised* Lagrangian density

$$\mathcal{L} = \frac{1}{2}\partial_\mu\phi\,\partial^\mu\phi - \frac{1}{2}m_R^2\phi^2 - \frac{1}{4!}\lambda_R\phi^4 + \delta\mathcal{L}_m + \delta\mathcal{L}_{\text{int}} \tag{15.38}$$

and the resulting theory is free of divergences up to and including 1-loop diagrams. It involves the addition of two new terms, $\delta\mathcal{L}_m$ and $\delta\mathcal{L}_{\text{int}}$, which change the coefficients of the ϕ^2 and ϕ^4 terms of the original Lagrangian. These terms are usually called *counterterms*. The procedure we just outlined represents an explicit realisation, in terms of the dimensional regularisation cut-off parameter ϵ, of the process we showed in equations (15.4) and (15.5). They provide the correct definition of the Lagrangian density up to this order of perturbation, by removing the short-distance ambiguities inherent in the local expressions $\phi^2(x)$ and $\phi^4(x)$.

Going to higher loops brings only one new element. At the 2-loop order we encounter the diagram of Figure 15.4. It contributes to the 2-point function $\Gamma^{(2)}(p^2)$, it is quadratically divergent, but contrary to the 1-loop diagram of Figure 15.2, it is not a constant and has a non-trivial dependence on p^2. The explicit calculation is left as an exercise (see Problem 15.4). It is instructive because it offers a non-trivial example of a multi-loop diagram. It yields divergent contributions which are constant, independent of the external momentum p and can be absorbed in Z_m, and a new one, which contributes a $1/\epsilon$ divergent term proportional to $ip^2 C\lambda_R^2$ with C a numerical constant. Such a term cannot be absorbed in a redefinition of either Z_m or Z_λ and requires a new renormalisation of the 2-point function. In the same spirit we change now the coefficient of the kinetic term $\mathcal{L}_k = (\partial_\mu \phi)^2/2$ in the Lagrangian. It is convenient to define a renormalised field by

$$\phi(x) = \sqrt{Z_\phi}\, \phi_R(x) = \left(1 + \frac{C\lambda_R^2}{2}\frac{1}{\epsilon} + O(\lambda_R^3)\right)\phi_R(x) \tag{15.39}$$

which introduces a new counterterm in (15.38), called *wave function renormalisation*, of the form

$$\delta\mathcal{L}_k = \frac{1}{2\epsilon}C\lambda_R^2 \partial_\mu \phi_R(x)\partial^\mu \phi_R(x) \tag{15.40}$$

This process of removing the ambiguities by introducing counterterms in the original Lagrangian can be extended to all orders of perturbation. The proof is rather complicated but, essentially, elementary. No new ideas are necessary. We must prove that, at any order, the terms appear with the correct combinatoric factor, even in the cases in which sub-diagrams are divergent to which counterterms have already been assigned. At the end, all Green functions of a renormalisable theory are well defined and calculable.

Before closing this section, let us see the price we had to pay for this achievement. It can be better seen in the example of the 4-point function. Looking back at the expression (15.25) we make the following two observations: First, as we noticed already, the finite part seems to depend on a new arbitrary parameter μ with the dimensions of mass. Second, the definition of Z_λ in (15.36) seems also arbitrary. We could add to it any term of the form $A\lambda_R$, with A an arbitrary constant independent of ϵ. Such an addition would change the value of the coupling constant at the one loop order. The two observations are not unrelated. Indeed, changing the parameter μ from μ_1 to μ_2 in (15.25) adds a constant term proportional to $\lambda \ln(\mu_1/\mu_2)$ which, as we just saw, can be absorbed in a redefinition of Z_λ and, thus, of the value of the coupling constant. We conclude that all arbitrariness of the renormalisation programme consists in assigning prescribed values to three quantities of the theory. The power counting argument shows that two should refer to the 2-point function and one to the 4-point function. A convenient choice is given by two conditions of the form

$$\Gamma^{(2)}(p^2 = m_{\text{ph}}^2) = 0 \tag{15.41}$$

$$\left.\frac{\partial \Gamma^{(2)}(p^2)}{\partial p^2}\right|_{p^2 = M^2} = i \tag{15.42}$$

and a third one given by

$$\Gamma^{(4)}(p_1, \ldots, p_4)\Big|_{\text{point } M} = -i\lambda_R^{(M)} \tag{15.43}$$

The first one, equation (15.41), defines the physical mass as the pole of the full propagator. Although this choice is the most natural for physics, from the purely technical point of view, we could use any condition assigning a prescribed value to $\Gamma^{(2)}(p^2)$ at a point $p^2 = M_1^2$, provided it is a point at which $\Gamma^{(2)}(p^2)$ is regular. Similarly, in the second one M^2 is some conveniently chosen value of p^2. Finally, in equation (15.43), by "point M" we mean some point in the space of the four momenta p_a, $a = 1, \ldots, 4$, provided it is a point at which $\Gamma^{(4)}$ is regular. To give an example, for a massive theory the point $p_a = 0$ is an acceptable choice. Once these conditions are imposed, all Green functions at any finite order of perturbation theory are well defined, calculable and independent of μ. For historical reasons the point M in momentum space is called the *subtraction point*.

If our purpose is to use these Green functions in the reduction formula (12.18), in order to compute scattering amplitudes, we must use the so-called *on-mass-shell renormalisation*, i.e. choose $M = m_{\mathrm{ph}}$ in condition (15.42). In this case the connected 2-point function $G_c^{(2)}(p^2)$, which is given by minus the inverse of $\Gamma^{(2)}(p^2)$, takes the form

$$G_c^{(2)}(p^2) = \frac{i}{(p^2 - m_{\mathrm{ph}}^2)\left[1 + (p^2 - m_{\mathrm{ph}}^2)\Sigma^{(2)}(p^2)\right]} \tag{15.44}$$

with $\Sigma^{(2)}(p^2)$ receiving only loop contributions, which has a pole at $p^2 = m_{\mathrm{ph}}^2$ and drops out from the reduction formula.

Finally, let us give the explicit 1-loop expressions of $G_c^{(2)}(p^2)$ and $\Gamma^{(4)}(p_1, .., p_4)$, which satisfy the conditions (15.44) and (15.43).

For the 2-point function at 1-loop we have $\Sigma^{(2)}(p^2)|_{\text{1-loop}} = 0$ because the contribution of the diagram of Figure 15.2 is a constant, independent of p^2. The non-zero values of $\Sigma^{(2)}(p^2)$ start at 2-loops with the diagram of Figure 15.4. So, up to and including one loop diagrams, we have

$$G_c^{(2)}(p^2)\Big|_{\text{1-loop}} = \frac{i}{p^2 - m_{\mathrm{ph}}^2} \tag{15.45}$$

We repeat, however, that this simple form is an accident of the ϕ^4 theory.

For the 4-point function we must compute the finite part, which we omitted in (15.25), and enforce the condition (15.43). We use the expansions

$$A^\epsilon = 1 + \epsilon \ln A + \mathcal{O}(\epsilon^2) \tag{15.46}$$

$$\Gamma\left(\frac{\epsilon}{2}\right) \simeq \frac{2}{\epsilon} - \gamma + \mathcal{O}(\epsilon) \tag{15.47}$$

with $\gamma \simeq 0.577216$ the Euler–Mascheroni constant, and obtain

$$\Gamma_1^{(4)}(p_1, \ldots, p_4) = -\frac{i\lambda_R^2}{32\pi^2}\left[\gamma - \ln(4\pi) - 2\ln\mu^2 + \int_0^1 dz \ln F^2\right] + \text{crossed} \tag{15.48}$$

where we used the expression (15.36) and added the contribution of the 1-loop counterterm $\delta\mathcal{L}_{\text{int}}$ to absorb the $1/\epsilon$ term. F^2 is given by

$$F^2 = m_{\mathrm{ph}}^2 - P^2 z(1-z) = m_{\mathrm{ph}}^2 - sz(1-z) \tag{15.49}$$

In the spirit of perturbation theory we have replaced m_{ph} for m. The s is the Mandelstam variable $s = (p_1+p_2)^2 = P^2$. The other two are $t = (p_1-p_3)^2$ and $u = (p_1-p_4)^2$

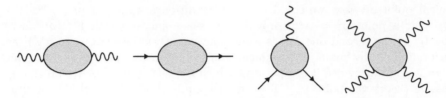

Fig. 15.5 The primitively divergent 1PI Green's functions of quantum electrodynamics. The last one, the light-by-light scattering, is in fact convergent as a consequence of current conservation.

and satisfy $s + t + u = p_1^2 + p_2^2 + p_3^2 + p_4^2$. In terms of these variables the crossed terms are obtained by adding the expressions with $s \to t$ and $s \to u$.

We want to use the freedom we have to choose the value of μ in order to enforce the renormalisation condition (15.43) at some point in the space of the four external momenta defined by a mass parameter M. A convenient choice is the following: (i) choose all four external momenta satisfying $p_a^2 = M^2$, $a = 1, \ldots, 4$, and, (ii) choose $s = t = u = 4M^2/3$. Notice that the 0-loop 4-point function $\Gamma_0^{(4)}(p_1, \ldots, p_4)$ satisfies the condition (15.43). Therefore, we must choose μ such that $\Gamma_1^{(4)}(p_1, \ldots, p_4)$ at the point M vanishes, i.e.

$$\gamma - \ln(4\pi) - 2\ln\mu^2 + \int_0^1 dz \ln\left(m_{\mathrm{ph}}^2 - \frac{4M^2}{3}z(1-z)\right) = 0 \qquad (15.50)$$

Using this equation we can trade μ for M and $\Gamma^{(4)}(p_1, \ldots, p_4)$, up to and including 1-loop diagrams, can be written as

$$\Gamma^{(4)}(p_1, \ldots, p_4) = -i\lambda_R - \frac{i\lambda_R^2}{32\pi^2} \int_0^1 dz \left[\ln \frac{m_{\mathrm{ph}}^2 - sz(1-z)}{m_{\mathrm{ph}}^2 - \frac{4M^2}{3}z(1-z)} + (s \to t) + (s \to u)\right]$$
$$(15.51)$$

It satisfies the renormalisation condition (15.43). In Problem 15.6 we ask the reader to perform the integrations over z and obtain the full answer for the 1-loop 4-point function. The answer is interesting because it shows the analyticity properties of the 4-point function in the complex s, t or u planes.

15.2.3.2 Renormalisation of quantum electrodynamics. As a second example, we shall present the renormalisation of quantum electrodynamics at the 1-loop level. The method is exactly the same and yields "renormalised" values of the various terms, which appear in the QED Lagrangian. According to the power counting formula (15.9), the only 1PI graphs, that can be divergent at 1-loop are those of Figure 15.5. A simple calculation gives:

- The contribution of the 1-loop diagram to the photon self-energy is

$$\Gamma_{(1)\mu\nu}^{(2,0)}(q) = \frac{2i\alpha}{3\pi} \frac{1}{\epsilon}\left(q_\mu q_\nu - q^2 \eta_{\mu\nu}\right) + \text{finite terms} \qquad (15.52)$$

where $\alpha = e^2/4\pi$ is the fine structure constant.

- Similarly, the electron self-energy is

$$\Gamma^{(0,2)}_{(1)}(p) = \frac{i\alpha}{2\pi}\frac{1}{\epsilon}\slashed{p} - \frac{2i\alpha}{\pi}\frac{1}{\epsilon}m + \text{finite terms} \tag{15.53}$$

where we have suppressed spinor indices. We see that in (15.52), as well as (15.53), the divergent parts are polynomials in the external momenta.

- The vertex function is

$$\Gamma^{(1,2)}_{(1)\mu}(p,p') = \frac{i\alpha}{2\pi}\frac{1}{\epsilon}e\gamma_\mu + \text{finite terms} \tag{15.54}$$

As before, all these divergences can be absorbed in the definition of renormalised quantities according to

$$A^\mu(x) = Z_3^{1/2}A_R^\mu(x) = \left(1 - \frac{\alpha}{3\pi}\frac{1}{\epsilon} + O(\alpha^2)\right)A_R^\mu(x) \tag{15.55}$$

$$\psi(x) = Z_2^{1/2}\psi_R(x) = \left(1 - \frac{\alpha}{4\pi}\frac{1}{\epsilon} + O(\alpha^2)\right)\psi_R(x) \tag{15.56}$$

$$m = Z_m m_R = \left(1 - \frac{2\alpha}{\pi}\frac{1}{\epsilon} + O(\alpha^2)\right)m_R \tag{15.57}$$

$$\Gamma^{(1,2)}_\mu(p,p') = -ieZ_1\gamma_\mu + \ldots = -ie\gamma_\mu\left(1 - \frac{\alpha}{2\pi}\frac{1}{\epsilon} + O(\alpha^2)\right) + \ldots \tag{15.58}$$

where we have included the tree-level terms. As we have noticed already, in QED the counterterms corresponding to the kinetic terms of the electron and the photon already appear at the 1-loop order. An important remark concerns the tensor structure in the right-hand side of (15.52). In general we would expect two independent divergent terms, one proportional to $\eta_{\mu\nu}$ and another proportional to $q_\mu q_\nu$. Were this the case, we would need two counterterms, the one shown in equation (15.55) and another one proportional to $A^\mu A_\mu$; in other words we would have generated a photon mass term. We could still recover massless quantum electrodynamics by imposing the vanishing of the physical photon mass as a renormalisation condition, but we would have to carry this term in all intermediate computations. It is the virtue of dimensional regularisation, which respects the gauge invariance of the theory, to guarantee the absence of such terms.

Let us introduce a charge renormalisation counterterm Z_e and write the interaction Lagrangian as

$$- e\bar\psi\gamma_\mu\psi A^\mu = -Z_e Z_2 Z_3^{1/2} e_R \bar\psi_R\gamma_\mu\psi_R A_R^\mu \tag{15.59}$$

The cancellation of the $1/\epsilon$ terms in the 3-point function determines the charge renormalisation constant Z_e as

$$Z_e Z_2 Z_3^{1/2} = Z_1 \tag{15.60}$$

By comparing (15.58) and (15.56), we see that, at least at this order, we have

$$Z_1 = Z_2 \tag{15.61}$$

which implies that the entire charge renormalisation is determined by the photon self-energy diagram. This property is known as *Ward identity* and we can show that it is valid to all orders of perturbation theory as a consequence of the conservation of the electromagnetic current. We will come back to this point in Chapter 18. It is the same property which guarantees that the last diagram of Figure 15.5, when computed using dimensional regularization which respects the conservation of the current, is in fact finite.

As we did for the ϕ^4 theory, we can choose any subtraction point M to define the Green functions but, for scattering amplitudes, we must use on-mass-shell renormalisation conditions by choosing $M = m_e$, the electron mass, for the electron lines and $M = 0$ for the photon lines. However, as we have discussed already, this last choice will lead to divergences, which we called "infrared divergences", and which we will discuss in Chapter 17.

This completes a very sketchy discussion of renormalisation theory. It only needs straightforward calculations to adapt it to any renormalisable theory and to any order in the perturbation expansion.

15.3   The Renormalisation Group

In the previous section we presented the renormalisation programme. We saw that, for every renormalisable field theory, there exists a precise prescription which makes it possible, order by order in perturbation theory, to obtain well defined expressions for all Green functions as formal power series in the expansion parameter.

In order to be specific, let us consider the ϕ^4 theory we studied earlier. A renormalisation scheme is defined through three renormalisation conditions like the ones shown in (15.41), (15.42) and (15.43). If we are interested in the calculation of physical scattering amplitudes we must choose the subtraction point in the conditions (15.41) and (15.42) to be the physical mass $M = m_{\mathrm{ph}}$.[3] However, if we only want to define the field theory and not to compute scattering amplitudes, any three renormalisation conditions are perfectly adequate. In order to simplify the discussion, we will choose (15.41) which defines the physical mass, but we will leave M as a free parameter in the other two. These conditions uniquely determine, order by order in perturbation theory, the three counterterms Z_ϕ, Z_m and Z_λ. The renormalisability of the theory then guarantees that all Green functions, for all n, are finite and calculable as formal power series in the renormalised coupling constant. The 1PIs are of the form $\Gamma^{(n)}\left(p_1, \ldots, p_n; m_R, \lambda_R^{(M)}; M\right)$, with the external momenta p_a assumed to flow inwards, and thus subject to the condition $\Sigma_{a=1}^n p_a = 0$. The dependence on the two parameters, the mass and the coupling constant, is expected. Different values of these parameters correspond to different physical theories. The subtraction point M, on the other hand, was introduced only in order to remove the ambiguities in the finite parts of the divergent diagrams. If we change M from one value M_1 to another M_2, we will obtain a different set of Green functions. In this section we want to determine under which conditions all these families of Green functions describe the same physical theory, that is under which conditions physical quantities are independent of M.

---

[3]The coupling constant in a ϕ^4 theory is not directly measurable, so there are no physical requirements that impose a unique renormalisation condition for the 4-point function.

Let us consider the transformation

$$M_1 \longrightarrow M_2 = \eta M_1 \tag{15.62}$$

with η a numerical factor, which can take any value in a domain including the value $\eta = 1$ and such that all resulting points M_2 belong to the domain of analyticity of $\Gamma^{(2)}$ and $\Gamma^{(4)}$ and yield well-defined sets of Green functions. For a renormalisable theory this transformation implies different choices for the counterterms Z_ϕ and Z_λ.[4] The two sets of Green functions will be physically equivalent if we can find functions $Z_\phi(m_R, \lambda_R^{(M_1)}, M_1, M_2)$ and $\lambda_R^{(M_2)}(m_R, \lambda_R^{(M_1)}, M_1, M_2)$ satisfying

$$\begin{aligned} Z_\phi(m_R, \lambda_R^{(M_1)}, M_1, M_2) &= 1 + O(\lambda_R^{(M_1)}) \\ \lambda_R^{(M_2)}(m_R, \lambda_R^{(M_1)}, M_1, M_2) &= \lambda_R^{(M_1)} + O((\lambda_R^{(M_1)})^2) \end{aligned} \tag{15.63}$$

and, in addition,

$$\begin{aligned} \Gamma^{(n)}(p_1, \ldots, p_n; m_R, \lambda_R^{(M_1)}; M_1) = \\ Z_\phi^{-n}(m_R, \lambda_R^{(M_1)}, M_1, M_2)\, \Gamma^{(n)}(p_1, \ldots, p_n; m_R, \lambda_R^{(M_2)}; M_2) \end{aligned} \tag{15.64}$$

in other words, if the change $M_1 \to M_2$ can be absorbed in a change of the values of the coupling constant and the normalisation of the field.

The transformations (15.64) form a group, called *the renormalisation group*.[5] It describes the change in the Green functions induced by a change of the subtraction point. It is instructive to write also the corresponding infinitesimal transformations with η close to one. For that we take the derivative of (15.64) with respect to M_2, set $M_1 = M_2 \equiv M$ and obtain

$$\left[M\frac{\partial}{\partial M} + \beta\frac{\partial}{\partial \lambda} - n\gamma \right] \Gamma^{(n)}(p_1, ..., p_n; m, \lambda; M) = 0 \tag{15.65}$$

where we have dropped the subscripts R and the reference to the subtraction point on λ. The functions β and γ are defined by

$$\beta\left(\lambda, \frac{m}{M}\right) = M\frac{\partial\lambda_R^{(M_2)}}{\partial M_2}\bigg|_{M_1=M_2=M} \tag{15.66}$$

$$\gamma\left(\lambda, \frac{m}{M}\right) = \frac{M}{2}\frac{\partial\ln Z_\phi}{\partial M_2}\bigg|_{M_1=M_2=M}$$

These functions are dimensionless and can only depend on the ratio m/M. By applying the general equation (15.65) for $n = 1$ and 2 and using the renormalisation

---

[4]We assume that m_R is chosen to remain the same, so we do not need to consider the counterterm Z_m.

[5]The concept of the renormalisation group was introduced formally in quantum field theory by E.C.G. Stückelberg and A. Petermann, *Helv. Phys. Acta* **26**, 499 (1953). The first application to the asymptotic behaviour of Green functions is due to M. Gell-Mann and F.E. Low, *Phys. Rev.* **95**, 1300 (1954).

conditions, we can express them as combinations of Green functions, or derivatives of Green functions, thus proving that they have a well defined expansion in powers of the coupling constant.

For $\Gamma^{(2)}$, upon differentiation with respect to p^2 and evaluating for $p^2 = M^2$, we obtain

$$\gamma\left(\lambda, \frac{m}{M}\right) = -\mathrm{i}\frac{M}{2}\frac{\partial}{\partial M}\frac{\partial\Gamma^{(2)}}{\partial p^2}\bigg|_{p^2=M^2} \tag{15.67}$$

while the equation for $\Gamma^{(4)}$ evaluated at the "point M" leads to

$$\beta\left(\lambda, \frac{m}{M}\right) = 2\lambda\gamma - \mathrm{i}M\frac{\partial\Gamma^{(4)}}{\partial M}\bigg|_{\text{point } M} \tag{15.68}$$

At lowest order β and γ vanish because there is no dependence on M at that order. At 1-loop order, using (15.45), we still find

$$\gamma\left(\lambda, \frac{m}{M}\right) = 0 \tag{15.69}$$

while, for the β-function, using (15.51) and performing the z integration, we obtain, for example, for $3m^2 \leq M^2$

$$\beta\left(\lambda, \frac{m}{M}\right) = \frac{3\lambda^2}{16\pi^2} - \frac{9\lambda^2}{32\pi^2}\frac{m^2/M^2}{\sqrt{1-3m^2/M^2}}\ln\left(\frac{1-\sqrt{1-3m^2/M^2}}{1+\sqrt{1-3m^2/M^2}}\right) \tag{15.70}$$

As we will show presently, the physically interesting values are those of the massless theory, for which we have

$$\gamma(\lambda) = 0 \quad\text{and}\quad \beta(\lambda) = \frac{3\lambda^2}{16\pi^2} \tag{15.71}$$

It is clear how to generalise this analysis to renormalisable theories with several fields ϕ_i, $i = 1, \ldots, N$, and several coupling constants λ_j, $j = 1, \ldots, k$. We shall need N γ- and k β-functions. In general, they will all depend on the k coupling constants and N mass ratios.

Equation (15.65) is the equation of the renormalisation group. It describes the response of the system under a rescaling of the subtraction point. One could argue that it is of limited interest because it only solves a problem we have created for ourselves by not making the physical choice $M = m_{\text{ph}}$. This is undoubtedly correct and we need some more input in order to extract physically interesting information out of this equation.

The important remark is that (15.65) describes the response of the system under the change of a *dimensionful* constant. Were M the only such constant, ordinary dimensional analysis would imply that the Green functions would depend on the ratios p_a/M, and the dependence on M could then be translated into non-trivial information about the dependence on the external momenta. So the question is under what conditions we can neglect the dependence on all physical masses m_{ph}. Intuitively we guess that this should be possible at very high energies.

Let us rescale all momenta p_a by a real parameter ρ. We can prove that, for generic values of the external momenta,[6] the asymptotic behaviour of the Green functions at large ρ is that of the corresponding massless theory. This is a consequence of a theorem known as the Kinoshita–Lee–Nauenberg theorem, which can be proven order by order in perturbation theory.

We can now use dimensional analysis to trade the derivative with respect to M for that with respect to ρ. The $\Gamma^{(n)}$ has dimension[7] $[M]^{4-n}$ and it can be written as

$$\Gamma^{(n)}_{\text{as}}(\rho p_a; m = 0, \lambda; M) = M^{4-n} F^{(n)} \left(\frac{\rho p_a}{M}; \lambda \right) \tag{15.72}$$

Therefore, the function $\Phi^{(n)} = \rho^{n-4} \Gamma^{(n)}$ satisfies asymptotically the equation

$$\left[-\rho \frac{\partial}{\partial \rho} + \beta(\lambda) \frac{\partial}{\partial \lambda} - n\gamma(\lambda) \right] \Phi^{(n)}_{\text{as}} = 0 \tag{15.73}$$

This equation can be solved using Monge's standard method. We define $\bar{\lambda}$ as a function of λ and ρ by

$$\left[-\rho \frac{\partial}{\partial \rho} + \beta(\lambda) \frac{\partial}{\partial \lambda} \right] \bar{\lambda}(\lambda, \rho) = 0 \ , \quad \bar{\lambda}(\lambda, \rho = 1) = \lambda \tag{15.74}$$

which is equivalent to, see Problem 15.8,

$$\rho \frac{\partial \bar{\lambda}}{\partial \rho} = \beta(\bar{\lambda}) \tag{15.75}$$

The general solution of (15.73) is now given by

$$\Phi^{(n)}_{\text{as}}(\rho p_1, ..., \rho p_n; \lambda) = \Phi^{(n)}_{\text{as}}(p_1, ..., p_n; \bar{\lambda}) \exp \left\{ -n \int_1^\rho d\rho' \gamma[\bar{\lambda}(\lambda, \rho')] \right\} \tag{15.76}$$

The physical meaning of (15.76) is clear. Scaling all momenta of a Green function by a common factor ρ and taking ρ large has the following effects: (i) It multiplies every external line by the scale dependent factor $\exp\left(-\int_1^\rho d\rho' \gamma(\bar{\lambda}(\lambda, \rho')) \right)$,[8] and (ii) it replaces the physical coupling constant λ by an effective one, $\bar{\lambda} = \bar{\lambda}(\lambda, \rho)$, which is the solution of (15.75). In the asymptotic region the effective strength of the interaction is not determined by the physical coupling constant λ, but instead by $\bar{\lambda}$, which is a function of λ and ρ, i.e. it depends on the initial value λ and on how deep in the asymptotic region we are. For this reason $\bar{\lambda}$ is often called *running* coupling constant.

The fact that the effective coupling constant of a renormalisable field theory depends on the scale of the external momenta can be understood by looking at Feynman diagrams. A Feynman diagram of nth order is proportional to λ^n. On the other hand,

---

[6]In practice this means that we must keep all scalar products $p_a \cdot p_b$ different from zero.

[7]We remind that for the ϕ^4 theory $\Gamma^{(2n+1)} = 0$.

[8]Naïvely, since the $\Phi^{(n)}$ are dimensionless, rescaling its arguments should not rescale them. Nevertheless, they get rescaled as if they had dimensions related to $\gamma(\lambda)$, which accordingly is often called "anomalous dimension" of the field ϕ.

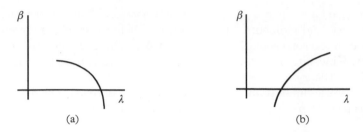

Fig. 15.6 The two kinds of zeros of the β-function.

if it is a multi-loop diagram, it contains logarithms of the external momenta of degree up to $\ln^n(k^2/M^2)$, where k^2 is the square of a typical momentum transfer. Let us call $\hat{\lambda} = \lambda \ln(k^2/M^2)$. We can reorganise the asymptotic expansion of some quantity, e.g. the 4-point function $\Gamma^{(4)}$, as

$$\Gamma^{(4)} = C_0^{(1)}\lambda + \lambda\left[C_1^{(1)}\hat{\lambda} + C_0^{(2)}\lambda\right] + \lambda\left[C_2^{(1)}\hat{\lambda}^2 + C_1^{(2)}\lambda\hat{\lambda} + C_0^{(3)}\lambda^2\right] + \ldots$$

$$= \lambda \sum_n C_n^{(1)}\hat{\lambda}^n + \lambda^2 \sum_n C_n^{(2)}\hat{\lambda}^n + \ldots \tag{15.77}$$

where the Cs are numerical constants. The first sum contains the leading logarithms in every term of the perturbation expansion, the second the next-to-leading ones, and so on. We can show, although we will not do it here, that solving equation (15.75) with the β-function computed at one loop is equivalent to summing the leading logs in (15.77).[9]

This re-summation induces a dependence of the effective coupling constant on the momentum scale.

The differential equation (15.75) provides a formal way to take into account this dependence. It shows that if $\beta > 0$, $\bar{\lambda}$ increases with increasing ρ and it will continue to increase as long as β remains positive. The limit of $\bar{\lambda}$ when $\rho \to \infty$ will be the first zero of β on the right of the initial value λ. Similarly, for $\beta < 0$, $\bar{\lambda}$ decreases with increasing ρ and, for $\rho \to \infty$, it tends to the first zero of $\beta(\bar{\lambda})$ for $\bar{\lambda} < \lambda$. Finally, if $\beta(\lambda) = 0$, $\bar{\lambda}$ is independent of ρ.

This analysis shows that we can classify the zeros of β in two classes: Those of Figure 15.6 (a) are ultraviolet (UV) attractive fixed points, starting anywhere in their neighbourhood, $\bar{\lambda}$ approaches them as ρ increases. Those of Figure 15.6 (b) are UV repulsive fixed points, $\bar{\lambda}$ moves further away from them with increasing ρ. The conclusion is that the asymptotic behaviour of a field theory depends on the position and nature of the zeros of its β-functions.

As long as perturbation theory is our only guide, we cannot say anything about the properties of $\beta(\lambda)$ for arbitrary λ. We do not know whether it has any non-trivial zeros, let alone their nature. The only information perturbation theory can hopefully provide is the behaviour of $\beta(\lambda)$ in the vicinity of $\lambda = 0$. We know that $\beta(0) = 0$, because $\lambda = 0$ is a free field theory. The nature of this zero (attractive or repulsive),

---

[9]Including the 2-loop term in the expansion of the β-function amounts to summing also the next-to-leading logs, etc.

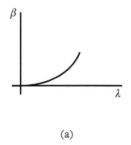

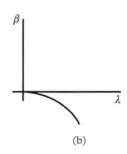

(a)   (b)

Fig. 15.7 The nature (repulsive or attractive) of the free field theory zero of the β-function. $\lambda = 0$ is an UV repulsive fixed point in (a), and UV attractive in (b).

will depend on the sign of the first non-vanishing term in the expansion of $\beta(\lambda)$ in powers of λ. But this expansion is precisely perturbation theory. If β starts as in Figure 15.7 (a) from positive values, the origin is repulsive. The effective coupling constant will be driven away to larger values as we go deeper and deeper into the asymptotic Euclidean region. On the contrary, if the first term of β is negative, Figure 15.7 (b), the origin is attractive. If we start somewhere between the origin and the next zero of β, the effective coupling constant will become smaller and smaller and it will vanish in the limit. A theory whose couplings exhibit such a behaviour is called *asymptotically free*.

And now we can state the following, very important theorem:

Out of all renormalisable field theories, only the non-Abelian gauge theories are asymptotically free.

The proof consists of a simple, straightforward computation. We compute the first term in the expansion of the β-function for all renormalisable field theories, i.e. ϕ^4, Yukawa, spinor electrodynamics, scalar electrodynamics, non-Abelian Yang–Mills theories and any combination of the above. Only for the pure non-Abelian Yang–Mills theory the β-function is negative. If we add any other, independent pieces with new coupling constants, we lose asymptotic freedom. In Chapter 21 we will show that an $SU(3)$ Yang–Mills theory describes the strong interactions among quarks.

15.3.1   Renormalisation group in dimensional regularisation

In Problem 15.7 we show that the first two terms in the expansion of the β-function (15.66) in powers of λ are independent of the renormalisation scheme we used to compute it. For practical calculations the most efficient method turns out to be the use of dimensional regularisation. In the notation we used in (15.35), we write the renormalised coupling constant as

$$\lambda_R = \lambda\mu^{-\epsilon}Z_\lambda^{-1}(\lambda_R, \epsilon) \tag{15.78}$$

In dimensional regularisation the β-function is defined as the derivative of λ_R with respect to μ keeping λ and ϵ fixed

$$\beta(\lambda_R, \epsilon) = \mu\frac{\partial\lambda_R}{\partial\mu}\bigg|_{\lambda,\epsilon} \tag{15.79}$$

This equation can be rewritten as

$$\left[\beta + \epsilon\lambda_R + \beta\lambda_R \frac{\partial}{\partial\lambda_R}\right] Z_\lambda = 0 \tag{15.80}$$

When $\epsilon \to 0$, Z_λ is singular and has an expansion in powers of ϵ^{-1}, i.e.

$$Z_\lambda = \sum_{n=0}^{\infty} \frac{Z_\lambda^{(n)}(\lambda_R)}{\epsilon^n} \tag{15.81}$$

with $Z_\lambda^{(0)}(\lambda_R) = 1$. On the other hand, $\beta(\lambda_R, \epsilon)$, which is regular when $\epsilon \to 0$, admits an expansion

$$\beta(\lambda_R, \epsilon) = \sum_{n=0}^{\infty} \beta^{(n)}(\lambda_R)\epsilon^n \tag{15.82}$$

Since (15.80) is an identity in ϵ, we can compute the coefficients $\beta^{(n)}(\lambda_R)$ by matching the powers of ϵ. We see that $\beta^{(1)}(\lambda_R) = -\lambda_R$ and all $\beta^{(n)}$s with $n \geq 2$ vanish. Dropping the index 0 and the subscript R, let us define $\beta(\lambda) = \beta(\lambda, \epsilon) + \epsilon\lambda$. We obtain the recursive relation

$$\lambda^2 \frac{\partial Z_\lambda^{(n+1)}}{\partial\lambda} = \beta(\lambda)\frac{\partial Z_\lambda^{(n)}}{\partial\lambda} \tag{15.83}$$

This is a very important relation. First, it makes it possible to compute $\beta(\lambda)$ order by order in perturbation. Writing (15.83) for $n = 0$ we obtain

$$\beta(\lambda) = \lambda^2 \frac{\partial Z_\lambda^{(1)}}{\partial\lambda} \tag{15.84}$$

which means that $\beta(\lambda)$ is entirely determined by the residues of the simple poles of Z_λ. Comparing with (15.36), we find

$$\beta(\lambda) = b_0\lambda^2 + b_1\lambda^3 + \dots \quad , \quad b_0 = \frac{3}{16\pi^2} \tag{15.85}$$

The advantage of this method is now obvious: in order to compute the β-function we only need the residues of the $1/\epsilon$ poles. The finite parts of the Green functions are not needed. A second consequence of the recursive relation (15.83) is that although in the nth order of perturbation theory we encounter singularities up to ϵ^{-n}, only the simple poles are new. The residues of the higher order poles are determined in terms of those of lower order.

15.4   Problems

Problem 15.1 Generating functionals of Green functions. We have defined in (10.25) the generating functional $Z[j]$ of Green functions for the ϕ^4 theory. Prove that

1. $W[j] = \ln Z[j]$ is the generating functional of connected Green functions.
2. Its Legendre transform $\Gamma[\phi_c] = W[j] - \int \mathrm{d}^4x j(x)\phi_c(x)$, where $\phi_c(x) = \delta W[j]/\delta j(x)$, is the generating functional of the one-particle irreducible (1PI) Green functions.

Problem 15.2 The power counting in various field theories.

1. Repeat the power counting computation and prove that the four-dimensional field theories mentioned in section 15.2.1 are indeed renormalisable.

2. Prove that a four-fermion theory with interaction Lagrangian density $\mathcal{L}_{\text{int}} \sim G_F(\bar{\psi}\psi)^2$, like the one proposed by Fermi for the weak interactions, is non-renormalisable.

3. Find in all these theories the dimensionality of the coupling constants.

4. Prove by power counting that a theory with an interaction Lagrangian density $\mathcal{L}_{\text{int}} \sim g\bar{\psi}\gamma_\mu\psi A^\mu$, where A^μ is a massive spin-one field, despite the fact that it has a dimensionless coupling constant, is non-renormalisable.

5. Prove that in a two-dimensional space-time all field theories involving only scalar fields with non-derivative polynomial interactions are super-renormalisable.

6. Are there any renormalisable theories in a six-dimensional space-time?

Problem 15.3 A simple way to regularise the ultraviolet behaviour of a quantum field theory is to introduce higher derivative terms in the quadratic part of the Lagrangian density. Consider the example of a single scalar field ϕ with $\mathcal{L}$ given by

$$\mathcal{L} = \frac{1}{2}\partial_\mu\phi\partial^\mu\phi - \frac{1}{2}m^2\phi^2 - \frac{\xi}{2}\partial_\mu\phi\Box\partial^\mu\phi - \frac{\lambda}{4!}\phi^4$$

with ξ a constant with dimensions $[m^{-2}]$.

1. Show that in this theory all Feynman diagrams are finite.

2. Show that this has been achieved at the expense of introducing degrees of freedom with negative energy.

Hint: Show that the propagator in this theory is equivalent to that of a theory with two scalar fields ϕ_1 and ϕ_2, one having negative kinetic energy.

Problem 15.4 Compute the $1/\epsilon$ part of the two loop diagram of Figure 15.4 and isolate the term proportional to p^2.

Problem 15.5 Consider a model with two neutral scalar fields, $\phi(x)$ and $\Phi(x)$ whose interaction is described by the Lagrangian density

$$\mathcal{L}_1 = \frac{1}{2}\partial_\mu\phi\partial^\mu\phi + \frac{1}{2}\partial_\mu\Phi\partial^\mu\Phi - \frac{1}{2}m^2\phi^2 - \frac{1}{2}M^2\Phi^2 - \frac{\lambda_1}{4!}\phi^4 - \frac{\lambda_2}{4!}\Phi^4 - \frac{2g}{4!}\phi^2\Phi^2$$

We consider the limit $M \to \infty$ keeping m and the interaction energy finite. Show that in this limit all effects of the Φ field in the dynamics of ϕ can be absorbed in the renormalisation conditions of the effective Lagrangian obtained from $\mathcal{L}_1$ by putting $\Phi = 0$

$$\mathcal{L}_{eff} = \frac{1}{2}\partial_\mu\phi\partial^\mu\phi - \frac{1}{2}m^2\phi^2 - \frac{\lambda_1}{4!}\phi^4$$

This is an example of the "decoupling theorem". It states that, if we have a renormalisable field theory and we take the limit in which some degrees of freedom become infinitely massive, the resulting theory is described by an effective field theory obtained by setting all heavy fields equal to zero, provided the resulting theory is also renormalisable.

Problem 15.6 Perform the z-integration in equation (15.51) and determine the analytic properties of the four-point function $\Gamma^{(4)}(p_1, \ldots, p_4)$ of Figure 15.2 as a function of the external momenta.

Problem 15.7 Consider two different renormalisation schemes for the ϕ^4 theory yielding coupling constants λ_1 and λ_2, and the corresponding β-functions $\beta_1(\lambda_1)$ and $\beta_2(\lambda_2)$. Using the fact that both schemes describe the same theory, prove that the first two terms in the expansion of the β-functions are scheme independent.

Hint. The fact that the two schemes describe the same theory implies that each coupling constant can be expressed as a power series in the other of the form $\lambda_1 = \lambda_2 + \sum_{n=2}^{\infty} C_n \lambda_2^n$.

Problem 15.8 1. Solve (15.75) with $\beta(\bar{\lambda}) = \bar{\lambda}^2$ for $\bar{\lambda}(\rho)$ and $\bar{\lambda}(1) = \lambda$.
2. Prove that the solution you found also satisfies (15.74).
3. Show for arbitrary $\beta(\bar{\lambda})$ the equivalence of (15.74) and (15.75).

16
The Electromagnetic Interactions

16.1 Introduction

In June 2–4 1947 the National Academy of Sciences sponsored the first American post-war Conference on the Foundations of Quantum Mechanics. It was held at Long Island's Shelter Island at Ram's Head Inn and gathered 24 participants. The chairman of the conference was K.K. Darrow, a theoretical physicist from the Bell Labs and long-time secretary of the American Physical Society; but the man who was really running the show was J.R. Oppenheimer. With the aura of the Manhattan Project, he was considered as one of the founding fathers of the American School of theoretical physics. Regarding the conference, Oppenheimer later declared it was the most successful scientific meeting he had ever attended. Indeed, in retrospect Shelter Island is a landmark of modern theoretical physics.

The most important contributions, which were presented in the conference, were not theoretical breakthroughs but two experimental results: (i) W.E. Lamb Jr., from Columbia University, presented the results, obtained together with his PhD student R.C. Retherford, on the fine structure of the hydrogen atom. (ii) I.I. Rabi, also from Columbia University, announced measurements by his colleagues P. Kusch and H.M. Foley of the anomalous magnetic moment of the electron. Both these results were important because, for the first time, they showed beyond any possible doubt that the predictions based on the Dirac equation for the electron are only approximately correct.

Lamb received his PhD at Berkeley in 1938 under the direction of Oppenheimer and started his career as a theorist. During the war he worked in radar and became an expert in microwave technology. After the war he used this expertise at Columbia to obtain very accurate measurements of hydrogen's fine structure. We recall here the results we obtained in Chapter 8: The relativistic corrections on the hydrogen spectrum computed by the Dirac equation split the degeneracy obtained by the Schrödinger equation and gave rise to the fine structure. However, the two levels $^2S_{1/2}$ and $^2P_{1/2}$ should remain degenerate, as we can see in equation (8.44). Lamb and Retherford proved instead that there is an energy difference of about 1000 MHz of the S level above the P level. This is the *Lamb shift*. Similarly, we showed in section 8.3.4 that the Dirac equation reproduces the non-relativistic Pauli equation with the particular value of the gyromagnetic ratio $g = 2$. The first measurement by Foley and Kusch gave a value $g = 2.00229 \pm 0.00008$.[1]

---

[1] The present value is $(g - 2)/2 = (1159.65218091 \pm 0.00000026) \times 10^{-6}$.

Elementary Particle Physics. John Iliopoulos and Theodore N. Tomaras, Oxford University Press.
© John Iliopoulos and Theodore N. Tomaras (2021). DOI: 10.1093/oso/9780192844200.003.0016

Both these results showed unambiguously that there is physics beyond the Dirac equation. As S. Weinberg has put it "...it was not so much that they forced theorists to change their theories, as that they forced them to take them seriously." Indeed, the formalism of quantum field theory existed already for both bosons and fermions. But the enormous prestige of the Dirac theory on the one hand, and the absence of a clear physical motivation on the other, discouraged theorists from seriously facing the problem of the divergences of perturbation theory. The fact that the theoretical ideas existed already in a subconscious form is witnessed by the fact that it took practically no time to develop them. The first estimation of the Lamb shift in a non-relativistic approximation was done by Bethe in the train that brought him back from New York to Ithaca. In the following months Feynman and Schwinger, independently, using apparently different formulations, set up the programme for the renormalised perturbation expansion of quantum electrodynamics and Schwinger gave the first calculation of $g - 2$. As it turned out, similar results were obtained independently in Japan by Sin-Itiro Tomonaga, who also obtained the first complete calculation of the Lamb shift. The equivalence of all these approaches was formally shown by F. Dyson in 1948. Rare are the examples in physics in which so much progress was accomplished in such a short time.[2]

16.2   Quantum Electrodynamics

In the previous chapters we have derived the Lagrangian density for quantum electrodynamics, the theory which describes the interaction among electrons (and, of course, also positrons) and photons

$$\mathcal{L} = \overline{\psi}(x)\,(i\slashed{\partial} - m)\,\psi(x) - \frac{1}{4}F_{\mu\nu}(x)F^{\mu\nu}(x) - e\overline{\psi}(x)\slashed{A}(x)\psi(x) \tag{16.1}$$

and in Chapter 15 we studied its renormalisation properties. Using these rules we can compute any process at any order in the perturbation expansion. As an example, let us indicate how we can compute the electron anomalous magnetic moment at one loop.

We consider the scattering of an electron by an external electromagnetic potential $A_\mu^{\mathrm{cl}}(x)$ at first order in A_μ^{cl}. For example, this could be the scattering of an electron from a heavy nucleus at small momentum transfer, where the recoil of the nucleus can be neglected. The interaction Lagrangian density can be written as

$$\mathcal{L}_{\mathrm{int}} = e\left[Z\mathcal{J}_\mu(x) - \overline{\psi}(x)\gamma_\mu\psi(x)\right]A^\mu(x) \tag{16.2}$$

where $\mathcal{J}_\mu(x)$ is a conserved electromagnetic current created by a charge Ze. In our approximation we can treat $\mathcal{J}_\mu(x)$ as a classical source.[3] The incident electron sees

---

[2]The Shelter Island conference was meant to be the first in a series on the same subject. Indeed a follow up conference was organised in Pocono (Pennsylvania) in 1948 and a third one in Oldstone (New York) in 1949. After that the programme of the Foundations of Quantum Mechanics was declared complete. A rather optimistic view, of course.

[3]For a static nucleus only $\mathcal{J}_0(x)$ is different from zero.

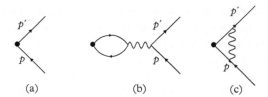

Fig. 16.1 The diagrams contributing to the scattering of an electron from an external electromagnetic field. The blob represents the external field. Diagram (a) represents the tree approximation and diagrams (b) and (c) the 1-loop corrections.

a classical electromagnetic potential which, in the gauge $\partial_\mu A^\mu(x) = 0$, satisfies the equation

$$\Box A_\mu^{\mathrm{cl}}(x) + eZ \mathcal{J}_\mu^{\mathrm{cl}}(x) = 0 \tag{16.3}$$

By Fourier transformation we obtain

$$\widetilde{A}_\mu^{\mathrm{cl}}(q) = eZ \widetilde{\mathcal{J}}_\mu^{\mathrm{cl}}(q) \frac{1}{q^2} \tag{16.4}$$

At lowest order the scattering is represented by the diagram of Figure 16.1(a) and the amplitude, suppressing for simplicity the polarisation index of the initial and final spinors, is given by

$$i\mathcal{M} = -ie\widetilde{A}_\mu^{\mathrm{cl}}(q)\bar{u}(p')\gamma^\mu u(p) \tag{16.5}$$

where $q = p - p'$ and the spinors $u(p)$ and $u(p')$ satisfy the free Dirac equation $(\not{p} - m)u(p) = (\not{p}' - m)u(p') = 0$ with $p^2 = p'^2 = m^2$. At 1-loop order we must add the diagrams of Figure 16.1(b) and (c). We recall that, since we are computing a scattering amplitude, we should not include the self-energy diagrams in the external electron lines.

Before proceeding to the explicit calculation, let us first examine the general structure of the amplitude. By Lorentz invariance, it can be written as

$$i\mathcal{M} = -ie\widetilde{A}_\mu^{\mathrm{cl}}(q)J^\mu(p,p')\,, \quad \text{with } J^\mu(p,p') = \bar{u}(p')\Gamma^\mu(p,p')u(p) \tag{16.6}$$

$\Gamma^\mu(p',p)$ is a 4×4 matrix, a superposition of the 16 matrices we found in Problem 7.3. Using parity conservation, the conservation of the electromagnetic current which implies the relation $q_\mu J^\mu(p,p') = 0$ and the Gordon decomposition of the current which we proved in Problem 8.2, we can write the most general form of J^μ as

$$J^\mu = \bar{u}_{s'}(p')\left[\gamma^\mu F_1(q^2) + \frac{i\sigma^{\mu\nu}q_\nu}{2m}F_2(q^2)\right]_{s's}u_s(p) \tag{16.7}$$

where F_1 and F_2 are scalar functions of q^2, which is the only Lorentz invariant variable. They are called *form factors*. We want to compute the form factors at 1-loop order. The zeroth order result, Figure 16.1(a), is given by equation (16.5). The low energy limit, which gave the Pauli equation in section 8.3.4, gives the values $F_1(0) = 1$ and $F_2(0) = 2$. These are the values predicted by the Dirac equation.

The 1-loop contribution is given by the two diagrams of Figure 16.1(b) and (c), in which we include the renormalisation counterterms we discussed in equations (15.52) and (15.54). Let us compute the contribution of each one of them.

Figure 16.1(b) gives the 1-loop correction to the photon 2-point function. As shown in equation (15.52), its contribution to $\mathcal{M}$ is of the general form

$$i\mathcal{M}_{(b)} = e^3 \, \widetilde{A}^{\text{cl}}_\mu(q)(\eta^{\mu\nu}q^2 - q^\mu q^\nu)\Pi(q^2)\frac{1}{q^2}\bar{u}(p')\gamma_\nu u(p) \tag{16.8}$$

$\Pi(q^2)$ is the finite renormalised value of the diagram containing the fermion loop. Since we have used on-shell renormalisation conditions, Π should vanish at $q^2 = 0$. Furthermore, $q^\nu \bar{u}(p')\gamma_\nu u(p) = 0$, because the spinors u satisfy the free Dirac equation. We conclude that this diagram contributes only to $F_1(q^2)$ but it is equal to zero at $q^2 = 0$. In fact, we can compute this contribution as a function of q^2 and show that it is both ultraviolet and infrared finite.

We are left with Figure 16.1 (c).

$$J^\mu_{(c)} = -ie^2 \int \frac{d^dk}{(2\pi)^d} \frac{1}{k^2 - m_\gamma^2} \bar{u}(p')\gamma_\nu \frac{1}{\not{p}' + \not{k} - m}\gamma^\mu \frac{1}{\not{p} + \not{k} - m}\gamma^\nu u(p) \tag{16.9}$$

where the iϵ prescription for the three Feynman propagators is understood. In writing (16.9) we used dimensional regularisation for the ultraviolet singularity and we introduced a fictitious mass m_γ in the photon propagator in order to avoid infrared divergences. At the end, the limit $m_\gamma \to 0$ should be taken. We have already computed the term which is singular when d equals 4 in equation (15.54) and we determined the counterterm, which is required to absorb it. Here we will need also the finite part. We get

$$J^\mu_{(c)} = -ie^2 \int \frac{d^dk}{(2\pi)^d} \frac{\bar{u}(p')\gamma_\nu(\not{p}' + \not{k} + m)\gamma_\mu(\not{p} + \not{k} + m)\gamma^\nu u(p)}{(k^2 - m_\gamma^2)(k^2 + 2p' \cdot k)(k^2 + 2p \cdot k)} \tag{16.10}$$

We use the Feynman formula (15.20) for $l = 3$, shift the integration variable $k \to K - z_2 p - z_1 p'$, and drop the terms linear in K. We obtain

$$J^\mu_{(c)} = -2ie^2 \int \frac{d^dK}{(2\pi)^d} \int_0^1 dz_1 dz_2 dz_3 \delta(z_1 + z_2 + z_3 - 1) \left[\frac{\bar{u}(p')\gamma_\nu \not{K}\gamma^\mu \not{K}\gamma^\nu u(p)}{(K^2 - R^2)^3} \right.$$

$$\left. + \frac{\bar{u}(p')\gamma_\nu((1 - z_1)\not{p}' - z_2\not{p} + m)\gamma^\mu((1 - z_2)\not{p} - z_1\not{p}' + m)\gamma^\nu u(p)}{(K^2 - R^2)^3} \right] \tag{16.11}$$

where $R^2 = m^2(z_1 + z_2)^2 + m_\gamma^2 z_3 - q^2 z_1 z_2$.

In the first term the K integration is logarithmically divergent and requires the subtraction we computed in equation (15.54). In the second term the K integration is convergent and the result, after we perform the Wick rotation, is given in equation (15.22). The expression in the numerator can be simplified using the identities we proved in Problem 7.6. In the end we find two types of terms: (i) terms proportional to γ_μ, which contribute only to F_1, and (ii) terms proportional to $(p_\mu + p'_\mu)$ which, with the help of the Gordon identity, we can trade for $\sigma_{\mu\nu}q^\nu$. They are the ones which contribute to F_2. We end up with

$$
J^\mu_{(c)} = -\frac{i\alpha}{\pi} \int_0^1 dz_1 dz_2 dz_3 \delta(z_1 + z_2 + z_3 - 1) \bar{u}(p') \left\{ \gamma^\mu \left[\frac{1}{\epsilon} + \ln \frac{R^2}{\mu^2} + C.T. \right. \right.
$$

$$
\left. \left. + \frac{q^2(1 - z_1)(1 - z_2) + m^2(1 - 4z_3 + z_3^2)}{4R^2} \right] + \frac{i\sigma^{\mu\nu}q_\nu}{2m} \frac{m^2 z_3(1 - z_3)}{R^2} \right\} u(p) \qquad (16.12)
$$

where μ is the mass parameter of the dimensional regularisation and $C.T.$ stands for the counterterm we introduced in section 15.2.3. It has a double role. It should: (i) absorb the $1/\epsilon$ ultraviolet divergence, and (ii) enforce the renormalisation condition $F_1(0) = 1$.[4] Therefore, it is given by

$$
C.T. = -\frac{1}{\epsilon} + \ln \frac{\mu^2}{m^2(z_1 + z_2)^2 + m_\gamma^2 z_3} - \frac{m^2(1 - 4z_3 + z_3^2)}{4[m^2(z_1 + z_2)^2 + m_\gamma^2 z_3]} \qquad (16.13)
$$

As expected, the μ^2 dependence disappears from the final result. The remaining $F_1(q^2)$ for $q^2 \neq 0$ is infrared divergent and we will discuss the $m_\gamma \to 0$ limit in Chapter 17.

The last term in equation (16.12) contributes to the form factor $F_2(q^2)$. A simple calculation gives

$$
F_2(0) = \frac{\alpha}{\pi} \int_0^1 dz_1 dz_2 dz_3 \delta(z_1 + z_2 + z_3 - 1) \frac{z_3(1 - z_3)}{(z_1 + z_2)^2} = \frac{\alpha}{2\pi} \qquad (16.14)
$$

It is the result first found by Schwinger. Note that $F_2(0)$ has neither ultraviolet nor infrared singularities and this value can be directly compared to experiment. It has been computed up to order α^5 with the result: $(g - 2)/2 = (1159.652181643 \pm 0.000000764) \times 10^{-6}$, to be compared with the experimental value of $(g - 2)/2 = (1159.65218091 \pm 0.00000026) \times 10^{-6}$. It gives one of the most precise tests of quantum electrodynamics.

To finish the calculation we must take the $m_\gamma \to 0$ limit. It cannot be taken directly in equations (16.12) and (16.13) because the resulting expressions are divergent. It is the problem of the infrared divergences that we have encountered several times already and it will be discussed in Chapter 17.

In this simple calculation we see all the properties of a renormalisable quantum field theory. The fact that we must enforce the condition $F_1(0) = 1$ order by order in perturbation theory means that we cannot compute the value of the electron charge e.

---

[4]This is the calculation of the scattering amplitude of an electron in an external electromagnetic field. According to our discussion in section 12.3 we have dropped the self-energy diagrams on the external electron lines. Had we computed the general 3 point Green function we would have to include them and we would have found that the equality $Z_1 = Z_2$ of equation (15.61) extends also to the finite parts, and the condition $F_1(0) = 1$ comes out automatically. See Problem 16.1.

The same holds for the electron mass m. These two basic parameters of the theory should be taken from experiment. However, all other quantities, such as the magnetic moment, or any measurable scattering cross section, appropriately defined in order to eliminate the infrared divergences as we will explain in Chapter 17, are finite and calculable as power series in α.

In this chapter we have considered the interaction of electrons with photons. Obviously, in order to write the complete Lagrangian density of the electromagnetic interactions of all the fundamental matter particles we should add to (16.1) the terms representing the kinetic energies and the interaction terms for every elementary charged particle. We will do this in Chapter 19.

16.3   Problems

Problem 16.1 Prove that the quantity $(p-p')^{\mu}\Gamma_{\mu}^{(1,2)}(p,p')$ can be expressed at 1-loop order in terms of the 2-point functions $\Gamma^{(0,2)}(p)$ and $\Gamma^{(0,2)}(p')$.

Problem 16.2 Compute the QED β-function at 1-loop order.

Problem 16.3 Assume that the β-function of QED has a non-trivial zero at some value of the running coupling constant $\bar{\alpha} = \alpha^* : \beta(\alpha^*) = 0$. Compute the asymptotic behaviour of the photon propagator $G_{\mu\nu}^{(20)}(q)$ at large q^2 for this value of the coupling constant.

Problem 16.4 Write the four-photon Green function at 1-loop (Figure 15.5) and show that it is finite.

Problem 16.5 In section 5.7.5 we noted that there were two alternative definitions of the parity operator acting on the Fock space of states: $\mathcal{P}_1$, such that $\mathcal{P}_1^2 = \mathbf{1}$, and $\mathcal{P}_2$, $\mathcal{P}_2^2 = (-)^F$ with F the fermion number of the state.

1. Show that quantum electrodynamics is invariant under either of them.

2. Prove that they both provide equivalent representations of the parity operator, in the sense that they are related by $\mathcal{P}_1 = U\mathcal{P}_2$, where U is a unitary operator corresponding to an internal symmetry of the theory.

17

Infrared Effects

17.1 Introduction

We have encountered the problem of infrared divergences several times in this book. It
was always associated with the computation of a scattering amplitude in the presence
of zero mass particles. Already in Chapter 4, where we formulated the scattering theory in non-relativistic quantum mechanics, we noted in section 4.3.2 that a divergence
appears in the case of the Coulomb potential, which does not vanish fast enough at
infinity. The same problem appeared in Chapter 12, because we had to assume the
absence of massless particles in order to define the asymptotic states and asymptotic
fields–see for example equation (12.11). Also, in explicit calculations of quantum electrodynamics, we found that the amplitude becomes singular at certain kinematical
regions because of photon's zero mass, as in equations (12.52), (12.53) and (16.12).
The prescription we adopted in these cases was to introduce a fictitious mass m_γ for
the photon, and we promised to discuss the limit $m_\gamma \to 0$ later. This discussion is the
purpose of the present chapter.

  In elementary particle physics the only known massless particles are gauge bosons
associated to gauge symmetries. It is the case of the photon in QED, but also, as we
shall see in Chapter 21, the particles we called "gluons", and which mediate the strong
interactions. Had we included also the study of gravitational interactions, we would
have encountered the same phenomenon of infrared divergences because a quantum
theory of gravity would imply the presence of a massless graviton. Massless particles
make it possible to consider excitations with arbitrarily low frequency, thus preventing
the appearance of a mass gap in the energy spectrum. In this sense, the infrared divergences have a completely different origin than the ultraviolet divergences we studied in
Chapter 15. The latter were due to a mathematical mistake, namely the multiplication
of the fields, which are operator valued distributions, at the same space-time point.
They were correctly handled order by order in perturbation theory by the process of
renormalisation. On the other hand, the infrared divergences have a physical origin, so
we do not expect a technical solution, but a physical one: we must correctly define the
physically measurable quantities in the presence of long range forces. In this chapter
we will indicate explicitly how this redefinition should be done in some simple cases
and how the results compare with experiment.

17.2 Examples of Infrared Divergent Terms in Perturbation Theory

We have found already infrared divergent diagrams in quantum electrodynamics. They
have various origins, and in this section we want to distinguish some specific cases.

Elementary Particle Physics. John Iliopoulos and Theodore N. Tomaras, Oxford University Press.
© John Iliopoulos and Theodore N. Tomaras (2021). DOI: 10.1093/oso/9780192844200.003.0017

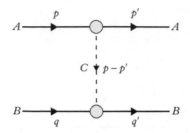

Fig. 17.1 Diagram exhibiting a pole singularity at $t = m_C^2$ of the $A + B \rightarrow A + B$ scattering amplitude. The blobs may include any type of corrections in the $A - C - A$, or $B - C - B$ vertices.

(i) *One particle exchange diagrams.* Let us consider the elastic scattering $A(p) + B(q) \rightarrow A(p') + B(q')$ and isolate the term representing the exchange of a virtual particle C. It is given by diagrams of the general topology shown in Figure 17.1. Energy-momentum conservation and the mass shell conditions give $p + q = p' + q'$, $p^2 = p'^2 = m_A^2$ and $q^2 = q'^2 = m_B^2$. If $t = (p-p')^2 = (q-q')^2$ is the momentum transfer square between the initial and the final particles, the contribution of the diagram of Figure 17.1 to the scattering amplitude is proportional to $1/(t - m_C^2)$, due to the C propagator. The physical region for the process $A + B \rightarrow A + B$ is $t < 0$, as we have shown in section 12.2. It follows that, for $m_C \neq 0$, the pole at $t = m_C^2$ lies outside the physical region and the physical amplitude has no singularity. However, when $m_C \rightarrow 0$ the pole moves in the complex t plane and reaches the edge of the physical region. This is the kind of singularities we found in the Bhabha or Møller scattering amplitudes of equations (12.52) and (12.53).

(ii) *Singularities due to integrations over loop momenta.* We encountered a singularity of this type in computing the diagram of Figure 16.1 (c). In equation (16.9), taking into account the mass shell conditions $\not{p} = \not{p}' = m$, we find that the amplitude for the scattering of an electron from an external electromagnetic field contains the integral $\int d^4k/k^2(k^2 - m_\gamma^2)$, which is infrared divergent for $m_\gamma = 0$. It is easy to verify that singularities of this type can be found at any order in the perturbation expansion and, as we see in the example of Figure 16.1(c), they occur when the massless propagator is attached to external lines which are on the mass shell, i.e. when we compute scattering amplitudes.

(iii) *Singularities due to soft photon emissions from external lines.* Let us consider the radiation of a photon by an external electron line which can be either entering the graph or leaving it. An example is shown in Figure 17.2. The electron propagator of the line before the emission is given by $1/(\not{p} + \not{k} - m) \sim 1/2p \cdot k$, where we assumed the electron line to be on the mass shell. Since the mass of the photon is zero, there is no lower limit for the energy of the emitted photon, and diagrams of this kind develop infrared singularities.

It is clear from these examples that in a theory which contains massless particles, infrared divergences occur only when we attempt to compute either a scattering amplitude, or a Green function in very special kinematical configurations, like the zero momentum transfer one of Figure 17.1. A general Green function, with the external

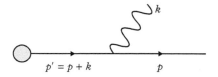

Fig. 17.2 A photon radiated from an external electron line.

lines off the mass shell and generic values of the momenta, is free of infrared singularities. In the following sections we shall show how, in the presence of zero mass
particles, we should modify our scattering theory in order to obtain finite and well
defined results, which can be compared with experiment.

17.3 Infrared Phenomena in QED

Quantum electrodynamics is the first quantum field theory to be formulated and analysed theoretically and it is in this framework that the problem of infrared divergences
was first addressed and solved. In this section we shall present the solution for the
simple example of electron Coulomb scattering which we studied in section 16.2.

17.3.1 Tree-level ($\mathcal{O}(\alpha^2)$) electron Coulomb scattering

In the tree approximation the process is represented by the diagram of Figure 16.1(a),
and the amplitude is given in (16.5). In the special case of Coulomb scattering, $\widetilde{A}^\mu_{\text{ext}}(q)$
has vanishing spatial components, while $\widetilde{A}^0_{\text{ext}}(q) = Ze/q^2$ is the Fourier transform of
the Coulomb potential $A^0_{\text{ext}}(x) = Ze/4\pi r$. The initial and final electron four-momenta
are $p^\mu = (E, \boldsymbol{p})$ and $p'^\mu = (E, \boldsymbol{p}')$, respectively, while the momentum transfer $q^\mu =
(0, \boldsymbol{q})$, with $\boldsymbol{q} = \boldsymbol{p}' - \boldsymbol{p}$, satisfies $-q^2 = \boldsymbol{q}^2 = 4\boldsymbol{p}^2 \sin^2(\theta/2) = 4E^2 v^2 \sin^2(\theta/2)$, where
θ is the scattering angle and v the speed of the electron. Note that in the static
approximation for the external field there is no energy transfer and $p^0 = p'^0 = E$.

The corresponding differential cross section for the scattering of an initial unpolarised electron beam summed over the electron polarisations in the final state is

$$\left(\frac{d\sigma}{d\Omega}\right)^{(0)}_C (p \to p') = \frac{Z^2 \alpha^2}{4E^2 v^4} \frac{1 - v^2 \sin^2(\theta/2)}{\sin^4(\theta/2)}$$

$$= \left(\frac{d\sigma}{d\Omega}\right)_R (p \to p') \left(1 - v^2\right) \left(1 - v^2 \sin^2 \frac{\theta}{2}\right) \qquad (17.1)$$

This is the Mott formula for Coulomb scattering of spin-$\frac{1}{2}$ particles. The expression
$(d\sigma/d\Omega)_R = Z^2 \alpha^2 / [4m^2 v^4 \sin^4(\theta/2)]$ is the classical Rutherford differential cross section, and coincides with the quantum mechanical one for the Coulomb scattering of
spinless non-relativistic particles. The expression (17.1) is singular at $\theta = 0$ and this
singularity corresponds to the $\boldsymbol{q}^2 = 0$ one of the external potential. We find the same
behaviour when we compute the cross section for the Bhabha or Møller scattering of
equations (12.52) and (12.53). The first is singular at $t = 0$ and the second at both
$t = 0$ and $u = 0$, due to the one-photon exchange diagrams.

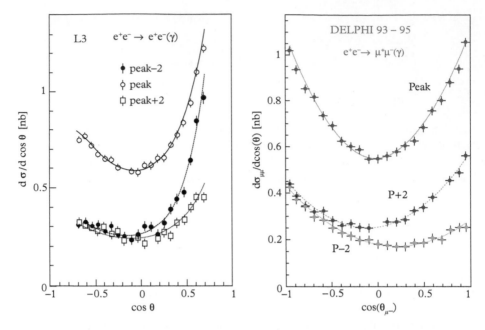

Fig. 17.3 The differential cross sections for $e^- + e^+$ annihilation as measured at LEP. The three curves correspond to the cross section at the peak of the Z pole and at ± 2 GeV above and below it. On the left we see the Bhabha scattering with the steep rise in the forward direction. For comparison, we show on the right the same quantity for the $e^+ e^- \rightarrow \mu^+ \mu^-(\gamma)$ process, in which there is no one-photon exchange contribution in the crossed channel and, as a result, there is no such phenomenon. *(Source: CERN, LEP Collaborations, L3 (left), DELPHI (right)).*

Before discussing the physical significance of such singularities, we want to point out that the above expressions are confirmed experimentally. Figure 17.3 shows the measured e^- e^+ cross section both for $e^- + e^+$ and $\mu^- + \mu^+$ final states[1] in the vicinity of the Z pole. Note the rise of the cross section at small angles. The theoretical curves are the ones given in equations (12.52) and (12.50), but they also include the radiative corrections. It is because of such corrections in the initial and final states that the curves do not look symmetric. We see that the agreement between theory and experiment is perfect.

Let us return to the Mott formula (17.1). The $\theta = 0$ singularity is an *infrared divergence*, since it is due to the long range nature of the Coulomb interaction and it is the same divergence we found in the non-relativistic Coulomb scattering in Chapter 4. Small scattering angle corresponds to large impact parameter, for which we still have non-trivial scattering due to the infinite range of the electromagnetic interaction. On the other hand, in actual experiments, in which an electron beam scatters off a target with charged particles, values of the scattering angle arbitrarily close to zero are not physically accessible. Indeed, as we explained in Chapter 4, there is no conceivable

---

[1]There are no good measurements for the Møller amplitude at very small angles.

experiment which could tell the difference between a particle scattered at zero angle and one which was not scattered. Therefore, placing a detector directly against the beam would not be meaningful because it will be saturated, and probably burned, by the incoming unscattered particles. The forward scattering cross section can only be reached by extrapolation. It follows that every experiment has a lower cut-off $\theta_0 > 0$ on the scattering angle. With the above in mind, for the total cross section we integrate over $\Omega(\theta, \phi)$ with $\theta_0 \leq \theta \leq \pi$ and $0 \leq \phi \leq 2\pi$, to obtain

$$\sigma_C^{(0)} = \frac{16\pi^3 Z^2 \alpha^2}{E^2 v^4} \left(\cot^2 \frac{\theta_0}{2} + 2v^2 \ln \sin \frac{\theta_0}{2} \right) \tag{17.2}$$

in perfect agreement with experiment.

17.3.2 One-loop corrections to the electron Coulomb scattering

The $\theta_0 \neq 0$ cut-off does not solve all the infrared problems because, as we saw in section 16.2, the 1-loop corrections to the Coulomb scattering bring new singularities. These are due to the diagram of Figure 16.1(c) which represents the one-photon correction to the interaction vertex. As we showed in section 16.2, the infrared divergent terms contribute only to the F_1 form factor, which is given in equations (16.12) and (16.13). Let us introduce the dimensionless ratios $\kappa^2 = q^2/m^2 \leq 0$ and $l^2 = m_\gamma^2/m^2$ and rewrite the Figure 16.1(c) contribution to the $F_1(q^2)$ form factor as

$$F_1^{(c)}(q^2) = \frac{\alpha}{2\pi} \int_0^1 \mathrm{d}z_1 \mathrm{d}z_2 \mathrm{d}z_3 \delta(z_1 + z_2 + z_3 - 1)$$

$$\times \left[\ln \left(\frac{(1 - z_3)^2 + l^2 z_3}{(1 - z_3)^2 - \kappa^2 z_1 z_2 + l^2 z_3} \right) + \frac{z_3^2 - 4z_3 + 1 + \kappa^2 (z_3 + z_1 z_2)}{(1 - z_3)^2 - \kappa^2 z_1 z_2 + l^2 z_3} \right.$$

$$\left. - \frac{z_3^2 - 4z_3 + 1}{(1 - z_3)^2 + l^2 z_3} \right] \tag{17.3}$$

where we used the fact that the sum of the three z_is is equal to 1. We want to isolate the term, which diverges when $l^2 \to 0$ and write $F_1^{(c)}(q^2)$ in the form

$$F_1^{(c)}(q^2) \simeq \frac{\alpha}{2\pi} \ln l^2 \, \mathcal{J}(\kappa^2) + \dots \tag{17.4}$$

where the dots stand for terms which are finite in the limit $l^2 \to 0$. It is easy to show that $\mathcal{J}(\kappa^2)$ equals the positive function[2]

$$\mathcal{J}(\kappa^2) = \int_0^1 \mathrm{d}s \frac{2 - \kappa^2}{2 - 2\kappa^2 s(1 - s)} - 1 = \frac{1}{2} \left[\frac{1 - \frac{2}{\kappa^2}}{\sqrt{1 - \frac{4}{\kappa^2}}} \ln \left(\frac{1 - \frac{2}{\kappa^2} + \sqrt{1 - \frac{4}{\kappa^2}}}{1 - \frac{2}{\kappa^2} - \sqrt{1 - \frac{4}{\kappa^2}}} \right) - 2 \right] \tag{17.5}$$

As we have seen, equation (17.4) contains all the infrared singular terms at the 1-loop order. Consequently, in the approximation in which we keep only infrared divergent terms, the total amplitude for the electron scattering process is

---

[2]Incidentally, the quantity $2\mathcal{J}(\kappa^2)$ can be written as $2\mathcal{J}(\kappa^2) = (1/v) \ln[(1+v)/(1-v)] - 2$, where v is the relative speed of the particles with four-momenta p_μ and p'_μ, given by $v = \sqrt{1 - 4/\kappa^2}/(1 - 2/\kappa^2)$.

Fig. 17.4 The diagrams representing the scattering of an electron from an external electromagnetic field with a photon radiated from the final, or the initial, state.

$$i\mathcal{M}_C^{(0)} + i\mathcal{M}_C^{(1)} \simeq i\mathcal{M}_C^{(0)} \left(1 + F_1^{(c)}(q^2) \right) \tag{17.6}$$

which implies that, up to order α^3 inclusive, the elastic cross section of Coulomb scattering of electrons is

$$\left(\frac{d\sigma}{d\Omega} \right)_C (p \to p') \simeq \left(\frac{d\sigma}{d\Omega} \right)_C^{(0)} \left| 1 + F_1^{(c)}(q^2) \right|^2 \simeq \left(\frac{d\sigma}{d\Omega} \right)_C^{(0)} \left(1 + 2F_1^{(c)}(q^2) \right)$$

$$\simeq \left(\frac{d\sigma}{d\Omega} \right)_C^{(0)} (p \to p') \left[1 + \frac{\alpha}{\pi} \ln \left(\frac{m_\gamma^2}{m^2} \right) \mathcal{J}(\kappa^2) \right] + \dots \tag{17.7}$$

In writing equation (17.7) we have kept the square of the lowest order term, which is of order α^2, and the interference between the lowest order and the 1-loop correction, which is of order α^3. The dots stand again for terms which are finite when m_γ goes to zero.

As anticipated, this expression diverges when we take the limit of vanishing photon mass. In addition to the tree-level infrared divergence present in $(d\sigma/d\Omega)_{(0)}$ for forward scattering $\theta = 0$, which we found in equation (17.1), loop corrections have made things worse, because of the infrared divergence in $F_1(k^2)$ due, at this order, to the exchange of a single virtual soft photon in Figure 16.1(c). The presence of a massless photon makes the scattering amplitude between initial and final electron Fock states nonsensical.

This sounds like a catastrophic result. We have often emphasised that most of our knowledge in particle physics comes from scattering experiments and the infrared divergences of equation (17.7) show that scattering amplitudes for theories, like QED, which contain massless particles, do not exist. The resolution of the apparent puzzle was found in 1937 by F. Bloch and A.T. Nordsieck who pointed out that, in the limit $m_\gamma \to 0$, the expression we computed in equation (17.7) does not correspond to a physically measurable quantity. Indeed, in the presence of the external field, the final and/or the initial electron may radiate photons. It is our familiar bremsstrahlung process. The key point is that every detector has a finite resolution in energy, i.e. it cannot detect photons with an energy smaller than some threshold δ. In other words, our experimental set up does not measure the cross section for the process of Figure 16.1 but rather the incoherent sum of all processes representing the electron scattering with any number of soft photons.

In perturbation theory every emitted photon carries a factor of the coupling e, so the $\mathcal{O}(\alpha)$ amplitude is the one of Figure 17.4 with just one soft photon emitted. As a result the order $\mathcal{O}(\alpha^3)$ transition probability, which is measured in our apparatus, is given by

$$\left(\frac{d\sigma}{d\Omega}\right)_C^{\text{measured}} = \left(\frac{d\sigma}{d\Omega}\right)_C (p \to p') + \int_k \frac{d\sigma}{d\Omega} (p \to p' + \gamma(k^0 < \delta)) \tag{17.8}$$

The first term on the right-hand side is given by (17.7) while, as indicated, the second involves an integration over the soft photon momenta $k^0 \in [0, \delta)$. As we will explicitly demonstrate presently, at order α^3, the quantity $(d\sigma/d\Omega)_{\text{measured}}$ is free of the logarithmic infrared divergence present in (17.7). The latter is exactly cancelled by the opposite contribution in the cross section of the soft photon emission process, to which we turn next.

17.3.3 Infrared divergence of the photon emission amplitude

Our goal in this section is the computation of the amplitude of electron Coulomb scattering with the emission of a single soft photon, represented by the diagrams of Figure 17.4. We obtain

$$i\mathcal{M}_{e \to e\gamma} = -ie^2 \bar{u}(p', \alpha') \left(\gamma^\mu \frac{\not{p}' + \not{k} + m}{(p' + k)^2 - m^2 + i\varepsilon} \gamma^\nu \right.$$
$$\left. + \gamma^\nu \frac{\not{p} - \not{k} + m}{(p - k)^2 - m^2 + i\varepsilon} \gamma^\mu \right) u(p, \alpha) \tilde{A}_{\text{ext}}^\nu(q) \epsilon_\mu^*(k) \tag{17.9}$$

At leading order for $k^0 \to 0$ it simplifies to the divergent expression

$$i\mathcal{M}_{e \to e\gamma} \simeq i\mathcal{M}_C^{(0)} e \left(\frac{p' \cdot \epsilon^*(k)}{p' \cdot k} - \frac{p \cdot \epsilon^*(k)}{p \cdot k} \right) \tag{17.10}$$

with $\mathcal{M}_C^{(0)}$ the tree level amplitude of elastic Coulomb scattering given in equation (16.5).

The cross section of interest to us according to (17.8), for an unpolarised electron beam and after summation over the final electron and photon polarisations, is

$$\int_k d\sigma(p \to p' + \gamma(k^0 < \delta)) \simeq d\sigma_C^{(0)}(p \to p')$$
$$\times e^2 \int_{|k| = m_\gamma}^\delta \frac{d^3\mathbf{k}}{(2\pi)^3} \frac{1}{2|\mathbf{k}|} \left(2 \frac{p \cdot p'}{p \cdot k \, p' \cdot k} - \frac{m^2}{(p \cdot k)^2} - \frac{m^2}{(p' \cdot k)^2} \right) \tag{17.11}$$

The lower end of the integral over the magnitude $|\mathbf{k}|$ of the photon momentum is set by the photon mass m_γ, while the upper limit is the characteristic energy δ given by the resolution of our detector.

We next denote by $\hat{\mathbf{k}}$ the photon momentum unit vector, we write the photon four-momentum $k^\mu = |\mathbf{k}|(1, \hat{\mathbf{k}}) = |\mathbf{k}| \, \hat{k}^\mu$, and perform the integration over $|\mathbf{k}|$ to obtain

$$\int_k d\sigma(p \to p' + \gamma(k^0 < \delta)) \simeq d\sigma_C^{(0)}(p \to p')$$
$$\times \frac{\alpha}{2\pi} \ln\left(\frac{\delta^2}{m_\gamma^2}\right) \int \frac{d\Omega_{\hat{\mathbf{k}}}}{4\pi} \left(2 \frac{p \cdot p'}{p \cdot \hat{k} \, p' \cdot \hat{k}} - \frac{m^2}{(p \cdot \hat{k})^2} - \frac{m^2}{(p' \cdot \hat{k})^2} \right) \tag{17.12}$$

It is straightforward to show that the last integration over $\Omega_{\hat{\mathbf{k}}}$ gives[3]

$$\int \frac{d\Omega_{\hat{\mathbf{k}}}}{4\pi} \left(2\frac{p \cdot p'}{p \cdot \hat{k}\, p' \cdot \hat{k}} - \frac{m^2}{(p \cdot \hat{k})^2} - \frac{m^2}{(p' \cdot \hat{k})^2} \right) = 2\mathcal{J}(\kappa^2) \tag{17.13}$$

so that the photon emission cross section in electron Coulomb scattering becomes

$$\int_k d\sigma(p \to p' + \gamma(k^0 < \delta)) \simeq d\sigma_C^{(0)}(p \to p') \frac{\alpha}{\pi} \ln\left(\frac{\delta^2}{m_\gamma^2} \right) \mathcal{J}(\kappa^2) \tag{17.14}$$

17.3.4 The $\mathcal{O}(\alpha^3)$ measured cross section is infrared finite

Combining (17.7), (17.8) and (17.14) we obtain for the physical cross section

$$\left(\frac{d\sigma}{d\Omega} \right)_C^{\text{measured}} \simeq \left(\frac{d\sigma}{d\Omega} \right)_C^{(0)} (p \to p') \left[1 - \frac{\alpha}{\pi} \mathcal{J}(\kappa^2) \ln\left(\frac{m^2}{\delta^2} \right) \right] \tag{17.15}$$

which, as promised, is well defined. The infrared divergences due to the virtual and the emitted photons nicely cancelled each other.

V.V. Sudakov has remarked that this expression simplifies considerably at large momentum transfer $-q^2 \gg m^2$. We first note that in equation (17.7) we can trade the $\ln(m_\gamma^2/m^2)$ factor for $\ln(-q^2/m_\gamma^2)$ without affecting its infrared behaviour. Second, at $-q^2 \gg m^2$ we have $\mathcal{J}(\kappa^2) \simeq \ln(-q^2/m^2)$. Combining the two we obtain for the measured cross section the *Sudakov double logarithm* expression

$$\left(\frac{d\sigma}{d\Omega} \right)_C^{\text{measured}} \simeq \left(\frac{d\sigma}{d\Omega} \right)_C^{(0)} (p \to p') \left[1 - \frac{\alpha}{\pi} \ln\left(\frac{-q^2}{m^2} \right) \ln\left(\frac{-q^2}{\delta^2} \right) \right] \tag{17.16}$$

17.4 The Summation to All Orders

Interestingly, the above cancellation of infrared divergences among virtual exchanged and real emitted soft photons is true to all orders in the perturbation expansion. Specifically, let us sum first over any number of virtual soft photons. The result is the exponentiation of the expression in the square brackets of the 1-loop expression (17.7), i.e.

$$d\sigma_C(p \to p') \simeq d\sigma_C^{(0)}(p \to p') \left(\frac{m_\gamma^2}{m^2} \right)^{\alpha \mathcal{J}(\kappa^2)/\pi} \tag{17.17}$$

which *vanishes*(!!) in the limit $m_\gamma \to 0$. To reiterate our previous conclusion, the presence of massless photons makes the scattering amplitudes between initial and final Fock states nonsensical. From the leading order result we know that we have to include

---

[3]Within our approximation of soft photon emission, we use $2p \cdot p' = 2m^2 - q^2$. Furthermore, the second and third terms of the integral are -1 each. As for the computation of the first integral, it is convenient to use Feynman's formula to write the integrand $1/(p \cdot \hat{k}\, p' \cdot \hat{k}) = \int_0^1 ds/[p \cdot \hat{k}s + p' \cdot \hat{k}(1-s)]^2$. It is then simple to perform the integration over $\Omega_{\hat{\mathbf{k}}}$ and verify (17.13).

processes involving the emission of soft photons. According to (17.14), including the contribution of at most one emitted soft photon to (17.17) we obtain

$$d\sigma_C(p \to p')_{\text{measured}} \simeq d\sigma_C^{(0)}(p \to p') \left(\frac{m_\gamma^2}{m^2} \right)^{\alpha \mathcal{J}(\kappa^2)/\pi} \left(1 + \frac{\alpha}{\pi} \mathcal{J}(\kappa^2) \ln \left(\frac{\delta^2}{m_\gamma^2} \right) \right)$$

(17.18)

Summation over any number of soft photons with total energy equal to δ exponentiates the expression in the last parenthesis of (17.18), and leads to the final answer for the measured Coulomb cross section

$$d\sigma_C(p \to p')_{\text{measured}} \simeq d\sigma_C^{(0)}(p \to p') \left(\frac{\delta^2}{m^2} \right)^{\alpha \mathcal{J}(\kappa^2)/\pi}$$

(17.19)

The above conclusions generalise to all scattering processes in QED or quantum gravity. The basic structure of equation (17.19) is common to all of them. The differences between the various processes are encoded in the detailed form of the tree level scattering amplitude and in the expression for the quantity $\mathcal{J}$, which is kinematical and depends on the specific process.

17.5 The QED Asymptotic Theory Revisited

We have shown that physically measurable quantities, taking into account the finite energy resolution in any experiment, are infrared finite. However, strictly speaking, the formalism we developed in Chapter 12 to describe scattering phenomena is incorrect in the presence of massless particles and we would like, if only for purely aesthetic reasons, to have a scattering theory in which the finite energy resolution of the detector is built in from the beginning. We will present in this section the main steps of such a formalism following the ideas introduced in 1970 by L.D. Faddeev and P.P. Kulish.[4]

17.5.1 The Faddeev-Kulish dressed electron states

In the Fock space the one electron state is strictly localised on the mass hyperboloid $p^2 = m^2$. It contains an electron with momentum p and nothing else. However, as we explained already, this state is experimentally indistinguishable from a multitude of others in which the electron is accompanied by an arbitrary number of photons, provided their total energy is less than the experimental resolution δ. This led Faddeev and Kulish to introduce the concept of *the dressed electron state*. It is a smeared state in which the energy is defined with a precision of order δ.[5] We see that such a formalism will contain three energy scales: m_γ is the auxiliary mass of the photon which will go to zero at the end of the calculation, δ is the energy resolution which will be kept finite and m is the mass of the electron. We shall choose $m_\gamma < \delta < m$. The first is

---

[4]For a more complete exposition of the Faddeev-Kulish approach see at T. N. Tomaras and N. Toumbas, "IR dynamics and entanglement entropy", *Phys.Rev. D* **101** 6, 065006 (2020), arXiv: hep-th 1910.07847.

[5]Already in Chapter 12 we introduced a smearing function in equation (12.11). But this was a technical step which allowed us to isolate the one-particle state. Here the smearing has a physical meaning.

obvious, since the ratio m_γ/δ will go to zero, while the second ensures that the one-electron dressed state will not overlap with multi-electron states. More precisely, the FK dressing is effected via the action of the unitary operator e^{-R_f}, where

$$R_f = \int \frac{\mathrm{d}^3\boldsymbol{p}}{(2\pi)^3}\, \hat\rho(\boldsymbol{p}) \int_{m_\gamma}^{\delta} \frac{\mathrm{d}^3\boldsymbol{k}}{(2\pi)^3}\, \frac{1}{(2\omega_{\boldsymbol{k}})^{1/2}}\, \left(f(\boldsymbol{k}, \boldsymbol{p}) \cdot a^\dagger(\boldsymbol{k}) - \text{h.c.}\right) \qquad (17.20)$$

Note that R_f is anti-Hermitian and e^{-R_f} is unitary. The notation is the following: $\hat\rho(\boldsymbol{p}) = \sum_s b_s^\dagger(\boldsymbol{p})b_s(\boldsymbol{p}) - d_s^\dagger(\boldsymbol{p})d_s(\boldsymbol{p})$ is the charge density operator. $b_s^\dagger(\boldsymbol{p})$ $(d_s^\dagger(\boldsymbol{p}))$ creates an electron (positron) with momentum $\boldsymbol{p}$ and polarisation s. $f(\boldsymbol{k}, \boldsymbol{p}) \cdot a^\dagger(\boldsymbol{k})$ stands for:

$$f(\boldsymbol{k}, \boldsymbol{p}) \cdot a^\dagger(\boldsymbol{k}) = \sum_r f^\mu(\boldsymbol{k}, \boldsymbol{p})\, \epsilon_{r\mu}^*(\boldsymbol{k})\, a_r^\dagger(\boldsymbol{k}) \qquad (17.21)$$

with $a_r^\dagger(\boldsymbol{k})$ the creation operator of a photon with momentum $\boldsymbol{k}$ and polarisation index r and $\epsilon_r^\mu(\boldsymbol{k})$, $r = 0, \ldots, 3$ the corresponding polarisation vector. We see that R_f is a linear superposition of photon creation operators, so the dressed state is a coherent state like the ones we introduced in Problem 2.3. The function $f^\mu(\boldsymbol{k}, \boldsymbol{p})$ is called *the dressing function* and its choice determines the content and the energy profile of the photon cloud, which surrounds the electron. Let us choose to describe free photons using the Lorenz gauge $\partial^\mu A_\mu = 0$. The physical photon states are the ones we introduced in section 10.4. They contain the transversely polarised photons, but also the combinations of longitudinal and scalar photons, which satisfy the Gupta-Bleuler condition (10.57). These last ones are zero norm states and we could drop them in this calculation. We recall that the cloud is supposed to describe the soft photons emitted from the moving electron through bremsstrahlung radiation. The latter are physical photons, so we choose $f^\mu(\boldsymbol{k}, \boldsymbol{p})$ to be transverse: $f \cdot k = 0$. Further constraints on f are imposed by various physical requirements: for example, it should vanish when the speed of the electron goes to zero because an electron at rest does not emit any bremsstrahlung radiation. Last but not least, it should be simple enough to make explicit calculations possible.

A simple choice satisfying these conditions is

$$f^\mu(\boldsymbol{k}, \boldsymbol{p}) = e\left(\frac{p^\mu}{p \cdot k} - c^\mu\right), \quad c^\mu = \left(\frac{1}{2k^0}, -\frac{\boldsymbol{k}}{2(k^0)^2}\right) \qquad (17.22)$$

c^μ is a null vector, $c^2 = 0$, satisfying $c \cdot k = 1$, thus ensuring the transversality of f. Notice that the dressing function $f^\mu(\boldsymbol{k}, \boldsymbol{p})$ is singular as the photon momentum $\boldsymbol{k}$ vanishes.

Therefore, the Faddeev-Kulish dressed one-electron state has the form

$$|\boldsymbol{p}, s\rangle_{\mathrm{d}} = |\boldsymbol{p}, s\rangle \times e^{-\int_{m_\gamma}^{\delta} \frac{\mathrm{d}^3\boldsymbol{k}}{(2\pi)^3} \frac{1}{2\omega_{\boldsymbol{k}}} \left(f(\boldsymbol{k},\boldsymbol{p}) \cdot a^\dagger(\boldsymbol{k}) - \text{h.c.}\right)} |0\rangle \qquad (17.23)$$

with

$$|\boldsymbol{p}, s\rangle = b_s^\dagger(\boldsymbol{p}) |0\rangle \qquad (17.24)$$

the Fock space one-electron state.[6] Thus, the charged particle is accompanied by a photon cloud described by a normalised coherent state. For finite non-zero m_γ, the coherent state can be rewritten in the following useful form:

$$|f_{\boldsymbol{p}}\rangle = \mathcal{N}_{\boldsymbol{p}}\ e^{-\int_{m_\gamma}^{\delta} \frac{d^3\boldsymbol{k}}{(2\pi)^3}\ \frac{1}{(2\omega_{\boldsymbol{k}})^{1/2}}\ f(\boldsymbol{k},\boldsymbol{p})\cdot a^\dagger(\boldsymbol{k})}|0\rangle \tag{17.25}$$

with the normalization factor $\mathcal{N}_{\boldsymbol{p}}$ given by

$$\mathcal{N}_{\boldsymbol{p}} = e^{\frac{1}{2}\int_{m_\gamma}^{\delta} \frac{d^3\boldsymbol{q}}{(2\pi)^3}\ \frac{1}{2\omega_{\boldsymbol{q}}}\ f^\mu(\boldsymbol{q},\boldsymbol{p})f_\mu(\boldsymbol{q},\boldsymbol{p})} \tag{17.26}$$

The exponent can be easily computed

$$\frac{1}{2}\int_{m_\gamma}^{\delta} \frac{d^3\boldsymbol{q}}{(2\pi)^3}\ \frac{1}{2\omega_{\boldsymbol{q}}}\ f^\mu(\boldsymbol{q},\,\boldsymbol{p})f_\mu(\boldsymbol{q},\,\boldsymbol{p}) = -\frac{\alpha}{2\pi}\ \ln\left(\frac{\delta}{m_\gamma}\right) I(v) \tag{17.27}$$

where $v = |\boldsymbol{p}|/p^0$ is the speed of the electron and

$$I(v) = -2 + v^{-1}\ln\left(\frac{1+v}{1-v}\right) \tag{17.28}$$

is a non negative kinematical factor. In particular, for small v, $I(v) = 2v^2/3 + \dots$ and vanishes at $v = 0$. As $v \to 1$, $I(v)$ diverges logarithmically. Setting

$$\mathcal{A}_{\boldsymbol{p}} = \frac{\alpha}{2\pi}\ I(v) \tag{17.29}$$

we get

$$\mathcal{N}_{\boldsymbol{p}} = \left(\frac{m_\gamma}{\delta}\right)^{\mathcal{A}_{\boldsymbol{p}}} \tag{17.30}$$

and so $\mathcal{N}_{\boldsymbol{p}}$ vanishes in the limit $m_\gamma \to 0$ (in which case (17.25) cannot be used).

Let us compute the number of transverse photons in this state. The photon number operator is $\hat{N}_{\text{photons}} = \sum_{l=1}^{2}\int d\Omega_0(q) a^{l\,\dagger}(\boldsymbol{q})\, a^l(\boldsymbol{q})$. Using the results we obtained in Problem 2.3, we get:

$$\langle f_{\boldsymbol{p}}|N_{ph}|f_{\boldsymbol{p}}\rangle = -\int_{m_\gamma}^{\delta} \frac{d^3\boldsymbol{q}}{(2\pi)^3}\ \frac{1}{2\omega_{\boldsymbol{q}}}\ f^\mu(\boldsymbol{q},\,\boldsymbol{p})f_\mu(\boldsymbol{q},\,\boldsymbol{p}) = \frac{\alpha}{\pi}\ln\left(\frac{\delta}{m_\gamma}\right) I(v) \tag{17.31}$$

Thus the cloud contains an infinite number of soft photons in the limit $m_\gamma \to 0$. On the other hand, the energy of the state is given by

$$\langle f_{\boldsymbol{p}}|H_{ph}|f_{\boldsymbol{p}}\rangle = -\frac{1}{2}\int_{m_\gamma}^{\delta} \frac{d^3\boldsymbol{q}}{(2\pi)^3}\ f^\mu(\boldsymbol{q},\,\boldsymbol{p})f_\mu(\boldsymbol{q},\,\boldsymbol{p}) = \frac{\alpha}{\pi}\ I(v)\ (\delta - m_\gamma) \tag{17.32}$$

For generic values of the electron velocity, this is a small fraction of the infrared scale δ.

---

[6]Strictly speaking the momenta $\boldsymbol{p}$ in the dressed and the Fock space states are not the same because the former includes also the momenta carried by the photons in the cloud. However, since by construction, their difference is smaller than δ and not measurable, we will use the same symbol for both.

The mean value of the cloud momentum is also interesting. It is given by

$$
\begin{aligned}
\langle f_{\boldsymbol{p}} | \boldsymbol{P}_{ph} | f_{\boldsymbol{p}} \rangle &= - \int_{m_\gamma}^{\delta} \frac{d^3 q}{(2\pi)^3} \frac{\boldsymbol{q}}{2\omega_q} \, f^\mu(\boldsymbol{q}, \boldsymbol{p}) f_\mu(\boldsymbol{q}, \boldsymbol{p}) \\
&= \frac{\alpha}{2\pi} (\delta - m_\gamma) \left[\frac{3}{v} I(v) - v I(v) - 2v \right] \hat{\boldsymbol{p}} \qquad (17.33)
\end{aligned}
$$

As the electron velocity approaches the speed of light, the energy and the magnitude of the cloud momentum grow logarithmically and become equal. Notice that both the energy and the momentum remain appreciably smaller than the energy and the momentum of the electron. As $v \to 0$, they become vanishingly small, albeit the momentum approaches zero faster.

Using the general equation (17.20) we can construct multi-electron dressed states, but we will never use them in this book.

17.5.2 Coulomb scattering of a dressed electron

In the Faddeev - Kulish formalism the measured cross-section is obtained from the scattering amplitude of dressed states. Let us apply this formalism to derive the Coulomb scattering amplitude of a dressed electron. We shall show that this is finite, free of infrared divergences and gives at order $\mathcal{O}(\alpha^3)$ the cross section we derived in the previous section. For that purpose we need the initial and final dressed electron states up to $\mathcal{O}(e)$, which according to (17.23) and (17.25) are

$$
|\boldsymbol{p}, s\rangle_{\mathrm{d}} = |\boldsymbol{p}, s\rangle \times \mathcal{N}_{\boldsymbol{p}} \left(1 - \int_{m_\gamma}^{\delta} \frac{d^3 k}{(2\pi)^3} \frac{1}{(2\omega_{\boldsymbol{k}})^{1/2}} \, f(\boldsymbol{k}, \boldsymbol{p}) \cdot a^\dagger(\boldsymbol{k}) + \dots \right) |0\rangle \qquad (17.34)
$$

for the initial electron, and similarly for the final one. The S-matrix element between the incoming/outgoing dressed states is then:

$$
\widetilde{S}_C(p \to p') = \mathcal{N}_{p'} \mathcal{N}_p \widetilde{M}_C \qquad (17.35)
$$

with

$$
\widetilde{M}_C = \langle \boldsymbol{p}', s' | \left(1 - \int_{m_\gamma}^{\delta} \frac{d^3 q}{(2\pi)^3} \frac{f_{p'}(\boldsymbol{q}) \cdot a(\boldsymbol{q})}{(2\omega_q)^{1/2}} \right) S \left(1 - \int_{m_\gamma}^{\delta} \frac{d^3 k}{(2\pi)^3} \frac{f_p(\boldsymbol{k}) \cdot a^\dagger(\boldsymbol{k})}{(2\omega_{\boldsymbol{k}})^{1/2}} \right) |\boldsymbol{p}, s\rangle
$$
$$
(17.36)
$$

Let us for the moment ignore the factor $\mathcal{N}_{p'} \mathcal{N}_p$ in $\widetilde{S}_C$ and focus on $\widetilde{M}_C$. This matrix element is the sum of four terms.

The simplest is the elastic Coulomb scattering amplitude of the bare electron, i.e.

$$
\widetilde{M}_C^{(0)} = \langle \boldsymbol{p}', s' | S | \boldsymbol{p}, s \rangle \equiv S_C(p \to p')
$$

Next, we have the two terms linear in the photon creation or annihilation operator. For the one linear in $a_r(\boldsymbol{q})$ we observe that $(2\omega_q)^{1/2} \langle \boldsymbol{p}', s' | a_r(\boldsymbol{q}) S | \boldsymbol{p}, s \rangle = \langle \boldsymbol{p}', s'; \boldsymbol{q}, r | S | \boldsymbol{p}, s \rangle$ is the amplitude for emission of a soft photon $(\boldsymbol{q}, r)$, which according to the *leading order soft theorem* (17.10), is related to the elastic Coulomb scattering amplitude $S_C(p \to p')$ by

$$\lim_{|q| \to 0} (2\omega_q)^{1/2} \langle p', s' | a_r(q) S | p, s \rangle = S_C(p \to p') e \left(\frac{p' \cdot \epsilon_r^*(q)}{p' \cdot q} - \frac{p \cdot \epsilon_r^*(q)}{p \cdot q} \right) \quad (17.37)$$

Similarly, for the term linear in the photon creation operator we have

$$\lim_{|k| \to 0} (2\omega_k)^{1/2} \langle p', s' | S a_r^\dagger(k) | p, s \rangle = -S_C(p \to p') e \left(\frac{p' \cdot \epsilon_r(k)}{p' \cdot k} - \frac{p \cdot \epsilon_r(k)}{p \cdot k} \right) \quad (17.38)$$

since the left hand side is the amplitude of absorption of a soft photon from the incoming photon cloud, either by the initial or by the final electron.

We use (17.37) and sum over the photon polarisations to obtain

$$\lim_{|q| \to 0} (2\omega_q)^{1/2} \sum_r \langle p', s' | f^\mu(p', q) \epsilon_{r\mu}(q) a_r(q) S | p, s \rangle$$

$$= S_C(p \to p') e f^\mu(p', q) \left(\frac{p'^\nu}{p' \cdot q} - \frac{p^\nu}{p \cdot q} \right) \sum_r \epsilon_{r\mu}(q) \epsilon_{r\nu}^*(q)$$

$$= -S_C(p \to p') f^\mu(p', q) (f_\mu(p', q) - f_\mu(p, q))$$

and similarly for the term of (17.36) containing the matrix element of $S f_p(k) \cdot a^\dagger(k)$. In summing over polarisations we used the relations: $\sum_{r=1}^2 \epsilon_{r\mu}(q) \epsilon_{r\nu}^*(q) = -\eta_{\mu\nu} + q_\mu c_\nu + q_\nu c_\mu$ and $f(p, q) \cdot q = 0 = f(p', q) \cdot q$, valid for the physical transverse photons. Taken together, the total contribution of the two terms linear in $a_r(q)$ or $a_r^\dagger(q)$ in the matrix element under discussion is

$$\widetilde{M}_C^{(1)} = -S_C(p \to p') \int_{m_\gamma}^\delta \frac{d^3 q}{(2\pi)^3 2\omega_q} (f_\mu(p', q) - f_\mu(p, q))^2 \quad (17.39)$$

This term is of $\mathcal{O}(\alpha)$.

Finally, there is an additional $\mathcal{O}(\alpha)$ contribution coming from a soft photon of the incoming cloud, which goes through to the outgoing cloud without interaction with the initial or the final electron. It is obtained from the terms of the form $(4\omega_q \omega_k)^{1/2} \langle p', s' | a_r(q) S a_z^\dagger(k) | p, s \rangle = \langle p', s'; q, r | S | p, s; k, z \rangle$ with $r = z$ and $q = k$, which gives

$$\langle p', s'; q, r | S | p, s; k, z \rangle = \langle p', s' | S | p, s \rangle \langle q, r | k, z \rangle = \delta^{rz} (2\pi)^3 2\omega_q \delta^{(3)}(q - k) S_C(p \to p') \quad (17.40)$$

and contributes to our matrix element the amount

$$\widetilde{M}_C^{(2)} = -S_C(p \to p') \int_{m_\gamma}^\delta \frac{d^3 q}{(2\pi)^3 2\omega_q} f_\mu(p', q) f^\mu(p, q) \quad (17.41)$$

We substitute all the above into (17.36) and take into account that, according to (17.26) to the order of interest here,

$$\mathcal{N}_p \simeq 1 + \frac{1}{2} \int_{m_\gamma}^\delta \frac{d^3 q}{(2\pi)^3 2\omega_q} f_\mu(p, q) f^\mu(p, q) \quad (17.42)$$

to write the Coulomb scattering amplitude of the dressed electron in the form

$$\widetilde{S}_C(p \to p') \simeq S_C(p \to p') \left(1 - \frac{1}{2} \int_{m_\gamma}^{\delta} \frac{d^3q}{(2\pi)^3 2\omega_q} \left(f_\mu(p', q) - f_\mu(p, q) \right)^2 \right) \qquad (17.43)$$

The integral inside the parenthesis was computed in (17.10) − (17.14). Substituting the result obtained there, we have

$$\widetilde{S}_C(p \to p') \simeq S_C(p \to p') \left(1 + \frac{\alpha}{2\pi} \ln \left(\frac{\delta^2}{m_\gamma{}^2} \right) \mathcal{J}(\kappa^2) \right) \qquad (17.44)$$

But, according to (17.6) and (17.7), the Coulomb scattering amplitude of the bare electron up to one-loop is given by

$$S_C(p \to p') \simeq S_C^{(0)} \left(1 + F^{(c)}(q^2) \right) = S_C^{(0)}(p \to p') \left(1 + \frac{\alpha}{2\pi} \ln \left(\frac{m_\gamma^2}{m^2} \right) \mathcal{J}(\kappa^2) \right)$$
$$(17.45)$$

and upon substitution in (17.44) our final result for the Coulomb scattering amplitude of dressed electrons is

$$\widetilde{S}_C(p \to p') \simeq S_C^{(0)}(p \to p') \left(1 - \frac{\alpha}{2\pi} \ln \left(\frac{m^2}{\delta^2} \right) \mathcal{J}(\kappa^2) \right) \qquad (17.46)$$

As promised, the Coulomb scattering amplitude of the dressed electron is free of infrared divergences and the corresponding cross section is in agreement with (17.15).

It is straightforward to repeat the calculation to all orders of perturbation theory and derive all the results of section 17.4 in this formalism.

17.6   Problems

Problem 17.1 Verify the tree-level expression (17.1) for the differential cross section of electron Coulomb scattering.

Problem 17.2 Compute the amplitude for an electron of momentum p to emit two soft photons with momenta q_1 and q_2.

Problem 17.3 Compute the energy emitted in soft photons with energy smaller than the threshold δ of the measuring apparatus during the Coulomb scattering of an electron.

18
The Weak Interactions

18.1 Introduction

The years which followed immediately after the Shelter Island conference were years of great enthusiasm. It was expected that the extension of the renormalisation programme would yield equally successful results for the other two interactions among elementary particles, to wit the strong interactions, which at that time were essentially the pion–nucleon interactions, and the weak interactions, which were described by the Fermi theory. Alas, the expectations soon proved to be illusory for both of them, although for different reasons.

As we have shown in Problem 15.2, the renormalisation programme was shown to be formally applicable to the Yukawa interactions of equations (6.14) or (6.15), but the results were useless for practical computations. The reason is a big difference in the strength of strong interactions as compared to the electromagnetic ones. The perturbation expansion is a formal power series in a dimensionless parameter, the renormalised coupling constant. Its numerical value is determined by experiment. For quantum electrodynamics, this value turns out to be roughly equal to 1/137, the celebrated fine structure constant α. The smallness of this number is responsible for the success of the theory, since higher order terms soon become negligible and a good estimate of the result can be obtained by keeping only the first few. The corresponding number for pion–nucleon interactions is instead on the order of 10 (see Problem 21.1) and the very idea of a power series expansion becomes meaningless. This is, of course, the reason why strong interactions are strong! The dynamics is often dominated by resonance production, a phenomenon completely outside the scope of perturbation theory. For a few years this caused considerable confusion among theorists, which was soon resolved by abandoning field theory techniques in strong interactions.

What about the weak interactions? Since, experimentally, they are much weaker than the electromagnetic ones, the previous objection did not apply. We remind the reader that the coupling constant in the Fermi theory is of order $10^{-5}/M_{\mathrm{p}}^2$ –see equation (6.17) –and this made the interactions very weak at the energies available at that time. However, a new theoretical problem appeared there: in Problem 15.2 we show that Fermi's current×current theory is non-renormalisable, which means that the theory is in fact lacking any predictive power. This double failure soon tarnished the glory of quantum electrodynamics. The general disappointment was such that renormalisation theory became an essentially esoteric topic in particle physics. The subject was not even taught in many universities. It took more than two decades before it was resuscitated in the form of non-Abelian gauge theories. In the process many

Elementary Particle Physics. John Iliopoulos and Theodore N. Tomaras, Oxford University Press.
© John Iliopoulos and Theodore N. Tomaras (2021). DOI: 10.1093/oso/9780192844200.003.0018

fundamental physical concepts were introduced, which we will describe in this chapter. The final form, which is a gauge theory based on the group $SU(2) \times U(1)$ incorporating the BEH mechanism of spontaneous symmetry breaking, will be presented in Chapter 19.

18.2   The Fermi Theory

We presented in section 6.6 the Fermi theory for β-decay. The early versions involved the point interaction among four fermions which, in general, involves the five types of bilinears we found in Problem 7.3. It took many years and a great experimental and theoretical effort to establish the particular form we know today. This achievement went through the introduction of new fundamental concepts and led to many important discoveries. It would make for a passionate story, but history is not the main purpose of this book. We will only sketch the most essential steps, not always in the right historical order.

The most general form of the Fermi theory is given in equation (6.18)

$$\mathcal{H}_I = \frac{G_F}{\sqrt{2}} \sum_{i=1}^{5} C_i \overline{\Psi}_p(x) O_i \Psi_n(x) \, \overline{\psi}_e(x) O_i \psi_\nu(x) \tag{18.1}$$

where the index i runs over the five bilinears we found in the Dirac theory, which we called S for "scalar" (with the corresponding matrix $O_S = \mathbf{1}$), P for "pseudo-scalar" (with the matrix $O_P = \gamma_5$), V for "vector" (with $O_V = \gamma_\mu$), A for "axial vector" (with $O_A = \gamma_\mu \gamma_5$), and T for "tensor" (with $O_T = \sigma_{\mu\nu} = -\frac{i}{4}[\gamma_\mu, \gamma_\nu]$).[1] Experiments were supposed to determine the values of the numerical coefficients C_i and, in particular, to find out if all five were needed, or if some could be put equal to zero. In the early days data were coming from β-decays of various nuclei of the form

$$N_i \rightarrow N_f + e^- + \bar{\nu}(\text{or, } e^+ + \nu) \tag{18.2}$$

The recoil of the final nucleus is very small, so we expect the nucleon part to be well described by a non-relativistic approximation. Keeping only the large components of the spinors we introduced in Chapter 8, we can find the non-relativistic limits of the five bilinears, see Problem 18.1. The results are shown in Table 18.1 so we see that, in the non-relativistic limit, we have two possible forms, either $\mathbf{1}$ for S and V, or $\boldsymbol{\sigma}$ for A and T. The possible presence of a pseudo-scalar term is not detectable in the non-relativistic approximation. All terms, which vanish in this limit, contribute to processes which are expected to have much smaller probabilities and are called *forbidden*. The transitions corresponding to the matrix $\mathbf{1}$ are called *Fermi transitions* and those corresponding to the matrix $\boldsymbol{\sigma}$ *Gamow–Teller transitions*. They are easily distinguishable because they obey different selection rules. Since the momentum transfer to the leptons is very small, the total angular momentum carried by the lepton pair is $J = 0$ for a Fermi transition and $J = 1$ for a Gamow–Teller one. If S_i and S_f are the initial and final nuclear spins, we define $\Delta S = |S_i - S_f|$ and we get:

---

[1]Before the discovery of parity violation, superpositions such as V and A, or S and P, were not considered.

Table 18.1 The five Dirac bilinears, their non-relativistic limits and their contribution to β-decay transitions

Operator	N-R limit	Transition
1	1	Fermi
$i\gamma_5$	0	—
γ_μ	$\gamma_0 \to 1$	Fermi
	$\gamma_i \to 0$	—
$\gamma_\mu \gamma_5$	$\gamma_i \gamma_5 \to \sigma_i$	Gamow–Teller
	$\gamma_0 \gamma_5 \to 0$	—
$\frac{-i}{4}[\gamma_\mu, \gamma_\nu]$	$[\gamma_i, \gamma_j] \to \epsilon_{ijk}\sigma_k$	Gamow–Teller
	$[\gamma_i, \gamma_0] \to 0$	—

- *Fermi transitions:* $\Delta S = 0$, no parity change.
- *Gamow–Teller transitions:* $\Delta S = 1$ or 0, but no $0 \to 0$; no parity change.

Experimentally both kinds of transitions have been observed: for example, the $O^{14} \to N^{14} + e^+ + \nu$ is a pure Fermi transition ($0 \to 0$), while the $He^6 \to Li^6 + e^- + \bar{\nu}$, or the $N^{12} \to C^{12} + e^+ + \nu$ are pure Gamow–Teller ones ($\Delta S = 1$). The decay of a free neutron is a mixture of the two. These results are given in Table 18.1. So, the existence of both Fermi and Gamow–Teller transitions can be translated for the coefficients C_i into: C_S and/or C_V is non-zero, *and* C_A and/or C_T is non-zero. Using only the non-relativistic results we cannot say anything about C_P.

In order to go further, we look at the leptonic part. In Problem 18.2 we ask the reader to prove the following statement: S and T give *the same* helicity for e^- and $\bar{\nu}$ in β-decay, while V and A give opposite helicities. Experimentally it was determined that fast β-decay electrons were maximally polarised with negative helicity in both Fermi and Gamow–Teller transitions. So, the final choice could be made by looking at the polarisation of the neutrino.[2] But before going into it, we present a revolutionary result coming from β-decay, namely the discovery of parity violation.

18.3 Parity Violation

Classical physics is invariant under space inversion, or parity: $\boldsymbol{x} \to -\boldsymbol{x}$. The fact that opposite parity energy levels in both atomic and nuclear physics do not mix, shows that this invariance remains true in quantum physics, at least as far as the strong and the electromagnetic interactions are concerned. For many years physicists were convinced that parity conservation was an exact law of nature.

The first hint against this belief came from the decay properties of the strange mesons, which in section 6.10 we called kaons. It was found experimentally that kaons have spin equal to 0 and they decay through weak interactions. For the charged kaons the main decay modes are leptonic $K^\pm \to \mu^\pm + \nu_\mu$ with the neutrino being ν or $\bar{\nu}$ depending on the lepton charge. A secondary decay mode is hadronic with either two or three pions:[3] $K^\pm \to \pi^\pm + \pi^0$, or $K^\pm \to \pi^\pm + \pi^+ + \pi^-$ and $K^\pm \to \pi^\pm + \pi^0 + \pi^0$. The

---

[2]We could also look at the $e^- - \bar{\nu}$ angular correlation in β-decay, see Problem 18.3, or the pion decay, see section 12.5.

[3]Energy conservation does not allow for final states with more than three pions.

spin and parity J^P of the pions were determined to be 0^-. It follows that the state of two pions coming from the decay of a spin-0 boson has positive parity. On the other hand, a careful analysis of the momentum distribution of the three pion state showed that all pairs of pions were predominantly in relative s states.[4] Therefore the parity of the three pion state is negative. Since parity was assumed to be conserved, people introduced two kaons with opposite parity. They were called $\theta^\pm$ with $J^P = 0^+$ and $\tau^\pm$ with 0^-. The first was supposed to decay into two pions and the second into three. The trouble was that in all other respects these two particles appeared to have identical properties: same mass, same lifetimes, same production cross sections, same leptonic decays. No other example of such *parity doubling* was known in particle physics. This was the famous *τ–θ puzzle.*

Today parity violation as a solution to the puzzle sounds as an obvious possibility, but it was not so in 1956. Although discussions in this direction did occur, the first serious study was the paper by two young Chinese physicists, T.D. Lee and C.N. Yang. Their great merit was not to "predict" parity violation,[5] but to present instead a clear scientific analysis of the situation. Their results were two-fold: First, they examined all available experimental data and concluded that there existed no evidence for parity conservation in weak interactions. Second, they proposed possible experiments to settle the issue.

Before the discovery of strange particles, weak interaction phenomena were manifest mainly in nuclear β-decay, but also in the decays of muons and pions. In all cases the measurements were limited to the energy of the final charged lepton. Let us look at β-decay. If $\boldsymbol{P}$ is the momentum of the initial nucleus and $\boldsymbol{k}_e$ that of the final electron, the decay spectrum in the rest frame of the nucleus is given by the mean value of the observable $< \boldsymbol{P} \cdot \boldsymbol{k}_e >$. This, being a scalar quantity, is insensitive to the possible presence of parity violating terms in the interaction Hamiltonian. Lee and Yang observed that, in order to detect such terms, we should measure a non-zero value of a *pseudo-scalar* quantity and, as such, they proposed $< \boldsymbol{S} \cdot \boldsymbol{k}_e >$, where $\boldsymbol{S}$ is a spin variable which is a pseudo-vector. Therefore, they proposed to measure the distribution of electrons coming from the β-decay of a polarised nucleus. For the latter, they suggested Co^{60}, which was known to be unstable under β-decay and, furthermore, it

---

[4]In the 1953 Bagnères-de-Bigorre conference R.H. Dalitz proposed a simple method to analyse the kinematics of three-particle decay processes. It is based on the following observation: In the rest frame of the decay particle the three-momenta of the final particles lie on a plane and their sum vanishes. Therefore, each point in the phase space can be represented by a point inside an equilateral triangle, (see Problem 18.7) with the distances of the point from the three sides being the energies of the three final pions. Equal areas in the triangle correspond to equal volumes in the phase space. The amplitude depends on two variables, (which can be chosen to be two angles among two pairs of momenta) and can be expanded in the corresponding Legendre polynomials. It follows that, from the distribution of the points in the triangle, we can infer the spin of the decaying particle as well as the relative angular momenta among pairs of the final particles. This is the *Dalitz plot*. The method was first applied to the decay $K \rightarrow 3\pi$. In 1953 Dalitz had 13 points in his plot, all coming from cosmic rays. At the 1955 Rochester Conference he had 53, 42 still coming from cosmic rays. The following year he had over 600, practically all coming from accelerators. As the plot thickened, the evidence for a uniform distribution became stronger, which implied zero spin for the decaying particle and zero relative angular momenta among the final particles.

[5]A. Pais reports that, in the train going back to Princeton from the 1956 Rochester Conference, he *and* C.N. Yang made each a one-dollar bet with J.A. Wheeler claiming that τ and θ were two distinct particles. As a result, the following year Wheeler won two dollars.

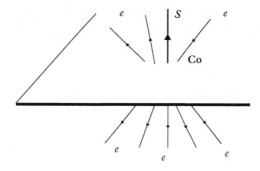

Fig. 18.1 The electrons from β-decay of polarised Co^{60} are emitted preferentially in the direction opposite to that of the nuclear spin.

has $S \neq 0$ and could be polarised. The first experimentalist to meet the challenge was C.S. Wu from Columbia University. The principle of the experiment is shown in Figure 18.1. In the rest frame of the polarised cobalt nucleus let us choose the z-axis parallel to the spin direction. The experiment consists in measuring a quantity named *up-down asymmetry* $\mathcal{A}$ defined as

$$\mathcal{A} = \frac{N^{(-)} - N^{(+)}}{N^{(-)} + N^{(+)}} \tag{18.3}$$

where $N^{(\pm)}$ denote the number of electrons emitted above, or below, the $x - y$ plane, respectively. It is clear that $\mathcal{A} \neq 0$ implies a non-zero value for $< \boldsymbol{S} \cdot \boldsymbol{k}_e >$, with $\boldsymbol{S}$ the spin of the initial cobalt nucleus. The spin and parity of Co^{60} is $J^P = 5^+$ and decays to a 4^+ excited state of Ni, which in turn decays into the ground state 0^+ with the emission of gamma rays. So, the overall decay scheme is $_{27}$Co$^{60} \rightarrow {}_{28}Ni^{60} + e^- + \bar{\nu} + 2\gamma$.

Although in principle the experiment sounds simple, in practice it was very complicated, especially with the 1956 technology. The first problem is due to the technical difficulty in obtaining an adequate degree of polarisation because of the small values of nuclear magnetic moments. The sample, consisting of a cobalt enriched cerium-magnesium nitrate, should be cooled to very low temperature, otherwise thermal fluctuations would wash out any polarisation. The cooling procedure is known as *the Gorter–Rose method* and is based on the phenomenon of adiabatic demagnetisation.[6] Columbia University did not have the required cryogenic facilities and C.S. Wu had to move her apparatus to Maryland, the headquarters of the National Bureau of Standards. The second problem was to obtain the desired nuclear polarisation. A vertical solenoid was introduced to the cryostat and switched on to align the cobalt spins in the positive, or negative, direction. This method worked because the salt of cerium-magnesium nitrate carrying the cobalt nuclei has a very anisotropic Landé factor, very small in the vertical direction to prevent appreciable re-heating and very large in the

---

[6]A magnetic material is cooled down to liquid helium temperature in the presence of a magnetic field. Then the field is switched off and this results into the formation of magnetic domains which get disoriented. If the material is kept thermally isolated this process absorbs thermal energy and the temperature drops. In Wu's experiment it reached 0.003 K.

horizontal one to allow for an efficient cooling. The resulting polarisation was moni-
tored by measuring the anisotropy of the emitted gamma rays, which were detected
using NaI scintillation counters. There was one counter in the near vertical direction
and one in the horizontal plane. Photon emission is due to electromagnetic interactions, which are known to conserve parity, so any difference in the counts of these
two detectors was a signature for a non-zero polarisation since, in the absence of it,
the distribution is expected to be isotropic. As we see in Figure 18.2, the polarisation
remained substantial for a period on the order of 5 minutes, after which it fell to zero
due to the rise in temperature. The third problem had to do with the detection of the
β-decay electrons. The counter should be placed inside the demagnetisation cryostat
and the scintillation photons should be transmitted to an outside photomultiplier.
The experiment was repeated several times often reversing the sign of the external
field. The final result is shown in Figure 18.2. For large times, when polarisation has
dropped to zero, the up-down asymmetry $\mathcal{A}$ vanishes, as expected. This shows that
there are no systematic biases. On the other hand, for non-zero polarisation, the electrons are emitted preferentially in the direction opposite to that of the nuclear spin,
thus indicating parity violation.

The article by Wu and her collaborators E. Ambler, R.W. Hayward, D.D. Hoppes
and R.P. Hudson, from the National Bureau of Standards, was published as a Letter
to the Editor in the Physical Review with submission date January 15, 1957. With
practically the same submission date, two more papers appeared also as Letters to
the Editor, reporting results on parity violation. The first was by R.L. Garwin, L.M.
Lederman and M. Weinrich from Columbia University and the second by J.I. Friedman
and V.L. Telegdi from Chicago. They both studied the decay chain

$$\pi^+ \to \mu^+ + \nu_\mu \text{ followed by } \mu^+ \to e^+ + \nu_e + \bar{\nu}_\mu \tag{18.4}$$

As Lee and Yang had pointed out, a sign of parity violation would be a non-zero
polarisation of the muons produced by the decay of pions at rest, and this could be
detected as a forward-backward asymmetry relative to the muon polarisation in the
distribution of electrons, see Problem 18.5. This is precisely what was measured. For
this work T.D. Lee and C.N. Yang were awarded the 1957 Physics Nobel Prize.

In the months which followed, several other measurements showing parity violation
in weak interactions were reported, so by the end of 1957 the phenomenon was clearly
and indisputably established. It was clearly a great shock[7] although, in retrospect,
some early signs had already appeared. In 1928 three experimentalists from New York[8]
attempted to measure a possible polarisation in an electron beam produced by β-decay.
Using today's language they found that the electrons were predominantly left-handed
and this effect was manifest mainly for the fastest electrons. In discussing their results
they did not make any connection with parity violation and nobody seems to have
paid any attention to it.

---

[7]It seems that when Pauli first heard the results of the Columbia–National Bureau of Standards
experiment, he burst into a shout "that's total nonsense".

[8]R.T. Cox, C.G. McIlwraith and B. Kurrelmeyer, "Apparent evidence of polarization in a beam of
β-rays", Proc. N.A.S. **14** 544, (1928)

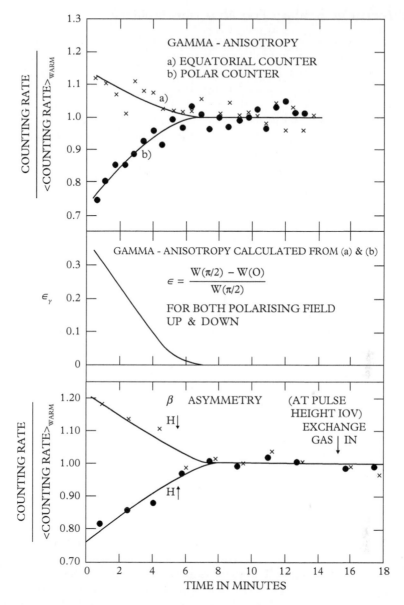

Fig. 18.2 The measurements which established parity violation. The upper two curves show the anisotropy in the γ-ray emission which implies nuclear polarisation. As expected, it changes sign with the applied field and remains non-zero for about five minutes. The lower curve shows that the anisotropy in the emitted electrons is opposite to the magnetic field. The vertical axis gives the ratio of the counting rates with and without polarisation (marked "warm" in the figure) for the two directions of the applied field. The ratio goes to one when the nuclear polarisation vanishes, implying that the electron anisotropy goes to zero. *(Source: C.S. Wu et al. Phys. Rev.* **105**, *1413 (1957)).*

18.4   Vector vs Scalar: The Neutrino Helicity

The discovery of parity violation showed that in the Fermi Hamiltonian of equation (18.1) we could combine terms with opposite parity, such as V and A or S and P. We also noted that the final choice could be made by a measurement of neutrino helicity.

In normal β-decay experiments the neutrinos go undetected. So, how can we measure the helicity of a particle we do not see? It sounds like an impossible task. Yet such a measurement was performed by M. Goldhaber, L. Grodzins and A.W. Sunyar in 1957. It is one of the most subtle experiments of weak interactions and we will describe here a simplified version.

They studied electron capture in $_{63}\mathrm{Eu}^{152}$ according to the scheme

$$_{63}\mathrm{Eu}^{152} + e^- \rightarrow _{62}\mathrm{Sm}^{152*} + \nu \tag{18.5}$$

The europium nucleus has spin and parity 0^- and the electron capture yields a nucleus of samarium in an excited state with spin-parity 1^-, i.e. (18.5) is a Gamow–Teller transition. In turn, $_{62}\mathrm{Sm}^{152*}$ decays into the ground state $_{62}\mathrm{Sm}^{152}$ (spin-parity 0^+) with the emission of a photon

$$_{62}\mathrm{Sm}^{152*} \rightarrow _{62}\mathrm{Sm}^{152} + \gamma \tag{18.6}$$

Several points make the measurement of the neutrino helicity possible using this chain of reactions:

• The electron capture occurs predominantly from an s-state. Indeed, we can estimate the probability for an electron to be captured directly from an outer shell and compare it with the probability of the electron to first fall to an s-state and then be captured. The first is much smaller.

• The neutrino energy equals 840 keV, which means that the recoil of the samarium nucleus is appreciable.

• The lifetime of $_{62}\mathrm{Sm}^{152*}$ is very short, of order 10^{-14} s. This means that it decays before it has time to stop and, as a result, we can select the photons in the direction of $_{62}\mathrm{Sm}^{152*}$. This selection is achieved by nuclear resonant scattering of the photons on $_{62}\mathrm{Sm}^{152}$ nuclei at rest.

In the decaying europium nucleus rest frame, the angular momentum analysis of the chain of reactions (18.5) and (18.6) is given in Figure 18.3 for the two possibilities of the neutrino helicity, left or right. The analysis is based on the fact that, for photons in the direction of motion of samarium, the entire process is collinear and the projection of the total angular momentum on the direction of motion is entirely due to the projection of the spins. The initial capture occurs from an s-state, therefore we have that the total angular momentum $J_{\mathrm{in}} = \frac{1}{2} = J_{\mathrm{fin}}$. Let us first assume that the neutrino is left-handed. Then, with the direction of the z-axis chosen to be that of the neutrino momentum, we have $m_z(\text{neutrino}) = -\frac{1}{2}$. This means that the possible values of the projection of the Sm^* spin are $m_z(\mathrm{Sm}^*) = 1, 0$. Therefore, since the photons are collinear and the spin of the ground state of samarium is zero, $m_z(\gamma) = 1$, i.e. the helicity of the photon is negative. The same reasoning, applied to the case of a right-handed neutrino, yields a positive helicity photon. We conclude that in both cases the neutrino and the photon

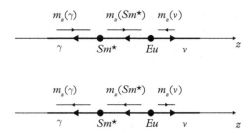

Fig. 18.3 The balance of J_z in the collinear process $\text{Eu}+e^- \to \nu+\text{Sm}+\gamma$. It shows that the helicities of the neutrino and the photon always have the same sign.

have same sign helicities. This was the magic: the helicity of the elusive neutrino can be measured by measuring that of a γ-ray. The latter can be measured by scattering in the presence of a magnetic field. The experiment gave a negative sign for the photon helicity, therefore, *the neutrino of β-decay is left-handed.* Going back to the analysis of section 18.2 we conclude that out of the five possible terms of the Fermi Hamiltonian (18.1), we are left with a superposition of vector V and axial vector A. Furthermore, the fact that we observe both Fermi and Gamow–Teller transitions, tells us that both are present.

18.5   The *V − A* Theory

By the late 1950s the Fermi theory of the weak interactions had taken its final form. It involves an operator $J_\lambda(x)$, the "weak current", which is the analogue of the familiar electromagnetic current. In terms of this, we build an effective Lagrangian density of the form

$$\mathcal{L}_{\text{weak}} = \frac{G_F}{\sqrt{2}} J^\lambda(x) J_\lambda^\dagger(x) \tag{18.7}$$

where † means hermitian conjugation.

This is the famous current×current theory, and G_F is the Fermi coupling constant, equal to $10^{-5} m_{\text{proton}}^{-2}$. The weak current $J_\lambda(x)$ is the sum of two parts, a leptonic part $l_\lambda(x)$ and a hadronic one $h_\lambda(x)$,

$$J_\lambda(x) = l_\lambda(x) + h_\lambda(x) \tag{18.8}$$

They are both of the *V-A* form, i.e. they can be written as a particular superposition of a vector and an axial vector part. These expressions are only formal because, in order to give a precise meaning to the current operators, we must first define the Hilbert space of the physical states in which these operators act. The total Hamiltonian can be written as a sum of the form

$$H = H_{\text{kinetic}} + H_{\text{strong}} + H_{\text{em}} + H_{\text{weak}} \tag{18.9}$$

where the various terms represent the kinetic energy and mass terms, the strong, the electromagnetic and the weak interactions, respectively. In order to write H_{kinetic} in terms of free fields we must make an assumption regarding the particles we may consider as "elementary". We will distinguish two cases: For the particles that have no strong interactions, such as leptons, photons, etc., the natural choice is to assign a field to each known particle, to wit the electron, the muon, the tau and their associated neutrinos as well as the photon and the intermediate bosons of the weak interactions. For the hadrons the corresponding assumption in Fermi's times was to take the proton and the neutron, but today there is a general consensus in favour of quarks and gluons. Ideally, the states of the Hilbert space we are looking for should be the eigenstates of H, but we can only find them in a perturbation expansion. Therefore, we need to split the Hamiltonian into an unperturbed part H_0 and a perturbation H_P. In all previous applications we have chosen H_0 to be H_{kinetic} and treated the entire interaction Hamiltonian as a perturbation. Such an assumption, however, does not make physical sense for the strong interactions, because we know of no perturbation expansion which would produce the observed hadron states, protons, neutrons, pions etc., starting from free quarks and gluons. So we will adopt a phenomenological strategy: we choose

$$H = H_0 + H_P; \quad H_0 = H_{\text{kinetic}} + H_{\text{strong}}, \quad H_P = H_{\text{em}} + H_{\text{weak}} \tag{18.10}$$

This choice sounds strange, because a basic assumption in all perturbation expansions is that we can compute the spectrum of the unperturbed Hamiltonian H_0, something impossible for H_{strong}. This is the reason why we called this strategy *phenomenological*: we will not compute the spectrum of H_0, we will take it from experiment. We will assume an equation of the form $H_0 |n\rangle = E_n |n\rangle$, where $|n\rangle$ are the observed physical states. Our Hilbert space $\mathcal{H}$ will be the space spanned by the vectors $|n\rangle$, taking into account all the superselection rules that we know from experiment. From first principles we know only a part of $\mathcal{H}$: since only hadrons have strong interactions, a state containing no hadrons will be an eigenstate of H_{kinetic} and we know that these states form our familiar Fock space of free particles, i.e. $\mathcal{H}_{\text{non-had}} = \mathcal{F}_{\text{non-had}}$. We do not have an analogous expression for the space of hadrons $\mathcal{H}_{\text{had}}$ and we do not expect to be able to write it as a Fock space starting from free quarks. So we take $\mathcal{H}_{\text{had}}$ to be the space of the observed hadrons, baryons and mesons. The only additional assumption we will make is that $\mathcal{H}_{\text{had}}$ has a unique state, which we will call *the hadronic vacuum state* $|0\rangle_{\text{had}}$, which is a zero energy eigenstate of the hadronic part of H_0. The entire Hilbert space is the direct product of the hadronic and non-hadronic Hilbert spaces

$$\mathcal{H} = \mathcal{H}_{\text{non-had}} \otimes \mathcal{H}_{\text{had}} \tag{18.11}$$

It is in this space that the operators, which form H_{em} and H_{weak}, are supposed to act. We will apply perturbation theory which, to first order, amounts to computing the matrix elements $\langle n| H_{\text{em}} |n'\rangle$ and $\langle n| H_{\text{weak}} |n'\rangle$. In this section we will study the latter, taking for H_{weak} the Hamiltonian we obtain from the Fermi theory (18.7). At first order of the perturbation expansion the product of two currents will give rise to three types of processes:

- Purely leptonic weak processes from $l^\lambda l_\lambda^\dagger$
- Semi-leptonic weak processes from $l^\lambda h_\lambda^\dagger$+h.c.
- Purely hadronic weak processes from $h^\lambda h_\lambda^\dagger$.
  We will study each one of them separately.

18.6 Leptonic Interactions

It is straightforward to write the leptonic current in terms of the fields of known leptons as

$$l_\lambda(x) = \bar{e}(x)\gamma_\lambda(1+\gamma_5)\nu_e(x) + \bar{\mu}(x)\gamma_\lambda(1+\gamma_5)\nu_\mu(x) \tag{18.12}$$

where we have used a notation in which the particle symbols denote also the corresponding Dirac fields; i.e. $e(x)$, $\nu_e(x)$, $\mu(x)$ and $\nu_\mu(x)$ are the Dirac fields of the electron, the electron–neutrino, the muon and the muon–neutrino, respectively. In 1960 the leptonic current had only these two terms; today we know that we must add a third one to describe the τ family: $\bar{\tau}(x)\gamma_\lambda(1+\gamma_5)\nu_\tau(x)$. The $V-A$ structure we mentioned above is shown by the presence of the $1+\gamma_5$ projector. The vector and the axial currents enter with exactly the same coefficient. This property, equal strength of the currents for all flavours, is often called "universality". As we saw in Chapter 8, $1+\gamma_5$ projects the four-component Dirac spinors into two-component left-handed Weyl spinors. In other words, $V-A$ implies that only the left-handed leptons participate in the weak interactions.

Another remarkable property of the $V-A$ current is that it generates an $SU(2)$ algebra. The origin of this algebra is simple to trace. Let us introduce a compact notation, which will also be useful later. We put every charged lepton, together with the associated neutrino, into a doublet

$$\Psi^i_{L,R}(x) = \begin{pmatrix} \nu_i(x) \\ \ell^-_i(x) \end{pmatrix}_{L,R} \tag{18.13}$$

where the index i runs over the three families, $\nu_i = \nu_e$, ν_μ and ν_τ and $\ell^-_i = e^-$, μ^- and τ^- for $i = 1, 2$ and 3, respectively. L and R denote the left- and right-handed spinors $\Psi_{L,R} = \frac{1}{2}(1 \pm \gamma_5)\Psi$.

For each family we can construct eight currents, four involving only Ψ_L and four only Ψ_R,

$$j^i_{\lambda L,R}(x) = \overline{\Psi^i_{L,R}}(x)\gamma_\lambda\Psi^i_{L,R}(x), \quad \boldsymbol{j}^i_{\lambda L,R}(x) = \overline{\Psi^i_{L,R}}(x)\gamma_\lambda\boldsymbol{\tau}\Psi^i_{L,R}(x) \tag{18.14}$$

where $\boldsymbol{\tau}$ are the Pauli matrices. All currents we have used so far can be written as linear combinations of the currents (18.14). For example, the electromagnetic current of the leptons is

$$j^{\text{em}}_\lambda = \sum_i \overline{\Psi^i_L}\gamma_\lambda\frac{1-\tau^3}{2}\Psi^i_L + \sum_i \overline{\Psi^i_R}\gamma_\lambda\frac{1-\tau^3}{2}\Psi^i_R \tag{18.15}$$

Similarly, the charged weak current (18.12) is given by

$$l_\lambda = \sum_i \overline{\Psi^i_L}\gamma_\lambda\tau^-\Psi^i_L \tag{18.16}$$

Two remarks are in order here: First, equations (18.15) and (18.16) show that electromagnetic and weak interactions, being vectorial, do not mix right- and left-handed components. Second, in 1960 the charged current (18.16) described all weak interactions known at the time. There was no evidence for a need to introduce neutral currents involving the unit matrix, or the τ^3 matrix in (18.16).

Since leptons have no strong interactions, the complete leptonic Lagrangian, in the limit of vanishing lepton masses, could be written as

$$\mathcal{L}_{\text{leptons}} = \sum_i \overline{\Psi_L^i}\, i\partial\!\!\!/\Psi_L^i + \sum_i \overline{\Psi_R^i}\, i\partial\!\!\!/\Psi_R^i - \frac{1}{4}F_{\mu\nu}^2 - ej_\lambda^{\text{em}} A^\lambda + \frac{G_F}{\sqrt{2}} l_\lambda l^{\lambda\dagger} \tag{18.17}$$

The quadratic part of (18.17) is invariant under $U(2)_L \times U(2)_R$ transformations, which rotate separately right- and left-handed doublets, as well as several $U(1)$ phase transformations, which correspond to the conservation of the various lepton numbers. A mass term, which is proportional to $\overline{\Psi}_L\Psi_R + \overline{\Psi}_R\Psi_L$, breaks the separate L and R symmetries into the diagonal subgroup leaving only the vector currents conserved. Notice also that, if neutrinos are massless, the right-handed neutrino fields are free fields and can be omitted. As we shall see later, recent experimental data show that neutrinos have non-zero masses and the three lepton numbers are not separately conserved. Neutrinos oscillate among the different species, but let us ignore this complication for the moment.

The generators of the $U(2)_L \times U(2)_R$ symmetry are precisely the space integrals of the time components of the $V - A$ and $V + A$ leptonic currents

$$Q_{L,R}^{(l)a} = \sum_i \int \mathrm{d}^3\boldsymbol{x}\, j_{0\,L,R}^a(x) \tag{18.18}$$

where $a = 0, 1, 2, 3$ denotes the $U(2)$ index. These charges satisfy the $U(2) \times U(2)$ algebra, i.e. the two $U(1)$ charges, corresponding to the value $a = 0$ of the index, commute among themselves as well as with all other charges, and the six $SU(2)_{L,R}$ ones satisfy

$$\left[Q_{L,R}^{(l)a},\, Q_{L,R}^{(l)b} \right] = \mathrm{i}\,\epsilon_{abc} Q_{L,R}^{(l)c}, \quad \left[Q_L^{(l)a},\, Q_R^{(l)b} \right] = 0\,, \quad a, b = 1, 2, 3 \tag{18.19}$$

Current conservation guarantees the time-independence of the charges $Q^{(l)a}$. In section 8.3.8 we introduced the term *chirality* to denote the eigenvalues of γ_5. By analogy, we call the transformations generated by the charges $Q_{L,R}^{(l)a}$ *chiral transformations* and the resulting symmetry *chiral symmetry*.

18.6.1   Muon decay

There are several purely leptonic weak processes in which the validity of the Lagrangian density (18.17) can be tested. Let us start with the most important one, both experimentally as well as historically, which is muon decay

$$\mu^-(p, r_1) \longrightarrow e^-(q, r_2) + \bar{\nu}_e(k_2, s_2) + \nu_\mu(k_1, s_1) \tag{18.20}$$

with the corresponding one for μ^+, where all particles are replaced by their antiparticles. Before studying this process, let us point out several important features of muon decay:

• Experimentally, we can only measure the parameters associated to the electron, but the study of its energy spectrum shows that we are dealing with a three-body decay.

• Anticipating the discussion, which will come later, we wrote two distinct neutrinos in the final state. We will come back to this point when we discuss neutrino scattering experiments in Chapter 20.

• Based only on the conservation of electric charge and angular momentum, we could expect the existence of the two-body decay mode

$$\mu^- \to e^- + \gamma \tag{18.21}$$

Experimentally we have only the upper bound $\Gamma(e\gamma)/\Gamma(\text{tot}) < 4.2 \times 10^{-13}$ and this was interpreted as an indication for a separate conservation of electron and muon numbers. In fact, it was the absence of this mode which prompted Schwinger to postulate the existence of a separate neutrino species ν_μ. As we will see in Chapter 20, we know today that this conservation is not exact, but the very stringent experimental limit shows that it holds to a very good approximation. We will postpone further discussion until Chapter 20.

Coming back to the decay mode (18.20), we see that in lowest order perturbation theory the amplitude is given, in an obvious notation, by

$$\mathcal{M} = \frac{G_F}{\sqrt{2}} \left[\overline{u}_{\nu_\mu}(k_1, \alpha_1)\gamma_\lambda(1+\gamma_5)u_\mu(p, \beta_1) \right] \left[\overline{u}_e(q, \beta_2)\gamma^\lambda(1+\gamma_5)v_{\nu_e}(k_2, \alpha_2) \right]$$
$$= \frac{G_F}{\sqrt{2}} \left[\overline{u_L}_{\nu_\mu}(k_1, \alpha_1)\gamma_\lambda u_{L\mu}(p, \beta_1) \right] \left[\overline{u_L}_e(q, \beta_2)\gamma^\lambda v_{L\nu_e}(k_2, \alpha_2) \right] \tag{18.22}$$

In Problem 18.5 we ask the reader to compute $|\mathcal{M}|^2$ and integrate over the phase space. It is convenient to reorder the amplitude $\mathcal{M}$ and put the spinors corresponding to the two neutrinos in the same bracket. This is possible by virtue of the *Fierz reordering theorem*, which we ask the reader to prove in Problem 18.4. A particular property of the $V - A$ amplitude (see Problem 18.4) is that it keeps this form under the exchange of the second and fourth spinor. The result is that $\mathcal{M}$ can be written also as

$$\mathcal{M} = \frac{G_F}{\sqrt{2}} \left[\overline{u_L}_{\nu_\mu}(k_1, \alpha_1)\gamma_\lambda v_{L\nu_e}(k_2, \alpha_2) \right] \left[\overline{u_L}_e(q, \beta_2)\gamma^\lambda u_{L\mu}(p, \beta_1) \right] \tag{18.23}$$

Three quantities can be measured experimentally in muon decay:

• The total decay rate. In the limit of vanishing electron mass it is

$$\Gamma_{\text{muon}} = \frac{G_F^2 m_\mu^5}{24(2\pi)^3} \tag{18.24}$$

The experimental lifetime is $\tau_{\text{muon}} = (2.1969811 \pm 0.0000022) \times 10^{-6}$ s. It is used to determine the value of G_F. With such a great experimental accuracy, it is important to evaluate also all sorts of corrections to equation (18.24) due to the non-vanishing electron mass, as well as electromagnetic (virtual photon exchange, or the emission and absorption of very soft photons) and even hadronic contributions. These calculations have been performed, but we will not present them here. The result, including the combined uncertainty, is

$$G_F = 1.1663787(6) \times 10^{-5} \text{GeV}^{-2} \qquad (18.25)$$

• The electron energy spectrum. It was computed already in 1950 by L. Michel. It was before the discovery of parity violation but, as we have already explained, this makes no difference for an energy spectrum. In fact, Michel computed it for an arbitrary combination of the five bilinears S, P, V, A and T. So we expect the result to depend on the parameters C_i of equation (18.1). The surprising result is that, in the limit $m_e \to 0$, the spectrum depends on only one parameter ρ, called *the Michel parameter*, and

$$\frac{d\Gamma}{\Gamma} = 4x^2 \left[3(1 - x) + 2\rho \left(\frac{4}{3}x - 1 \right) \right] dx \qquad (18.26)$$

where $x = 2E_e/m_\mu$ is the energy carried by the electron as a fraction of its maximum value. For the particular case of the $V - A$ interaction, we find $\rho = 3/4$ (see Problem 18.5). Experimentally we find $\rho = 0.74979 \pm 0.00026$, another confirmation of the $V - A$ structure.

• We can also measure the angular distribution of the electrons coming from the decay of polarised muons. The calculation is straightforward and the result is in full agreement with $V - A$.

18.6.2   Other purely leptonic weak processes

Today we know other purely leptonic weak processes beyond muon decay. They are:
• The leptonic decays of the next sequential lepton, the τ. It is heavier than the muon with mass $m_\tau = 1776.86 \pm 0.12$ MeV and, therefore, it can have the two leptonic decay modes

$$\tau^- \to e^- + \bar{\nu}_e + \nu_\tau \quad \text{and} \quad \tau^- \to \mu^- + \bar{\nu}_\mu + \nu_\tau \qquad (18.27)$$

Each one is described by the analogue of the amplitude (18.22) with the obvious replacements. Since the ratios $(m_\mu/m_\tau)^2$ and $(m_e/m_\tau)^2$ are very small, we expect the ratio of the purely leptonic modes $\Gamma(\tau \to \mu)/\Gamma(\tau \to e)$ to be very close to one. This is a test of *electron–muon universality*. Experimentally we have

$$\frac{\Gamma(\tau^- \to \mu)}{\Gamma(\tau^- \to \text{all})} = (17.39 \pm 0.04)\% \quad \text{and} \quad \frac{\Gamma(\tau^- \to e)}{\Gamma(\tau^- \to \text{all})} = (17.82 \pm 0.04)\% \qquad (18.28)$$

which, taking into account the small $e - \mu$ mass difference, is in excellent agreement with universality. A further test is provided by the Michel parameter ρ measured in either the electron or the muon decay mode. We find $\rho(e) = 0.747 \pm 0.01$ and $\rho(\mu) = 0.763 \pm 0.02$.

As we see, the purely leptonic modes amount to roughly one third of the total τ width. The rest is given by semi-leptonic modes of the form

$$\tau \to \nu_\tau + \text{hadrons} \qquad (18.29)$$

where "hadrons" means mainly pions and kaons.

(a)   (b)

Fig. 18.4 Semi-leptonic process. (a) A hadron decay producing a pair of leptons. h_2 may be the hadronic vacuum state. (b) A lepton–hadron scattering. h_1 and h_2 represent hadronic systems. For the semi-leptonic decays of the τ lepton h_1 is the vacuum state.

• The elastic neutrino–electron scattering. It played a crucial role in the development of the theory of weak interactions and we will study it in Chapter 20.

18.7 Semi-Leptonic Interactions

They are described by the product of the leptonic and the hadronic parts of the weak current: $h^\lambda(x)l^\dagger_\lambda(x)+l^\lambda(x)h^\dagger_\lambda(x)$. We have already written the explicit form of $l^\lambda(x)$ in terms of the fields of known leptons in equation (18.12), but we have no way to do that for $h^\lambda(x)$. Even if we assume that the quarks are the elementary hadronic fields, still we can only use experimental data which involve hadrons. So our first problem is to determine $h^\lambda(x)$ as an operator acting in the hadronic Hilbert space $\mathcal{H}_{\text{had}}$. We remind the reader that we have assumed some knowledge of the non-perturbative effects of strong interactions, since we included it in the unperturbed part of the Hamiltonian H_0, and, therefore, here to "determine" means to identify $h^\lambda(x)$ with a known operator of the strong interactions. It took more than ten years to solve this problem and it was one of the most demanding, but also among the most rewarding enterprises in theoretical particle physics. The solution required the introduction of new concepts and ideas, whose importance transcends the domain of weak interactions and covers many aspects of modern theoretical physics. It is this story, which we will present briefly in this section.

As we explained already, we will follow a phenomenological approach. A semi-leptonic weak process is of the general form shown in Figure 18.4. It is either a decay, in which a hadron h_1 gives a hadronic state h_2 and a lepton pair (see Figure 18.4 (a)), or a scattering process in which a lepton l_1 scatters off a hadron h_1 and gives a hadronic state h_2 and a lepton l_2 (Figure 18.4 (b)). For conceptual and pedagogical purposes we will distinguish two cases: (i) *strangeness conserving*, or $\Delta S = 0$ processes, in which the states h_1 and h_2 have the same strangeness, and (ii) *strangeness violating* processes, in which $S(h_1) \neq S(h_2)$.

18.7.1 Strangeness conserving semi-leptonic weak interactions

18.7.1.1 Neutron β-decay. The first manifestation of weak interactions, namely nuclear β-decay, belongs to this class. Its simplest form is the decay of a free neutron.

$$n \longrightarrow p + e^- + \bar{\nu}_e \tag{18.30}$$

The leading order contribution to the amplitude is given by

$$i T_{fi} = i \frac{G_F}{\sqrt{2}} \int d^4x \, \langle p(q_2, \beta_2), e^-(k_1, \alpha_1), \bar{\nu}_e(k_2, \alpha_2) | \, l^\mu(x) \, h_\mu^\dagger(x) \, | n(q_1, \beta_1) \rangle$$

$$= i \frac{G_F}{\sqrt{2}} (2\pi)^4 \delta^4(q_1 - q_2 - k_1 - k_2)$$

$$\times \langle e^-(k_1, \alpha_1), \bar{\nu}_e(k_2, \alpha_2) | \, l^\mu(0) \, | 0 \rangle \, \langle p(q_2, \beta_2) | \, h_\mu^\dagger(0) \, | n(q_1, \beta_1) \rangle \qquad (18.31)$$

where q_1, q_2, k_1, k_2 are the momenta of the four particles and $\beta_1, \beta_2, \alpha_1$ and α_2 their polarisation indices. We used the fact that at this order the amplitude factorises into a leptonic and a hadronic part, and the fact that for any local operator $\mathcal{O}(x)$ we have $\mathcal{O}(x) = e^{iP \cdot x} \mathcal{O}(0) e^{-iP \cdot x}$, with P^μ the four-momentum operator.

Using (18.12) we can write the leptonic part in (18.31) explicitly in terms of the electron and neutrino wave functions, but we can only give a formal expression for the hadronic part. The invariant amplitude becomes

$$\mathcal{M} = \frac{G_F}{\sqrt{2}} \left[\bar{u}_e(k_1, \alpha_1) \gamma^\lambda (1 + \gamma_5) v_{\nu_e}(k_2, \alpha_2) \right] \langle p(q_2, \beta_2) | \, h_\lambda^\dagger(0) \, | n(q_1, \beta_1) \rangle \qquad (18.32)$$

and it is the hadronic factor on the right that we want to study in this section. Let k be the four-momentum carried by the lepton pair: $k = k_1 + k_2$. By energy-momentum conservation, it is also equal to the momentum transfer between the nucleons, i.e. $k = q_1 - q_2$. The hadronic current $h_\lambda^\dagger(x)$ is the sum of a vector current $V_\lambda(x)$ and an axial vector one $A_\lambda(x)$. Let us start with the vector part. The most general form of its matrix element consistent with Poincaré invariance can be written as

$$\langle p | V^\lambda(0) | n \rangle = \bar{u}_p(q_2, \beta_2) \left[g_V(k^2) \gamma^\lambda + g_T(k^2) \sigma^{\lambda\mu} k_\mu + g_S(k^2) k^\lambda \right] u_n(q_1, \beta_1) \qquad (18.33)$$

where u_p and u_n are the Dirac spinors for the proton and the neutron. A similar expression holds for the matrix element of $A^\lambda(0)$, i.e.

$$\langle p | A^\lambda(0) | n \rangle = \bar{u}_p(q_2, \beta_2) \left[g_A(k^2) \gamma^\lambda + \tilde{g}_T(k^2) \sigma^{\lambda\mu} k_\mu + g_P(k^2) k^\lambda \right] \gamma_5 u_n(q_1, \beta_1) \qquad (18.34)$$

The form factors $g(k^2)$ are functions of the only Lorentz invariant variable, the nucleon momentum transfer square. These expressions can be simplified by imposing several invariance properties on the matrix element but, in fact, we will not need them here. In the rest frame of the decaying neutron the components of k are bounded by the neutron–proton mass difference, i.e. $k^0, |\mathbf{k}| \leq \delta m = m_n - m_p \simeq 1.3$ MeV. It follows that in writing the general form for the hadronic matrix element $\langle p | h_\lambda^\dagger(0) | n \rangle$ we can restrict ourselves to $k = 0$. Terms proportional to k^μ will give contributions of order $\delta m / m \sim 10^{-3}$, comparable to or smaller than electromagnetic corrections, which we have neglected. Under this assumption the most general form of the hadronic matrix element becomes

$$\langle p | h_\lambda^\dagger(0) | n \rangle = \bar{u}_p(q_2, \beta_2) \gamma_\lambda \left[g_V(0) + g_A(0) \gamma_5 \right] u_n(q_1, \beta_1) \qquad (18.35)$$

with $g_V(0)$ and $g_A(0)$ two real numbers to be determined by experiment.[9] If we factor out the Fermi coupling constant determined by muon decay, see equation (18.25), we obtain the values

$$g_V(0) \simeq 1, \quad g_A(0) \simeq 1.25 \tag{18.36}$$

We want to use these values in order to guess the properties of the current $h^\lambda(x)$ as an operator acting on the space of hadron states. Having just two real numbers does not sound like a great help and, indeed, were they two arbitrary numbers, we wouldn't be able to draw any conclusions. It is the special value $g_V(0) \simeq 1$, which will give us the clue to the solution.

18.7.1.2 The conserved vector current (CVC) hypothesis. In order to develop some intuition, we will imagine a very naïve toy model, in which the proton and the neutron are "elementary" particles described by quantum Dirac fields Ψ_p and Ψ_n, which, at zeroth order perturbation theory, can be expanded in creation and annihilation operators. Let us assume that, to this order, the hadronic weak current responsible for β-decay has the same simple form as the leptonic current (18.12), just replacing lepton with nucleon fields, i.e.

$$h_\lambda(x) = \overline{\Psi}_n(x)\gamma_\lambda(1 + \gamma_5)\Psi_p(x) \tag{18.37}$$

In writing $h_\lambda(x)$ we have assumed that the same coupling strength G_F describes both β-decay and μ-decay. This assumption generalises the notion of *universality* which we introduced for the leptonic weak interactions. In our toy model it is expressed by the form we have assumed for $h_\lambda(x)$ in equation (18.37), which amounts to assuming that the form factors of equation (18.35) take the values

$$g_V(0) = 1, \quad g_A(0) = 1 \tag{18.38}$$

So far we worked at zeroth order in the strong interactions. To study their effect we assume some simple form of the strong interaction Hamiltonian, for example the isospin invariant pion–nucleon coupling (6.15)

$$\mathcal{L}_{\pi N} = g_{\pi NN}\overline{\Psi}(x)\,\tau^a\,\gamma_5\Psi(x)\,\pi^a(x)$$

with $g_{\pi NN}$ the pion–nucleon coupling constant and $\pi^a(x)\,(a = 1, 2, 3)$ the pion triplet quantum fields, for which we assume that, at zeroth order, perturbation theory can be expanded in pion creation and annihilation operators. This is only a toy model because, if we try to describe the observed pion–nucleon interactions with this Lagrangian in leading order perturbation theory, we find $g_{\pi NN} > 1$, which implies that an expansion in powers of it is not meaningful. Nevertheless, we will continue with this model, because our purpose is not to obtain reliable numerical results, but only to guess some properties, which we will prove are model independent.

---

[9]If we relax the zero momentum transfer approximation, we obtain the contribution of the additional form factors such as the tensor $g_T(k^2)\sigma_{\lambda\mu}k^\mu$ of equation (18.33). It is called "weak magnetism" and contributes to some nuclear transitions, which are highly forbidden by angular momentum conservation for the other form factors.

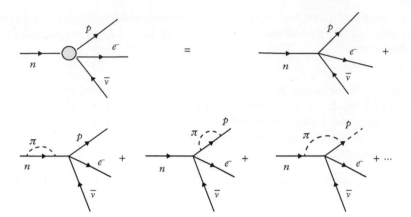

Fig. 18.5 Neutron β-decay. Strong interactions are represented by pion exchange graphs. These are the 1-loop diagrams. The dots stand for the infinite series with more virtual pions.

We can formally write the higher order corrections due to pion exchange diagrams shown in Figure 18.5. We obtain for $g_V(0)$ and $g_A(0)$ the formal power series

$$g_V(0) = 1 + \sum_{n=1}^{\infty} g_{\pi NN}^{2n} g_V^{(n)}(0) \tag{18.39}$$

$$g_A(0) = 1 + \sum_{n=1}^{\infty} g_{\pi NN}^{2n} g_A^{(n)}(0) \tag{18.40}$$

while, obviously, no such corrections exist for the leptonic current $l^{\lambda}(x)$, since leptons have no strong interactions. Comparing with the experimental values (18.36), we conclude that in this model the strong interaction corrections for the axial form factor sum up to 0.25, while those for the vector one vanish. Although we cannot compute the sum, a priori, we do not expect the strong interaction corrections to be so small, but the big surprise is related to the vector form factor, which seems to receive no corrections at all from strong interactions. In the physicists' jargon we say that *the vector current coupling constant does not get renormalised*. This rings a bell: recall that, in computing the renormalisation properties of quantum electrodynamics, we found that the corrections due to the diagrams of the electron self-energy and the vertex function cancel, see equation (15.61) and Problem 16.1. In section 15.2.3 we argued that this is a consequence of the conservation of the electromagnetic current, and it is valid to all orders of perturbation theory. Here we want to present a simple heuristic proof of this property and point out some very important physical implications, one of which will be the precise definition of the weak hadronic current.

Let us consider a general field theory with N fields $\phi^i(x)$, $i = 1, 2, \dots, N$, described by a Lagrangian density $\mathcal{L}(\phi^i(x), \partial_{\mu}\phi^i(x))$. The spin values of the fields will not be important in this discussion. We assume that $\mathcal{L}$ is invariant under a group $\mathcal{G}$ of internal symmetry transformations, under which the fields transform linearly

$$\phi^i \to \phi^i + \delta\phi^i, \delta\phi^i = i\epsilon_{\alpha}T_{\alpha}^{ij}\phi^j \tag{18.41}$$

where, in the usual notation, $\epsilon_\alpha (\alpha = 1, 2, \ldots, m)$ are m infinitesimal parameters, m is the dimension of the Lie algebra of $\mathcal{G}$, and T_α are m $N \times N$ Hermitian matrices, representing the generators of $\mathcal{G}$. According to Noether's theorem this invariance implies the existence of m conserved currents $j_\alpha^\mu(x)$, $\partial_\mu j_\alpha^\mu(x) = 0$. The corresponding charges

$$Q_\alpha = \int d^3\mathbf{x}\, j_\alpha^0(x) = \mathrm{i} \int d^3\mathbf{x}\, \pi^i(t, \mathbf{x}) T_\alpha^{ij} \phi^j(t, \mathbf{x}) \tag{18.42}$$

are conserved, $\dot{Q}_\alpha = 0$, and belong to the adjoint representation of the algebra of $\mathcal{G}$, that is, they satisfy

$$[Q_\alpha, Q_\beta] = \mathrm{i} f_{\alpha\beta\gamma} Q_\gamma \tag{18.43}$$

with $f_{\alpha\beta\gamma}$ the structure constants of $\mathcal{G}$. Their action on the quantum fields ϕ^i is given by:

$$[Q_\alpha, \phi^i] = T_\alpha^{ij} \phi^j \tag{18.44}$$

The important property of this relation is that it is homogeneous in the fields but inhomogeneous in the charges. We have seen that going from the classical to the quantum theory, we must introduce renormalisation counterterms. In particular, each field will acquire a multiplicative wave function renormalisation $\phi^i = \sqrt{Z} \phi_R^i$ and the invariance under $\mathcal{G}$ implies that the renormalisation constants Z are the same for all fields belonging to the same irreducible representation of $\mathcal{G}$. Therefore, Z will drop out from the relation (18.44); in other words, the commutation relation remains intact when we replace unrenormalised with renormalised fields. However, this is not true for the charges. If we had to renormalise the current with a multiplicative constant $\hat{Z}$ as $Q = \hat{Z} Q_R$, we would modify the commutation relation. We conclude that, *if the renormalisation procedure respects a symmetry, the corresponding conserved currents do not get renormalised.* Appropriately generalised, this gives the proof of the Ward identity we found in Chapter 15.

An immediate and very important application of this theorem concerns the value of the electric charge. Let us consider quantum electrodynamics applied to protons and electrons. We know experimentally that, to a very great accuracy $|Q_p| = |Q_e|$.[10] Therefore, in writing the Lagrangian of quantum electrodynamics, we must impose the condition that the unrenormalised values of the charges are the same. But, contrary to electrons, protons have strong interactions described, for example, by the pion-nucleon interaction (6.15). Nevertheless, the non-renormalisation theorem guarantees that the series represented by the diagrams shown in Figure 18.6 sums up to zero and, as a result, the renormalised electric charges of the proton and the positron remain equal.

Let us come back to the problem of the weak hadronic current. There also we found out that the equality of the vector current coupling constants between hadrons and leptons, as expressed by the relation $g_V(0) = 1$, implies that the strong interaction corrections shown in the series of equation (18.39) vanish. This strongly suggests that the vector hadronic current of the weak interactions must be a conserved current of the strong interactions, which we should identify with the Noether current of a symmetry.

---

[10]The fact that a macroscopic piece of matter appears to be electrically neutral imposes a relation of the form $|Q_p| = |Q_e| \pm \mathcal{O}(10^{-20})$.

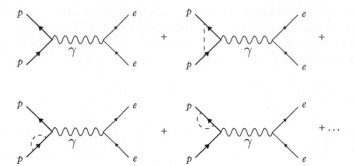

Fig. 18.6 The non-renormalisation of the proton electric charge relative to that of the electron. In the notation of Chapter 15, the one pion exchange contributions give $Z_1 = Z_2$ and the conservation of the electromagnetic current guarantees that this Ward identity remains true to all orders in the strong interactions. As a result the relation $|Q_p| = |Q_e|$ remains valid.

S.S. Gershtein and Y.B. Zeldovich in 1956, and, independently, R.P. Feynman and M. Gell-Mann in 1957, formulated the *conserved vector current* (CVC) hypothesis, according to which *the vector part of $h^\lambda(x)$ belongs to the triplet of the conserved isospin currents of strong interactions.*

Specifically, if we ignore strange particles, strong interactions are invariant under an $SU(2) \times U(1)$ group of internal symmetry transformations. The $U(1)$ part corresponds to the conservation of baryon number and the $SU(2)$ part to isospin. This invariance leads to four conserved currents $j_0^\lambda(x)$ and $\boldsymbol{j}^\lambda(x) = (j_1^\lambda(x), j_2^\lambda(x), j_3^\lambda(x))$. We can illustrate the transformation properties of these currents using the proton–neutron isospin doublet, but it will be more convenient for later generalisations to adopt the quark notation restricting ourselves to the first two flavours u and d. Let q denote an isospin doublet. Taking into account the fact that the quarks have baryon number equal to $\frac{1}{3}$ and electric charges $\frac{2}{3}$ and $-\frac{1}{3}$, respectively, the four currents can be written as[11]

$$q(x) = \begin{pmatrix} u(x) \\ d(x) \end{pmatrix} ; \quad j_0^\lambda(x) = \frac{1}{3}\bar{q}(x)\gamma^\lambda q(x), \quad \boldsymbol{j}^\lambda(x) = \bar{q}(x)\gamma^\lambda \frac{\boldsymbol{\tau}}{2} q(x) \tag{18.45}$$

The electromagnetic current is given by

$$j_{\text{em}}^\lambda(x) = \frac{1}{2}j_0^\lambda(x) + j_3^\lambda(x) = \frac{2}{3}\bar{u}(x)\gamma^\lambda u(x) - \frac{1}{3}\bar{d}(x)\gamma^\lambda d(x) \tag{18.46}$$

and, according to CVC, the vector part of the weak strangeness conserving hadronic current by

$$V^\lambda(x) = j^{\lambda-}(x) = \bar{q}(x)\gamma^\lambda \tau^+ q(x) \tag{18.47}$$

with $\tau^\pm = (\tau^1 \pm \mathrm{i}\,\tau^2)/2$. This completes the CVC picture and solves one-half of our problem: we have determined the vector part of the hadronic current.

---

[11]All these expressions are supposed to be summed over the three colours, although we did not write it explicitly.

Before closing this section let us note that the CVC hypothesis, besides determining the weak vector current, has some precise physical consequences, which can be compared with experiment. For example, let us look at the β-decay matrix element, equation (18.33), and impose the conservation equation $\partial_\lambda V^\lambda(x) = 0$. The first term gives a vanishing contribution in the limit $m_p = m_n$ (remember that CVC assumes the isospin invariance of the strong interactions and is expected to receive electromagnetic corrections) and the tensor term vanishes by symmetry. So, $\partial_\lambda V^\lambda(x) = 0$ implies $g_S(k^2) = 0$, i.e. there is no induced scalar form factor. This is not surprising, since such a term should be proportional to the scalar part of the current $\partial_\lambda V^\lambda(x)$. We can test this result in nuclear physics, where we find that transitions induced by this scalar term are highly suppressed, compatible with electromagnetic corrections.

More precise tests can be obtained by using the fact that V^λ is assumed to belong to the same multiplet as the isovector part of the electromagnetic current. In Problem 18.9 we ask the reader to establish the relation

$$g_T(0) = \frac{\mu_n - \mu_p}{2m} \tag{18.48}$$

where g_T is the weak magnetic form factor of equation (18.33) and μ_n (μ_p) is the anomalous magnetic moment of the neutron (proton). The m is the common nucleon mass. The experimental measurement of this quantity has a complicated history, but the final result confirms the CVC hypothesis. Similar relations can be obtained between the neutrino scattering amplitudes and the corresponding ones in photon or electron scattering.

Finally, let us remark that, since V^λ is part of the isospin current, its matrix elements between states belonging to different isomultiplets vanish. For example, if we look at the decay $\Sigma^- \to \Lambda + e^- + \bar\nu_e$, we expect the contribution of the vector part to be suppressed and the decay to be dominated by the axial current. Therefore, it must be nearly parity conserving, in agreement with the data.

18.7.1.3 Partial conservation of the axial current, spontaneous breaking of chiral symmetry and charged pion decay. This section has a very long title. The reason is that it will address several, apparently distinct, fundamental physics questions.

Encouraged by our success with the vector part of the hadronic current, we turn our attention to the axial part. Could $A^\lambda(x)$ be conserved? The answer is no, but it is instructive to understand the reasons.

- The conservation equation $\partial_\lambda A^\lambda(x) = 0$ would imply $g_A(0) = 1$, in contradiction with experiment, see equation (18.36).
- It is more enlightening to look at the amplitude for the charged pion decay $\pi^+ \to \mu^+ + \nu_\mu$. This is a semi-leptonic process, so the invariant amplitude factors into a hadronic and a leptonic part

$$\mathcal{M} = \frac{G_F}{\sqrt{2}} \langle 0| \, h_\lambda(0) \, |\pi^+(q)\rangle \left[\bar{u}_{\nu_\mu}(k_2, s_2)\gamma^\lambda(1 + \gamma_5)v_\mu(k_1, s_1) \right] \tag{18.49}$$

Since the pion is pseudo-scalar, only the axial part of the current contributes. By Lorentz invariance we can write its matrix element as

$$\langle 0| A_\lambda^\dagger(0) | \pi^+(q) \rangle = f_\pi \, q_\lambda \tag{18.50}$$

with f_π a constant. In section 12.5 we showed how to compute the decay rate and the result is

$$\Gamma(\pi^+ \to \mu^+ + \nu_\mu) = \frac{G_F^2 f_\pi^2 m_\mu^2 (m_\pi^2 - m_\mu^2)^2}{8\pi m_\pi^3}$$

Inserting the measured values of the pion lifetime, the masses and the Fermi coupling constant, this relation yields

$$f_\pi \simeq 130 \, \mathrm{MeV} \simeq 0.93 \, m_\pi$$

for the pion decay constant.

On the other hand, from (18.50) and noting that $q^2 = m_\pi^2$, we obtain for the divergence of the axial current

$$\langle 0| \partial^\lambda A_\lambda^\dagger(0) | \pi^+(q) \rangle = \mathrm{i} f_\pi m_\pi^2 \tag{18.51}$$

We conclude that a conserved axial current would imply a massless and/or stable pion.

So we see that we have at least two pieces of evidence against the conservation of the axial current. However, a striking feature is common to both: they are both numerically weak. As we have noted already, in the scale of strong interactions the change from 1 to 1.25 is a rather small renormalisation, and m_π^2 is very small compared to the square of a typical hadron mass, which is of order 1 GeV^2. So the question is: could we assume that the axial current is *approximately* conserved? If this is the case, we should enlarge the internal symmetry of the strong interactions from $SU(2)$ (isospin) to $SU(2) \times SU(2)$, in which case the strangeness conserving hadronic current would satisfy approximately an algebra similar to the one we found for the leptonic current. We could define the right- and left-handed components as $V \pm A$ and we would obtain the algebra

$$[Q_{L,R}^a, Q_{L,R}^b] = \mathrm{i}\,\epsilon_{abc}\, Q_{L,R}^c, \qquad [Q_L^a, Q_R^b] = 0 \tag{18.52}$$

In the free quark toy model notation we used for the vector current in equations (18.45) to (18.47), the axial current is given by the same expressions with an additional γ_5, so the complete hadronic current $h_\lambda(x)$ can be written as

$$h_\lambda(x) = \bar{q}(x)\gamma_\lambda(1 + \gamma_5)\,\tau^- q(x) \tag{18.53}$$

Naturally, in the real world, where the axial current is not exactly conserved, the corresponding charges will not be time independent and we expect $\dot{Q}_5^a \neq 0$. Nevertheless, we will assume that the algebra of (18.52) remains exact, provided we compute the commutators involving axial charges at equal times.

The trouble is that there is no trace of such a chiral symmetry in the spectrum of known hadrons. The proton and the neutron are approximately degenerate, but they have no partners with similar masses and opposite parity. If the strong interactions are approximately invariant under chiral transformations, this is not manifest in the spectrum of hadrons.

Here the analysis we presented in Chapter 14 comes in handy. We saw that the invariance of the equations of motion under a group of transformations may be realised in at least two different ways. In the first, which we call *Wigner realisation*, the eigenstates of the Hamiltonian form irreducible representations of the symmetry group with degenerate eigenvalues. This is the case of the isospin symmetry. The proton and the neutron form a doublet with near degenerate masses, the three pions form a triplet, and so on. However, we saw that there exists an alternative way, called *the Nambu–Goldstone realisation*. We called such a symmetry *spontaneously broken* and saw that it manifests itself by the presence of a zero mass excitation, *the Goldstone particle*, with quantum numbers the ones of the divergence of the corresponding conserved current. We want to speculate here that this phenomenon occurs in the strong interactions. To be precise, we assume that the strong interaction Hamiltonian can be written in the form

$$H_{\text{strong}} = H_{\text{symmetric}} + H_{\text{breaking}} \tag{18.54}$$

where $H_{\text{symmetric}}$ is invariant under chiral $SU(2)_L \times SU(2)_R$ transformations and H_{breaking} breaks this invariance, but its effects are relatively small. We will give precise estimations of these breaking effects presently, and in Chapter 19 we will explain their physical origin. Furthermore, we assume that $SU(2)_L \times SU(2)_R$ is spontaneously broken to the diagonal subgroup $SU(2)_L \times SU(2)_R \to SU(2)_V$, with $SU(2)_V$ involving only vector currents, which we identify with isospin. In the absence of H_{breaking} this spontaneous breaking would lead to three massless pseudo-scalar Goldstone bosons. Switching on H_{breaking} turns these bosons into "pseudo-Goldstone bosons" with small, non-zero masses. We assume that they can be identified with the pions. A convenient way to express the pseudo-Goldstone character of the pions is to promote (18.51) into an operator equation and write

$$\partial^\lambda A_\lambda^a(x) = i \frac{f_\pi}{\sqrt{2}} m_\pi^2 \phi_\pi^a(x) \tag{18.55}$$

where a is an $SU(2)$ index and $\phi_\pi^a(x)$ is a field operator normalised to create a one-pion state out of the vacuum: $\langle \pi^a(k)| \phi^b(x) |0\rangle = \delta^{ab} e^{ik \cdot x}$. The only non-trivial part of this relation is the coefficient $f_\pi m_\pi^2$, which fixes the normalisation. Indeed, as we explained in section 12.2, any operator with the right quantum numbers, appropriately normalised, can be used as a field operator in the reduction formula.

This theoretical scheme is known as PCAC for *partially conserved axial current*. It has several observable consequences, and in the rest of this section we will present some of them, together with the experimental evidence which supports them, but, before doing so, we want to stress once again the remarkable fact, that the quest for understanding the nature of the weak interactions led us to discover an unsuspected hidden symmetry of the strong interactions.

We start with a very general relation first obtained by S. Adler. It involves the amplitude for emission or absorption of a low momentum Goldstone or pseudo-Goldstone boson, and it is based on the fact that a Goldstone particle has the quantum numbers of the divergence of the corresponding current. Although we will derive it for pions,

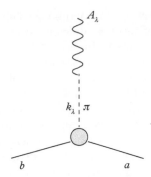

Fig. 18.7 The one-pion dominance of the axial current matrix elements.

it applies to any spontaneously broken symmetry. We consider the matrix element $M_\lambda(k)$ of the axial current between two hadronic states $|a\rangle$ and $|b\rangle$

$$M_\lambda(k) = \int \mathrm{d}^4 x e^{-ik\cdot x} \langle a| A_\lambda^\dagger(x) |b\rangle \tag{18.56}$$

where we have suppressed the $SU(2)$ indices. If we compute $k^\lambda M_\lambda(k)$ we obtain the matrix element of the divergence of the axial current. Were the symmetry not spontaneously broken, the result would be zero. But now we have an almost massless Goldstone particle, the pion, whose quantum numbers are precisely those of the divergence of the current. Using the PCAC relation (18.55) we obtain

$$k^\lambda M_\lambda(k) = f_\pi m_\pi^2 \int \mathrm{d}^4 x e^{-ik\cdot x} \langle a| \phi_\pi(x) |b\rangle \tag{18.57}$$

The matrix element on the right-hand side can be approximated at low momenta k by the one-pion intermediate state represented by the diagram of Figure 18.7. If we call $M_\pi(k)$ the invariant amplitude for emission $a \to b + \pi$, or absorption $a + \pi \to b$, of a pion, we get

$$k^\lambda M_\lambda(k) = \frac{\mathrm{i} f_\pi m_\pi^2}{k^2 - m_\pi^2} M_\pi(k) \tag{18.58}$$

If m_π^2 is very small, this diagram gives the dominant contribution for low k because of the almost vanishing denominator. The expression (18.58) is Lorentz invariant and we can choose to compute it in the frame $\boldsymbol{k} = 0$. In the same approximation of very small m_π we can replace $\boldsymbol{k} \to 0$ with $k \to 0$ and obtain

$$\lim_{k \to 0} M_\pi(k) = \mathrm{i} f_\pi^{-1} \lim_{k \to 0} k^\lambda M_\lambda(k) \tag{18.59}$$

This relation is called *Adler's condition*. The amplitude for emission, or absorption, of a Goldstone particle vanishes at low energy linearly with the momentum of the Goldstone particle.

We now specialise the matrix element of β-decay. The graph is a special case of the one shown in Figure 18.7 with a proton and a neutron replacing the states $|a\rangle$ and

$|b\rangle$. Using equation (18.34), the matrix element of the divergence of the axial current between a proton and a neutron can be written as

$$- \mathrm{i} \langle p| \partial^\lambda A_\lambda^\dagger(0) |n\rangle = k^\lambda M_\lambda(k) = \left[2M g_A(k^2) + k^2 g_P(k^2)\right] \bar{u}_p \gamma_5 u_n \qquad (18.60)$$

where u_p and u_n are the free Dirac spinors for the proton and the neutron, M is the nucleon mass and g_A and g_P are two form factors, called "axial vector" and "induced pseudo-scalar". For $k^2 \to 0$ it gives $k^\lambda M_\lambda(k) \to 2M g_A(0) \bar{u}_p \gamma_5 u_n$. We use again the PCAC relation (18.55) and approximate the matrix element of ϕ by the one-pion exchange diagram. This leads to $\lim_{k\to 0} M_\pi(k) = \mathrm{i}\sqrt{2} g_{\pi NN} \bar{u}_p \gamma_5 u_n$. From (18.59) we then obtain

$$g_A(0) = \frac{g_{\pi NN} f_\pi}{\sqrt{2} M} \qquad (18.61)$$

where $g_{\pi NN}$ is the pion–nucleon coupling constant. This is the *Goldberger–Treiman* relation, whose study played an important role in the development of the theory of weak interactions. Using the values for $g_{\pi NN} \simeq 13.15$ and $f_\pi \simeq 130$ MeV, we find $g_A(0) \simeq 1.3$, in good agreement with experiment. The discrepancy of a few per cent, gives the order of magnitude of the expected difference between the actual world and the hypothetical one with $m_\pi^2 \simeq 0$.

As a last step we derive a relation which tests the entire chiral algebra (18.52), thus obtaining experimental evidence to the hypothesis that the axial current, which satisfies the commutation relations (18.52), is the one which describes neutron β-decay. This relation was derived independently by S. Adler and W. Weisberger in 1965, and was the first confirmation of the chiral symmetry of strong interactions.[12] We start with the quantity

$$T^{ab}(s,t,k^2,k'^2) =$$

$$= \frac{(k^2 - m_\pi^2)(k'^2 - m_\pi^2)}{f_\pi^2 m_\pi^4} \int \mathrm{d}^4x \mathrm{d}^4y \mathrm{e}^{\mathrm{i}k\cdot x} \mathrm{e}^{-\mathrm{i}k'\cdot y} \langle p'| T(\partial^\mu A_\mu^a(x) \partial^\nu A_\nu^b(y)) |p\rangle \qquad (18.62)$$

where $s = (p+k)^2$, $t = (p-p')^2$. The $|p\rangle$ and $|p'\rangle$ are one-nucleon states and a and b are $SU(2)$ indices. Using the reduction formula we see that $T^{ab}(s,t,m_\pi^2,m_\pi^2)$ equals the physical pion–nucleon scattering amplitude. By partial integration this equation becomes

$$T^{ab}(s,t,k^2,k'^2) =$$

$$= \frac{(k^2 - m_\pi^2)(k'^2 - m_\pi^2)}{f_\pi^2 m_\pi^4} \int \mathrm{d}^4x \mathrm{d}^4y \mathrm{e}^{\mathrm{i}k\cdot x} \mathrm{e}^{-\mathrm{i}k'\cdot y} \Big\{ - \mathrm{i}k^\mu \langle p'| T(A_\mu^a(x) \partial^\nu A_\nu^b(y)) |p\rangle$$

$$- \langle p'| [A_0^a(x), \partial^\nu A_\nu^b(y)] |p\rangle \delta(x^0 - y^0) \Big\}$$

$$= \frac{(k^2 - m_\pi^2)(k'^2 - m_\pi^2)}{f_\pi^2 m_\pi^4} \int \mathrm{d}^4x \mathrm{d}^4y \mathrm{e}^{\mathrm{i}k\cdot x} \mathrm{e}^{-\mathrm{i}k'\cdot y} \Big\{ k^\mu k'^\nu \langle p'| T(A_\mu^a(x) A_\nu^b(y)) |p\rangle$$

$$+ \delta(x^0 - y^0) \left(\mathrm{i}k^\mu \langle p'| [A_0^b(y), A_\mu^a(x)] |p\rangle - \langle p'| [A_0^a(x), \partial^\nu A_\nu^b(y)] |p\rangle \right) \Big\} \qquad (18.63)$$

The equal-time commutators come from the derivatives of the θ functions implicit in the time-ordered symbols. We intend to derive a low energy theorem keeping terms

---

[12]We will present here a derivation proposed by S. Weinberg in 1966.

Fig. 18.8 The two diagrams which could give a $1/k$ singularity. If k and k' are the momenta associated with the two current operators, the nucleon propagator of the diagram on the left is proportional to $(\not{p} + \not{k} - M)^{-1}$ and that of the one on the right is proportional to $(\not{p} - \not{k'} - M)^{-1}$.

up to, and including, first order in k and k'. In this limit $k^2 \simeq k'^2 \simeq t \simeq 0$ and we are left with only one invariant $s \simeq M^2 + 2p \cdot k \simeq M^2 + 2p \cdot k'$. We will then argue that $T^{ab}(s,0,0,0)$ is a good approximation, up to terms of order m_π^2/M^2, of the physical scattering amplitude at threshold, namely $T^{ab}(s,0,0,0) = T^{ab}(s,0,m_\pi^2,m_\pi^2) + \mathcal{O}(m_\pi^2/M^2)$.

Let us dispose of the first term. It contains two k factors, so it will give a vanishing contribution, unless the matrix element has a singularity like $1/k$. In field theory such singularities come from vanishing propagators and, in this matrix element, they correspond to the two diagrams of Figure 18.8. They can be computed explicitly. For example, the first one, apart from numerical factors, gives

$$\bar{u}(p')\not{k'}\gamma_5 \frac{\not{p} + \not{k} + M}{k^2 + 2p \cdot k}\not{k}\gamma_5 u(p) = \bar{u}(p')\not{k'}\frac{\not{p} + \not{k} - M}{k^2 + 2p \cdot k}\not{k}u(p) \qquad (18.64)$$

We want to evaluate this expression at the physical threshold $\boldsymbol{p} = \boldsymbol{p'} = \boldsymbol{k} = \boldsymbol{k'} = 0$ and $k_0 = k'_0 = m_\pi$. Using the Dirac equation and neglecting terms of second order, we find

$$\bar{u}(p')\not{k'}\left[1 - \frac{\not{k}}{k_0}\right]u(p) \sim \mathcal{O}(k^2) \qquad (18.65)$$

i.e. the contribution of this diagram at threshold, despite the apparent pole term, is of second order. The same is true with the second diagram. This is not surprising because, by parity conservation, the pole diagrams contribute to the p-wave pion–nucleon scattering at threshold, see Problem 18.12.

We are thus left with the two equal-time commutator terms, which we want to evaluate at $\boldsymbol{k} = \boldsymbol{k'} = 0$. The first one gives, to first order in k^0,

$$\frac{ik_0}{f_\pi^2}\int \mathrm{d}^4x\,\mathrm{d}^4y\,\mathrm{e}^{ik_0x_0}\,\mathrm{e}^{-ik'_0y_0}\,\langle p'|\,[A_0^b(y), A_0^a(x)]\,|p\rangle\,\delta(x_0 - y_0)$$

$$= \frac{ik_0}{f_\pi^2}\int_{-\infty}^{+\infty}\mathrm{d}x_0\,\mathrm{e}^{i(k_0-k'_0)x_0}\,\langle p'|\,[Q_5^b(x_0), Q_5^a(x_0)]\,|p\rangle$$

$$= \frac{k_0}{f_\pi^2}\epsilon^{abc}\int_{-\infty}^{+\infty}\mathrm{d}x_0\,\mathrm{e}^{i(k_0-k'_0)x_0}\,\langle p'|\,Q^c\,|p\rangle \qquad (18.66)$$

where Q_5^a is the charge corresponding to the axial current and we have used the commutation relations (18.52). Q^c is the isospin generator whose matrix element between two nucleon states can be written as

$$\langle p' | Q^c | p \rangle = I^c \bar{u}(p') u(p) (2\pi)^3 \delta^3(\boldsymbol{p} - \boldsymbol{p}') \qquad (18.67)$$

with I^c a numerical coefficient denoting the isospin part of the matrix element. Putting all this together, and noting that the time integration gives an extra δ-function, we obtain the contribution of the second term

$$\frac{k_0}{f_\pi^2} \epsilon^{abc} I^c \bar{u}(p') u(p) (2\pi)^4 \delta^4(p + k - p' - k') \qquad (18.68)$$

This expression is the direct consequence of the $SU(2) \times SU(2)$ algebra. We now turn to the third term. The same analysis leads to the matrix element of the equal-time commutator $\langle p' | [Q_5^b(x_0), \dot{Q}_5^a(x_0)] | p \rangle$. This commutator is not part of the algebra (18.52), but we can make two remarks. First, since the equal-time commutator of the two axial charges gives a vector charge which is time-independent, it is straightforward to show that the commutator $[Q_5^b(x_0), \dot{Q}_5^a(x_0)]$ is symmetric in the indices a and b. Therefore, the part of the pion–nucleon scattering amplitude at threshold, which is antisymmetric in the two pion isospin indices, is completely determined by the $SU(2) \times SU(2)$ algebra. The threshold amplitude is parametrised by the s-wave scattering lengths $a_{1/2}$ and $a_{3/2}$ for the isospin combinations $\frac{1}{2}$ and $\frac{3}{2}$ respectively and equation (18.68) gives

$$a_{\frac{1}{2}} - a_{\frac{3}{2}} = \frac{3m_\pi}{2\pi f_\pi^2} \simeq 0.22 m_\pi^{-1} \qquad (18.69)$$

This is the Adler–Weisberger relation in the simple form obtained by Weinberg.[13] We can go one step further and evaluate the commutator involving the time derivative of the axial charge. A simple argument is that the non-vanishing time derivative is due to the non-conservation of the axial current whose divergence, according to the PCAC equation (18.55), is of order $\mathcal{O}(m_\pi^2)$. So, in our approximation we must put it equal to zero. In a more refined analysis we should look for possible singular contributions and it is easy to see that there is none. This is due to the fact that this commutator has the quantum numbers of 0^+ and there is no scalar hadron with a small mass like that of the pion.[14] Putting this commutator equal to zero gives a second relation for the scattering lengths of the symmetric combination of the amplitude

$$a_{\frac{1}{2}} + 2 a_{\frac{3}{2}} = 0 \qquad (18.70)$$

The values of the scattering lengths are obtained by a fit of the measured low energy s-wave scattering amplitude extrapolated to threshold. They depend slightly on the model used for the fit and this is reflected in an uncertainty on the order of a few per cent. The average values are $a_{1/2} = 0.175 \, m_\pi^{-1}$ and $a_{3/2} = -0.082 \, m_\pi^{-1}$, in good agreement with equations (18.69) and (18.70).

---

[13] Adler and Weisberger expressed the threshold amplitude in terms of an integral over the total pion–nucleon cross sections and used the Goldberger–Treiman relation (18.61) to replace f_π by g_A and the pion–nucleon coupling constant.

[14] This term is called the σ-term in the literature from the corresponding field in the σ-model we studied in Problem 14.2.

18.7.1.4   Other strangeness conserving semi-leptonic weak processes.   There are several other strangeness conserving weak transitions, which have been studied both theoretically and experimentally. They include:

(i) The $\Sigma^0 \to \Lambda$ transition we have mentioned already, to which, as we pointed out, the vector part of the current contributes very little.

(ii) We have studied already the charged pion decays into μ-ν or e-ν. We can also consider the β-decay of charged pions of the form $\pi^+ \to \pi^0 + e^+ + \nu_e$. Because of limited phase space the branching ratio is very small $\Gamma_\beta / \Gamma_{\text{tot}} \approx 10^{-8}$.

(iii) The τ-lepton semi-leptonic decays of the form: $\tau \to \nu_\tau$ +hadrons, where "hadrons" means mainly pions and kaons. They can be considered as the analogues, by crossing symmetry, of the pion β-decay.

(iv) Another important process is the neutrino–nucleon scattering, which we will discuss in Chapter 20.

18.7.2   Strangeness violating semi-leptonic weak interactions

Strange particles were identified through their weak decays, so strangeness violation has been understood to be a property of weak interactions since the early 1950s. The study of these processes led to many fundamental discoveries, such as parity violation, but we will not follow here the historical evolution.

In the previous section we found that the remarkable algebraic properties of the strangeness conserving hadronic current can be encoded in a simple free quark model as shown in equation (18.53). The discovery of strange particles led us to introduce a third quark s with charge $-\frac{1}{3}$, so it is natural to try to add to $h_\lambda(x)$ of equation (18.53) a strangeness changing piece proportional to $\bar{u}(x)\gamma_\lambda(1+\gamma_5)s(x)$. This way the total weak current should be a superposition of the form

$$h_\lambda^\dagger(x) = \bar{u}(x)\gamma_\lambda(1+\gamma_5)(c_1 d(x) + c_2 s(x))$$
$$= \sqrt{c_1^2 + c_2^2}\,\bar{u}(x)\gamma_\lambda(1+\gamma_5)[\cos\theta d(x) + \sin\theta s(x)] \qquad (18.71)$$

where we have defined $\cot\theta = c_1/c_2$. The principle of universality, i.e. equal couplings of the leptonic and the hadronic current, imposes the condition $\sqrt{c_1^2 + c_2^2} = 1$ leading to the form proposed by N. Cabibbo in 1963.[15] θ is called *the Cabibbo angle*, and it is a parameter which should be determined by experiment. So,

$$h_\lambda^\dagger(x) = \bar{u}(x)\gamma_\lambda(1+\gamma_5)[\cos\theta d(x) + \sin\theta s(x)] \qquad (18.72)$$

Note that in 1963 the quark model had not yet been introduced and Cabibbo postulated that the current operator was a member of an octet of $SU(3)$. Before checking the experimental consequences, let us make the following remarks:

• Having only one quark with charge $+2/3$ and two with charge $-1/3$, equation (18.72) is the most general expression we can write. This means that the orthogonal combination

$$\tilde{h}_\lambda^\dagger(x) = \bar{u}(x)\gamma_\lambda(1+\gamma_5)[\cos\theta s(x) - \sin\theta d(x)] \qquad (18.73)$$

---

[15]The introduction of a parameter like an angle to relate the strengths of the $\Delta S = 0$ and $\Delta S = 1$ parts of the weak current was first proposed by M. Gell-Mann and M. Lévy in 1960.

will not participate in the weak interactions. We will come back to this point in Chapter 19.

• The angle θ has a simple physical meaning. In section 6.11 we presented an ordering of the interactions among elementary particles following their strength and the symmetries they respect. The very strong interactions conserve $SU(3)$ and do not distinguish any particular direction in $SU(3)$ space. It is the medium strong interactions which, by breaking $SU(3)$, specify a particular direction in it and define what we call strangeness. Weak interactions also break $SU(3)$ but they choose a different direction. The Cabibbo angle is the difference between these two directions. So, saying that "the weak interactions do not conserve strangeness" is not accurate. The correct expression is that "weak interactions do not conserve the strangeness chosen by the medium strong interactions". Were medium strong interactions absent, we would have chosen the Cabibbo-rotated basis in quark space (u, $d_c = \cos\theta d(x) + \sin\theta s(x)$ and $s_c = \cos\theta s(x) - \sin\theta d(x)$) and there would be no strangeness violation.

• In the previous section we studied the properties of the strangeness conserving piece of the current, and formulated the principles of CVC and PCAC. We can extend these concepts to the strangeness changing part and obtain the algebra of chiral $SU(3) \times SU(3)$, but the results will not be as reliable as the ones based on $SU(2)$. The breaking of $SU(3)$ is much harder than that of isospin, so the conservation of the vector current is not expected to be a very good approximation. Furthermore, the mass of the K-mesons ($m_K \simeq 494$ MeV) is not that small and this makes the assumption of approximate chiral symmetry with spontaneous breaking questionable.

Let us now examine quantitatively some experimental consequences of the current given by equation (18.72).

• *The $\Delta S < 2$ rule.* It is clear that, to first order in the weak interactions, the form of the current (18.72) does not allow for semi-leptonic weak decays which change strangeness by more than one unit. This is confirmed experimentally by the absence of decays of the form $\Xi^0 \to p + e^- + \bar{\nu}$, or $\Xi^- \to n + e^- + \bar{\nu}$. The present bounds on their partial decay widths are on the order of 10^{-3}.

• *The $\Delta S = \Delta Q$ rule.* Similarly, the strangeness changing processes described by a current belonging to an $SU(3)$ octet, such as the $\bar{q}q$ form we used before, satisfy always $\Delta S = \Delta Q$. Therefore $\Delta S = -\Delta Q$ transitions are forbidden. The best evidence is the absence of the decay $\Sigma^+ \to n + e^+ + \nu$ at the 5×10^{-6} level.[16]

• *The determination of the Cabibbo angle.* In his original 1963 paper Cabibbo attempted a determination of the angle θ by comparing the rates of $K^+ \to \mu^+ + \nu$ and $\pi^+ \to \mu^+ + \nu$. Apart from kinematical factors due to the mass difference between pions and kaons, their ratio is proportional to $\tan^2\theta$, but also to the ratio of the two constants f_K^2/f_π^2, for which Cabibbo assumed the value 1, the value given by exact $SU(3)$. This gives $\theta \simeq 0.26$. As an independent check he looked at the ratio between the rates of $K^+ \to \pi^0 + e^+ + \nu$ and $\pi^+ \to \pi^0 + e^+ + \nu$ with the same result for θ. More precise estimates give a ratio $f_K/f_\pi \simeq 1.2$ due to $SU(3)$ breaking effects, thus

---

[16]In 1962, there appeared an erroneous experiment claiming evidence for a $\Sigma^+ \to n + \mu^+ + \nu$ decay and caused considerable confusion.

reducing the value of θ. We will postpone a detailed discussion of the present situation until we obtain a more complete picture in Chapter 19.

• *Corrections to* CVC. An immediate consequence of the form (18.72) is that the vector form factor g_V of β-decay changes to $g_V \cos\theta$. This implies that the actual value of $g_V(0)$ we used in equation (18.36) should be corrected. On the other hand, we remind the reader that $g_V(0) \simeq 1$ was the basis of our discussion, which led to the conserved vector current hypothesis. Leaving a more accurate discussion for Chapter 19, we only remark here that the actual value of $g_V(0)$ is close to 0.97, i.e. slightly smaller than 1, in agreement with the Cabibbo theory. In retrospect, we can say that we were lucky that the early determinations of $g_V(0)$ were not very precise, otherwise we could have missed the conservation of the vector current.

18.8 Purely Hadronic Weak Interactions

In the current $\times$ current theory they are described by the product $h^\lambda(x)h_\lambda^\dagger(x)$, with $h_\lambda^\dagger(x)$ given in equation (18.72). The strangeness conserving transitions are often masked by the strong interactions with the exception of some small parity violating effects in nuclear physics. The most important processes are the non-leptonic K and hyperon decays. In Chapter 6 we presented some of their properties.

• *The $\Delta S < 2$ rule.* With the form (18.72) of the weak current any $\Delta S \geq 2$ hadronic transition must be of higher order in the weak interactions and, therefore, highly suppressed. This is confirmed by the absence of the decays $\Xi^0 \to p + \pi^-$ and $\Xi^- \to n + \pi^-$ at the 10^{-5} level, or $\Omega^- \to \Lambda + \pi^-$ at the 10^{-6} level. An even stronger indication comes from the tiny value of the $K_L^0 - K_S^0$ mass difference, as we explained in section 6.10.

• *The $\Delta I = \frac{1}{2}$ rule.* The strangeness conserving piece of the current is an isovector operator and the strangeness changing piece is an isospinor. It follows that the product $h^\lambda(x)h_\lambda^\dagger(x)$ will induce transitions with $\Delta I = \frac{1}{2}$ and $\Delta I = \frac{3}{2}$. Both are observed. However, it is an empirical fact that the $\Delta I = \frac{1}{2}$ transitions are considerably enhanced. The amplitude for the decay $K^+ \to \pi^+ + \pi^0$, which is a pure $\Delta I = \frac{3}{2}$ decay (see Problem 6.7), is around 20 times smaller than the $K_S^0 \to \pi^+ + \pi^-$ one. We have no convincing dynamical explanation of such a large enhancement factor, although some estimates go in the right direction.

18.9 The Intermediate Vector Boson (IVB) Hypothesis

With the establishment of the current $\times$ current theory (18.7) weak interactions share a common feature with the electromagnetic ones, namely, they both involve vector currents. We can push the analogy much further with the hypothesis of an intermediate vector boson. We replace the current $\times$ current Lagrangian $\mathcal{L}_F$ by a new $\mathcal{L}_W$

$$\mathcal{L}_F = \frac{G_F}{\sqrt{2}} J^\lambda(x) J_\lambda^\dagger(x) \quad \Rightarrow \quad \mathcal{L}_W = g J^\lambda(x) W_\lambda(x) + \text{h.c.} \qquad (18.74)$$

where $W_\lambda(x)$ is a complex vector field describing an electrically charged spin one particle $W^\pm$ with mass m_W satisfying the Proca equation (7.101). The g is a dimensionless coupling constant, and $\mathcal{L}_W$ splits the four-fermion interaction of equation (18.7) and,

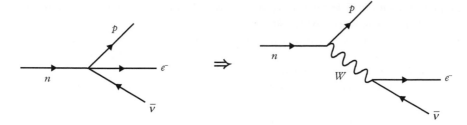

Fig. 18.9 Neutron β-decay in the Fermi four-fermion theory and according to the intermediate vector boson hypothesis.

e.g. for beta decay at lowest order, we obtain the diagrams of Figure 18.9. If m_W is larger than the typical energy or momentum transfer of the process, the predictions of the two theories in (18.74) agree, provided

$$\frac{G_F}{\sqrt{2}} = \frac{g^2}{m_W^2} \tag{18.75}$$

The intermediate vector bosons $W^\pm$ remained as hypothetical particles for more than a quarter of a century and their existence was only established in 1983. Written in the form of $\mathcal{L}_W$, the weak interactions look quite similar to the electromagnetic ones, modulo some important differences:

1. The electromagnetic interactions are long range – the photon is massless. Weak interactions give rise to short range forces – the Ws must be very massive.
2. The electromagnetic current is conserved, the weak current is not. $(\partial_\lambda J^\lambda(x) \neq 0)$.
3. The photon is neutral, while W is charged.

These physical differences imply several technical ones, the most important of which concerns the vector boson propagator. We know that the photon propagator is given in the Feynman gauge by $-i\eta_{\mu\nu}/k^2$ and, therefore, behaves at large momenta like k^{-2}. The W-propagator, on the other hand, is given by equation (7.103) and tends asymptotically to a non-vanishing constant. In Problem 15.2 we proved that this behaviour implies that the resulting theory is non-renormalisable. The IVB hypothesis did not make Fermi's theory mathematically consistent.

18.10 Charged and Neutral Currents

In the previous sections we introduced the notion of the weak current $J_\lambda(x)$ as a sum of a hadronic and a leptonic piece and showed that the Hamiltonian in the form of a product $J^\lambda J_\lambda^\dagger$ describes, in a phenomenological way, a large class of weak processes. The question we want to address here is whether this form describes *all* weak interactions.

An important property of $J_\lambda(x)$ is that it induces a change of the electric charge by one unit. It describes transitions such as neutron to proton, neutrino to electron, u quark to d or s. For this reason we call $J_\lambda(x)$ *a charged current*. In the language of the IVB this led to the introduction of charged intermediaries ($W^\pm$). Question: are there processes for which we need to introduce *neutral weak currents?*, in other words, do there exist weak transitions like $p \rightarrow p$, or $\nu \rightarrow \nu$, or $d \rightarrow d$ or s, etc. Phrased differently, do there exist neutral intermediate vector bosons? Today we know that the answer is yes, but it is instructive to describe the quest, both theoretical and experimental, which led to their discovery.

Examples of physical processes which would require neutral currents are:

1. Parity violation effects in atomic physic or in elastic electron–proton scattering. The amplitude for the process $e^- + p \rightarrow e^- + p$ is dominated by the electromagnetic interactions, which conserve parity. If a hypothetical weak neutral current violates parity like the charged one, we would expect small parity violating effects, which could be detected. Indeed, such effects have been measured at SLAC in high energy $e^- - p$ scattering in 1979 and at the Ecole Normale Supérieure in atomic physics experiments in 1982.

2. Elastic, or quasi-elastic, neutrino scattering. They are reactions such as

$$\nu + N \rightarrow \nu + X \,, \quad \nu_\mu + e^- \rightarrow \nu_\mu + e^- \tag{18.76}$$

where N denotes a nucleon and X any hadronic system. It was in reactions of this type that the existence of neutral currents was first established in 1973 and we will describe the relevant experiments in Chapter 20.

3. A very clean signal for a neutral current with no important background was provided by decays of K mesons like

$$K^0 \rightarrow \mu^+ + \mu^- \quad \text{or} \quad K^+ \rightarrow \pi^+ + e^- + e^+ \tag{18.77}$$

Such decays were known to be highly suppressed. The present results for the corresponding branching ratios are $(6.84 \pm 0.11) \times 10^{-9}$ and $(3.00 \pm 0.09) \times 10^{-7}$, respectively. These values are compatible with the absence of neutral currents and with the possibility that the transitions are due to second order charged current weak interactions. For many years this suppression was interpreted as evidence against the very existence of neutral currents, even though the above results refer only to strangeness changing transitions. We will show in Chapter 19 that their correct interpretation led to the discovery of fundamental new physics.

18.11   *CP* Violation

In the analysis of the $K^0 - \overline{K^0}$ system, which we presented in section 6.10, we assumed that the weak interactions were invariant under the operation CP, the product of charge conjugation and parity. By the CPT theorem this is equivalent to the invariance under time reversal T. Indeed, until 1964, all experimental evidence confirmed this assumption. This invariance was used in equations (6.31) and (6.32), in order to construct the states $|K_L^0\rangle$ and $|K_S^0\rangle$ as eigenstates of CP. The most direct experimental evidence for this conservation was the absence of the decay $K_2^0 \rightarrow 2\pi$, as we explained in section 6.10. Therefore, it came as a great surprise the discovery in 1964 by J.H. Christenson, J.W. Cronin, V.L. Fitch and R. Turlay, that such decays did in fact occur. The principle of the experiment is rather simple. It is based on the fact, which we explained in section 6.10, that the two eigenstates of the $K^0 - \overline{K^0}$ system have very different lifetimes, namely 0.9×10^{-10} s for the short-lived component and 5×10^{-8} s for the long-lived one. Their corresponding $c\tau$ values are 2.7 cm and 15 m, respectively. Therefore, a beam of neutral kaons, after a certain distance, consists entirely of the long-lived component. In the Cronin–Fitch experiment this distance was 57 ft (17.37 m) and no K_S^0 was expected to survive. Still, they found 45 ± 9 $\pi^+ - \pi^-$ decay events in a sample of 22700 K^0 decays. As a further check, they placed a tungsten plate just before the decay region in order to regenerate the short-lived component according to the Pais–Piccioni scheme we presented in section 6.10. They checked that the 2π events they observed in the absence of the plate had exactly the same profile with the ones obtained from the K_S^0 decays after regeneration. For this discovery Cronin and Fitch were awarded the 1980 Physics Nobel Prize.

Since both long- and short-lived components of the neutral kaons decay to two pions, CP is not conserved and the simple analysis that we presented in section 6.10 must be repeated. Let us consider a state $|\Phi\rangle$, which is a superposition of $|K^0\rangle$ and $|\overline{K^0}\rangle$

$$|\Phi\rangle = C_1 |K^0\rangle + C_2 |\overline{K^0}\rangle \tag{18.78}$$

Its time evolution is given by a matrix equation of the form

$$i\frac{d}{dt} \begin{pmatrix} C_1(t) \\ C_2(t) \end{pmatrix} = \mathcal{M} \begin{pmatrix} C_1(t) \\ C_2(t) \end{pmatrix} \tag{18.79}$$

where $\mathcal{M}$ is a 2×2 matrix called *the mass matrix*. Among all states, which satisfy the evolution equation (18.79), there exist only two which have simple exponential time dependence of the form $|\Phi\rangle = e^{i\lambda t} |\Phi\rangle_0$. They are the eigenstates of $\mathcal{M}$ with eigenvalues λ

$$\mathcal{M} \begin{pmatrix} p \\ q \end{pmatrix} = \lambda \begin{pmatrix} p \\ q \end{pmatrix} \tag{18.80}$$

Due to the presence of the decay channels, $\mathcal{M}$ is not Hermitian, otherwise it would have had real eigenvalues. It is convenient to separate the Hermitian and anti-Hermitian parts of $\mathcal{M}$ and write

$$\mathcal{M} = M - \frac{i}{2}\Gamma, \quad M = \frac{1}{2}(\mathcal{M} + \mathcal{M}^\dagger), \quad \Gamma = i(\mathcal{M} - \mathcal{M}^\dagger) \tag{18.81}$$

Thus the study of the $K^0 - \overline{K^0}$ system has been reduced to that of determining the matrix elements of the Hermitian matrices M and Γ. This can only be done phenomenologically.

 • *The unitarity relations.* The probability as a function of time for the decay $|\Phi\rangle \rightarrow |F\rangle$, where $|F\rangle$ is a specific final state, will be given by the square of the transition matrix element

$$|\langle F|T|\Phi\rangle|^2 = |C_1 \langle F|T|K^0\rangle + C_2 \langle F|T|\overline{K^0}\rangle|^2 \tag{18.82}$$

The total decay rate $\Gamma_{\text{tot}}(\Phi)$ of $|\Phi\rangle$ is given by the sum over all possible final states

$$\Gamma_{\text{tot}}(\Phi) = |C_1|^2 \sum_F |\langle F|T|K^0\rangle|^2 + |C_2|^2 \sum_F |\langle F|T|\overline{K^0}\rangle|^2$$
$$+ (C_1^* C_2 \sum_F \langle F|T|K^0\rangle^* \langle F|T|\overline{K^0}\rangle + c.c.) \tag{18.83}$$

Conservation of probabilities implies that, for any state $|\Phi\rangle$, that is for any C_i, $\Gamma_{\text{tot}}(\Phi)$ must be equal to the rate of decrease $-d|\langle\Phi|\Phi\rangle|^2/dt$ of its norm, which using (18.78–18.81) is given by

$$-\frac{\mathrm{d}}{\mathrm{d}t}|\langle\Phi|\Phi\rangle| = C_i^* \Gamma_{ij} C_j = |C_1|^2 \Gamma_{11} + |C_2|^2 \Gamma_{22} + C_1^* C_2 \Gamma_{12} + C_2^* C_1 \Gamma_{21} \tag{18.84}$$

Therefore, we obtain for the matrix elements of Γ

$$\Gamma_{11} = \sum_F |\langle F|T|K^0\rangle|^2, \quad \Gamma_{22} = \sum_F |\langle F|T|\overline{K^0}\rangle|^2,$$

$$\Gamma_{21} = \Gamma_{12}^* = \sum_F \langle F|T|\overline{K^0}\rangle^* \langle F|T|K^0\rangle \tag{18.85}$$

These equations are called *unitarity relations*, because they follow from the conservation of probabilities. They relate the matrix elements of the anti-Hermitian part of $\mathcal{M}$ with the various decay rates. We remark that they do not give any information on the elements of M because the summations in equation (18.85) extend only over all physical transitions, i.e. those which conserve energy and momentum. The elements of M depend also on all possible virtual transitions, such as nucleon–anti-nucleon pairs, arbitrary numbers of pions, etc., which do not contribute to the conservation of probabilities.

• *The CPT relations.* We presented in section 11.2.1 the *CPT* theorem and explained its very general validity. In Problem 12.1 we derived the relation $T(A \to B) = T(\bar{B}' \to \bar{A}')$, where $T(A \to B)$ is the transition amplitude for the process $A \to B$, "bar" denotes replacement of particles with anti-particles and vice versa, and prime denotes reversal of spin directions. Applied to the transitions $K^0 \to K^0$ and $\overline{K^0} \to \overline{K^0}$, the above relation yields

$$\mathcal{M}_{11} = \mathcal{M}_{22}, \quad \text{that is} \quad M_{11} = M_{22} \text{ and } \Gamma_{11} = \Gamma_{22} \tag{18.86}$$

• *Long- and short-lived kaons.* Let us now solve, formally, the eigenvalue problem of equation (18.80) and obtain the two eigenstates, which will be the "true particles" in the presence of weak interactions, i.e. the particles which have well-defined lifetimes. Since equation (18.80) is linear and homogeneous in p and q, we can only determine the ratio $p_\pm/q_\pm = \pm\sqrt{\mathcal{M}_{12}/\mathcal{M}_{21}}$, with eigenvalues $\lambda_\pm = \mathcal{M}_{11} \pm \sqrt{\mathcal{M}_{12}\mathcal{M}_{21}}$, respectively. We will use the convention

$$p^2 = i\mathcal{M}_{12}, \quad q^2 = i\mathcal{M}_{21} \tag{18.87}$$

and write the eigenvectors and eigenvalues as

$$\mathcal{M}\begin{pmatrix} p \\ \pm q \end{pmatrix} = \lambda_\pm \begin{pmatrix} p \\ \pm q \end{pmatrix}, \quad \lambda_\pm = \mathcal{M}_{11} \mp \mathrm{i}pq = \mathcal{M}_{11} \pm \sqrt{\mathcal{M}_{12}\mathcal{M}_{21}} \tag{18.88}$$

We therefore define the two states

$$\left| K_S^0 \right\rangle = \frac{p \left| K^0 \right\rangle + q \left| \overline{K^0} \right\rangle}{\sqrt{|p|^2 + |q|^2}} \text{ and } \left| K_L^0 \right\rangle = \frac{p \left| K^0 \right\rangle - q \left| \overline{K^0} \right\rangle}{\sqrt{|p|^2 + |q|^2}} \tag{18.89}$$

which satisfy

$$\left| \left\langle K_L^0 \middle| K_S^0 \right\rangle \right| = \left| \frac{|p|^2 - |q|^2}{|p|^2 + |q|^2} \right| \tag{18.90}$$

Their time evolution is

$$\left| K_S^0 \right\rangle \to \mathrm{e}^{-\mathrm{i}\lambda_+ t} \left| K_S^0 \right\rangle, \quad \left| K_L^0 \right\rangle \to \mathrm{e}^{-\mathrm{i}\lambda_- t} \left| K_L^0 \right\rangle$$

$$\lambda_+ \equiv m_S - \frac{\mathrm{i}}{2}\gamma_S, \quad \lambda_- \equiv m_L - \frac{\mathrm{i}}{2}\gamma_L \tag{18.91}$$

where m_L (m_S) and γ_L (γ_S), given above in terms of the matrix elements of $\mathcal{M}$, are the mass and decay rate of the long- (short-)lived component. Note that, if CP is conserved, $p = q$, so that, with an appropriate change of phase, we recover the eigenstates of section 6.10.

• *The phenomenological analysis.* The masses and lifetimes have been determined experimentally. So are the amplitudes for the various decay modes, which enter in the unitarity relations (18.85). Combining all this information, we can solve the equations and determine the parameters p and q. As an illustration, we do it here under the oversimplified assumption that the only final states which contribute in the sums (18.85) are the two pion states, namely the 2π with isospin $I = 0$, $|\pi\pi, I = 0\rangle$, and the 2π with $I = 2$, $|\pi\pi, I = 2\rangle$.

Let us start by giving a more precise characterisation of the two-pion states. Pions interact strongly among themselves and therefore we cannot assume that the state contains just two free, non-interacting particles. We do not know how to compute the two-pion state as an exact eigenstate of the strong interaction Hamiltonian and we will adopt a phenomenological description which we expect to be valid for low energy pions. It is based on two assumptions: First, we shall truncate the full pion–pion interaction by keeping only the pion–pion elastic scattering. Second, we shall approximate the latter using equation (4.28), which expresses the scattering amplitude in terms of the phase shifts, as we derived for non-relativistic collisions. Therefore, we can write

$$|\pi\pi, I\rangle = \mathrm{e}^{2\mathrm{i}\delta_I} |I\rangle, I = 0, 2 \tag{18.92}$$

where δ_0 and δ_2 are the s-wave pion–pion phase shifts for isospin 0 and 2, respectively, computed at a centre-of-mass energy m_K. The $|I\rangle$ denotes the state of two non-interacting pions with isospin I. The pion–pion phase shifts are not directly measurable and are extracted from a theoretical analysis of various processes in which two-pion states appear. A good example is provided by the charged kaon decay $K^+ \to \pi^+\pi^- e^+\nu_e$. A recent evaluation gives $\delta_0 \simeq 40$ and $\delta_2 \simeq -10$.

In order to proceed, we recall that the relative phase between the states $|K^0\rangle$ and $|\overline{K^0}\rangle$ is not determined by the strong interactions because $|K^0\rangle$ and $|\overline{K^0}\rangle$ correspond to different values of strangeness, which defines a superselection rule for strong and electromagnetic interactions. Here we shall adopt the following convention: we assume that the amplitude A_0, defined as

$$A_0 = \langle I = 0| T |K^0\rangle = \langle I = 0| T |\overline{K^0}\rangle \tag{18.93}$$

is real and positive. We also define the amplitude A_2 as: $A_2 = \langle I = 2| T |K^0\rangle$, but A_2 is not necessarily real.[17] In terms of the amplitudes for the decays of the physical states $|K_L^0\rangle$ and $|K_S^0\rangle$, we obtain

$$\frac{(p+q)A_0}{\sqrt{|p|^2 + |q|^2}} = \langle I = 0| T |K_S^0\rangle, \qquad \frac{(p-q)A_0}{\sqrt{|p|^2 + |q|^2}} = \langle I = 0| T |K_L^0\rangle$$

$$\frac{(pA_2 + qA_2^*)}{\sqrt{|p|^2 + |q|^2}} = \langle I = 2| T |K_S^0\rangle, \qquad \frac{(pA_2 - qA_2^*)}{\sqrt{|p|^2 + |q|^2}} = \langle I = 2| T |K_L^0\rangle \tag{18.94}$$

Experimentally we measure the following ratios

$$\eta_{+-} = \frac{\langle \pi^+\pi^-| T |K_L^0\rangle}{\langle \pi^+\pi^-| T |K_S^0\rangle}, \qquad \eta_{00} = \frac{\langle \pi^0\pi^0| T |K_L^0\rangle}{\langle \pi^0\pi^0| T |K_S^0\rangle} \tag{18.95}$$

where the two-pion states contain the interactions.

For the phenomenological analysis it is useful to define the following quantities

$$\epsilon = \frac{p-q}{p+q} = \frac{\langle I = 0| T |K_L^0\rangle}{\langle I = 0| T |K_S^0\rangle}, \qquad \omega = \frac{1}{\sqrt{2}} e^{2i(\delta_2 - \delta_0)} \frac{\langle I = 2| T |K_S^0\rangle}{\langle I = 0| T |K_S^0\rangle}$$

$$\epsilon' = \frac{1}{\sqrt{2}} e^{2i(\delta_2 - \delta_0)} \frac{\langle I = 2| T |K_L^0\rangle}{\langle I = 0| T |K_S^0\rangle} \simeq \frac{i}{\sqrt{2}} e^{2i(\delta_2 - \delta_0)} \frac{\mathrm{Im}A_2}{A_0} \tag{18.96}$$

Note that the quantities ϵ, ϵ' and ω, as well as η_{+-} and η_{00}, are very small because they all vanish in the limit of either CP conservation (ϵ, ϵ', η_{+-} and η_{00}) or exact $|\Delta I| = \frac{1}{2}$ (ϵ' and ω). In other words, the amplitude $\langle I = 0| T |K_S^0\rangle$ is much larger than any one of the others. We thus obtain

$$\eta_{+-} = \frac{\epsilon + \epsilon'}{1 + \omega} \simeq \epsilon + \epsilon' \quad , \quad \eta_{00} \simeq \epsilon - 2\epsilon' \tag{18.97}$$

It is straightforward to complete the analysis by including also the 3π and the semi-leptonic final states. The experimental values are

$$|\epsilon| \simeq |\eta_{00}| \simeq |\eta_{+-}| = (2.232 \pm 0.006) \times 10^{-3}, \mathrm{Re}\,(\epsilon'/\epsilon) = (1.66 \pm 0.23) \times 10^{-3} \tag{18.98}$$

and the phases

---

[17]We do not assume the validity of the $|\Delta I| = \frac{1}{2}$ rule because, as we have seen, it is valid only approximately and the $|\Delta I| = \frac{3}{2}$ amplitude is not negligible compared to the CP violating amplitude.

$$\phi_{+-} \simeq \phi_{00} \simeq \phi_\epsilon \simeq 43.5 \pm 0.05 \qquad (18.99)$$

Of particular interest are the semi-leptonic decay modes of K_L^0 in which we measure a CP violating parameter

$$A_L = \frac{\Gamma(\pi^- l^+ \nu_l) - \Gamma(\pi^+ l^- \bar{\nu}_l)}{\Gamma(\pi^- l^+ \nu_l) + \Gamma(\pi^+ l^- \bar{\nu}_l)} \qquad (18.100)$$

with l being either an electron or a muon. Both have been measured and they are essentially equal. Their weighted average is $A_L \simeq (0.332 \pm 0.006)\%$. It is worth noticing that the non-vanishing of A_L provides a convention-independent way to distinguish between electrons and positrons, in other words matter from anti-matter.

As we shall see later, with the introduction of heavy quark flavours, the problem of CP violation has been extended beyond the neutral kaon system and we shall study it in Chapter 22.

18.12 Problems

Problem 18.1 Find the non-relativistic limit of the five bilinears S, P, V, A and T, and establish the results shown in Table 18.1.

Problem 18.2 Prove that the couplings S, P and T give the same helicity for the electron and the anti-neutrino in β-decay, while V and A give opposite helicities.

Problem 18.3 Compute the e^- - $\bar{\nu}$ angular correlations in β-decay for each one of the couplings in eq. (18.1).

Problem 18.4 *The Fierz reordering theorem.* It applies to the four fermion theory (18.1). Let Ψ_a, $a = 1, \ldots, 4$, be four Dirac spinors and O_i $i = 1, \ldots, 5$, combinations of the five independent 4×4 matrices $\mathbf{1}$, γ_5, γ_μ, $\gamma_\mu \gamma_5$ and $\sigma_{\mu\nu}$. Equation (18.1) is written as $\sum_{i=1}^5 C_i \overline{\Psi}_1 O_i \Psi_2 \, \overline{\Psi}_3 O_i \Psi_4$. For most applications in weak interactions there is a natural pairing among the four spinors, for example the one we choose for β-decay in equation (18.1) is: (p, n) and (e, ν_e). But there are cases in which it is convenient to change the initial pairing and write the expression in an alternative form, e.g. $\sum_{i=1}^5 C_i' \overline{\Psi}_1 O_i \Psi_4 \, \overline{\Psi}_3 O_i \Psi_2$ with a new set of constants C_i'.

1. Compute the 5×5 transformation matrix $\mathbf{R}$ such that

$$\begin{pmatrix} S \\ P \\ V \\ A \\ T \end{pmatrix} [(\overline{\Psi}_1 \Psi_2)(\overline{\Psi}_3 \Psi_4)] = \mathbf{R} \begin{pmatrix} S \\ P \\ V \\ A \\ T \end{pmatrix} [(\overline{\Psi}_1 \Psi_4)(\overline{\Psi}_3 \Psi_2)] \qquad (18.101)$$

or, explicitly,

$$(\overline{\Psi}_1 O_i \Psi_2)(\overline{\Psi}_3 O_i \Psi_4) = \sum_{j=1}^5 R_i^j (\overline{\Psi}_1 O_j \Psi_4)(\overline{\Psi}_3 O_j \Psi_2) \quad \text{with}$$

$$\mathbf{R} = -\frac{1}{4} \begin{pmatrix} 1 & 1 & 1 & 1 & 1 \\ 4 & -2 & 0 & 2 & -4 \\ 6 & 0 & -2 & 0 & 6 \\ 4 & 2 & 0 & -2 & -4 \\ 1 & -1 & 1 & -1 & 1 \end{pmatrix}$$

The overall minus sign comes from the anti-commuting properties of the spinors.

2. Prove that the combination $V - A$ is stable under Fierz reordering.

Problem 18.5 1. Compute the amplitude for μ-decay and prove that, for the $V - A$ interaction, the Michel parameter has the value $\rho = 0.75$. Obtain equation (18.24) for the decay rate.

2. The main decay mode of a charged pion is (18.4). Consider a pion decaying at rest and show that the produced muons are polarised.

3. Study the decay of the produced muon given in (18.20). Show that in the $V - A$ theory there is a forward–backward asymmetry in the angular distribution of the final electrons relative to the polarisation of the intermediate muons.

Problem 18.6 *The neutron decay parameters.*

1. Compute the energy spectrum of the emitted electron in neutron β-decay in the vicinity of the "end point", i.e. in the region in which the electron takes almost all the available energy. Show that the form of the spectrum is very sensitive to a possible non-vanishing value of the neutrino mass.

2. In the following we will assume $m_e \simeq m_\nu \simeq 0$. We consider the decay of a polarised free neutron and compute the up-down asymmetry of the electron A_e and that of the anti-neutrino A_ν. Experimentally we have $A_e \approx -0.12$ and $A_\nu \approx 0.98$. Explain the large difference.

3. Explain why in C.S. Wu's experiment of the decay of polarised Co^{60} she found a much larger value for A_e.

Problem 18.7 The Dalitz plot. Consider the three-body decay $K \to 3\pi$. In order to simplify the calculations we make the following assumptions: (i) We shall neglect the mass difference between charged and neutral pions. (ii) Since the average kinetic energy of the final pions is low, $\approx 25\text{MeV}$, we shall assume the three pions to be non-relativistic. Let $\boldsymbol{p}_a$, $a = 1, 2, 3$, be the three-momenta of the final pions

1. Prove that each point in the phase space can be represented by a point inside an equilateral triangle with the distances of the point from the three sides being the energies of the three final pions. Determine the boundary of the kinematically allowed region.

2. Prove that equal areas in the triangle correspond to equal volumes in the phase space.

3. Show that if all pairs of pions have zero relative angular momentum, the distribution of points in the plot is uniform.

4. For comparison, find the distribution for the hypothetical case in which the decay is a two-stage one of the form: $K \to A + \pi \to 3\pi$, with A some particle of mass m_A and width Γ_A decaying into two pions.

Problem 18.8 In the leptonic weak interactions (18.17) ignore the electromagnetic coupling. Using canonical anti-commutation relations, derive the chiral symmetry algebra (18.19).

Problem 18.9 Prove that the CVC hypothesis implies the relation (18.48) which relates the weak interaction form factor g_T to the electromagnetic properties of the nucleons.

Problem 18.10 Consider the theory of free quarks with N flavours $q = (q_1, \ldots, q_N)^T$.

$$\mathcal{L} = \overline{q}(x)\mathrm{i}\partial\!\!\!/q(x) - \overline{q}(x)Mq(x)$$

with M an orthogonal $N \times N$ matrix. Determine the group of continuous transformations, other than Poincaré transformations, which leave this theory invariant in the following cases 1. $M = 0$, 2. $M \propto \mathbf{1}$, 3. M an arbitrary, non-singular orthogonal matrix.

Problem 18.11 Spontaneous symmetry breaking in the σ-model. We continue the study of the model we started in Problem 14.2.

We add to $\mathcal{L}$ the term $\mathcal{L}_{\mathrm{br}} = c\sigma(x)$, where c is a constant with dimensions $[\mathrm{M}]^3$, which breaks $SU(2) \times SU(2)$ to $SU(2)_V$. In order to reach the minimum of the potential, we shift σ.

1. Keeping only terms of first order in c, study the spectrum of the resulting theory in both cases, $M^2 > 0$ and $M^2 < 0$. Show that when there is no underlying spontaneous symmetry breaking, the shift of the field goes to zero together with c. The same happens to all symmetry breaking effects, such as the fermion mass. On the other hand, if $M^2 < 0$, the shift is $\sigma' = \sigma + v - c/2M^2$ with $v^2 = -6M^2/\lambda$. In this case, the symmetry breaking effects remain, even when c goes to zero. Find the masses of the pions in this case and show that they are what we called *pseudo-Goldstone bosons*. We believe that such a model provides a good description for real pions.

2. Compute, using Noether's theorem, the triplet of axial currents, i.e. the currents which correspond to the spontaneously broken generators of $SU(2) \times SU(2)$. Show that they are conserved in the absence of c. Show that they satisfy the divergence equation

$$\partial^\mu j^5_\mu \sim c\boldsymbol{\pi}$$

in other words, the divergence of the current is proportional to the field of the pseudo-Goldstone boson. We called this relation PCAC, for "partially conserved axial current" and it has played an important role in the development of the theory of both strong and weak interactions.

Problem 18.12 Using the results about the action of parity established in Problem 8.1, prove that the pole diagrams of Figure 18.8 contribute to the p-wave pion–nucleon scattering at threshold, thus explaining the suppression factor in equation (18.65).

19

A Gauge Theory for the Weak and Electromagnetic Interactions

19.1 Introduction

We want to apply all the powerful machinery of gauge theories to the real world and construct a realistic gauge theory of the fundamental interactions. Following the historical order, we start in this chapter with the weak and electromagnetic interactions. The idea of putting these two fundamental forces together and treating them on an equal footing was not obvious. Indeed, as we just saw, the weak interactions violate parity and involve $V - A$ currents, while the electromagnetic ones conserve parity and involve only vector currents. The first suggestion along these lines goes back to the work of S.L. Glashow of 1961, but it took several years before a realistic theory could be built. A remarkable fact of this achievement is that it had a theoretical, even an aesthetic, motivation. As we explained in Chapter 18, at the phenomenological level Fermi theory offered a perfectly satisfactory model. It fitted all data that were available at the time. The only unsatisfactory aspect was theoretical; it was not a renormalisable quantum field theory. It was the quest for mathematical consistency which motivated theorists, and their success has been a triumph of abstract theoretical thought. The main field theory result is due to G. 't Hooft and M. Veltman who proved that a Yang–Mills theory, with or without the BEH mechanism, is renormalisable. For this work they shared the 1999 Nobel Prize.

All these discoveries brought geometry into physics and culminated in what is known as "The Standard Model of Particle Physics". The road has been long and circuitous and many a time it gave the impression of leading to a dead end. No detailed history of this revolution has yet been written and we shall not attempt here to follow the historical order. This would have led to too lengthy a presentation, mainly because the theoretical developments often anticipated the experimental results. In many cases theorists had only their intuition for a guide and their deductions were necessarily speculative. Nevertheless, both theory and experiment made spectacular discoveries, which were parallel and complementary. It would have been instructive to follow in detail the historical order, precisely in order to show this close cooperation, but it would lengthen the exposition considerably. Therefore, we choose to go directly to the model in its final form, as it has been established today, and we shall only comment on the highlights of the interrelation between theory and experiment as they appear.

Elementary Particle Physics. John Iliopoulos and Theodore N. Tomaras, Oxford University Press.
© John Iliopoulos and Theodore N. Tomaras (2021). DOI: 10.1093/oso/9780192844200.003.0019

19.2 The Rules of Model Building

In this section we follow a step-by-step approach from the available experimental data to a model based on gauge invariance. The essential steps are:

1. Choose a gauge group G.

2. Choose the fields of the "elementary" particles whose interactions you want to describe, and assign them to representations of G. Include scalar fields to allow for the Brout–Englert–Higgs (BEH) mechanism.

3. Write the most general renormalisable Lagrangian invariant under G. At this stage, gauge invariance is still manifest and all gauge vector bosons are massless.

4. Choose the parameters of the BEH potential so that spontaneous symmetry breaking occurs.

5. Translate the scalar fields and rewrite the Lagrangian in terms of the shifted fields. Choose a suitable gauge and quantise the theory.

A remark: Gauge theories provide only the general framework, not a detailed model. The latter will depend on the particular choices made in steps (1) and (2). These choices will be inspired by the available experimental data.

19.3 The Lepton World

We start with the lepton world. This is also what happened historically, for reasons we will explain shortly. The leptonic part of the model was first proposed by S. Weinberg and A. Salam in 1967 and 1968. We will construct it following the five-step programme of section 19.2.

Step 1: Looking at the Table 6.6 we see that, for the combined electromagnetic and weak interactions, we have four gauge bosons, namely $W^{\pm}$, Z and the photon. As we explained in Chapter 13, each one of them corresponds to a generator of the group G. The only non-trivial group with four generators is $U(2) \cong SU(2) \times U(1)$. Here comes the first remark about the theory–experiment connection: When the model was proposed, the three intermediate vector bosons were still out of reach of any accelerator. Furthermore, the weak neutral currents, such as $\bar{e}\gamma_{\lambda}(1 \pm \gamma_5)e$, or $\bar{\nu}\gamma_{\lambda}$ $(1 \pm \gamma_5)\nu$, namely the sources of Z, had not yet been observed and their very existence was in doubt. Consequently, several attempts were made to build models which avoided them. None was particularly attractive and the subsequent discovery of the neutral currents at CERN by the Gargamelle group in 1973 established the $SU(2) \times U(1)$ model unambiguously and offered the first triumph of these ideas. We will discuss this discovery in Chapter 20. A naïve identification would have been to assign the photon to the generator of $U(1)$ and the other three to those of $SU(2)$. Glashow was the first to remark that a much richer algebraic structure is obtained if you allow for a mixing between the two neutral generators, that of $U(1)$ and the third component of $SU(2)$. Following the notation which was inspired by the hadronic physics, we call T_a, $a = 1, 2, 3$, the generators of $SU(2)$ and Y the one of $U(1)$. Then, the electric charge operator Q will be a linear combination of T_3 and Y. By convention, we write

$$Q = T_3 + \frac{Y}{2} \tag{19.1}$$

The coefficient in front of Y is arbitrary and only fixes the normalisation of the $U(1)$ generator relatively to those of $SU(2)$. This ends our discussion of the first step.

Step 2: Leptons have always been considered as elementary particles, both in 1967, when the model was initially proposed, and today. Their number has increased from four to six with the discovery of the τ and its associated neutrino ν_τ, so we must look for $SU(2)$ representations of dimension six. However, as we have already noticed, a striking feature of the data is the phenomenon of family repetition. We do not understand why nature chooses to repeat itself three times, but the simplest way to incorporate this observation into the model is to use three times the same representations, one for each family. This leaves $SU(2)$ doublets and/or singlets as the only possible choices. A further experimental input we shall use is the fact that the charged vector bosons couple only to the left-handed components of the lepton fields, in contrast to the photon, which couples with equal strength to both right and left. These considerations lead us to assign the left-handed components of the lepton fields to doublets of $SU(2)$. In the notation of section 18.6, we write

$$\Psi_L^i(x) = \frac{1}{2}(1 + \gamma_5) \begin{pmatrix} \nu_i(x) \\ l_i^-(x) \end{pmatrix}, \quad i = 1, 2, 3 \tag{19.2}$$

The right-handed components are assigned to singlets of $SU(2)$

$$\nu_{iR}(x) = \frac{1}{2}(1 - \gamma_5)\nu_i(x) \quad (?), \quad \ell_{iR}^-(x) = \frac{1}{2}(1 - \gamma_5)\ell_i^-(x) \tag{19.3}$$

The question mark next to the right-handed neutrinos means that the presence of these fields is not confirmed by the data. We shall drop them in this chapter, but we shall come back to this point later. We shall also simplify the notation and put $\ell_{iR}^-(x) = R_i(x)$. The resulting transformation properties under local $SU(2)$ are

$$\Psi_L^i(x) \rightarrow e^{i\boldsymbol{\theta}(x)\cdot\mathbf{t}} \Psi_L^i(x), \quad R_i(x) \rightarrow R_i(x) \tag{19.4}$$

with $t^a = \tau^a/2$ the 2×2 generators of $SU(2)$. This assignment and the Y normalisation given by equation (19.1) also fix the $U(1)$ charge, and therefore, the transformation properties, of the lepton fields. For all $i = 1, 2, 3$, we find

$$Y(\Psi_L^i) = -1, \quad Y(R_i) = -2 \tag{19.5}$$

If a right-handed neutrino exists, it has $Y(\nu_{iR}) = 0$, which shows that it is not coupled to any gauge boson.

We are left with the choice of the BEH scalar fields. One of them has already been discovered and, although its detailed properties have not yet been fully studied, the data seem to favour the minimal solution initially proposed by S. Weinberg. We must give masses to three vector gauge bosons and keep the fourth one massless. The latter will be identified with the photon. We recall that, for every vector boson acquiring mass, a scalar with the same quantum numbers decouples. At the end, we shall remain with at least one physical, neutral, scalar field. It follows that the minimal number to start with is four, two charged and two neutral. We choose to put them in a complex doublet of $SU(2)$,

$$\Phi = \begin{pmatrix} \phi^+ \\ \phi^0 \end{pmatrix}, \quad \Phi(x) \to e^{i\boldsymbol{\theta}(x)\cdot\mathbf{t}}\,\Phi(x) \tag{19.6}$$

with the conjugate fields ϕ^- and ϕ^{0*} forming $\Phi^\dagger$. The $U(1)$ charge of Φ is $Y(\Phi) = 1$.

This ends our choices for the second step. At this point the model is complete. All further steps are purely technical and uniquely defined.

Step 3: What follows is straightforward algebra. We write the most general, renormalisable Lagrangian, involving the fields (19.2), (19.3) and (19.6) invariant under gauge transformations of $SU(2) \times U(1)$. We shall also assume separate conservation of the three lepton numbers, but we shall further discuss this point in Chapter 20. The requirement of renormalisability implies that all terms in the Lagrangian are monomials in the fields and their derivatives and their canonical dimension is smaller than or equal to four. The result is

$$\mathcal{L} = -\frac{1}{4} W^\alpha_{\mu\nu} W^{\alpha\mu\nu} - \frac{1}{4} B_{\mu\nu} B^{\mu\nu} + |D_\mu \Phi|^2 - V(\Phi)$$

$$+ \sum_{i=1}^{3} \left[\overline{\Psi^i_L}\, i\slashed{D} \Psi^i_L + \overline{R_i}\, i\slashed{D} R_i - G_i \left(\overline{\Psi^i_L} R_i \Phi + \text{h.c.} \right) \right] \tag{19.7}$$

If we call $W^a_\mu, a = 1, 2, 3$ and B_μ the gauge fields associated to $SU(2)$ and $U(1)$ respectively, the corresponding field strengths $W^a_{\mu\nu}$ and $B_{\mu\nu}$ appearing in (19.7) are given by (13.26) and (2.7), i.e.

$$W^a_{\mu\nu} = \partial_\mu W^a_\nu - \partial_\nu W^a_\mu + g\,\epsilon^{abc} W^b_\mu W^c_\nu, \quad B_{\mu\nu} = \partial_\mu B_\nu - \partial_\nu B_\mu \tag{19.8}$$

Similarly, the covariant derivatives in (19.7) are determined by the assumed transformation properties of the fields, as shown in (13.19)

$$D_\mu \Psi^i_L = \left(\partial_\mu - i g\, t^a\, W^a_\mu + i\frac{g'}{2}\, B_\mu \right) \Psi^i_L, \quad D_\mu R_i = (\partial_\mu + i g' B_\mu)\, R_i$$

$$D_\mu \Phi = \left(\partial_\mu - i g\, t^a\, W^a_\mu - i\frac{g'}{2} B_\mu \right) \Phi \tag{19.9}$$

The two coupling constants g and g' correspond to the groups $SU(2)$ and $U(1)$, respectively. The most general invariant potential $V(\Phi)$ for the scalar fields compatible with the transformation properties of the field Φ is

$$V(\Phi) = \mu^2 \Phi^\dagger \Phi + \lambda (\Phi^\dagger \Phi)^2 \tag{19.10}$$

The last term in (19.7) is a Yukawa coupling term between the scalar Φ and the fermions. Since we have assumed the absence of right-handed neutrinos, this is the most general term which is invariant under $SU(2) \times U(1)$. As usual, h.c. stands for "hermitian conjugate". G_i are three arbitrary coupling constants.

A final remark: As expected, the gauge bosons $W^a_\mu, a = 1, 2, 3$, and B_μ appear to be massless. The same is true for all fermions. This is not surprising because the different transformation properties of the right- and left- handed components forbid the appearance of a Dirac mass term in the Lagrangian. Furthermore, the conservation of the three lepton numbers forbids the appearance of a Majorana mass term. In fact, the only dimensionful parameter in (19.7) is μ^2, the parameter in the BEH potential (19.10). Therefore, the mass of every particle in the model is expected to be proportional to $|\mu|$.

Step 4: The next step of our programme consists in choosing the parameter μ^2 of the scalar potential negative, in order to trigger the phenomenon of spontaneous symmetry breaking and the BEH mechanism. The minimum of the potential occurs at a point $\Phi^\dagger\Phi = v^2 = -\mu^2/\lambda$. As we explained in Chapter 14, we can choose the direction of the breaking to be along the real part of ϕ^0. As we have built the model, out of all the generators t^a and Y of the $SU(2) \times U(1)$, only the linear combination $Q = t^3 + Y/2$ annihilates the vacuum. In the physicist's jargon we have a breaking of $SU(2) \times U(1) \to U(1)_{\text{em}}$ and according to the BEH mechanism three of the gauge bosons will acquire masses, while one will remain massless and will be identified with the photon.

Step 5: Shifting the BEH field by its value at the minimum

$$\Phi \to \Phi + \frac{1}{\sqrt{2}}\begin{pmatrix} 0 \\ v \end{pmatrix}, \quad v^2 = -\frac{\mu^2}{\lambda} \tag{19.11}$$

transforms the Lagrangian and generates new terms, as was explained in Chapter 14. Let us look at some of them:

(*i*) *Fermion mass terms.* These are obtained from the Yukawa terms in (19.7), upon substitution of ϕ^0 by $v/\sqrt{2}$, and lead to the lepton masses

$$m_e = \frac{1}{\sqrt{2}}G_e v, \quad m_\mu = \frac{1}{\sqrt{2}}G_\mu v, \quad m_\tau = \frac{1}{\sqrt{2}}G_\tau v, \quad m_{\nu_i} = 0, \, i = 1,2,3 \tag{19.12}$$

Since we had three arbitrary constants G_i, we can fit the three observed lepton masses.

(*ii*) *Gauge boson mass terms.* These arise from the $|D_\mu\Phi|^2$ term in the Lagrangian. Substitution of (19.11) leads to the following quadratic terms among the gauge boson fields

$$\frac{1}{8}v^2 \left[g^2(W_\mu^1 W^{1\mu} + W_\mu^2 W^{2\mu}) + (g'B_\mu - gW_\mu^3)^2\right] \tag{19.13}$$

The masses of the charged vector bosons

$$W_\mu^\pm = \frac{1}{\sqrt{2}}(W_\mu^1 \mp iW_\mu^2) \tag{19.14}$$

are

$$m_W = \frac{1}{2}vg \tag{19.15}$$

The neutral gauge bosons B_μ and W_μ^3 have a 2×2 non-diagonal mass matrix. With $\tan\theta_W = g'/g$, the mass eigenstates are

$$Z_\mu = \cos\theta_W W_\mu^3 - \sin\theta_W B_\mu \quad \text{and} \quad A_\mu = \sin\theta_W W_\mu^3 + \cos\theta_W B_\mu \tag{19.16}$$

with masses

$$m_Z = \frac{v(g^2 + g'^2)^{1/2}}{2} = \frac{m_W}{\cos\theta_W} \quad \text{and} \quad m_A = 0 \tag{19.17}$$

respectively.

The BEH mechanism breaks the original symmetry according to $SU(2) \times U(1) \rightarrow U(1)_{\text{em}}$ and θ_W is the angle between the original $U(1)$ and the one left unbroken. It is the parameter first introduced by S.L. Glashow and it is often referred to as the "Weinberg angle".

(*iii*) *Physical scalar mass.* Three out of the four real fields of the Φ doublet will be absorbed by the BEH mechanism in order to allow for the three gauge bosons $W^{\pm}$ and Z to acquire their masses. The fourth one, which corresponds to $(|\phi^0\phi^{0*}|)^{\frac{1}{2}}$, remains physical. Its mass is given by the coefficient of the quadratic part of $V(\Phi)$ after the shift (19.11) and is equal to

$$m_H = \sqrt{-2\mu^2} = \sqrt{2\lambda}\, v \tag{19.18}$$

Notice that the mechanism of spontaneous symmetry breaking is at the origin of the creation of all masses of elementary particles, fermions as well as gauge bosons, with the exception of the remaining physical scalar.

In addition, various coupling terms are generated, which we shall present, together with the hadronic ones, in the next section.

19.4 Extension to Hadrons

Introducing the hadrons into the model presents some novel features. They are mainly due to the fact that the individual quark flavour numbers are not separately conserved, and we encounter the phenomenon of flavour mixing. As regards to *Step 2*, today there is consensus regarding the choice of the "elementary" constituents of matter: besides the six leptons, there are also six quarks. They are fractionally charged and each flavour comes in three "colours".

Let us pause here for a moment and make a second historical comment: In the 1960s only three quark flavours were known, the ones we present in Table 6.6 as u, d and s. Their electric charges are $+\frac{2}{3}, -\frac{1}{3}$ and $-\frac{1}{3}$, respectively, and the weak current, which was known at the time, was the one given in (18.72). Trying to extend the gauge theory ideas to this three-quark hadronic world we faced the following difficulty: The commutator of h and $h^{\dagger}$, which gives the neutral component of the $SU(2)$ current, contains pieces of the form $\bar{d}s$ and $\bar{s}d$, i.e. flavour-changing neutral currents. Their presence would induce decays of the type: $K^0 \rightarrow \mu^+\mu^-$ or $K^0 \rightarrow \nu\bar{\nu}$, both of which, as we explained in section 18.10, were absolutely excluded experimentally. The solution to this puzzle was found in 1970 and consisted in postulating the existence of a fourth quark flavour, named c for "charm", which made possible the addition of a second piece to the charged weak current

$$h_{\lambda}^{\dagger} = \bar{u}_L\gamma_{\lambda}(\cos\theta\ d_L + \sin\theta\ s_L) + \bar{c}_L\gamma_{\lambda}(\cos\theta\ s_L - \sin\theta\ d_L) \tag{19.19}$$

Now, it is easy to check that the resulting neutral current is diagonal in flavour space. The introduction of the fourth quark implied the prediction of the existence of

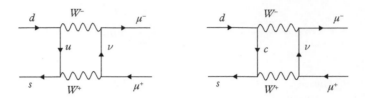

.**Fig. 19.1** The 1-loop contributions to a strangeness changing neutral current process.

an entire new hadronic sector, namely the hadrons containing c and or $\bar{c}$ among their constituents. Since, experimentally, the value of $\cos\theta$ is close to one, the "charmed" particles are predicted to decay predominantly to strange particles. We could even go a step further: in the limit of exact flavour symmetry, when all quarks have the same mass, the processes involving flavour changing neutral currents were forbidden by the symmetry. But we know that flavour symmetry is broken and the quarks do not have the same mass. If we look at these processes at higher orders we find, for example, the square diagrams of Figure 19.1. They describe amplitudes for processes such as $K^0 \to \mu^+\mu^-$, or the $\Delta S = 2$ transition $K^0\text{-}\overline{K^0}$, which gives the $K_S\text{-}K_L$ mass difference. The u and c quark contributions have opposite signs, so the resulting amplitudes are proportional to $(m_c^2 - m_u^2)/m_W^2$. This property implied the translation of the experimental upper bounds on these processes to an upper bound on m_c, and consequently, to an upper bound of a few GeV on the masses of the charmed particles.[1] The subsequent experimental discovery of these new particles in the predicted mass range and with precisely the correct decay signature, was a second great success of these ideas.

Coming back to the present, we note that, in order to explore the lepton–hadron universality property, we must use doublets and singlets for the quarks too. The first novel feature we mentioned above is that all quarks appear to have non-vanishing Dirac masses, so we must introduce both right-handed singlets for each family. A naïve assignment would be to write the analogue of (19.2) and (19.3) as

$$Q_L^i(x) = \begin{pmatrix} U^i(x) \\ D^i(x) \end{pmatrix}_L ; \quad U_R^i(x), \quad D_R^i(x) \tag{19.20}$$

with $U^i = u, c, t$ and $D^i = d, s, b$ for $i = 1, 2, 3$, respectively.[2] This assignment determines the $SU(2)$ transformation properties of the quark fields. It also fixes their Y charges. Using (19.1), we find

$$Y(Q_L^i) = \frac{1}{3}, \quad Y(U_R^i) = \frac{4}{3}, \quad Y(D_R^i) = -\frac{2}{3} \tag{19.21}$$

The presence of the two right-handed singlets has an important consequence. Already for a single family we can write two distinct Yukawa terms coupling the quarks to the scalar fields, namely

---

[1]This is known as *the GIM mechanism*, for Glashow, Iliopoulos, Maiani.
[2]An additional index a, running also through 1, 2 and 3 and denoting the colour, is understood.

$$\mathcal{L}_{\text{Yuk}} = G_d(\overline{Q_L} D_R \Phi + h.c.) + G_u(\overline{Q_L} U_R \widetilde{\Phi} + h.c.) \tag{19.22}$$

where $\widetilde{\Phi} = i\tau_2 \Phi^*$ is the doublet made out of ϕ^{0*} and ϕ^-. $\widetilde{\Phi}$ is also a doublet of $SU(2)$ with $Y(\widetilde{\Phi}) = -Y(\Phi) = -1$.

Were there only one family, this would have been the end of the story. The hadron Lagrangian would be the same as (19.7) with quark fields replacing leptons, plus the extra term (19.22). The complication we alluded to before arises with the addition of more families. In this case, the total Lagrangian is not just the sum over the family index. The physical reason is the non-conservation of the individual quark quantum numbers we mentioned previously. In writing (19.20), we implicitly assumed a particular pairing of the quarks in each family, namely u with d, c with s, and t with b. In general, we could choose any basis in family space and, since we have two Yukawa terms, we will not be able to diagonalise both of them simultaneously. It follows that the most general Lagrangian will contain a matrix with non-diagonal terms which mix the families. By convention, we attribute it to a different choice of basis in the d-s-b space. It follows that the correct generalisation of the Yukawa Lagrangian (19.22) to many families is given by

$$\mathcal{L}_{\text{Yuk}} = \sum_{i,j} \left[(\overline{Q_L^i}\, G_D^{ij} D_R^j \Phi + h.c.) \right] + \sum_i \left[G_U^i (\overline{Q_L^i}\, U_R^i \widetilde{\Phi} + h.c.) \right] \tag{19.23}$$

where the Yukawa coupling constant G_D has become a matrix in family space. After shifting the scalar field, we shall produce masses for the up quarks given by $m_u = G_U^1 v/\sqrt{2}$, $m_c = G_U^2 v/\sqrt{2}$ and $m_t = G_U^3 v/\sqrt{2}$, as well as the three-by-three mass matrix $M_D^{ij} = G_D^{ij} v/\sqrt{2}$ for the charge-$(-\frac{1}{3})$ quarks. We usually prefer to work in a field space where the masses are diagonal, so we change our starting D^i basis to the physical $\widetilde{D}^i = d, s, b$, in which G_D^{ij} is diagonal with elements G_D^i. This is done with a three-by-three transformation $D^i = V^{ij} \widetilde{D}^j$ such that $V^\dagger G_D V = \text{diag}(m_d, m_s, m_b)$. In the simplest example of only two families, it is easy to show that the most general such matrix, after using all freedom for field redefinitions and phase choices, is a real rotation

$$V_{\text{C}} = \begin{pmatrix} \cos\theta & \sin\theta \\ -\sin\theta & \cos\theta \end{pmatrix} \tag{19.24}$$

with θ being our familiar Cabibbo angle. For three families we can show (see Problem 19.1) that the matrix has four parameters, three angles, the three Euler angles, $\theta_{12}, \theta_{23}, \theta_{13}$, and one arbitrary phase δ. It is conventionally written in the form

$$V_{\text{CKM}} = \begin{pmatrix} c_{12}c_{13} & s_{12}c_{13} & s_{13}\,e^{-i\delta} \\ -s_{12}c_{23} - c_{12}s_{23}s_{13}\,e^{i\delta} & c_{12}c_{23} - s_{12}s_{23}s_{13}\,e^{i\delta} & s_{23}c_{13} \\ s_{12}s_{23} - c_{12}c_{23}s_{13}\,e^{i\delta} & -c_{12}s_{23} - s_{12}c_{23}s_{13}\,e^{i\delta} & c_{23}c_{13} \end{pmatrix} \tag{19.25}$$

with the notation $c_{kl} = \cos\theta_{kl}$ and $s_{kl} = \sin\theta_{kl}$, $k, l = 1, 2, 3$. The three angles θ_{kl} can be chosen to lie in the first quadrant, so that $c_{kl}, s_{kl} \geq 0$. The novel feature is the possibility of introducing the phase δ. This means that a six-quark model has a natural source of CP, or T, violation, while a four-quark model does not. This brings us to a third historical remark: It has been known since 1965 that weak interactions are not invariant under time reversal. The absence of a natural source for such violation in a two-family model prompted M. Kobayashi and T. Maskawa in 1973 to postulate the existence of a third family, a prediction which was again verified with the discovery of τ, b and t. So we see that the consistency and the simplicity of the electroweak theory repeatedly provided important information about the spectrum of elementary particles.

According to our present understanding of the theory, the four angles of the V_{CKM} matrix – the so-called Cabibbo–Kobayashi–Maskawa matrix – are fundamental parameters and should be determined by experiment. In principle this can be done in the way we determined the Cabibbo angle in section 18.7.2, namely by comparing the decay rates of various flavoured particles, but we will postpone this discussion until section 22.4.

The Lagrangian density before shifting the scalar field, is now

$$
\mathcal{L} = -\frac{1}{4} W^a_{\mu\nu} W^{a\mu\nu} - \frac{1}{4} B_{\mu\nu} B^{\mu\nu} + |D_\mu \Phi|^2 - V(\Phi)
$$

$$
+ \sum_{i=1}^{3} \left[\overline{\Psi^i_L}\, i\gamma^\mu D_\mu \Psi^i_L + \overline{R_i}\, i\gamma^\mu D_\mu R_i - G_i \left(\overline{\Psi^i_L} R_i \Phi + h.c. \right) \right.
$$

$$
+ \overline{Q^i_L}\, i\gamma^\mu D_\mu Q^i_L + \overline{U^i_R}\, i\gamma^\mu D_\mu U^i_R + \overline{D^i_R}\, i\gamma^\mu D_\mu D^i_R + G^i_U \left(\overline{Q^i_L} U^i_R \widetilde{\Phi} + h.c. \right) \Big]
$$

$$
+ \sum_{i,j=1}^{3} \left[\left(\overline{Q^i_L}\, G^{ij}_D D^j_R \Phi + h.c. \right) \right] \tag{19.26}
$$

The covariant derivatives on the quark fields are

$$
D_\mu Q^i_L = \left(\partial_\mu - i g\, t^a\, W^a_\mu - i\frac{g'}{6} B_\mu \right) Q^i_L
$$

$$
D_\mu U^i_R = \left(\partial_\mu - i\frac{2g'}{3} B_\mu \right) U^i_R
$$

$$
D_\mu D^i_R = \left(\partial_\mu + i\frac{g'}{3} B_\mu \right) D^i_R \tag{19.27}
$$

The classical Lagrangian (19.26) contains seventeen arbitrary real parameters. They are

1. The two gauge coupling constants g and g'.
2. The two parameters λ and μ^2 of the BEH potential.
3. Three Yukawa coupling constants for the three lepton families, $G_{e,\mu,\tau}$.
4. Six Yukawa coupling constants G^i_U and G^i_D for the three quark families.
5. Four parameters of the Cabibbo–Kobayashi–Maskawa V_{CKM} matrix, the three angles and the phase δ.

A remark: Fifteen out of these seventeen parameters are directly connected with the BEH sector.

Shifting the field Φ according to equation (19.11) and diagonalising the resulting down quark mass matrix produces the mass terms for fermions and bosons, which we have discussed already, as well as several interaction terms. We shall write the latter here in the unitary gauge we introduced in section 14.3, in which only fields corresponding to physical particles appear.

(*i*) *The gauge boson–fermion couplings.* These are the ones which generate the known weak and electromagnetic interactions. The photon field A_μ is coupled to the charged fermions through the usual electromagnetic current, i.e.

$$\frac{gg'}{(g^2 + g'^2)^{1/2}} \left[-\bar{e}\gamma^\mu e + \sum_{a=1}^{3} \left(\frac{2}{3}\bar{u}_a\gamma^\mu u_a - \frac{1}{3}\bar{d}_a\gamma^\mu d_a \right) + \dots \right] A_\mu \tag{19.28}$$

where the dots stand for the contribution of the other two families $(e, u, d) \rightarrow (\mu, c, s), (\tau, t, b)$, while a is the colour index. Equation (19.28) shows that the electric charge e is given, in terms of g and g' by

$$e = \frac{gg'}{(g^2 + g'^2)^{1/2}} = g\sin\theta_W = g'\cos\theta_W \tag{19.29}$$

Similarly, the couplings of the charged vector bosons to the weak current are

$$\frac{g}{2\sqrt{2}} \left(\bar{\nu_e}\gamma^\mu(1 + \gamma_5)e + \sum_{a=1}^{3} \bar{u_a}\,\gamma^\mu(1 + \gamma_5)D_a^1 + \dots \right) W_\mu^+ + h.c. \tag{19.30}$$

As expected, only left-handed fermions participate. $D_a^i = V_{\text{CKM}}^{ij}\tilde{D}_a^j$ are linear combinations of the Dirac fields of the physical charge-$(-\frac{1}{3})$ quarks. Thus, the diagonalisation of the down quark mass matrix introduced off-diagonal terms in the hadronic current. When considering processes, like nuclear β-decay, or μ-decay, where the momentum transfer is very small compared to the W-mass, the W propagator can be approximated by im_W^{-2} and the effective Fermi coupling constant is given by

$$\frac{G_F}{\sqrt{2}} = \frac{g^2}{8m_W^2} = \frac{1}{2v^2} \tag{19.31}$$

Contrary to the charged weak current (19.30), the $Z-$fermion interaction terms in the Lagrangian involve both left- and right-handed fermions. The coupling terms of Z to the leptons are

$$\frac{e}{2} \frac{1}{\sin\theta_W \cos\theta_W} \left[\bar{\nu_L}\gamma^\mu\nu_L + (\sin^2\theta_W - \cos^2\theta_W)\bar{e_L}\gamma^\mu e_L \right.$$
$$\left. +2\sin^2\theta_W\bar{e_R}\gamma^\mu e_R + \dots \right] Z_\mu \tag{19.32}$$

while its couplings to the quarks are

$$-\frac{e}{2} \sum_{a=1}^{3} \left[\left(\frac{1}{3}\tan\theta_W - \cot\theta_W \right) \bar{u_L^a}\gamma^\mu u_L^a + \left(\frac{1}{3}\tan\theta_W + \cot\theta_W \right) \bar{d_L^a}\gamma^\mu d_L^a \right.$$
$$\left. +\frac{2}{3}\tan\theta_W \left(2\bar{u_R^a}\gamma^\mu u_R^a - \bar{d_R^a}\gamma^\mu d_R^a \right) + \dots \right] Z_\mu \tag{19.33}$$

Again, the summation is over the colour indices, and the dots stand for the contribution of the other two families. We verify in this formula the property of the weak neutral current to be diagonal in the quark flavour space. Another interesting feature is that the axial part of the neutral current is proportional to $[\bar{u}\gamma_\mu\gamma_5 u - \bar{d}\gamma_\mu\gamma_5 d]$. This particular form of the coupling is important for the phenomenological applications, such as the induced parity violating effects in atoms and nuclei.

(*ii*) *The gauge boson self-couplings.* One of the characteristic features of Yang–Mills theories is the particular form of the self-couplings among the gauge bosons. They come from the square of the non-Abelian curvature in the Lagrangian, which, in our case, is the term $-\frac{1}{4}W^a_{\mu\nu}W^{a\,\mu\nu}$. Expressed in terms of the physical fields, this term becomes

$$
\begin{aligned}
&- ig(\sin\theta_W A^\mu + \cos\theta_W Z^\mu)(W^{-\,\nu}W^+_{\mu\nu} - W^{+\,\nu}W^-_{\mu\nu}) \\
&- ig(\sin\theta_W F^{\mu\nu} + \cos\theta_W Z^{\mu\nu})W^-_\mu W^+_\nu \\
&- g^2(\sin\theta_W A^\mu + \cos\theta_W Z^\mu)^2 W^+_\nu W^{-\,\nu} \\
&+ g^2(\sin\theta_W A^\mu + \cos\theta_W Z^\mu)(\sin\theta_W A^\nu + \cos\theta_W Z^\nu)W^+_\mu W^-_\nu \\
&- \frac{g^2}{2}(W^+_\mu W^{-\,\mu})^2 + \frac{g^2}{2}(W^+_\mu W^-_\nu)^2
\end{aligned}
\tag{19.34}
$$

where we have used the following notation: $F_{\mu\nu} = \partial_\mu A_\nu - \partial_\nu A_\mu$, $W^\pm_{\mu\nu} = \partial_\mu W^\pm_\nu - \partial_\nu W^\pm_\mu$ and $Z_{\mu\nu} = \partial_\mu Z_\nu - \partial_\nu Z_\mu$. Let us focus on the two $\gamma W^+ W^-$ cubic couplings. If we forget, for the moment, about the $SU(2)$ gauge invariance, we can use different coupling constants for the two trilinear couplings in (19.34), say e for the first and $e\kappa$ for the second. For a charged, massive W, the magnetic moment μ and the quadrupole moment Q are given by (see Problem 19.2)

$$
\mu = \frac{(1+\kappa)e}{2m_W}, \quad Q = -\frac{e\kappa}{m_W^2}
\tag{19.35}
$$

Looking at (19.34), we see that $\kappa = 1$. Therefore, $SU(2)$ gauge invariance gives very specific predictions concerning the electromagnetic parameters of the charged vector bosons. The gyromagnetic ratio equals 2 and the quadrupole moment equals $-em_W^{-2}$.

(*iii*) *The scalar–fermion couplings.* These are given by the Yukawa terms in (19.7) and (19.26). The same terms generate the fermion masses through spontaneous symmetry breaking. It follows that the physical BEH scalar couples to quarks and leptons with strength proportional to the fermion mass. Therefore the prediction is that it will decay predominantly to the heaviest possible fermion compatible with phase space. For the recently discovered BEH particle this is the b quark. This property provides a typical signature for its identification.

(*iv*) *The scalar–gauge boson couplings.* These come from the covariant derivative term $|D_\mu\Phi|^2$ in the Lagrangian. In the unitary gauge only one neutral scalar field ϕ survives. Its interactions with the gauge fields can be extracted from

$$
\frac{1}{4}(v + \phi)^2 \left[g^2 W^+_\mu W^{-\mu} + \frac{1}{2}(g^2 + g'^2)Z_\mu Z^\mu \right]
\tag{19.36}
$$

(*v*) *The scalar self-couplings.* These are proportional to $\lambda(v+\phi)^4$. Equations (19.18) and (19.31) show that $\lambda = G_F m_S^2/\sqrt{2}$, so this coupling has been measured together with m_H, the BEH mass. The latter is around 125 GeV, so λ is on the order of 0.12,

Table 19.1 The numerical values of the parameters of the Standard Model

THE PARAMETERS OF THE S.M.				
g	g'	λ	v	$\lvert\mu\rvert$
0.63	0.34	0.12	255.17 GeV	88.4 GeV

appreciable, but according to our experience, still in the perturbative regime.[3] On the other hand this relation shows that, were the physical scalar particle very heavy, it would have been also strongly interacting and this sector of the model would have been non-perturbative.

The five-step programme is now complete for both leptons and quarks. Although the number of arbitrary parameters seems very large, we should not forget that they are all mass and coupling parameters, like the electron mass and the fine structure constant of quantum electrodynamics. The reason we have more of them is that the Standard Model describes in a unified framework a much larger number of particles and interactions.

If we leave the fermions aside, the purely bosonic part of the model contains four independent parameters: the two gauge coupling constants g and g' and the two parameters of the scalar potential μ and λ. Alternatively, we can choose the electric charge e, the Fermi coupling constant $G_F/\sqrt{2}$, the weak angle θ_W and the physical scalar mass m_H. The values of e and G_F were well known already before the construction of the model. The angle θ_W was measured in several independent ways in neutrino scattering experiments. As we will see in Chapter 22, the precise value of m_H had only a weak influence on the low energy measurements. In Table 19.1 we show the values of these parameters computed from the measured values of the electric charge e and the masses m_W, m_Z and m_H using the relations (19.15), (19.17), (19.18) and (19.29). These relations receive higher order radiative corrections, which affect the values shown in the table by a few per cent.

To these parameters we must add those connected to the fermions, masses and mixing angles. Leaving aside the neutrinos, the other fermion masses are shown in Table 19.2.

A few remarks are in order here:

• We see that the dimensionless coupling constants g and g' are of the same order of magnitude, the same as e. It follows that in the unbroken phase all these interactions have the same strength. Weak interactions appear weak at low energies because the symmetry breaking occurs at a large scale v and this leads to large values of the intermediate vector boson masses. At very high energies, weak and electromagnetic interactions are expected to become comparable.

• Even without the neutrinos, fermion masses are spread over a huge range. The ratio m_t/m_e is larger than 10^5. If we include the limits on the neutrino masses, the ratio m_t/m_{ν_e} exceeds 10^{11} and it may well be much bigger. This spread comes from

---

[3]Of course, the numerical value depends on the definition of the coupling constant. In section 11.5.1 we noticed that in the $\lambda\phi^4$ theory there are certain combinatoric factors, which occur naturally. For this reason, a redefinition of the coupling constant is often used which, for a complex field, amounts to the change $\lambda = \tilde{\lambda}/4$. The value of the new coupling constant $\tilde{\lambda}$ is now close to $\frac{1}{2}$.

Table 19.2 This shows the fermion masses in MeV. The lepton masses are directly measurable but those of the quarks are only indirectly estimated. This is particularly significant for the u and d quarks. The uncertainties, experimental or estimated, are also shown. For the electron and the muon the uncertainties are too small to be included

FERMION MASSES					
Leptons			Quarks		
e	μ	τ	u/d	c/s	t/b
			2.3 (+0.7, -0.5)	1275 ± 25	173070 ± 520
0.51100	105.6584	1776.82 ± 0.16	4.8 (+0.5, -0.3)	95 ± 5	4180 ± 30

the Yukawa coupling constants, which are chosen in order to reproduce the observed mass spectrum. On the other hand, it is hard to imagine that a fundamental theory contains parameters with so widely different numerical values.

• In the Standard Model all fermions are massless in the symmetric phase, and chiral symmetry is exact. The spontaneous breaking of the symmetry produces the fermion masses and gives the explicit breaking of chiral symmetry. In the quark sector, Table 19.2 shows that this breaking is very small for the first family and quite appreciable for the others. This explains the fact we presented in section 18.7, that PCAC is a good approximation for non-strange particles and questionable for the others.

• For the two heavy quark families the up-type quarks are heavier than the down-type ones. This pattern is reversed for the light quarks of the first family.

• If we restrict ourselves to the first family we see that the mass ratio of the two quarks is about 2, so, at the quark level, isospin symmetry is very badly broken. If it appears to be a good approximate symmetry in hadronic physics, it is because the quark masses contribute very little to the masses of hadrons. The scale of the first is a few MeV, while that of the second is 1 GeV. The global isospin symmetry seems to be accidental. For particles like a proton or a neutron the main part of their masses comes from the spontaneous breaking of the chiral symmetry we presented in section 18.7. The part which is due to the masses of their constituent quarks seems to be negligible. The situation is largely reversed for the heavier families.

• In spite of their small values, the mass pattern of the light quarks seems to have very important consequences. The fact that the down quark is much heavier than the up results, probably, into the neutron being slightly heavier than the proton. Remarkably, hydrogen stability, at the origin of all matter creation, is due to this accident.

Our confidence in this model is amply justified on the basis of its ability to accurately describe the bulk of our present day data and, especially, of its enormous success in predicting new phenomena. We have mentioned some of them already. Let us give here a brief summary:

• The discovery of weak neutral currents by Gargamelle in 1973

$$\nu_\mu + e^- \to \nu_\mu + e^-, \quad \nu_\mu + N \to \nu_\mu + X$$

Not only their existence, but also their detailed properties were predicted. In general, we would expect for every lepton and every quark flavour a parameter, which determines the strength of the neutral current relatively to the charged one, and another to fix the ratio of the vector and axial parts. In the Standard Model, in which the breaking comes through an $SU(2)$ doublet scalar field, they are all expressible in terms of the angle θ_W. This is impressively confirmed by the data. It has become customary to express this agreement by defining the dimensionless quantity ρ, which, for the Standard Model to lowest order in perturbation theory, equals 1:

$$\rho = \frac{m_W^2}{m_Z^2 \cos^2 \theta_W} = 1 \tag{19.37}$$

In Problem 19.4 we ask the reader to compute the corrections to this value induced by a second scalar field, which is an $SU(2)$ triplet and has non-zero vacuum expectation value.

• The discovery of charmed particles at SLAC in 1974–1976. Their presence was essential to ensure, through the GIM mechanism, the absence of strangeness changing neutral currents, that is the suppression of processes like $K^0 \to \mu^+ + \mu^-$. The characteristic property of charmed particles is to decay predominantly into strange particles.

• We can show–although we will not do it in this book–that a necessary condition for the consistency of the Model is that $\sum_i Q_i = 0$ inside each family. When the τ lepton was discovered, the b and t quarks were predicted with the right electric charges.

• The discovery of the W and Z bosons at CERN in 1983 involved a brilliant innovation in accelerator technology. The characteristic relation of the Standard Model with an isodoublet BEH mechanism $m_Z = m_W / \cos \theta_W$ is confirmed with very high accuracy (including radiative corrections).

• The t quark had been *seen* at LEP through its effects in radiative corrections before its actual discovery at Fermilab.

• The final touch: The discovery of the Brout–Englert–Higgs scalar at CERN, announced on 4 July 2012.

A brief comparison of the predictions of the Standard Model with all available experimental data will be presented later. All its seventeen parameters have been measured.

19.5 Problems

Problem 19.1 *The general Cabibbo–Kobayashi–Maskawa matrix.* Assuming n_f quark families in the Standard Model, find the number of physically measurable arbitrary parameters which appear in the resulting Cabibbo–Kobayashi–Maskawa matrix. Verify that, for $n_f = 3$, we obtain the three Euler angles and a phase.

Problem 19.2 Prove that in the coupling of a massive, charged, spin-1 boson to the electromagnetic field the three classical terms, charge, magnetic moment and quadrupole moment are parametrised as in equation (19.35). Use only Lorentz invariance and electromagnetic gauge symmetry, and do not require renormalisability.

Problem 19.3 *The elastic $\bar{\nu}_\mu e$ and $\nu_\mu e$ cross sections.*

1. In the framework of the Standard Model write the invariant amplitude for the processes $\nu_\mu e \to \nu_\mu e$ and $\bar{\nu}_\mu e \to \bar{\nu}_\mu e$.

2. Compute the corresponding cross sections at neutrino energies $m_e \ll E_\nu \ll M_Z$. Assume massless neutrino, unpolarised initial electrons and sum over its final polarisations. Evaluate the ratio $\sigma(\bar{\nu}_\mu e)/\sigma(\nu_\mu e)$.

Problem 19.4 *The symmetry breaking pattern of the Standard Model.* Find how the relations (19.17) or (19.37) are modified if we add to the Standard Model a second scalar field, belonging to a triplet of weak $SU(2)$ and taking a vacuum expectation value v'. How many physical scalar particles would remain?

Problem 19.5 *Symmetry of the Standard Model for $g' = 0$.* Consider the Standard Model for $g' = 0$ in the tree approximation. Show that, after spontaneous symmetry breaking, the bosonic sector remains invariant under a global $O(3)$ symmetry, which, in particular, explains the equality of the W and Z masses in this case.

Problem 19.6 *The normalisation of the $U(1)$ generator.* In the Standard Model the generator of $U(1)$ is normalised through (19.1). In Chapter 23 we will present theories in which the group of the Standard Model is embedded into a simple group G. Find the $U(1)$ normalisation necessary for such an embedding.

Problem 19.7 *A gauge theory of the weak and electromagnetic interactions without neutral currents.*

Weak neutral currents were discovered experimentally in 1973. Prior to this discovery there were attempts to build gauge theories without weak neutral currents. The simplest is due to Howard Georgi and Sheldon L. Glashow in 1972. They consider $O(3)$ as the gauge group with the photon identified with the neutral vector boson. For simplicity we shall consider only the leptonic sector of the model with only the electron family.

1. Prove that the model requires the introduction of new, as yet unobserved, leptons.

2. Following Georgi and Glashow, consider a model containing two new leptons carrying the electron leptonic number: a positively charged one X^+ and a neutral X^0. We build two $O(3)$ triplets $\psi_R = (X^+, X^0, e^-)_R$ and $\psi_L = (X^+, X^0 \cos\beta + \nu \sin\beta, e^-)_L$ with the remaining neutral components forming $O(3)$ singlets: $s_R = \nu_R$ and $s_L = (X^0 \sin\beta - \nu \cos\beta)_L$, where β is a new mixing angle. In addition we consider a real triplet of scalar fields ϕ^a $a = 1, 2, 3$.

Write the corresponding $O(3)$ invariant gauge theory.

3. Allow for spontaneous symmetry breaking. Find the spectrum of the resulting physical particles.

4. Show that the neutral component of the vector bosons remains massless and has the correct coupling to the electromagnetic current of the charged particles to be identified with the photon.

Problem 19.8 *A model with spontaneously broken parity.*

In the standard electroweak theory parity is broken explicitly because the right- and left-handed components of the fermion fields belong to different representations of the group. In this problem we study a model in which parity is broken spontaneously by the same BEH mechanism that breaks the internal symmetry.

1. We choose the gauge group to be $SU(2) \times U(1)$. For simplicity we limit ourselves to the leptonic sector with only the electron family. We introduce three Dirac fermion fields, the electron e, and two neutral ones N and S. We assume that N and e form a doublet D of $SU(2)$ and S is a singlet. Under $U(1)$ we assume $Y_D = 1$ and $Y_S = 0$. Write the gauge invariant and parity conserving Lagrangian density of the fields D and S with terms of dimension smaller than or equal to 4.

2. We introduce in addition two multiplets of complex scalar fields Φ_1 and Φ_2, each one having the same transformation properties under $SU(2) \times U(1)$ as the one we used in the Standard Model. In order to limit the number of possible terms in the Lagrangian, we impose the invariance under a $U(1)$ group of global transformations with parameter η of the form

$$S' = e^{i\gamma_5 \eta} S, \quad \Phi_1' = e^{-i\eta} \Phi_1, \quad \Phi_2' = e^{i\eta} \Phi_2$$

with all other fields remaining invariant. Write the complete Lagrangian density.

3. How do the fields Φ_1 and Φ_2 transform under parity?

4. We choose the parameters of the potential $V(\Phi_1, \Phi_2)$ such that

$$<\Phi_1> = v \neq 0 \quad \text{and} \quad <\Phi_2> = 0$$

After shifting the field Φ_1 write the resulting fermion mass terms. Which are the physical particles of the model?

5. As we did in the Standard Model, we define the vector bosons $W^\pm$, Z and γ. Write their couplings with the fermions. Are they compatible with the known experimental results?

6. Express the electromagnetic and weak coupling constants α and G_F in terms of the parameters of the model.

Problem 19.9 *A model with a left-right symmetric group.*

Another model in which parity is spontaneously broken is the following:

1. The gauge group is

$$G = SU(2)_L \times SU(2)_R \times U(1)_V$$

where the subscripts L and R denote that the corresponding groups act on left- and right-handed fermions, respectively, and $U(1)_V$ has only vector couplings. In order to simplify the calculation we consider only the first family and we write

$$E_L = \begin{pmatrix} \nu \\ e^- \end{pmatrix}_L, \quad E_R = \begin{pmatrix} \nu \\ e^- \end{pmatrix}_R, \quad q_L^i = \begin{pmatrix} u^i \\ d^i \end{pmatrix}_L, \quad q_R^i = \begin{pmatrix} u^i \\ d^i \end{pmatrix}_R$$

where $i = 1, 2, 3$ denotes the three colours. The doublets L and R transform according to the (2,1) and (1,2) representations of $SU(2)_L \times SU(2)_R$ and, if Q is the electric charge, the charge Y of the $U(1)_V$ group is given by $Q = T_L^3 + T_R^3 + Y/2$. We shall further impose the discrete symmetry $L \leftrightarrow R$.

Write the corresponding renormalisable and gauge invariant Lagrangian.

2. For the symmetry breaking we introduce the following scalar fields: (i) A complex doublet $\phi = \begin{pmatrix} \phi^+ \\ \phi^0 \end{pmatrix}$ which belongs to the (1,2) representation and (ii) a real quadruplet σ which belongs to the (2,2) representation. Complete the new terms in the Lagrangian.

3. We assume that σ and ϕ take non-zero vacuum expectation values $< \sigma >= v_\sigma$ and $< \phi >= v_\phi$ which are relatively real. Give the spectrum of physical particles which results after shifting the scalar fields. Can we keep the neutrino massless? What is the corresponding mass of the electron?

4. The experimental upper bound for the contribution of right-handed currents in weak processes is on the order of 1%. Find the corresponding limits for the ratio v_σ / v_ϕ.

20

Neutrino Physics

20.1 A Blessing and a Curse

Neutrinos are the only particles having only weak interactions, therefore they are the ideal probes for the study of the latter. For more than fifty years they have offered a rich collection of results, which have influenced profoundly our understanding of the fundamental forces. Furthermore, their weak interactions give them almost miraculous properties. Their absorption cross section in matter is so small that practically nothing stops them. In Problem 20.1 we estimate their penetration length. This property is both a blessing and a curse. A blessing because it makes them very efficient messengers to study the state of matter under extreme conditions. For example, using neutrinos we can "see" the interior of a massive star. It is also a curse because their weak interactions make them very hard to detect. In Chapter 6 we saw that it took almost a quarter of a century between the prediction of their existence and their experimental discovery. Their study today is one of the most challenging, and also one of the most exciting, fields of experimental high energy physics. In this chapter we will expose in an approximate chronological order the most important steps of this passionate story. We warn the reader that the field is evolving very rapidly with many experiments either taking data or planned for the near future. As a result the picture may change radically, and neutrino physics may have more surprises in store for us.

The Cowan and Reines historic experiment we presented in Chapter 6 used neutrinos produced by the β-decay of neutrons coming from a nuclear reactor. They were very low energy electron anti-neutrinos $\bar{\nu}_e$. In Problem 20.1 we compute the neutrino–nucleon cross section in the simple Fermi theory and we derive equation (6.19), which shows that the cross section is proportional to the square of the neutrino momentum in the centre-of-mass system. This result remains valid also in the gauge electroweak theory as long as the neutrino energy is small compared to the W-mass. In this simple calculation we see the need to build high energy neutrino beams. They would allow for a dramatic increase in detection efficiency.

There exists a plethora of physics goals, which can be assigned to such experiments and these were already identified in the late 1950s.

1. The obvious one is the study of weak interactions at various energy scales. As long as we are restricted to decay processes, the available energy is necessarily limited. Scattering experiments make it possible to measure the weak interaction cross section as a function of energy and to check the validity of the Fermi theory at higher energies.

2. A second goal is related to the intermediate vector boson $W^{\pm}$ for the weak interactions we introduced in section 18.9. In decay experiments, we could only set a

Elementary Particle Physics. John Iliopoulos and Theodore N. Tomaras, Oxford University Press.
© John Iliopoulos and Theodore N. Tomaras (2021). DOI: 10.1093/oso/9780192844200.003.0020

lower bound on its mass. With a high energy neutrino beam we could either discover the W, or improve the bound considerably. Since it decays semi-weakly, i.e. the decay amplitude is proportional to the coupling constant g of equation (18.75) rather than g^2, as is the case for the weak decays we studied so far, the lifetime is expected to be very short and it would leave no visible track in the detector. The corresponding reactions would be

$$\nu + N \rightarrow W^+ + l^- + X \tag{20.1}$$

followed by

$$W^+ \rightarrow e^+ + \nu_e \quad \text{or} \quad W^+ \rightarrow \mu^+ + \nu_\mu \tag{20.2}$$

where N is a nucleon, l^- denotes the charged lepton associated to the incident neutrino and X some hadronic state. The signature would be the presence in the final state of the pair of opposite charge leptons.

3. A third one is the possible existence of weak neutral currents we mentioned in section 18.10, which would manifest themselves as elastic or quasi-elastic collisions of neutrinos in matter.

4. Finally, an important question has to do with the identity of the neutrinos, which appear in various decays. We have always assumed that each lepton was associated to its own neutrino type. Let us consider the example of muon decay. In section 18.6 we wrote it in the form $\mu^- \rightarrow e^- + \bar{\nu}_e + \nu_\mu$ and we explained that the absence of the decay mode $\mu^- \rightarrow e^- + \gamma$ suggests the existence of a separate conserved lepton quantum number for electrons and muons, in which case the two neutrinos appearing in muon decay should be different. It was immediately realised that neutrino beams could definitely settle this issue experimentally. Indeed, if the charged leptons are associated to their own neutrinos, a ν_μ interacting with matter should produce only muons and a ν_e only electrons, according to

$$\nu_\mu + N \rightarrow \mu^- + X \quad \text{and} \quad \nu_e + N \rightarrow e^- + X \tag{20.3}$$

Therefore, if we could control the types of neutrinos in a beam, we would be able to predict the types and relative abundance of charged leptons that would be produced from their collisions with matter.

20.1.1 The first neutrino beams

Neutrinos are among the main decay products of charged pions and kaons which are copiously produced in proton–nucleon collisions. The corresponding decay modes are

$$\pi^\pm \rightarrow \mu^\pm \pm \nu_\mu \quad \text{and} \quad K^\pm \rightarrow \mu^\pm \pm \nu_\mu \tag{20.4}$$

where we have used the notation $-\nu_{\mu(e)} \equiv +\bar{\nu}_{\mu(e)}$. We must also point out that we have denoted the neutrino produced in these decays as $\nu_{\mu(e)}$, even though, when the first neutrino beams were designed, there was no clear evidence that there were different kinds of neutrinos. The important point is that, as we see in the decays (20.4), the lepton in the final state is predominantly a muon. Electrons are practically absent. Indeed, in section 12.5 we computed the branching ratio $B = \Gamma(\pi^+ \rightarrow e^+ + \nu_e)/\Gamma(\pi^+ \rightarrow \mu^+ + \nu_\mu)$ and found it (see equation (12.60)) to be proportional to $m_e^2/m_\mu^2 \sim \mathcal{O}(10^{-4})$.

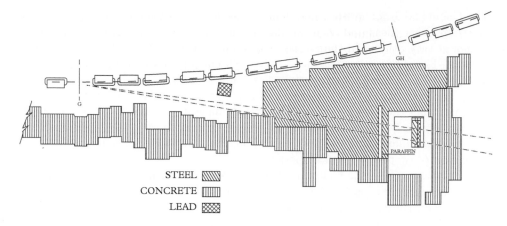

STEEL
CONCRETE
LEAD

Fig. 20.1 The layout of the first neutrino beam at Brookhaven. We see the beam, the iron shield and the detector. *(Source: G. Danby et al. Phys. Rev. Lett. 9, 36 (1962))*.

In fact, the decay mode (20.4) accounts for 99.99% of the charged pion decays, while for charged kaons it represents more than 90% of all leptonic and semi-leptonic modes. The only decay which yields an appreciable electron signal is the three-body decay $K^{\pm} \to \pi^0 + e^{\pm} \pm \nu_e$ which occurs with a branching ratio of 5.6%. If we take into account the relative abundance of kaons and pions, the resulting electron neutrinos are less than 1% of the total number of neutrinos.

The proposal to build a neutrino beam was made independently by B. Pontecorvo in 1959 in the Soviet Union and M. Schwartz in 1960 in the USA. The principle is simple. A proton beam of an accelerator hits a target and produces a large number of hadrons, the majority of which are pions and kaons. After a certain distance the charged pions and kaons decay in flight, predominantly according to (20.4). Recall that the $c\tau$ values for these particles are 7.8 m for $\pi^{\pm}$ and 3.7 m for $K^{\pm}$. Then the resulting beam goes through a thick shielding, which absorbs essentially all particles with the exception of the neutrinos. After the shielding we are left with a beam rich in neutrinos. A detector is placed immediately after the shielding.

The first neutrino beam was commissioned at Brookhaven in 1962 by a team from Columbia University, following this very simple design, see Figure 20.1. The primary proton energy was limited to 15 GeV in order to limit the energy of the produced particles and reduce the need for shielding. The most penetrating among the secondary particles, other than the neutrinos, are the muons, which are absorbed in matter through the electromagnetic interactions. Their energy sets the required amount of shielding. In the Brookhaven experiment a 13.5 m thick iron shield wall was placed at a distance of 21 m from a beryllium target. In fact, they used the steel from the disabled battleship Missouri. At an early stage of the design the Columbia University team was planning to place a bubble chamber behind the shield, but they switched to spark chambers following the proposal of Fukui and Miyamoto, as we explained in Chapter 6. They used 10 aluminium spark chambers weighting 1-ton each. It was the first use of such detectors in a major particle physics experiment.

The Columbia-Brookhaven experiment addressed most of the physics questions we listed earlier. They measured both the elastic and the inelastic neutrino–nucleon cross sections and they concluded that "...no large modification to the Fermi interaction is required...". The statistics was rather limited (29 elastic events and 22 inelastic ones were recorded)[1] but the measurements were compatible with the linear rise with the momentum square described by equation (6.19). They also looked for the intermediate vector boson, but their limited statistics did not make it possible to reach any meaningful conclusions. On the other hand, as we will explain shortly, the neutrino beams in all early experiments suffered from an irreducible neutron background, which made the detection of neutral currents impossible.

The great success of the Columbia-Brookhaven experiment was the confirmation of the distinct neutrino species. They observed 34 single muon events out of which five were estimated to be due to cosmic rays. In comparison, they had six candidate electron events, all compatible with the expected background. They concluded that neutrinos produced together with muons in pion or kaon decays, produced only muons (and not electrons) when they interacted with matter. For this discovery L.M. Lederman, M. Schwartz and J. Steinberger, the three senior members of the collaboration, received the Nobel Prize in 1988.

In parallel with the Brookhaven experiment, and in direct competition with it, a neutrino beam was built at CERN. In fact, the European project started earlier because the CERN AGS (alternating gradient synchrotron) was ready in September 1959, while the one in Brookhaven was commissioned during the summer of 1960. However, the project experienced many delays, in particular because the pion yield turned out to be smaller than what was estimated. As a result CERN completed the experiment later and could only confirm the results obtained at Brookhaven. This relative failure was counterbalanced by the fact that CERN delivered a much superior engineering project, which set the standard for all subsequent neutrino experiments. The main element was *the van der Meer horn,* a magnetic focusing device invented by the CERN engineer S. van der Meer. It consisted of two co-axial cones with a magnetic field between the two surfaces. Its purpose was to collect and focus the charged pions and kaons produced by the proton beam hitting the target. In fact, the problem was that the pions and the kaons were produced with a wide angular distribution, therefore only a small fraction of them were inside the solid angle formed by the point target and the distant detector. Therefore, the use of this focusing device produced a large increase in counting rates. An additional advantage of the horn was that it offered many possibilities. For example, one could choose to collect as many as possible of the produced pions and kaons and obtain a rich neutrino beam with a wide energy distribution, a "wide-band beam", or, alternatively, focus only particles of a given energy in order to obtain a "narrow-band beam" containing almost mono-energetic neutrinos. In addition, one could select positive or negative particles, thus obtaining a beam of neutrinos or anti-neutrinos, according to the decay modes (20.4).

The CERN experiment used a shielding made out of 4000 tons of concrete and 650 tons of steel. As detectors, the CERN experimentalists deployed both spark chambers

---

[1]By "elastic" they mean events of the form $\nu + n \rightarrow \mu^- + p$ or $\bar{\nu} + p \rightarrow \mu^+ + n$ and by "inelastic", events in which hadrons, mainly pions, were produced in the final state.

and a bubble chamber and confirmed the Brookhaven results with much higher statistics. For example, the bubble chamber reported 454 muon events and only three electron events, the latter being compatible with a small ν_e contamination of the beam coming from the decay $K^{\pm} \to \pi^0 + e^{\pm} \pm \nu_e$ we mentioned earlier.

20.1.2 Gargamelle and the neutral currents

The first neutrino scattering experiments answered conclusively two out of the four physics questions we listed in the first part of this section: they confirmed the existence of two distinct neutrino species and verified the validity of the Fermi theory up to energies of order a few GeV, the maximum energy of the first neutrino beams. As we saw in Chapter 19, the intermediate vector bosons turned out to have masses of order 80 and 90 GeV, which put them out of reach of the accelerators of the 1960s or 70s. In this section we want to explain the reasons for which the remaining question, namely the one concerning the existence of weak neutral currents, remained unanswered, and we will present the strategy which was adopted in order to overcome them and obtain a definite answer.

In a neutrino scattering experiment an event due to a charged current interaction has the form $\nu_{\mu} + N \to \mu^- + X$, with X a hadronic system, and it appears in the detector, bubble chamber or spark chamber, with a picture shown in Figure 20.2 (a): the long track is the muon and the short tracks are due to the hadrons. If neutral currents exist, they would produce two kinds of events: (i) $\nu_{\mu} + N \to \nu_{\mu} + X$, or (ii) $\nu_{\mu} + e^- \to \nu_{\mu} + e^-$, depending on whether the neutrino hits a nucleon or an atomic electron. The two events give images shown in Figure 20.2 (b) and (c), respectively. The one corresponding to neutrino–electron elastic scattering of Figure 20.2 (c) is easily identifiable because of the isolated fast electron track. However, because of the small electron mass, the centre-of-mass energy of this reaction is very small and so is the cross section. It follows that we need a very high neutrino flux and a large detector in order to observe events of this kind. On the other hand, events of the type shown in Figure 20.2 (b), which correspond to neutrino–nucleon scattering, have a much larger cross section, but suffer from a serious problem of neutron background. Even in the absence of weak neutral currents, the neutrino beam going through the shielding gives reactions of the type shown in Figure 20.2 (a), in which a slow neutron may be produced as part of the hadronic system X. Since neutral particles do not leave a visible track, when such a neutron enters the detector, it produces a hadronic reaction, which gives an image precisely like the one of Figure 20.2 (b), i.e. it mimics a neutral current. This was the neutron background problem, which haunted all early neutrino experiments and made the detection of neutral currents essentially impossible.

The mean free path of a 1 MeV neutron in matter is on the order of 70–80 cm. Therefore this problem could be solved if we had a detector several meters long. We could look at the events observed far away from the shielding in order to ensure that if an event of type 20.2 (b) is due to a neutron, the latter is produced inside the detector and can be detected. These arguments explain the drive for large detectors in the study of neutral currents.

In February 1964 a team of physicists under the direction of A. Lagarrigue from the Paris Ecole Polytechnique proposed to build a large heavy liquid bubble chamber,

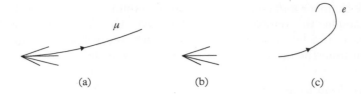

Fig. 20.2 Bubble chamber pictures of neutrino interactions. (a) A charged current interaction with a visible muon track. (b) A neutral current hadronic interaction with no visible lepton. (c) A neutral current leptonic interaction with an isolated electron track.

nicknamed *Gargamelle*,[2] for the study of neutrino interactions. It was built at Saclay and commissioned at CERN in December 1970. It was cylindrical and measured 4.8 m long and 2 m in diameter and was filled with 12 m^3 of heavy liquid (freon). In order to bend the tracks of charged particles and make possible their identification, the chamber was surrounded by a coil producing a magnetic field of 2 tesla. It was equipped with a very sophisticated optics system consisting of several cameras with fish-eye lenses. The entire installation weighed 1000 tons. It was operated by an international collaboration of 55 physicists coming from seven European institutions (Aachen, ULB Bruxelles, CERN, Ecole Polytechnique Paris, Universitá di Milano, LAL Orsay and University College London.) In fact, it was the first time that an experiment has been designed, run and analysed by such a multinational team. In this respect, Gargamelle set a precedent in establishing such collaborations,[3] as well as in defining their working rules. The run was supervised by all the members of the collaboration and the films were shared among the seven laboratories. Strict scanning and measuring rules ensured the same standards everywhere.

Gargamelle was taking data at CERN until 1979, the year it was decommissioned. Today, it is a museum piece in the CERN exhibition area. During those years it obtained a rich harvest of important results in neutrino physics, but its greatest achievement was the discovery of the weak neutral currents. Although initially this item was not among the priorities of the collaboration, they were convinced to launch an important search programme motivated by the rise of the electroweak gauge theory presented in Chapter 19. As we saw, this theory predicted the existence of strangeness-conserving weak neutral currents as part of the $SU(2) \times U(1)$ gauge symmetry, while, at the same time, it avoided the appearance of strangeness-violating ones through the introduction of the c-quark.

Using the van der Meer horn, Gargamelle was exposed separately to very intense neutrino and anti-neutrino beams. The strategy that was adopted for the discovery of the weak neutral currents was twofold:

(i) Try to identify events due to neutrino–electron elastic scattering, which we showed in Figure 20.2 (c). They were considered as "gold-plated" events because the

---

[2]Gargamelle was the mother of Gargantua, a hero of the popular book *Pantagruel* by François Rabelais, first published in France in 1532.

[3]For comparison, ATLAS, one of the large experiments at LHC, involves 3000 physicists coming from 182 institutions in 38 countries.

presence of the final electron would produce a rapidly curling track in the magnetic field, thus making the event relatively easy to identify. Furthermore, they were essentially background free because, in the absence of neutral currents, no known interaction could simulate such an event. However, the smallness of the expected cross section implied that their discovery could be a matter of luck. The collaboration was indeed lucky because in December 1972 the Aachen group found the event shown in Figure 6.8.[4] It was found in the anti-neutrino exposure and shows a very clean isolated electron track. No comparable event was found in the film taken with neutrinos, and this absence was used in order to obtain a first rough estimate of the value of θ_W in the $SU(2) \times U(1)$ electroweak gauge theory.

(ii) The most important challenge was the control of the neutron background for the hadronic neutral current events, and the Gargamelle collaboration had to show that they could use the large volume of their detector to succeed where all previous experiments had failed. Let us first present the basic idea in a simplified form. We call $N_n(z)$ the number of events of the type shown in Figure 20.2 (b) with z denoting the position where each event took place, as measured from the front end of the chamber. Obviously, we have $0 \leq z \leq L$, with $L = 4.8$ m, the total length of the chamber. $N_n(z)$ is the number of events, which were candidates for neutral current interactions. Assume that there are no hadronic neutral currents. In this case, all events of this type must be due to the neutron background. As we said earlier, the neutron mean free path is of order 70–80 cm, so the neutron responsible for each of these events was produced by a neutrino interaction at a distance of that order and, since the chamber was 4.8 m long, many of them should be visible in the chamber. Indeed, the collaboration found such events with the topology shown in Figure 20.3. Let us call them *associated events*, and the corresponding number $N_{AS}(z)$, where z denotes the position of the neutral current candidate event. Let us now define a quantity Δ as

$$\Delta(z) = N_n(z) - N_{AS}(z) \tag{20.5}$$

and plot Δ as a function of z. If all N_n events are in fact associated events, we expect Δ to decrease with z and go to zero after a few neutron mean free paths. This was not what Gargamelle observed. The value of Δ did decrease as a function of z at the beginning of the chamber, which shows that some of the $N_n(z)$ events were indeed due to neutrons, but it reached a finite value near the end of the chamber. This irreducible number cannot be explained as coming from neutrons and should correspond to genuine neutral current events.

As stated, this argument is simplified and the real situation was more complex. The chamber was 4.8 m long but only 2 m in diameter, so this analysis could not exclude neutrons produced in the steel surrounding the chamber and entering from the sides. For that, a much more sophisticated analysis was required, which involved a simulation of all neutrino interactions in the entire environment of the chamber. Such simulations had been used before in high energy physics experiments, but it was the first time that the behaviour of such a large and complicated detector had

---

[4]Two more events of the same type were found the following years providing a confirmation of the discovery.

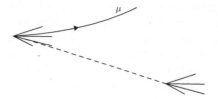

Fig. 20.3 A schematic view of a bubble chamber picture showing a neutrino "associated event": A charged current interaction produces a neutron which creates a hadronic event downstream.

been analysed numerically. They used the so-called *Monte Carlo algorithm*,[5] which has become the most important tool in data analysis. For the computer resources of the time it was a real "tour de force". The paper, signed by all the 55 members of the collaboration, appeared in July 1973. The existence of strangeness-conserving neutral currents was established, and a first evaluation of their relative strength compared to that of the charged currents was given.

20.1.3 Very high energy neutrino beams

During the 1970s a new generation of proton accelerators came into operation with energies of several hundred GeV. In the USA a high energy physics laboratory was established in Batavia, Illinois, later named *Fermilab*, and housed a four-mile ring proton synchrotron, which on 1 March 1972 reached the record energy of 200 GeV. In 1976, CERN commissioned a 400 GeV accelerator and in May of the same year Fermilab went to 500 GeV. The main purpose of these accelerators was the study of hadronic collisions, but, naturally, an important part of their beam time was devoted to the production of intense neutrino beams with energies of order 100 GeV or higher. It was also the time when bubble chambers became obsolete and were replaced by large electronic detectors. The main result of these experiments, which is relevant for weak interactions, is the measurement of the neutrino–(and anti-neutrino–)nucleon cross sections as shown in Figure 20.4. The linear rise with the neutrino laboratory energy is well verified. On the other hand, these high energies obtained highly inelastic events with a large number of hadrons produced in the final state. Their study offered

---

[5] Also known as *the Metropolis algorithm,* it was invented in the very early days of the first computers (1953) by N. Metropolis. Strictly speaking it is a very fast sampling algorithm, suitable for studying the landscape of multi-dimensional hyper surfaces. Technically it is a Markov chain Monte Carlo method, which provides efficient sampling from a given probability distribution for which direct methods are impossible, or extremely time-consuming. As such, it is suitable for the study of "the most probable" configuration among a multitude of possible ones and it is in this sense that it is used in high energy physics experiments. Alternatively, we can use it to estimate a multi-dimensional integral using the steepest descent method if we have a fast-decreasing integration measure, for example a Gaussian measure, and once the landscape of the integrand is found. We will see such applications later on in this book. In the course of the years the Monte Carlo method has grown into a full branch of computational science incorporating many rigorous results from numerical analysis and the theory of stochastic processes.

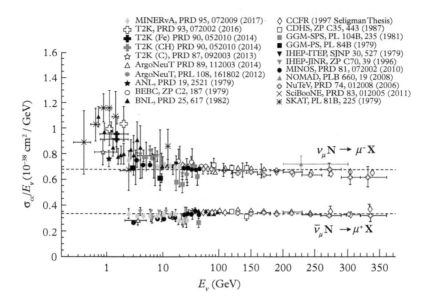

Fig. 20.4 Measurements of $\nu - N$ and $\bar{\nu} - N$ charged current cross sections divided by the energy of the incident neutrino in the laboratory frame. Note that $E_{\text{CM}} \sim \sqrt{2GeV \times E_{\text{lab}}}$. The energy scale in the Figure changes from logarithmic to linear at 100 GeV. Neutral current cross sections are not shown but they are generally smaller. *(Source: Particle Data Group, www.pdg.lbl.gov).*

new insights into the structure of the nucleon and the nature of strong interactions, which will be discussed in Chapter 21.

20.2 Neutrino Masses and Oscillations

20.2.1 Introduction

Our presentation of neutrino physics up to this point has been based on two implicit assumptions: (i) that there are only three distinct neutrino species, and (ii) that they are all massless. Here we want to review the experimental evidence on which these assumptions were based and see how our theory should be modified, were these assumptions found faulty.

All present evidence supports the view that neutrinos come in families accompanied by electrically charged leptons, but we have no theoretical indication of the total number of families. The Large Electron Positron (LEP) collider has set a model-independent lower bound of order 100 GeV, the maximum energy reached by each beam at LEP, on the masses of hypothetical heavy leptons L, provided they carry an appreciable fraction of the electron charge. The bound is based on the fact that no such lepton has been observed. LHC has pushed this limit to somewhere between 1 and 2 TeV, depending on assumptions about their decay modes.

However, we can constrain the number of species of such heavy leptons if we assume that they come in families together with their own neutrinos. Recall that a new neutrino even with very low, or even vanishing, mass would still remain undetectable if its companion charged lepton is sufficiently heavy beyond the reach of current experiments. For example, ν_τ was detected only after $\tau^\pm$ was produced. In other words, direct searches for neutral particles cannot give a bound on the number of even massless neutrinos. This problem was brilliantly solved and such a bound was obtained by the first LEP experiment. The clue was to measure very precisely the width of the Z boson. From (19.32) and (19.33) we can deduce the contribution to the Z width of each elementary fermion of the model. Its partial widths due to its decays into quarks and charged leptons can be measured separately, and their sum constitutes the visible part of the Z width.[6] The neutrino final states cannot be measured directly but their total contribution can be determined by the difference between the experimentally measured total and visible widths. In turn this influences the value of the cross section at the peak of the Z curve. Using (19.32) we can express the latter in terms of N_ν, the number of neutrino species contributing to the Z decay, i.e. those with masses less than half the mass of Z. In practice, and in order to reduce the errors, we define the "invisible ratio" $R^0_{\rm inv} = \Gamma_{\rm inv}/\Gamma_{l\bar{l}}$, i.e. the ratio of the invisible width over the purely leptonic width. We shall assume first that all invisible width is due to Standard Model neutrinos and second that the principle of universality holds for all families. Then we obtain

$$R^0_{\rm inv} = N_\nu \left(\frac{\Gamma_{\nu\bar{\nu}}}{\Gamma_{l\bar{l}}} \right)_{\rm SM} \tag{20.6}$$

where the ratio which multiplies N_ν is the one we compute in the Standard Model. We compare the measured value of $R^0_{\rm inv}$ with the Standard Model calculation and we obtain

$$N_\nu = 3 \tag{20.7}$$

We can visualise the result in Figure 20.5, which shows also the strong dependence of the hadronic peak cross section on the number of neutrinos. The great precision of the measurements allows to put very strict limits on any other conceivable contributions of unknown neutral particles in the Z decay width. If more families exist, their neutrinos are heavier than $m_Z/2$.

Let us now come to the three known neutrinos, which up to now have been assumed massless. This assumption is compatible with all direct measurements which have been performed so far, and have given only upper bounds on the neutrino masses. Not surprisingly, the most stringent limits concern $\bar{\nu}_e$.[7] They come from precise measurements of the shape of the electron energy spectrum in tritium β-decay near the end point, where the electron gets the maximum available energy. In Problem 18.6 we asked the reader to compute the dependence of the spectrum on the assumed neutrino mass. The upper limit is obtained by a fit of this curve. For the other two neutrino species the limits are obtained by an as-precise-as-possible determination of missing energy and momentum in various decays. The results, with 90–95% confidence level, are

---

[6]By "visible" we mean the decays whose final states are visible in the detector.

[7]We will come back shortly and give a more precise definition of the physical states we call ν_e, ν_μ and ν_τ in the light of the most recent experimental results.

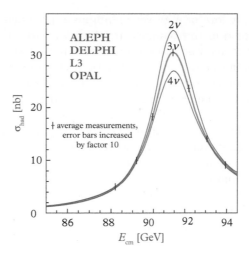

Fig. 20.5 The hadronic production cross section in the vicinity of Z. The predicted curves for two, three and four neutrino species are shown under the assumption of Standard Model couplings and negligible masses. Note that the error bars have been multiplied by ten for illustration purposes. *(Source: CERN, LEP Collaborations).*

$$m_{\nu_e} < O(1\text{eV})^8 \ , \quad m_{\nu_\mu} < 0.19 \text{ MeV} \ , \quad m_{\nu_\tau} < 18.2 \text{ MeV} \tag{20.8}$$

In the standard cosmological model a combination of observations gives a very strong bound on the sum of all light neutrino masses $\sum m < 0.2$ eV.

Let us note also that, if all three neutrinos are massless, each lepton number is separately conserved and no neutrino oscillations are possible.[9] This applies, in particular, to the electroweak theory we presented in Chapter 19. So it came as a great surprise when it was discovered experimentally that the three neutrino species do oscillate among themselves, which shows that they have non-zero unequal masses and the individual lepton numbers are not conserved. This ground-breaking experimental discovery was made initially using neutrinos of "natural" origin, namely solar and atmospheric, and it was later verified with "man-made" neutrinos produced in nuclear reactors and accelerators.

The experiments designed to study neutrino oscillations use a neutrino beam, whose intensity and composition are known. In practice they are either electron neutrinos coming from nuclear $\beta^\pm$-decay, or muon neutrinos coming from the decays of pions or kaons. A detector capable of detecting charged leptons produced by neutrino interactions is placed at a distance L from the origin of the beam. Such experiments are of two types:

1. *Disappearance experiments.* If the beam is made out of ν_l, l being the associated charged lepton, the detector at distance L measures fewer than expected leptons l,

---

[8]In fact, the various measurements present significant dispersion in their results. PDG quotes an estimation for the electron neutrino mass squared as: $(-0.6 \pm 1.9)(eV)^2$. Obviously, the negative sign is not meaningful, it only shows the order of the upper bound.

[9]The same holds true for massive neutrinos if all three masses are equal.

hence the term *disappearance*. A typical example is provided by a ν_e (or $\bar{\nu}_e$) beam made out of neutrinos coming from nuclear $\beta^{\pm}$-decay. Even if we use heavy radioactive elements, the maximum neutrino energy does not exceed 10–20 MeV, far below the threshold for producing any lepton other than electrons (positrons). In this case, detection of an electron (positron) deficit is a signal of neutrino oscillations. As we shall see presently, experiments of this kind provided the first evidence of neutrino oscillations.

2. *Appearance experiments.* In these, the beam is made initially of ν_l, but the detector detects a different lepton l'. A beautiful such experiment is OPERA. A high energy neutrino beam is produced at CERN. As we have said earlier, it consists mainly of ν_μs with a small admixture of ν_es, all coming from pion or kaon decays. This is controlled and measured via a nearby detector placed at the origin of the beam right after the shielding. Then the beam travels through the crust of the Earth to the underground laboratory situated at Gran Sasso in Italy, around 750 km away. The detector at Gran Sasso detects τ leptons. This means that ν_es and ν_μs of the beam were converted in flight to ν_τs.

20.2.2 Experimental evidence for neutrino oscillations

The first suggestion that neutrinos may undergo quantum mechanical oscillations among various quantum states was made by Pontecorvo in 1958. He was considering only electron neutrinos and proposed the possibility of $\nu - \bar{\nu}$ oscillations in analogy with the $K^0 - \overline{K^0}$ mixing we discussed already. Although he did not consider such oscillations as probable, he remarked that this could be easily checked experimentally since neutrinos and anti-neutrinos have different interactions in matter.[10] But the first evidence that oscillations may in fact occur came from attempts to detect solar neutrinos.

Already in 1920, A.S. Eddington suggested that nuclear fusion reactions in the interior of the stars were the source of stellar energy as well as the origin of heavy elements. In 1939, Bethe gave the first detailed chain of reactions, which was completed later for the heavier elements by F. Hoyle. The first and most important step in this chain is the fusion of hydrogen leading to the formation of ^4He

$$p + p \rightarrow d + e^+ + \nu_e, \quad p + d \rightarrow {}^3He + \gamma, \quad {}^3He + {}^3He \rightarrow {}^4He + p + p$$

It is followed by several other reactions, such as $^7Be + e^- \rightarrow {}^7Li + \nu_e$, some of which produce neutrinos in the final state. The rate of each one of them depends on the equation of state in the interior of the star, which, in turn, depends on the energy output of these reactions. Knowing the energy released in each one of them and the total energy radiated by the Sun, we can compute the neutrino flux and their energy spectrum. They are the messengers of the processes occurring in star's interior.

The first measurements of solar neutrinos were made by the team of R. Davis, who used Pontecorvo's reaction $\nu_e + {}^{37}Cl \rightarrow e^- + {}^{37}Ar$. They detected the argon produced

---

[10]In 1946, Pontecorvo had proposed the reaction $\nu_e + {}^{37}Cl \rightarrow e^- + {}^{37}Ar$ as a way to detect neutrinos. The proposal was judged impractical and he never published it. His 1958 suggestion for the $\nu - \bar{\nu}$ oscillation was triggered by a preliminary result, not confirmed, by R. Davis Jr. indicating that anti-neutrinos coming from a nuclear reactor, i.e. from neutron β-decay, could induce a nuclear reaction of the form $^{37}Cl \rightarrow {}^{37}Ar$.

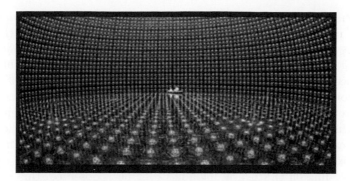

Fig. 20.6 The underground neutrino observatory "Super-Kamiokande". A 50,000 cubic metre tank with purified water and a dense network of photomultipliers. *(Source: Super-Kamiokande Observatory Archives).*

in a tank full of chlorine. It was the first detection of solar neutrinos, but the measured flux was found to be less than half of what was predicted by the standard solar model. The result was received with scepticism by the scientific community, with the exception of V. Gribov and B. Pontecorvo, who immediately interpreted it as a manifestation of $\nu_e - \nu_\mu$ oscillations.

We will not review here all the experiments which confirmed the solar neutrino deficit problem. They include gallium experiments, the experiments which were designed to detect proton decay, and, most important statistically, water Cherenkov experiments. The most significant of the latter was *Super-Kamiokande*, an experiment in Japan near the town of Mozumi. It is an underground neutrino observatory, whose detector consists of a tank filled with 50,000 tons of purified water, equipped with photomultipliers, see Figure 20.6. An electron neutrino interacting in the water through the charged current interaction will produce an electron, which can be detected via the resulting Cherenkov radiation. The results confirmed the neutrino deficit. For these observations R. Davis and M. Koshiba received the 2002 Physics Nobel Prize.

All these measurements, which were accumulated by the 1990s, showed clearly that something was fundamentally wrong concerning solar neutrinos: the experiments, the solar model, our theoretical ideas on neutrinos, or any combination of the above. The issue was brilliantly settled by two independent experiments, both involving underground neutrino observatories.

The first was the SNO (Sudbury Neutrino Observatory) experiment in Canada, which used a tank filled with heavy water. As a result they could detect the charged current interactions producing an electron, but also the neutral current interactions by detecting the nuclear recoil. The result was surprisingly simple: the charged current measurements confirmed the neutrino deficit, but the neutral current ones did not. In other words, the Sun was sending us all the neutrinos the solar model was predicting, but not all of them could produce electrons. Since the energy of solar neutrinos is less than 20 MeV, whichever neutrinos had changed nature to ν_μ, ν_τ or whatever other state exists did not have enough energy to be detected through charged current interactions, although they were perfectly well detectable through the neutral current ones.

The second experiment was Super-Kamiokande again, but this time they looked at neutrinos produced by cosmic rays in the high atmosphere. Like the accelerator neutrinos, they come from meson decays produced by the primary cosmic rays, mostly protons, colliding with the nuclei of the air. Therefore there is a sizable flux of ν_μs, which produce muons in the water. The subsequent Cherenkov light has the form of a cone, from which we can determine the direction of the incoming neutrino. In particular, we can distinguish between *downward going* neutrinos, generated in the upper-half atmosphere, and *upward going* ones produced on the other side of the Earth. If we call N_d and N_u the corresponding numbers detected as muon neutrinos, Super-Kamiokande found that $N_d \approx 2N_u$. Since no such significant dependence of the cosmic ray flux on the zenith angle has ever been observed, this result can only be attributed to the fact that N_u have travelled a longer distance going through the Earth, which implies neutrino oscillations and, furthermore, constrains the parameters which determine the characteristic oscillation length and amplitude. For this discovery T. Kajita and A.B. McDonald shared the 2015 Nobel Prize in Physics.

In Problem 20.2 we ask the reader to study the phenomenon of neutrino oscillations in vacuum, as well as in the presence of matter. Comparing with the same problem we studied for the K^0–$\overline{K^0}$ oscillations, we find two important differences: first, we have at least three neutrino species, so the corresponding matrix will be at least 3×3 and, second, neutrinos are not known to decay and therefore the mass eigenvalues should be real. If we call them m_1, m_2 and m_3, ignoring for the moment the possible existence of unknown neutrino species, we find that the measured oscillation amplitudes depend on two mass differences, which can be parametrised as $\Delta m_{21}^2 = m_2^2 - m_1^2$ and $\Delta m_{32}^2 = m_3^2 - m_2^2$. A fit of all oscillation results gives $\Delta m_{21}^2 = (7.53 \pm 0.18) \times 10^{-5}$ eV^2. For the second mass difference we find two possible values, either $\Delta m_{32}^2 = (2.51 \pm 0.05) \times 10^{-3}$ eV^2, or $\Delta m_{32}^2 = (2.56 \pm 0.04) \times 10^{-3}$ eV^2. We call the pattern corresponding to the first value *normal hierarchy*, i.e. $m_1 < m_2 < m_3$ and that corresponding to the second one *inverted hierarchy* $m_3 < m_1 < m_2$. An important goal of many planned neutrino experiments is to resolve the ambiguity and determine the pattern of neutrino masses.

20.2.3 The neutrino mass matrix

In formulating the electroweak gauge theory we assumed that all three neutrinos were massless. As a result we could use only the left-handed components of Dirac fields. The discovery of the phenomenon of neutrino oscillations, which, as we explained, implies non-zero neutrino masses, changed this picture. The minimal modification of the theory we presented in Chapter 19 amounts to keeping the assumption of having three neutrino species described by Dirac fields and adding their right-handed components. They are singlets of $SU(2)$ and carry no $U(1)$ charge, so none of the three ν_Rs couples to a gauge boson. They have only Yukawa couplings to the BEH scalars similar to the ones we wrote for the quarks in equation (19.23). As a result, spontaneous symmetry breaking produces a 3×3 mass matrix like the one shown in equation (19.25) and this implies the introduction of seven new parameters in the model, to wit three neutrino masses, three mixing angles which, in the notation we used in (19.25), we will denote by $\tilde{\theta}_{12}$, $\tilde{\theta}_{13}$ and $\tilde{\theta}_{23}$, and a new CP violating phase $\tilde{\delta}$.

At present, this simplest extension of the Standard Model fits all available experimental data.[11] Neutrino oscillations imply non-conservation of the individual lepton numbers, but the total lepton number is conserved. In this respect leptons behave similarly to quarks: baryon number is conserved although individual quark flavour numbers are not. The experimental evidence for baryon number conservation is the absence of any sign of proton decay. The current bounds on the mean lifetime of the proton, depending on the particular decay mode, are $\tau_p > 10^{31} - 10^{33}$ years. The corresponding bounds on the conservation of lepton number are weaker and we will discuss them shortly. Note that we cannot use the bounds on the electron lifetime to conclude anything about the level of lepton number violation, because the electron stability is guaranteed by the conservation of electric charge.

However, there is cosmological evidence that the conservation of baryon and/or lepton numbers may be violated. The reason is the apparent predominance of matter over anti-matter in the entire observable universe.[12] In the Standard Model of Cosmology this is attributed to the fact that in the primordial universe, when matter was created, baryon and lepton numbers were not absolutely conserved and the matter which is observed today is the result of the unbalance in the primordial matter–anti-matter creation rates. Assuming that the lepton number is not absolutely conserved, the right-handed neutrinos carry no conserved charge whatsoever and therefore could be described by Majorana fields. On the other hand, the left-handed neutrinos, which belong to $SU(2)$ doublets and carry non-zero $U(1)$ charges, are Weyl fields.

If neutrinos are Majorana particles the structure of the lepton mass matrix will differ from that of the quarks, equation (19.25). The reason is that Majorana fields are real and therefore we have no free parameters corresponding to the relative phases of the fields. As a result the 3×3 neutrino mass matrix contains three arbitrary phases. In the same notation we used for the Cabibbo–Kobayashi–Maskawa quark mixing matrix, the corresponding one for the leptons is

$$U_\nu = \begin{pmatrix} \tilde{c}_{12}\tilde{c}_{13} & \tilde{s}_{12}\tilde{c}_{13} & \tilde{s}_{13}\,\mathrm{e}^{-i\tilde{\delta}} \\ -\tilde{s}_{12}\tilde{c}_{23} - \tilde{c}_{12}\tilde{s}_{23}\tilde{s}_{13}\,\mathrm{e}^{i\tilde{\delta}} & \tilde{c}_{12}\tilde{c}_{23} - \tilde{s}_{12}\tilde{s}_{23}\tilde{s}_{13}\,\mathrm{e}^{i\tilde{\delta}} & \tilde{s}_{23}\tilde{c}_{13} \\ \tilde{s}_{12}\tilde{s}_{23} - \tilde{c}_{12}\tilde{c}_{23}\tilde{s}_{13}\,\mathrm{e}^{i\tilde{\delta}} & -\tilde{c}_{12}\tilde{s}_{23} - \tilde{s}_{12}\tilde{c}_{23}\tilde{s}_{13}\,\mathrm{e}^{i\tilde{\delta}} & \tilde{c}_{23}\tilde{c}_{13} \end{pmatrix}$$
$$\times \begin{pmatrix} 1 & & \\ & \mathrm{e}^{i\alpha_{12}/2} & \\ & & \mathrm{e}^{i\alpha_{13}/2} \end{pmatrix} \tag{20.9}$$

with $\tilde{\delta}$, α_{12} and α_{13} the three phases and $\tilde{c}_{12} \equiv \cos\tilde{\theta}_{12}$, etc. the new angles. This matrix implies a richer pattern of possible CP violating phenomena but we will discuss them in Chapter 22.

---

[11]As we said already, this may change due to ongoing experiments.

[12]There do not seem to exist any galaxies, even very distant ones, made out of anti-matter, since we do not observe any extended sources of very energetic γ-rays, which should be produced by annihilations in regions in which matter and anti-matter overlap. The rare traces of anti-protons found in cosmic rays are compatible with the expected production of anti-nucleons in high energy collisions in the interstellar medium.

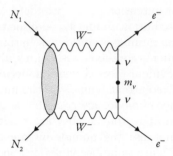

Fig. 20.7 Neutrinoless double β-decay. A nucleus N_1 gives a nucleus N_2 and two electrons. Charged current interactions conserve lepton number and this is shown by the arrows in the neutrino line. However, a Majorana mass term violates lepton number by two units and changes a neutrino to an anti-neutrino. As shown in the figure, the amplitude is proportional to the neutrino Majorana mass m_ν.

20.2.4 Neutrinoless double β decay

The most stringent limits regarding the possible non-conservation of lepton number come from the absence of neutrinoless double β-decay. This term refers to nuclear processes of the general form

$$(A, Z) \rightarrow (A, Z + 2) + e^- + e^- \tag{20.10}$$

They violate lepton number conservation by two units, and if neutrinos are described by a Majorana field, they are represented by the diagram of Figure 20.7. We see that the amplitude is proportional to the neutrino mass m, so the limits can be translated into upper limits on m_ν. The current best upper limits have been obtained by the KamLAND-Zen and GERDA phase II experiments searching for $(\beta\beta)0\nu$-decay of ^{136}Xe and ^{76}Ge, respectively. They are both of order 0.1 eV with quite large uncertainties, due, in particular, to the difficulty in obtaining an accurate estimation of the matrix elements of the weak current between the corresponding nuclear states. Current and planned experiments aim at improving by at least one order of magnitude the bound on m_ν.

20.3 Problems

Problem 20.1

1. Using the Fermi theory for charged current interactions, compute the electron-neutrino–nucleon cross section at lowest order perturbation theory and derive formula (6.19).

2. Estimate the mean free path of ν_e with energy a few MeV in lead.

Problem 20.2 *Neutrino oscillations.*

1. Start with a two-flavour model, ν_e and ν_μ. Assume that the neutrinos are massive Dirac particles and the flavour numbers are not separately conserved. Call the mass eigenstates $|\nu_1\rangle$ and $|\nu_2\rangle$ with mass eigenvalues m_1 and m_2, respectively. At $t = 0$ a ν_e is produced with momentum $\boldsymbol{p}$ and travels through empty space. Find the probability that it interacts like a ν_μ at time t. Show that it exhibits an oscillatory behaviour, which is absent if the masses m_1 and m_2 are equal. Show that the oscillation amplitude depends on a single mixing parameter.

2. Repeat the exercise for the physical case of three flavours. Find the number of parameters which determine the mixing, when the neutrinos are described by (a) Dirac fields, and (b) Majorana fields.

3. The solar neutrinos which were detected on Earth have travelled a long way inside the Sun and the approximation of vacuum propagation is not justified. Show that, in the Standard Model, electron neutrinos interact with matter differently from the other two species and draw the corresponding diagrams. What is the difference between neutrinos and anti-neutrinos? *For this question a qualitative answer will be sufficient.*

4. Let us make the simplifying assumption of a medium with a constant matter density. If H is the total Hamiltonian, which governs neutrino propagation, we can write $H = H_V + H_m$. The H_V is the part which describes propagation in the vacuum and contains the kinetic energy and mass terms. As we noted above, it is diagonal in the basis of the mass eigenstates $|\nu_1\rangle$, $|\nu_2\rangle$ and $|\nu_3\rangle$. Show that H_m, which represents the matter effects, is diagonal in the basis of flavour eigenstates $|\nu_e\rangle$, $|\nu_\mu\rangle$ and $|\nu_\tau\rangle$. Show that in this basis and in the Fermi theory approximation in which the vector boson propagator is replaced by a point interaction, it can be written, up to an overall factor, as $H_m \propto \mathrm{diag}\ (C + a\rho_e, C, C)$ with ρ_e the density of electrons, C a factor common to all flavours and a a constant affecting only electron neutrinos which changes sign between ν_e and $\overline{\nu}_e$.

Remark: The constant density approximation is not justified for the Sun where the density changes from a maximum in the centre to zero at the surface. The general case depends on the solar model but there is an approximation, which is quite realistic and allows for exact solutions. It is the adiabatic approximation valid when density variations are small over one oscillation length. In this case the problem can be solved by going to a variable basis of eigenstates, exactly analogous to the phenomenon of spin precession in a slowly varying magnetic field.

21

The Strong Interactions

21.1 Strong Interactions are Complicated

Strong interactions entered the physics scene in the first half of the twentieth century with the development of nuclear physics. It was immediately evident that they are short ranged – no macroscopic effects related to the strong force have ever been observed – and they are very strong, since they overcome the electrostatic repulsion among protons and lead to the formation of stable nuclei. In Chapter 6 we reviewed the early history and showed how their study revealed many fundamental principles of particle physics, such as the Yukawa theory of particle exchanges being the common mechanism of all forces and the Heisenberg concept of internal symmetries determining the form of the interaction. In fact, the study of nuclear structure has been the early source of information on the nature of the strong force.

In nuclear physics the non-relativistic approximation is quite adequate and the early theoretical calculations were based on phenomenological potentials describing the dynamics of an effective one-particle model, an approach inspired by the Hartree–Fock self-consistent method. With the increasing precision in the measurements of nuclear levels and the study of more and more complex nuclei, theoretical nuclear physics has become an independent field of research, which we will not attempt to present in this book.

The situation changed drastically in the 1940s with the extensive use of accelerators. They made possible the study of strong interactions over a broad energy range, in which relativistic effects are crucial and revealed the presence of more potential players beyond nucleons, such as mesons and baryon resonances.

The first field theory model for strong interactions was proposed by Kemmer in 1938. In modern notation it is the pion–nucleon isospin invariant theory we presented in equation (6.15). The complete Lagrangian, including the kinetic energy terms, is given by

$$\mathcal{L}_{\pi N} = \overline{\Psi}(x)(i\partial\!\!\!/ - M)\Psi(x) + \frac{1}{2}\partial^\mu \boldsymbol{\pi}(x) \cdot \partial_\mu \boldsymbol{\pi}(x) - \frac{1}{2}m^2 \boldsymbol{\pi}(x) \cdot \boldsymbol{\pi}(x)$$
$$+ g\overline{\Psi}(x)\boldsymbol{\tau} \cdot \boldsymbol{\pi}(x)\gamma_5 \Psi(x) - \lambda \left(\boldsymbol{\pi}(x) \cdot \boldsymbol{\pi}(x)\right)^2 \tag{21.1}$$

where M and m denote the nucleon and the pion mass, respectively, and we have included a four-pion interaction term with a new dimensionless coupling constant λ. In Problem 21.1 we ask the reader to use this Lagrangian in order to compute the proton–proton elastic cross section in the one-pion exchange approximation, compare with the experimental value and extract the numerical value of the coupling constant

Elementary Particle Physics. John Iliopoulos and Theodore N. Tomaras, Oxford University Press.
© John Iliopoulos and Theodore N. Tomaras (2021). DOI: 10.1093/oso/9780192844200.003.0021

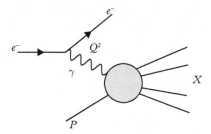

Fig. 21.1 Deep inelastic electron–nucleon scattering with the production of a final state $|X\rangle$.

g in this approximation. As we have said already, the result is $g > 1$, and this makes the perturbation expansion, which is a series in powers of g, meaningless. In fact, experiments show that the dynamics is dominated by the production of a large number of resonances, a phenomenon which cannot be described by perturbation theory. PDG lists more than 30 pion–nucleon resonances with masses between 1 and 2.5 GeV, with spins as high as five and a half, and it is clear that no simple quantum field theory could accommodate all of them. Strong interactions appeared to be very complicated.

We know today that this complexity shows only a superficial part of the picture. The interactions appear very complicated because the objects we were trying to study, the hadrons, are themselves complex. It is as if we were trying to discover quantum electrodynamics by looking at the interactions among complicated macromolecules.[1] By changing the perspective, we discovered a completely different picture.

21.2   Strong Interactions are Simple

The modern theory of strong interactions was developed as a response to certain intriguing experimental results. In 1966, a high energy linear electron accelerator was commissioned at Stanford in the laboratory which became known as *Stanford Linear Accelerator Center* (SLAC). It was 2 miles long and reached an energy of 50 GeV. By the late 1960s and early 1970s a series of experiments were carried out at SLAC, studying the high energy and large momentum transfer scattering, often called *deep inelastic scattering,* of electrons off nucleons.[2] The process is shown schematically in Figure 21.1 in the one-photon exchange approximation. It is an *inclusive process,* meaning that, in the final state, only the electron was measured. The state denoted by X was not. This choice was dictated partly by some theoretical considerations,[3] but mostly by several experimental constraints: (i) The detectors were not able to distinguish between a π^+ and a proton if their kinetic energy was much higher than 1 GeV. (ii) The forward and backward regions were not covered, so a complete identification of all particles in

---

[1] R.P. Feynman has given a nice analogy which accurately describes the efforts to understand strong interactions: imagine you want to study the mechanism of a fine Swiss watch and, to do that, you take two of them and smash them one against the other.

[2] If we use a hydrogen target we obtain the scattering on protons. If we combine the results with those on other nuclei, for example deuteron, we can extract the scattering cross section on neutrons.

[3] There have been theoretical suggestions by K. Johnson and F.E. Low, as well as by J.D. Bjorken that this kinematic region is indeed interesting.

X was not possible. (iii) The electrons in the beam were coming in bunches but the detectors were not able to resolve multiple collisions. Therefore, what was measured corresponds to a sum over all possible X.

We want to compute the amplitude of the process shown in Figure 21.1. If we look at it as a Feynman diagram, it is a product of three factors: the electron–photon vertex, the photon propagator and the vertex representing the interaction of the virtual photon with the incoming nucleon producing the hadronic state X. The first two factors are well known in quantum electrodynamics, but we have no explicit form for the third one. This is first because we do not know the state X and second, the blob in the lower part of Figure 21.1 includes the effects of the strong interactions for which our hadronic field theory models were found to be inadequate. As we have done already several times in this book, we go around this difficulty by writing a general parametric form for this vertex constrained only by symmetry principles, such as Lorentz invariance, current conservation etc. In an obvious notation, we write the transition matrix element to a particular final state $|X\rangle$ as

$$S_{fi} = -\frac{ie^2}{q^2} \bar{u}(k', s')\gamma^\mu u(k, s) \int d^4 y \, e^{iq\cdot y} \langle X| J_\mu^h(y) |P\rangle \tag{21.2}$$

where $q = k - k'$ is the momentum of the virtual photon and $J_\mu^h(y)$ is the operator representing the hadronic part of the electromagnetic current. We used the photon propagator in the Feynman gauge. The differential cross section will be given by the square of this expression. We assume that the initial proton and electron are unpolarised and we do not measure the polarisation of the final electron. Furthermore, since we are interested in the inclusive cross section, we sum over all final states $|X\rangle$. We obtain

$$\frac{d\sigma}{d\Omega'dE'} \sim \sum_X \frac{e^4}{q^4} \left| \bar{u}(k', \alpha')\gamma^\mu u(k, \alpha) \int d^4 y \, e^{-iq\cdot y} \langle X| J_\mu^h(y) |P\rangle \right|^2 \tag{21.3}$$

We parametrise this expression as

$$\frac{d\sigma}{d\Omega'dE'} \sim \frac{e^4}{q^4} L_{\mu\nu} W^{\mu\nu} \tag{21.4}$$

with $L_{\mu\nu}$ and $W^{\mu\nu}$ the leptonic and hadronic parts, respectively. In the deep inelastic regime we can for simplicity set $m_e = 0$. The spinors $u(k, \alpha)$ and $u(k', \alpha')$ satisfy the free Dirac equation. Then, summing over α' and averaging over α, gives the factor $L_{\mu\nu} = 2 \left(k_\mu k'_\nu + k_\nu k'_\mu - \eta_{\mu\nu} k \cdot k' \right)$.

We do not have such an explicit expression for $W^{\mu\nu}$ and we write it, averaged over the spin of the initial nucleon, as

$$W_{\mu\nu} = \frac{1}{2} \sum_X \int d^4 x \, d^4 y \, e^{iq\cdot y} e^{-iq\cdot x} \langle P| J_\mu^{h\dagger}(y) |X\rangle \langle X| J_\nu^h(x) |P\rangle \tag{21.5}$$

This expression can be further simplified by using the translation operator $e^{i\hat{P}\cdot x}$ and applying equation (5.198): $J_\nu^h(x) = e^{i\hat{P}\cdot x} J_\nu^h(0) e^{-i\hat{P}\cdot x}$

$$W_{\mu\nu} = \frac{1}{2} \sum_X \int d^4 y \ e^{iq \cdot y} \ \langle P | J_\mu^{h\dagger}(y) | X \rangle \langle X | J_\nu^h(0) | P \rangle (2\pi)^4 \delta^4(k + P - k' - P_X)$$

$$= \frac{1}{2} \int d^4 y \ e^{iq \cdot y} \ \langle P | J_\mu^{h\dagger}(y) J_\nu^h(0) | P \rangle \tag{21.6}$$

where, in the last step, we used the completeness relation in summing over the final states $|X\rangle$ consistent with energy and momentum conservation. In Problem 21.2 we ask the reader to prove that we can replace in this equation the product of the two current operators $J_\mu^{h\dagger}(y) J_\nu^h(0)$ by their commutator $[J_\mu^{h\dagger}(y), J_\nu^h(0)]$.

Since in (21.5) we sum over all possible intermediate states, $W_{\mu\nu}$ represents the total scattering cross section of a virtual photon of momentum q off a proton at rest. Using the optical theorem (12.9), it is given by the imaginary part of the forward elastic proton–virtual photon scattering.[4] Therefore,

$$W_{\mu\nu} \sim \text{Im} \, T_{\mu\nu} \ , \qquad T_{\mu\nu} = \int d^4 y \ e^{iq \cdot y} \ \langle P | T(J_\mu^{h\dagger}(y) J_\nu^h(0)) | P \rangle \tag{21.7}$$

The expression (21.6) is still formal. We have no way to compute the matrix elements of the hadronic electromagnetic current between hadron states. We only know that $W_{\mu\nu}$ is a two-index symmetric tensor, which depends on two four-vectors q and P. Taking into account the conservation of the electromagnetic current as well as invariance under parity (see Problem 21.3 for parity violating processes), we can write the most general form of $W_{\mu\nu}$ as

$$W_{\mu\nu} = -\left(\eta_{\mu\nu} - \frac{q_\mu q_\nu}{q^2}\right) W_1 + \frac{1}{M_N^2} \left(P_\mu - q_\mu \frac{q \cdot P}{q^2}\right) \left(P_\nu - q_\nu \frac{q \cdot P}{q^2}\right) W_2 \tag{21.8}$$

with M_N the nucleon mass, and $W_{1,2}$ two scalar functions, which can depend only on $\nu = q \cdot P$ and q^2. We can choose two dimensionless variables, $x \equiv Q^2/2\nu$ and M_N^2/Q^2, with $Q^2 \equiv -q^2$. From the inequality $P_X^2 \geq M_N^2$ we deduce that the physical region for x is $0 \leq x \leq 1$. It is customary to define the two dimensionless *structure functions* $F_1 = M_N W_1$ and $F_2 = \nu W_2/M_N$. We conclude that the inclusive differential cross section of the deep inelastic scattering of electrons off nucleons depends on two unknown functions of two dimensionless variables, x and M_N^2/Q^2.

The surprising result of the SLAC experiment was that, when both ν and Q^2 become large with their ratio x kept fixed, these structure functions were approximately functions only of x. Here "large" means large compared to the nucleon mass. This property became known as *scale invariance*; varying the energy scale of the experiment, Q^2 and ν with x fixed, does not affect the cross section.

This result is very interesting, because it is very easy to "understand" it using a naïve and wrong reasoning. When $Q^2 \to \infty$ with fixed x, the second variable M_N^2/Q^2 goes to zero. Naïvely, a function $F(x, M_N^2/Q^2)$ can be approximated by

$$F(x, M_N^2/Q^2) = F(x, 0) + \frac{M_N^2}{Q^2} F^{(1)}(x, 0) + \dots \tag{21.9}$$

where $F^{(1)}$ is the first derivative of F with respect to M_N^2/Q^2 keeping x fixed. So, when M_N^2/Q^2 goes to zero, we expect to be left with only the x dependence. This argument,

---

[4]In Problem 10.2 we proved such a relation between $\langle 0 | T(\phi(x)\phi(y)) | 0 \rangle$ and $\langle 0 | [\phi(x), \phi(y)] | 0 \rangle$ for a free scalar field.

however, is technically wrong because it assumes that the structure functions $F_{1,2}$ are analytic in the variable M_N^2/Q^2 around the point $M_N = 0$ and can be expanded in a power series. This assumption is false in quantum field theory, because, as we saw in Chapter 17, we often encounter infrared divergences when we attempt to set a mass parameter equal to zero. We remind the reader, for example, of the result of the 1-loop calculation we performed in Problem 15.6 for the ϕ^4 theory: The expression contains terms of the form $\ln(m_\phi)$ and makes no sense for $m_\phi = 0$. In fact, the naïve argument is based on the intuitive idea that at very high energies and momentum transfers, the masses should be unimportant and, consequently, the theory should exhibit scale invariance. Feynman had even built a simple model which implemented this scale behaviour. Let us assume that the target nucleon is made out of elementary constituents which interact with the incident photon as point particles. We shall call these constituents collectively *partons*. If we neglect all interactions among the partons we can easily reproduce this scaling property, see Problem 21.5. The trouble, of course, is that the assumption of no interaction among the partons does not seem to make sense. The partons cannot be free and, at the same time, bind strongly to form a nucleon. Nevertheless, it was such a schizophrenic behaviour that was implied by the data. The partons were almost free when probed by a virtual photon in the deep inelastic region and still very strongly bound in ordinary hadronic experiments. Strong interactions had this dual behaviour: very complicated with no perturbative expansion at the level of hadrons and very simple, approximated by a free field theory, at the level of partons.

21.3   Asymptotic Freedom

We can offer a simple geometrical picture, which exemplifies such a behaviour. Imagine the partons being hard spheres tied together with strings to form a hadron. When the partons are close together the strings are loose. So, when the photon hits a parton the latter responds as if it were free, because a loose string transmits little, or no force. On the other hand, with the partons at larger distances the strings are stressed and bind the partons strongly inside the hadron.

This picture rings a bell. It reminds us of the analysis we presented in section 15.3, where we studied the properties of a renormalisable quantum field theory under the transformations of the renormalisation group. We found that there is one class of theories, the unbroken non-Abelian Yang–Mills theories, which have the property of asymptotic freedom: the effective interaction is strong at large distances and becomes weak at short distances.

The moral of the story is simple: if we want to understand SLAC's results on deep inelastic electron–nucleon scattering in terms of a quantum field theory, we must assume that strong interactions are described by an unbroken non-Abelian Yang–Mills theory.

21.4   Quantum Chromodynamics

Deciding which Yang–Mills theory describes the strong interactions at the level of the hadronic elementary constituents has not been easy. Naturally, we assume the standard quark model in which we have six quark flavours coming in three colours each, and having fractional electric charges. We can write the quark fields as a 6×3 matrix of the form

$$
\begin{matrix}
u_1 & d_1 & c_1 & s_1 & t_1 & b_1 \\
u_2 & d_2 & c_2 & s_2 & t_2 & b_2 \\
u_3 & d_3 & c_3 & s_3 & t_3 & b_3
\end{matrix}
\tag{21.10}
$$

Therefore, there is a natural non-Abelian group $SU(6) \times SU(3)$ in which $SU(6)$ mixes quark fields having the same colour and different flavours, while $SU(3)$ does the opposite; it mixes colours and leaves flavours unchanged. We know experimentally that the flavour group is badly broken, because the quark masses are spread over a very wide range and there exist no massless vector bosons with flavour quantum numbers. As we have explained already, breaking of the symmetry destroys asymptotic freedom. This leaves the colour group $SU(3)$ as the natural choice for the strong interaction gauge group. The resulting theory is called *Quantum Chromodynamics*, or QCD, and was formulated by D. Gross and F. Wilczek, as well as D. Politzer, in 1973. For this work they shared the 2004 Nobel Prize.

QCD has eight (3^2-1) massless gauge bosons, which we call *gluons*. The Lagrangian density with the six flavours of quark colour triplets is given by

$$
\mathcal{L}_{\mathrm{QCD}} = -\frac{1}{2} Tr\, \mathcal{G}_{\mu\nu} \mathcal{G}^{\mu\nu} + \sum_{f=1}^{6} \sum_{i,j=1}^{3} \bar{q}_f^i \left(i \not{D}_{ij} - m_f \delta_{ij} \right) q_f^j
\tag{21.11}
$$

$q_f^i(x)$ is the Dirac field of the i-th colour of the f-th quark flavour with mass m_f, and $\bar{q}_f^i$ is its Dirac adjoint. The matrix valued field strength $\mathcal{G}_{\mu\nu}(x)$ can be written in terms of the eight gluon gauge fields $G_\mu^\alpha(x)$, $\alpha = 1, ..., 8$, as

$$
\mathcal{G}_\mu(x) = \sum_{\alpha=1}^{8} G_\mu^\alpha(x)\, T^\alpha \; ; \quad \mathcal{G}_{\mu\nu} = \partial_\mu \mathcal{G}_\nu - \partial_\nu \mathcal{G}_\mu + i g_s \left[\mathcal{G}_\mu, \mathcal{G}_\nu \right]
\tag{21.12}
$$

where $T^\alpha = \lambda^\alpha/2$ and λ^α the eight Gell-Mann matrices. The g_s is the strong interaction coupling constant. The covariant derivative $\mathcal{D}_\mu$ is a 3×3 matrix given by

$$
\mathcal{D}_\mu = \partial_\mu + i g_s \mathcal{G}_\mu(x)
\tag{21.13}
$$

It is straightforward, although rather lengthy, to compute the 1-loop β-function $\beta(g_s)$ of this theory, see Problem 21.4. It receives two contributions: The first comes from the non-Abelian self-coupling among the vector bosons to which we must add the couplings with the Faddeev–Popov ghosts we found in section 13.2. These diagrams result in a negative contribution, as expected. The second comes from the diagrams with one fermion loop. Since each flavour contributes independently, the result is proportional to N_f, the total number of flavours, and each gives a positive contribution to the β-function. For a general $SU(N)$ gauge group with N_f Dirac fermions in the fundamental representation, the 1-loop β-function of the gauge coupling is

$$
\beta(g_s) = -\frac{g_s^3}{(4\pi)^2} \left(\frac{11}{3} N - \frac{2}{3} N_f \right) = -b_0\, g_s^3
\tag{21.14}
$$

We see that, for $SU(3)$, we still have asymptotic freedom, provided $N_f < 17$ quark triplets. Following the analysis of section 15.3, we conclude that the effective strength of the interaction in the deep Euclidean region will be given by the running coupling constant $\bar{g}_s(t)$, solution of the equation

$$t\frac{\partial \bar{g}_s}{\partial t} = \beta(\bar{g}_s) \tag{21.15}$$

which, for t large enough, may become sufficiently small for perturbation theory to be applicable. Indeed, solving (21.15) with β given by (21.14), we obtain

$$\bar{g}_s^2(t) = \frac{g^2}{1 + 2b_0 g^2 t} \tag{21.16}$$

where $g = \bar{g}_s(0)$, the momenta are scaled according to $p_i \to \rho p_i$ and $t = \ln \rho$. As long as b_0 remains positive, i.e. $N_f < 17$, the denominator does not vanish. For large ρ, $\bar{g}_s$ tends to zero logarithmically. As in quantum electrodynamics, it is customary to define $\alpha_s(t) = \bar{g}_s^2(t)/4\pi$. If Q denotes the typical momentum which grows large and μ the initial subtraction point where $t = 0$, we have $2t = \ln(Q^2/\mu^2)$ and the relation (21.16) becomes

$$\alpha_s(Q^2) = \frac{\alpha_s(\mu^2)}{1 + 4\pi b_0 \alpha_s(\mu^2) \ln(Q^2/\mu^2)} \tag{21.17}$$

We introduce the parameter Λ with the dimensions of mass through the relation $\ln(\Lambda^2/\mu^2) = -[4\pi b_0 \alpha_s(\mu^2)]^{-1}$ and rewrite (21.17) as

$$\alpha_s(Q^2) = \frac{1}{4\pi b_0 \ln(Q^2/\Lambda^2)} \tag{21.18}$$

in which all reference to the initial value of the coupling constant $\alpha_s(\mu^2)$ has disappeared in favour of the scale parameter Λ. It shows clearly the behaviour of the effective coupling constant: At scales Q^2 much larger than Λ, $\alpha_s(Q^2)$ decreases in accordance with asymptotic freedom. On the other hand, when Q^2 decreases, $\alpha_s(Q^2)$ increases and it diverges when $Q = \Lambda$. Of course, the perturbation expansion, on which this analysis is based, cannot be trusted when α_s becomes large, but we expect (21.18) to represent a reliable description of the behaviour of the effective coupling strength of the theory for $Q^2 \gg \Lambda^2$. The value of Λ is determined by comparing equation (21.18) with the experimental data. Figure 21.2 shows such a comparison with $\Lambda \sim \mathcal{O}(200 \text{ MeV})$. With this choice the agreement between theory and experiment is impressive.

What about the region $Q^2 \lesssim \Lambda^2$? Perturbation theory breaks down, in particular because (21.11) predicts the existence of asymptotic states corresponding to free quarks and gluons, none of which has ever been seen in hadronic collisions. We get out of this difficulty by assuming that this breakdown of perturbation is not only quantitative, but also qualitative, in the sense that even the space of states is not correctly described by it. Specifically, as we mentioned in the Introduction of Chapter 14, only colour singlet states appear as free particles. This property, which we called "confinement", implies in particular that quarks and gluons cannot exist as asymptotic states. Proving the property of confinement for unbroken non-Abelian Yang–Mills theories remains one of the main unsolved problems in quantum field theory.[5] In the absence of a real proof of this property, we can only try to explain this concept a bit further, by analogy to similar phenomena in other areas of physics.

Let us go back to quantum electrodynamics. Consider an electron and a positron, created at point x_0 and pulled at distance L apart. For large L the electric field between them decreases like L^{-2}, because the field lines spread over all space. Imagine that we had a way to confine the electric field lines inside a thin tube starting at the position of the electron and ending at that of the positron. The field would remain constant, independent of L, and we would need an infinite amount of energy to separate the two particles. The QED vacuum does not have this property and, therefore, electric charges are not confined.

By contrast, recall that a fundamental property of a superconductor is that it exhibits the so-called *Meissner effect*. Once a piece of metal immersed in an external magnetic field is cooled below its characteristic critical temperature and becomes a superconductor, it expels from its interior the magnetic field lines. The magnetic field can only penetrate in a narrow surface layer of depth λ_s, called penetration depth, characteristic of the superconductor.

Imagine now, that there exist in nature magnetic monopoles and anti-monopoles with magnetic charge g and $-g$, respectively. They are sources and sinks of magnetic field, dual to the positive and negative electric charges, which are sources and sinks, respectively, of electric field.[6] Let us now place such a monopole–anti-monopole pair at a distance L from each other inside a superconductor. A magnetic flux $4\pi g$ will emerge from the monopole and end at the antimonopole, in an environment hostile to the presence of magnetic fields in its bulk. The natural compromise for the magnetic field is to be compressed inside a straight tube connecting the monopole and the anti-monopole, with characteristic cross section on the order of $\pi \lambda_s^2$. The magnitude of the magnetic field inside the tube is constant and roughly equal to $4g/\lambda_s^2$, and its energy density about $B^2/8\pi \simeq 2g^2/\pi \lambda_s^4$. Consequently, the potential energy of the monopole–anti-monopole will be proportional to their distance L. We conclude that the interaction between a monopole and an anti-monopole inside a superconductor is described by a linear potential, which means that it costs infinite energy to take them apart. It is a model for confinement. Incidentally, we shall see in Chapter 22 that a linear potential $U(r) \sim \sigma r$ at distances larger than $r \sim \mathcal{O}(0.2fm)$, which is consistent with quark confinement, is a good approximation for the phenomenological description of the spectrum of a non-relativistic quark–anti-quark pair.

Comparing QCD predictions with experiment is not as direct as with the theories we have studied so far. There are at least two reasons for that. First, the Lagrangian (21.11) is written in terms of quarks and gluons, but experiments are done with hadrons. Hadron formation is a phenomenon which happens in the strong coupling

---

[5]It is one of the "7 millenium problems", each one of which the Clay Mathematics Institute has endowed in 2000 with a prize of 1 million dollars.

[6]In fact, there are theoretical arguments which strongly suggest that such particles may well exist. In 1931, Dirac argued that if a magnetic monopole exists, then the electric charges in nature are quantised in agreement with observation. And conversely, all known theories which explain naturally the observed electric charge quantisation, predict the existence of magnetic monopoles.

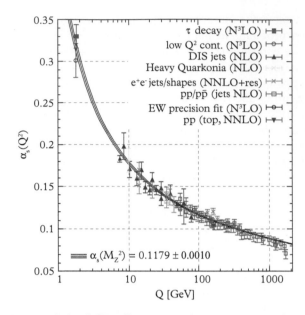

Fig. 21.2 The evolution of the QCD effective coupling constant and comparison with experimental measurements. The theoretical precision with which α_s has been extracted from the data is indicated in parenthesis: NLO means "next-to-leading order", NNLO "next-to-next-to-leading order", etc. The width of the line gives the QCD prediction including the theoretical uncertainties. *(Source: Particle Data Group)*.

regime of QCD, in a region in which perturbation theory does not apply. Second, Figure 21.2 shows that even in the region which we expect to be accessible to perturbation, the effective coupling constant α_s is still appreciable, of order 0.1 to 0.2, to be compared with $\alpha_{\mathrm{QED}} \simeq 1/137$. It follows that lowest order perturbation calculations will not give satisfactory answers and we must invent efficient methods to go beyond. In the remainder of this section we will only sketch the approach and the results.

21.4.1 Quantum chromodynamics in perturbation theory

21.4.1.1 The electon–positron annihilation into hadrons. We start with the example of the electon-positron annihilation into hadrons, i.e. $e^+ + e^- \rightarrow$ hadrons.

We want to compute the total cross section. The general Feynman diagram is shown in Figure 21.3 in the one-photon exchange approximation. The amplitude factors into the leptonic part, which is given by $-ie\bar{v}(k_2, \alpha_2)\gamma_\mu u(k_1, \alpha_1)$, the photon propagator $-i\eta^{\mu\nu}/Q^2$ with $Q = k_1 + k_2$, and the hadronic part, which is the amplitude for a virtual photon of momentum Q to give a hadronic state X. In order to isolate the hadronic part it is convenient to define the dimensionless ratio

$$R(Q^2) = \frac{\sigma(e^+ + e^- \rightarrow \text{hadrons})}{\sigma(e^+ + e^- \rightarrow \mu^+ + \mu^-)} \tag{21.19}$$

If Q^2 is sufficiently large, we expect the QCD coupling constant to be small and we can compute R perturbatively. At zeroth order, the hadronic system consists of a single

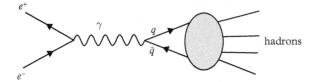

Fig. 21.3 The process $e^+ + e^- \to$ hadrons in the one-photon approximation.

quark–anti-quark pair and we must sum over all quark flavours, whose production is energetically possible. We thus obtain the very simple result

$$R(Q^2) = \sum_i e_i^2 \tag{21.20}$$

where e_i is the electric charge of the i-th type of quark measured in units of the electron charge. In this approximation R is a constant. The dependence on Q^2 comes from the higher order QCD corrections. In order to simplify the presentation we use the optical theorem we derived in Chapter 12. If T is the strong interaction transition matrix, the total cross section we are looking for can be written as $\sum_n \langle Q| T |n\rangle \langle n| T^\dagger |Q\rangle$ and, by the optical theorem, it is proportional to the imaginary part of the matrix element $\langle Q| T |Q\rangle$, which represents the photon 2-point function of momentum Q in QCD. The expansion of this quantity in powers of the QCD coupling constant α_s is shown in Figure 21.4. In fact, at this order the diagrams are the same as the corresponding diagrams in QED, which people had computed already in order to estimate higher order corrections to the Lamb shift or to the electron anomalous magnetic moment, so no new calculations were required. The result, which includes the contribution of the 2-loop diagrams of Figure 21.4 (b), (c) and (d), i.e. the $\mathcal{O}(\alpha_s)$ corrections to the parton model result of equation (21.20), is

$$R(Q^2) = \sum_i e_i^2 \left(1 + \frac{\alpha_s(Q^2)}{\pi} + \mathcal{O}(\alpha_s^2)\right) \tag{21.21}$$

At higher orders we must include diagrams with more gluons, which will involve also the three- and four-gluon vertices. Figure 21.5 shows the value of R, as measured in electron–positron colliders. To compare with (21.20) we should take into account the following: At very low values of $\sqrt{Q^2}$, e.g. below 1 GeV, the effective QCD coupling constant is large, and we do not expect perturbation theory to apply. Indeed, the data show large variations in R due to resonance production. Surprisingly, the asymptotic regime of equation (21.20) seems to be reached quite soon, at $\sqrt{Q^2} \sim 2$ GeV, although the corresponding value of α_s is not that small, see Figure 21.2. At this energy only u, d and s quarks can be produced and equation (21.20) gives $R = 3(4/9 + 1/9 + 1/9) = 2$ in rather good agreement with the data. At $\sqrt{Q^2} \sim 3$–4 GeV we cross the charm threshold. As expected, perturbation theory breaks down because, at this range, Q^2 cannot be considered larger than all relevant masses. We observe, instead, the charm–anti-charm resonances, such as the J/Ψ, the $\Psi(2S)$ etc. However, as soon as we pass

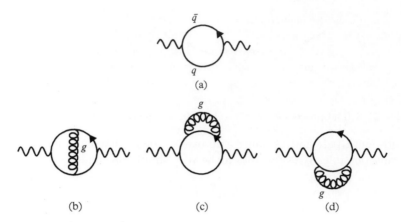

Fig. 21.4 The hadronic contributions to the photon propagator at zero order in α_s (a), and the one-gluon corrections (b), (c) and (d). In our case the last two diagrams give equal contributions.

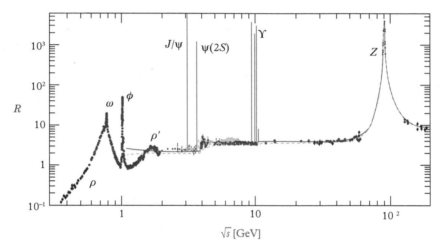

Fig. 21.5 The ratio R from low energies, up to and above the Z mass. The green curve is the parton model prediction and the red one includes QCD corrections. *(Source: Particle Data Group).*

the thresholds, R settles down to a new constant value, in agreement with the parton model result. The new value is obtained by adding the charm quark contribution, which equals $3 \times 4/9 = 4/3$, bringing the new value of R to $10/3$. The same happens again at $\sqrt{Q^2} \sim 10$ GeV when we cross the b–$\bar{b}$ threshold. Above it R increases by $3 \times 1/9 = 1/3$ to a total of $11/3$. We predict a new jump by $4/3$ when we cross the t–$\bar{t}$ threshold, but no electron–positron collider has reached this energy yet. At energies close to the Z mass the one-photon approximation breaks down and the total cross section is dominated by the Z pole.

21.4.1.2 Deep inelastic electron–nucleon scattering. The $e^+ + e^-$ total hadronic cross section was easy to analyse because, by applying the optical theorem, we expressed it in terms of a quantity, the photon 2-point function, which depends only on Q^2. The property of asymptotic freedom implies that α_s becomes small at large Q^2 and we can use perturbation theory. Our next example will be more complicated. It is the calculation of the QCD contributions to the deep inelastic electron–nucleon scattering shown in Figure 21.1. In equations (21.5) and (21.6) we expressed the cross section in terms of the matrix element of a product of two currents taken between one-proton states. The problem is, precisely, these proton states. We have noted already that perturbative QCD does not describe hadrons, so we do not know how to compute this matrix element. Technically, the problem is that, as we explained in section 21.2, the structure functions depend on two variables $F_i(x, M_N^2/Q^2)$, $i = 1, 2$, and the limit when the second variable approaches zero has infrared singularities. To get around this problem we will apply the lessons we learned in Chapter 17.

We start with a more precise formulation of the parton picture. The nucleon is a bag full of partons. Let k_i^μ denote the 4-momentum of the parton i in the rest frame of the nucleon. The momentum of the nucleon at rest is

$$P^\mu = \sum_i k_i^\mu = \left(\sum_i k_i^0, \mathbf{0} \right) = (M_N, \mathbf{0}) \tag{21.22}$$

We can perform a Lorentz boost in the z direction with parameter $\omega = \sqrt{(1 + v_z)/(1 - v_z)}$ and define $P^+ = (P^0 + P^z)/2$ and $P^- = P^0 - P^z$. Under the boost the momenta become

$$
\begin{aligned}
P^\mu &\longrightarrow (P^+, P^-, P^\perp) = \left(\tfrac{1}{2} M e^\omega, M e^{-\omega}, 0^\perp \right) \\
k_i^\mu &\longrightarrow (k_i^+, k_i^-, k_i^\perp) = \left(k_i^+ e^\omega, k_i^- e^{-\omega}, k_i^\perp \right)
\end{aligned}
\tag{21.23}
$$

where $\perp$ denotes the momentum in the x-y plane. The parton picture becomes simple in the infinite momentum frame, $\omega \to \infty$. In that limit we have

$$z = \frac{k^z}{P^z} \simeq \frac{k^+}{P^+} + \mathcal{O}\left(\frac{M_N}{P^+} \right) \tag{21.24}$$

Basic assumptions of the model are: (a) that $0 \leq z \leq 1$, in other words, at large ω there are no partons going in the opposite direction and, (b) that the parton transverse momenta $k_i^\perp$ are bounded.

In the parton picture the hadronic part of deep inelastic scattering is shown in Figure 21.6. The photon–nucleon interaction is described as the incoherent sum of photon–parton interactions. A key concept of the model is the *parton distribution function* $f_i(z)$, defined as the probability of finding inside the nucleon a parton of type i carrying a fraction of the nucleon's longitudinal momentum contained between z and $z+dz$. The set of $f_i(z)$s form a probability distribution, which means that summed over i and integrated over z, are normalised to one. Clearly, $f_i(z)$ are different for the various hadrons. In quantum chromodynamics the partons are the various quark flavours with

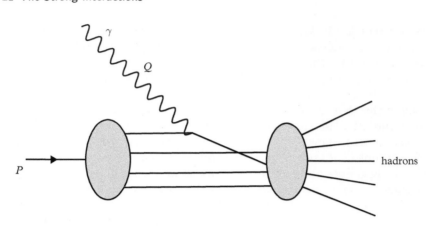

Fig. 21.6 The photon–nucleon interaction as the incoherent sum of photon–parton interactions.

distribution functions (summed over the three colours) $u(z)$, $d(z)$, $s(z)$, etc., the anti-quarks $\bar{u}(z)$, $\bar{d}(z)$, etc., and the gluons $G_a(z)$, $a = 1, \ldots, 8$. The normalisation condition takes the form

$$\int_0^1 dz\, z \left[\sum_{i=1}^{N_f} \left(q_i(z) + \bar{q}_i(z) \right) + \sum_{a=1}^{8} G_a(z) \right] = 1 \tag{21.25}$$

which is generic for all hadrons and expresses the fact that the momentum of the hadron is carried by its constituents quarks and gluons. Furthermore, specifically for the proton, its distribution functions satisfy the sum rules

$$\int_0^1 dz\, [u(z) - \bar{u}(z)] = 2\,, \qquad \int_0^1 dz\, [d(z) - \bar{d}(z)] = 1$$

$$\int_0^1 dz\, [s(z) - \bar{s}(z)] = 0\,, \qquad \int_0^1 dz\, [c(z) - \bar{c}(z)] = 0 \tag{21.26}$$

etc., for the heavier flavours

$$\sum_q \int_0^1 dz\, [q(z) - \bar{q}(z)] = 3 \tag{21.27}$$

where in equation (21.27) we have called $q(z)$ the distribution function of quark species q. Equation (21.25) expresses the conservation of the total momentum, and equations (21.26) and (21.27) refer to a proton target and express the fact that a proton has electric charge and baryon number equal to 1 and 0 strangeness, charm etc.

We can now compute the diagram of Figure 21.6. The gluons are neutral, so only quarks and anti-quarks interact with the incident virtual photon. The quantity $W_{\mu\nu}$ defined in (21.5) can be expressed in terms of the same quantity $W_{\mu\nu}^{(q)}$ in which the proton is replaced by the parton q with momentum k_q^μ, integrated over the momentum fraction z carried by the parton, and summed over all contributing partons. Thus, we write[7]

$$W_{\mu\nu}(x, Q^2) = \sum_q \int_x^1 \frac{\mathrm{d}z}{z} q(z)\, W_{\mu\nu}^{(q)}\left(\frac{Q^2}{2k_q \cdot q}, Q^2\right) \tag{21.28}$$

Note that $Q^2/2k_q \cdot q = x/z$, while the explicit $1/z$ factor takes into account the difference in the incident fluxes between a nucleon of momentum P and a parton of momentum k_q. From the optical theorem $W_{\mu\nu}^{(q)}$ is the imaginary part of the forward Compton scattering amplitude of a virtual photon of mass $-Q^2$ on a free on-shell quark, or anti-quark, satisfying the free Dirac equation. So, apart from an overall kinematical factor, we have

$$W_{\mu\nu}^{(q)} \sim \mathrm{Im} \sum_{\text{spins}} \frac{\bar{u}(k_q, \alpha_q)\gamma_\mu(\slashed{q} + \slashed{k}_q + m_q)\gamma_\nu u(k_q, \alpha_q)}{(q + k_q)^2 - m_q^2 + i\epsilon}.$$

$$\sim \delta(x - z)\mathrm{Tr}\left[\slashed{k}_q\gamma_\mu(\slashed{q} + \slashed{k}_q)\gamma_\nu\right] \tag{21.29}$$

For simplicity, we neglect the quark masses, since the probability of finding inside the nucleon a very heavy quark flavour is negligible. Replacing (21.29) into (21.28) and using (21.8), we finally obtain (see Problem 21.5)

$$F_1(x) = \sum_i e_i^2\, q_i(x) \quad \text{and} \quad F_2(x) = 2\sum_i e_i^2\, x\, q_i(x) \tag{21.30}$$

where e_i is the electric charge of the i-th quark. As expected, the structure functions depend only on x, which is the explicit manifestation of the scaling property of the parton model. Furthermore, we obtain an identity, the *Callan–Gross* relation:

$$2\, x\, F_1(x) = F_2(x) \tag{21.31}$$

a consequence of the fact that the charged partons have spin-$\frac{1}{2}$.

In the real world the quarks inside a nucleon are not free, and this simple picture will have to change. We want to develop a formalism allowing for the computation of such changes in the perturbation expansion in powers of α_s. For such an expansion to be meaningful, two conditions must be satisfied. First, α_s must be small and, because of the property of QCD to be asymptotically free, we expect this to be the case for large values of Q^2, i.e. at short distances. Second, we should disentangle the dependence of the structure functions on the second variable M_N^2/Q^2. As we have pointed out already, the dependence on this variable cannot be neglected, even at large values of Q^2, because of infrared singularities. The dependence on M_N brings a second scale, of order $\mathcal{O}(M_N^{-1})$, which involves large distances. From our discussion of the infrared singularities in Chapter 17, we can guess that this disentanglement will be possible

[7]We use the same symbol q to denote the photon momentum and the parton and its distribution function $q(z)$. It is clear which is which in the formulae below.

only for some carefully defined observables. Specifically, they must satisfy a property we shall call *factorisation*,[8] which takes the form of a convolution

$$F(x, M_N^2/Q^2) = \sum_{i=q,\bar{q},g} \mathcal{F}_i(\mu^2/Q^2, x/\xi, \alpha_s) * f_i(\xi, M_N^2/\mu^2, \alpha_s) + \dots$$

$$= \sum_{i=q,\bar{q},g} \int_x^1 \frac{d\xi}{\xi} \mathcal{F}_i(\mu^2/Q^2, x/\xi, \alpha_s) f_i(\xi, M_N^2/\mu^2, \alpha_s) + \dots \qquad (21.32)$$

where $\mathcal{F}_i$ are "short-distance" functions, which can be computed in perturbation for large Q^2, and f_i are "long-distance" functions, which depend on M_N and are non-perturbative. The dots stand for terms of order $\mathcal{O}(1/(Q^2)^p)$. This is, in particular, the structure of $W_{\mu\nu}$ in (21.28) with z, $W_{\mu\nu}^{(q)}$ and $q(z)$ in the roles of ξ, $\mathcal{F}$ and f, respectively. This analysis follows the one we made for the renormalisation group and neglects terms which vanish as inverse powers of Q^2 at large Q^2. The variable μ with dimensions of mass is the *factorisation scale* separating long- and short-distance phenomena. We can understand its physical meaning by looking at the relation (21.32) order by order in perturbation. At zero order in α_s, (21.32) reduces to (21.30) and describes the process shown in Figure 21.6. At first order in α_s, a parton may radiate a gluon with momentum k_g and reduce its share of momentum from x to ξ. If the transverse part $k_g^\perp$ of k_g is small, i.e. $k_g^\perp < \mu$, the emission is collinear, we call the process *soft* and we include it in the long-distance function f. If, on the other hand, $k_g^\perp > \mu$, we call it a *hard* process and we include it in the short-distance function $\mathcal{F}$. Otherwise, μ is arbitrary and, as it happens with the subtraction point in the renormalisation group, an observable should not depend on it.

$$\mu \frac{d}{d\mu} \ln F = 0 \qquad (21.33)$$

leads to

$$\mu \frac{df(\xi, \mu, \alpha_s)}{d\mu} = P(\xi/z, \alpha_s) * f(z, \mu, \alpha_s)$$

$$-\mu \frac{d\mathcal{F}(\mu^2/Q^2, \eta, \alpha_s)}{d\mu} = \mathcal{F}(\mu^2/Q^2, z, \alpha_s) * P(\eta/z, \alpha_s) \qquad (21.34)$$

The functions $P(z, \alpha_s)$ are called *splitting functions* and can depend only on variables, which are common to both $\mathcal{F}$ and f. The convolutions are integrals over the intermediate variable z. The appearance of $P(z, \alpha_s)$ reflects the fact that quarks and gluons can emit or absorb gluons.

   This set of equations was obtained by G. Altarelli and G. Parisi, as well as by Y.L. Dokshitzer, V.N. Gribov and L.N. Lipatov and they are known as APDGL equations. Their importance is clear: $\mathcal{F}$ depends only on short-distance variables and can be computed in perturbation. At zero order we have $\mathcal{F}_g^{(0)} = 0$ and $\mathcal{F}_q^{(0)} = e_q^2 \delta(x - z)$. At first order we must include the emission of hard gluons, etc. Knowing $\mathcal{F}$ we can

---

[8]For the structure functions of deep inelastic scattering the property of factorisation can be proven rigorously at any order in the QCD perturbation expansion.

compute P which, in turn, determines the evolution of f. We cannot compute f, but we can extract its value from experiments at a given scale Q_0, provided it is large enough for perturbation to be reliable. By choosing $Q_0^2 \sim \mu^2$, the evolution equation gives the value of f at any other scale.

Let us summarise: Perturbative QCD allowed us to compute the total e^+-e^- annihilation cross section in powers of α_s. When we considered deep inelastic scattering, we had to give up hope of actually computing the structure functions. We found instead that if we measure them at a given scale, we can compute their values at any other scale within the perturbative regime. Figure 21.7 shows the comparison between theory and experiment compiled by the HERA[9] collaborations.

We can generalise this picture to other inclusive processes for which the property of factorisation holds. For example, in proton–proton collisions we should visualise the process as a parton from proton A interacting with one of proton B, so the expression for the cross section will involve a double convolution. In order for the factorisation property to hold, we should not try to compute the amplitude for a specific final state but consider instead inclusive processes, such as a total cross section.

21.4.2 Quantum chromodynamics and hadronic physics

In Chapter 11 we noticed that a discrete space-time offers a way to define directly the path integral without reference to perturbation expansion. Therefore it makes it possible, in principle, to perform approximate calculations in the strong coupling regime, where perturbation theory is unreliable. In this section we shall try to use this approach for the physically interesting case of quantum chromodynamics. Figure 21.2 shows that at energies smaller than $\mathcal{O}(2\,\text{GeV})$ the effective coupling constant becomes strong. But, these are the characteristic energy scales at which quarks and gluons bind to form hadrons. So, an important test of the theory would be its ability to reproduce the observed spectrum of baryons and mesons.

In a very simplified form the strategy of the computation is the following: In section 13.3 we described briefly how to write the gauge fields on a four-dimensional Euclidean lattice. For quantum chromodynamics we need also to introduce the quarks. In the continuum the Lagrangian density is given by equation (21.11). On a finite lattice the functional integral becomes a multi-dimensional integral. Suppose, for example, that we want to find the spectrum of pseudo-scalar mesons. They have the quantum numbers of the local operators $\mathcal{O}_{ij}(x) = \overline{q}_i(x)\gamma^5 q_j(x)$. For a finite lattice we can compute the correlation function

$$F_{ij,i'j'}(\boldsymbol{x}-\boldsymbol{x}\prime) = \langle \mathcal{O}_{ij}(\boldsymbol{x})\mathcal{O}_{i'j'}(\boldsymbol{x}\prime)\rangle = \frac{\int \mathcal{D}[\Phi]\mathcal{O}_{ij}(\boldsymbol{x})\mathcal{O}_{i'j'}(\boldsymbol{x}')\,\mathrm{e}^{-S_{\mathrm{QCD}}[\Phi]}}{\int \mathcal{D}[\Phi]\mathrm{e}^{-S_{\mathrm{QCD}}[\Phi]}} \qquad (21.35)$$

where Φ denotes collectively quark and gluon fields and $S_{\mathrm{QCD}}[\Phi]$ is the QCD action. On the lattice we can evaluate the integrals numerically and study the behaviour of $F_{ij,i'j'}(\boldsymbol{x}-\boldsymbol{x}')$ as a function of the distance $r = |\boldsymbol{x}-\boldsymbol{x}'|$. If there exists a bound state with mass m having the correct quantum numbers to couple to both operators,

[9] *Hadron–Elektron–Ringanlage* was a e^- (or e^+)–proton collider operating in DESY near Hamburg, Germany. An electron, or positron, beam of 27.5 GeV was colliding with a 920 GeV proton beam. It was commissioned in 1992 and shut down in 2007 to be transformed into a synchrotron radiation source.

H1 and ZEUS

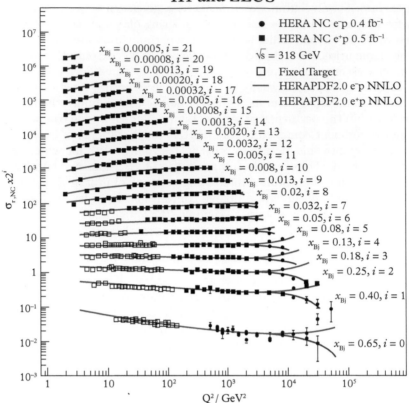

Fig. 21.7 The electron–proton and positron–proton cross sections for various values of the parameter x, from $x = 0.65$ ($i = 0$), to $x = 0.00005$ ($i = 26$), plotted against Q^2. In equation (21.3) we have computed the cross section in the one-photon exchange approximation. However, at values of $Q^2 \geq \mathcal{O}(10^4)$ GeV^2, the one-Z exchange contribution becomes important. It is included, together with the γ–Z interference term, in the calculation of the cross sections presented here. The interference term explains the difference between the e^- and e^+ cross sections for large Q^2. For illustration purposes the cross section for various x is multiplied by 2^i. Exact scaling would imply a constant σ, independent of Q^2. The data show instead significant scaling violating effects: the cross section is falling slowly for large values of x and rises sharply at small x, (note that both horizontal and vertical scales are logarithmic) with increasing Q^2. The pivot value of almost exact scaling is $x \sim 0.14$. The solid curves are QCD predictions computed up to and including next-to-next-to leading order effects. The agreement is remarkable. *(Source: HERA collaborations).*

equation (4.15) shows that $F_{ij,i'j'}(r)$ behaves, for large r, as e^{-mr}/r. We can repeat the calculation choosing different local operators and extract the complete spectrum of hadrons.

In practice this programme becomes complicated for various reasons, both theoretical and numerical. On the theoretical side the integration over the quark fields must be performed in a Grassmannian manifold following the Berezin rules we explained in section 10.3.2. This sounds like a very serious problem because computers do not know how to handle such integrals, but in fact this is only a numerical problem. All the theories we may be interested in, including quantum chromodynamics, are quadratic in the fermion fields, so the integrations can be done explicitly following equation (10.45), i.e.

$$\int \mathcal{D}[q]\mathcal{D}[\bar{q}] \, e^{-\int d_E^4 x(\bar{q} \, \mathcal{D} \, q)} = \det(-\mathcal{D}) \tag{21.36}$$

The resulting determinant of the covariant derivative depends only on the gauge fields, which are bosonic, but its numerical computation is very time-consuming.

A second theoretical problem has to do with the fact that the Lagrangian of fermion fields contains only first order derivatives. This creates some conceptual difficulties which we will not study in this book.

Let us now come to the practical problems of the numerical computations. In the real world we do not dispose of unlimited computing power and we must make approximations. In fact, it is not surprising that reliable results have been obtained only recently and the programme is still in progress. A simple order of magnitude estimation will convince you of its difficulty: Let us assume that the computation of a single integral from $-\infty$ to $+\infty$ requires K simple operations. For simple well-behaved functions, K may be of order 10^2 to 10^3. For a four-dimensional theory a rather low value for the number of lattice points N is of order 10^6, which gives around thirty points in each direction. The resulting number of elementary operations is K^N, clearly beyond any available, or even imaginable, computing system. In real computations we use various forms of the Metropolis algorithm we mentioned in section 20.1.2. An important bottleneck in the calculation has been the computation of the determinant of the huge matrix resulting from the integration over the fermion fields for each field configuration. For this reason, all early results from lattice simulations were treating quarks as external sources with no dynamical degrees of freedom. This was called *the quenched approximation*. It is only recently, with the increasing computing power, that the influence of fermion loops has been included.

Before going into the actual calculations, let us estimate the lattice parameters we should ideally need in order to simulate the physically interesting, non-perturbative regime of QCD. In units of mass, we want to cover the area from a few MeV, the order of magnitude of the light quark masses, to a few GeV, above which perturbation theory becomes reliable. The upper end tells us that we need to consider lattice spacings a smaller than one-tenth of a fermi.[10] The lower end requires a lattice size in each direction larger than 100 fermi, which means at least 1000 points. At present – 2021 – we still do not have the means to consider such large lattices and, consequently, the low mass region, where the spontaneous breaking of chiral symmetry is expected to occur, can be approached only by extrapolation. State-of-the-art computations use lattices with fewer than 100 points in each direction.

Figure 21.8 gives a general view of the spectrum of low-lying hadrons obtained in various lattice computations. The different colours refer to the different groups that

---

[10]1 fermi equals 10^{-13} cm, which, in our units, corresponds to $(200\text{MeV})^{-1}$.

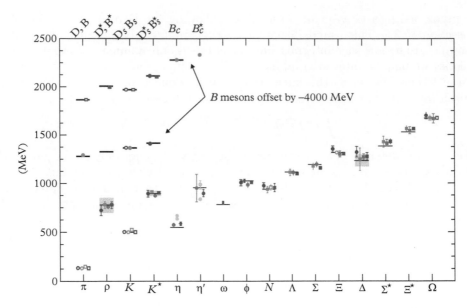

Fig. 21.8 The hadron spectrum. *(Source: A.S. Kronfeld).*

performed the calculations. The B-meson masses are displaced by 4 GeV in order to be included in the figure. Isospin symmetry has been assumed by taking the masses of the two light quarks m_u and m_d equal. The input parameters are the common mass $m_u = m_d$ and the masses of the s, c and b quarks. To them we have to add another one, related to the value of the strong interaction coupling constant, or, equivalently, a scale parameter similar to the one we introduced in the continuum in equation (21.18). The question of the spontaneous breaking of the chiral symmetry has been studied independently.

Putting all these results in the same figure may be misleading, because they come from different simulations, with typically different assumptions and input parameters. However, it offers a comprehensive picture and shows that the lattice calculations have reached maturity. The fit is quite impressive.

There are many more results which have been obtained in lattice QCD, such as the spectrum of gluon bound states, called *glueballs*, the hadronic matrix elements in weak interaction transitions, or the evidence for confinement for all couplings in non-Abelian theories. The work is still in progress, but the results presented above show already the power of the method.

21.5 Problems

Problem 21.1 The proton–proton elastic cross section near threshold is on the order of 20 mb. By using the model (21.1) at lowest order of perturbation theory, estimate the value of the coupling constant g. What do you conclude? Is the theory in the strong or the weak coupling regime?

Problem 21.2 Prove that in (21.6) we can replace the product of the two current operators by their commutator.

Hint: Insert between the two current operators a complete set of states satisfying energy and momentum conservation.

Problem 21.3 Write the analogue of (21.8) for the case of deep inelastic neutrino–nucleon scattering.

Problem 21.4 Compute the 1-loop β-function of an $SU(N)$ Yang–Mills theory with N_f multiplets of Dirac fermions in the fundament representation.

Problem 21.5 1. Consider a simple parton model in which the proton is a collection of free quarks. Show that it implies the scale independence of the structure functions.

2. Prove the relations (21.30) for the proton structure functions F_1 and F_2 in deep inelastic electron–proton scattering.

22

The Standard Model and Experiment

22.1 A Unified Picture

The electroweak gauge theory together with quantum chromodynamics form what we call the *Standard Model* of elementary particle physics. The unifying principle on which it is based is *gauge invariance*: all fundamental forces are described by gauge interactions. This principle covers also the gravitational forces, although we have not yet succeeded in incorporating them into the Standard Model. For the latter, the gauge group is $SU(3) \times SU(2) \times U(1)$. It is a rank four group and the corresponding Lie algebra has twelve generators: one for $U(1)$, three for $SU(2)$ and eight for $SU(3)$. Correspondingly we have twelve gauge bosons which, in the unbroken phase, are massless. The symmetry is spontaneously broken through the BEH mechanism to $SU(3) \times U(1)_{\text{em}}$

$$SU(3) \times SU(2) \times U(1) \to SU(3) \times U(1)_{\text{em}} \tag{22.1}$$

The scale of the phase transition is of order $\mathcal{O}(10^2)$ GeV. As a result, the three weak interaction gauge bosons, W^+, W^- and Z, acquire masses of that order, while the photon and the eight gluons remain massless.

As we have said already, this model describes essentially all available experimental results with great accuracy. In the following sections we will discuss some of the experiments which established the validity of the Standard Model and present the overall agreement. In fact, most of these experimental discoveries were made following the corresponding theoretical predictions. We often say that progress in physics occurs when an unexpected experimental result contradicts our current theoretical ideas. This forces theorists to elaborate new theoretical models. However, the evolution which brought geometry into physics and led to the construction of the Standard Model had a theoretical motivation, the quest for a renormalisable quantum field theory. This new theoretical paradigm made specific predictions, which occasionally seemed to go against the available experimental results. This triggered new experiments and led to important discoveries. We will present the most spectacular of them in the next section.

22.2 Classic Experiments

We have already presented the discovery of the weak neutral currents in the heavy liquid bubble chamber Gargamelle exposed to a neutrino beam at CERN. It was the first concrete evidence that with gauge theories we were in the right track. In this

Elementary Particle Physics. John Iliopoulos and Theodore N. Tomaras, Oxford University Press.
© John Iliopoulos and Theodore N. Tomaras (2021). DOI: 10.1093/oso/9780192844200.003.0022

section we will briefly describe a few more important experimental discoveries which consolidated this belief. We have chosen three of them: (i) the discovery of charmed particles, mainly at SLAC, (ii) that of the weak gauge bosons, W^+, W^- and Z, at CERN and (iii) the discovery of the BEH boson also at CERN. We will not describe the actual experiments in any detail; we will only explain why we consider each of these discoveries as being particularly important.

22.2.1 Experimental discovery of charmed particles

The necessity to introduce a fourth quark flavour, the charm quark, was discovered theoretically in 1970. We presented it in section 19.4 as the GIM mechanism, the mechanism to reconcile an $SU(2) \times U(1)$ gauge theory for the electroweak interactions with the absence of strangeness-violating neutral currents and large $\Delta S = 2$ weak transitions. It was also shown that the same mechanism predicted an upper bound for the charm quark mass of order a few GeV. A new quark implies the existence of new hadrons carrying this new quantum number and their characteristic signature would be their preferential decay into strange particles. It was also predicted that the charm quark would produce a new vector meson as a $\bar{c}c$ $J^P = 1^-$ bound state, the analogue of ρ, ω and ϕ we encountered in section 6.11.

In 1972, a new electron–positron collider was commissioned at SLAC with the acronym SPEAR for Stanford positron electron asymmetric ring. It reached an energy of 4 GeV and one of the first measurements was the value of R defined in equation (21.19). The results, as measured at the time, are shown in Figure 22.1. Note that the value of R is shown also in Figure 21.5, and 22.1 should be just an expanded version around the region of a few GeV. It does not look at all the same: in the old data R appears to be rising and this became a real problem, especially after the development of QCD which, as we saw in equation (21.21), predicted R approaching a constant value from above. Some people suspected that this could be due to the production of charmed particles,[1] but the situation turned out to be more complex and more exciting.

The measurements which gave Figure 22.1 were done with the machine energy increasing by steps of around 200 MeV. There were no deep reasons for this choice other than convenience, they were eager to arrive at the highest energy as soon as possible. In November 1974, following some vague hints, the SPEAR experimentalists decided to go back and sweep the region above 3 GeV in fine steps of 1 MeV. To their great surprise they obtained a totally different picture, namely the one shown in Figures 22.2 and 22.3. The first one, Figure 22.2, presents the SPEAR November 1974 measurements showing an extremely narrow resonance. It is known as J/Ψ and has a mass of 3096.9±0.006 MeV and a full width of 92.9±2.8 keV. Note that the cross section scale in the figure is logarithmic and the value at the resonance peak is one thousand times the background. Figure 22.3 shows the value of R in the energy region from 3 to 5 GeV. It contains an additional narrow resonance, called $\Psi(2S)$, with mass

---

[1]In the plenary session report on gauge theories in the 1974 International High Energy Physics Conference in London we read: "...the hadron production cross section, which absolutely refuses to fall, creates a serious problem. The best explanation may be that we are observing the opening of the charmed thresholds, in which case everything fits together very nicely."

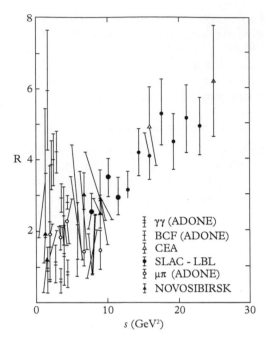

Fig. 22.1 Compilation of all early measurements of the ratio R, as presented in the 1974 London International Conference on High Energy Physics by Burton Richter.

of 3686.097±0.025 MeV and a full width of 294±8 keV.[2] Both J/Ψ and $\Psi(2S)$ decay predominantly to hadrons, yielding a very large number of possible final states. We understand now why they were missed in the earlier rough sweep: the experimental resolution was insufficient. Recall equation (2.60), which states that we cannot observe a resonance if the width is much smaller than the experimental resolution. Note also that J/Ψ was seen independently as an e^+e^- resonance in Brookhaven by a group led by S. Ting. They could not determine the width; they could only set an upper bound of order 20 MeV, their experimental resolution. For these discoveries B. Richter from Stanford and S. Ting from MIT shared the 1976 Nobel Prize in Physics.[3]

The significance of these resonances was at first a subject of controversy. What puzzled people were the extremely narrow widths, something totally unheard of for hadronic resonances, which should decay through strong interactions. Naïvely one would expect a width of several hundred MeV, given the phase space available and the large number of possible decay channels. In comparison, the ρ resonance we saw in section 6.10, which has mass of 775 MeV and decays almost exclusively into two pions, has width of 150 MeV, more than three orders of magnitude larger than that of the J/Ψ.

[2]These are the present values given by PDG. The values first obtained in 1974 had larger error bars.

[3]The Brookhaven team had called the 3.1 GeV resonance J, while that of SPEAR called it Ψ. Hence the double letter name J/Ψ. In the first SPEAR publications the 3.7 GeV state was called Ψ'. Today, we know that it is an excited state of J/Ψ and we use the terminology from the corresponding states of positronium. According to this rule J/Ψ should be called $\Psi(1S)$.

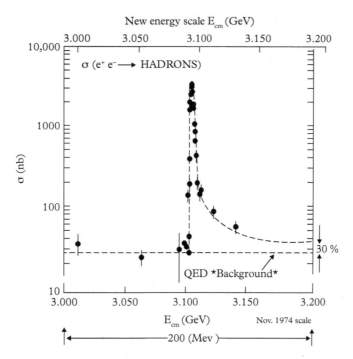

Fig. 22.2 The 10 November 1974 measurements, which established the J/Ψ resonance. The upper energy scale corresponds to a later recalibration of the machine energy. *(Source G. Goldhaber, The Rise of the Standard Model).*

The answer turned out to be simple, but subtle. In order to understand it we must look at the three known 1^- resonances, ρ, ω and ϕ. As we noted in the discussion which followed Table 6.5 of section 6.11, the first two are quark–anti-quark bound states of the u and d quarks and the third one ϕ, with mass of 1020 MeV, is an $\bar{s}s$ bound state. It has a total width of 4.25 MeV, quite narrow for its mass, but the most astonishing fact is that its main decay mode is into a K–$\overline{K}$ pair with a branching ratio of 85%, despite the fact that the phase space for this mode is tiny ($m_K \sim 495$ MeV). If we look at all the pionic modes, they make a width of only 650 keV. Even before the discovery of QCD, this experimental result gave rise to a phenomenological rule, known as *the Okubo–Zweig–Iizuka (OZI) rule*, which states that in a decay of a quark–anti-quark system, the modes which require the annihilation of the initial q–$\bar{q}$ pair are strongly suppressed.

When the rule was first formulated in the early 1960s it was purely empirical, but with QCD we can understand it better. The decay mode $\phi \to$ pions, which requires the annihilation of the initial $\bar{s}s$ quarks, has to go through an intermediate state of at least three gluons, because the one-gluon state is forbidden due to colour conservation and the two-gluon state does not couple directly to a state having $C = -1$. As a result, the decay probability is proportional to α_s^3. Even at 1 GeV α_s is substantially smaller than one, so the mode is suppressed. In fact, we can do better. We expect the lightest charmed particles to be $\bar{c}u$ or $\bar{c}d$ 0^- bound states, the analogues of pions and

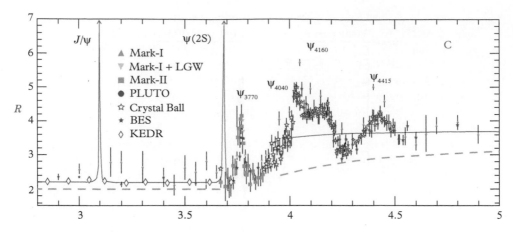

Fig. 22.3 The value of R for energies between 3 and 5 GeV as it is given today by the Particle Data Group.

kaons. We call them D mesons. In section 18.7.1 we saw that the pions are very light because they are approximately the Goldstone bosons of spontaneously broken chiral symmetry. As a result, the ρ meson is well above the two-pion threshold and the decay $\rho \to 2\pi$ is highly favoured. We also noted that the argument of spontaneous breaking is marginally applicable to kaons and not at all to the mesons of heavier flavours. Indeed, $m_K \gg m_\pi$ and the decay $\phi \to \bar{K}K$ is barely possible. By extrapolation we expect J/Ψ to be below the $\bar{D}D$ threshold and it should decay only via the three-gluon intermediate state. Indeed, the measured value of m_D is 1870 MeV, therefore neither J/Ψ nor $\Psi(2S)$ can decay into $\bar{D}D$. Figure 21.2 shows that between ϕ and J/Ψ, i.e. between 1 GeV and 3 GeV, α_s has dropped by a factor of 2, therefore dividing 650 keV, the ϕ pionic partial width, by 8, we obtain the correct value for the J/Ψ width.

Looking again at Figure 22.3, we see that after the narrow resonances J/Ψ and $\Psi(2S)$, we find a series of broad bumps, which should be interpreted as $\bar{c}c$ resonances lying above the $\bar{D}D$ threshold. Indeed, a detailed analysis of the hadrons in this region showed an abundance of $K–\pi$ resonances and made possible the first discovery of charmed mesons in 1976.[4]

As a final remark, an accidental degeneracy further complicated the analysis of the e^+e^- data: it turned out that the third sequential charged lepton, the τ, has a mass of 1777 MeV, so the threshold for $\tau^+\tau^-$ pair production roughly coincides with that of charm production. Since taus decay promptly, they were initially counted as part of the hadronic cross section, thus increasing the apparent value of R by one unit.

22.2.2 Stochastic cooling and the discovery of $W^\pm$ and Z

The most characteristic prediction of a gauge theory is the existence of vector gauge bosons. As we explained in Chapter 19, by the middle 1970s the measurement of θ_W and the knowledge of the values of the electric charge e and the Fermi coupling constant

---

[4]Evidence for charm particle production had been reported earlier, coming from cosmic rays as well as from neutrino experiments.

$G_F/\sqrt{2}$ led to very precise predictions for the masses of $W^\pm$ and Z, of order 80 and 90 GeV, respectively. These values were beyond reach for both CERN and Fermilab, which were operating fixed target proton accelerators in the 450 GeV range. The simple kinematical relation (6.26) shows that only a collider could reach such values. Hence the idea to convert the existing accelerators into proton–antiproton colliders. In Problem 22.1 we ask the reader to estimate the required minimum energy and luminosity using some realistic assumptions concerning production rates and detection efficiencies. A proposal along these lines was formulated in 1976 by Carlo Rubbia, Peter McIntyre and David Cline. It was first addressed to Fermilab, where the accelerator was already running, but it was turned down. In 1978, Rubbia obtained the approval of the CERN Council and the collider started operating in 1981. In Chapter 6, we presented briefly the principle of stochastic cooling which made this p–$\bar{p}$ collider possible.

In the eight years of operation, Sp$\bar{p}$S yielded many interesting physics results, but the most spectacular achievement was the discovery of the gauge bosons of the weak interactions. Several detectors were placed at collision points named after their location as UA1 to UA6 (for Underground Area 1 to 6, respectively) but only UA1 and UA2 were designed to look for Ws and Zs.

The weak vector bosons have a lifetime of order 10^{-24} s, too short to leave any visible track in a detector. Therefore we can detect them only through their decay products. A quick glance at the couplings given in equations (19.30), (19.32) and (19.33) shows that the decay modes, which could be seen at Sp$\bar{p}$S are

$$W^+ \to U + \overline{D} , \quad W^+ \to l^+ + \nu_l$$
$$Z \to U + \overline{U} , \quad Z \to \widetilde{D} + \overline{\widetilde{D}} , \quad Z \to l + \bar{l} \tag{22.2}$$

where, in the notation of section 19.4, U stands for a u or c quark, $D = V_{\text{CKM}}\widetilde{D}$ for the Cabibbo–Kobayashi–Maskawa combinations of d, s or b quarks and l for any of the charged leptons e, μ or τ. Obviously, since quarks do not exist as free particles, the quark decay modes will appear in the detector as hadronic systems. We have not included the ν–$\bar{\nu}$ decay modes of Z, because neutrinos are not visible in the detector.

At first sight, we would expect the hadronic modes to be the most promising. Because of the colour factor, they contribute about 70% of the decay rate and they do not involve invisible neutrinos. But this is ignoring the background. In a p–$\bar{p}$ collision we produce a very large number of hadrons through strong interactions, which by far outnumber the relatively rare events which involve Ws or Zs. Therefore the UA1 and UA2 collaborations opted for the leptonic modes. A weak vector boson of such high mass would be produced with relatively small kinetic energy and, as a result, we expect a substantial number of the leptons produced in the decays (22.2) to fly at large angles with respect to the beam. So the signal would be high transverse momentum isolated charged leptons. Specifically,

- For $W^+ \to l^+ + \nu_l$: The signal would be a large p_T isolated charged lepton[5] together with large missing transverse energy and momentum carried away by the elusive neutrino.
- For $Z \to l + \bar{l}$: The signal would be a high p_T $e^+ - e^-$, or $\mu^+ - \mu^-$ pair.

---

[5]By "isolated" we mean "not part of a hadronic cluster".

These considerations determine the necessary performances of the detectors. They must offer: (i) good identification of charged leptons, electrons and/or muons, especially in the central region, (ii) good measurement of their energy, and (iii) good determination of missing transverse energy and momentum.

UA1 was an all-purpose large magnetic detector, the most complete detector of its time, which became the prototype for most subsequent collider detectors. Its design resembles, in a smaller scale and simplified form, the one we presented in section 6.9 for ATLAS. It was a cylindrical detector surrounding the collision area and offering an almost 4π detection capability with good hermeticity, i.e. it had very few "dead angles". A dipole magnet produced a 0.7 T horizontal magnetic field perpendicular to the beam axis over a volume of $7{\times}3.5{\times}3.5$ m^3. Immersed in the magnetic field there was the central detector for tracking and vertex determination surrounded by an electromagnetic calorimeter for electron identification and measurement. The magnet return yoke contained the hadronic calorimeter and outside it there were layers of muon chambers. Two end cups with calorimeters and muon chambers were closing the cylinder at the two ends leaving only the beam pipe. UA2 in contrast was simpler, optimised for large angle electron identification and measurement. It had an electromagnetic calorimeter with very good granularity, which covered the central region but it had no end cups and no muon chambers.

In the 1982 run, the collider delivered a total luminosity of 18 nb^{-1}, the equivalent of 10^9 collisions. UA1 recorded 39 events in the central detector with evidence for an electron with large E_T (a signal in the electromagnetic calorimeter with $E_T > 15$ GeV containing an isolated high p_T track, no associated signal in the hadronic calorimeter, and with the energy measurement matching the one given by the inner detector). A careful analysis of these events showed that only five satisfied the criteria of being accompanied with missing E_T approximately matching the electron E_T and no hadronic jets. A similar analysis in the end cups yielded one more candidate. After a detailed study, the UA1 collaboration could exclude the possibility of these events being due to any conceivable background and announced their results at CERN on 20 January 1983. The following day, UA2 announced their observation of four more events of the same type. The results were published in Physics Letters B under the title "Experimental Observation of Isolated Large Transverse Energy Electrons with Associated Missing Energy at $\sqrt{s} = 540$ GeV".

No signal for a Z boson was found in the 1982 run, but this was expected because the $Z \to e^+ + e^-$ branching ratio is smaller than the $W \to e + \nu$ one. The signal came with the addition of the 1983 data. UA1 found four events consistent with $Z \to e^+ + e^-$ and one event with $Z \to \mu^+ + \mu^-$. UA2 had eight events of $Z \to e^+ + e^-$. In Figure 22.4 we show a reconstruction of the $Z \to \mu^+ + \mu^-$ event.

Both UA1 and UA2 used a display, known as *lego-plot*, in which the electromagnetic calorimeter is unfolded and the energy deposited in each cell is shown in the form of a tower. Figure 22.5 shows such plots for a $W \to e + \nu$ and a $Z \to e^+ + e^-$ event.

The combined measurements of UA1 and UA2 gave an estimate for the W and Z masses, namely

$$m_W = 82.1 \pm 1.7 \text{ GeV} , \quad m_Z = 93.0 \pm 1.7 \text{ GeV} \tag{22.3}$$

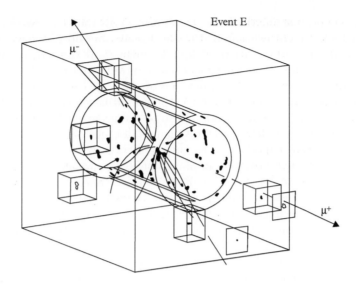

Fig. 22.4 A 3D reconstruction of the UA1 $Z \to \mu^+ + \mu^-$ event. *(Source: CERN, UA1 collaboration).*

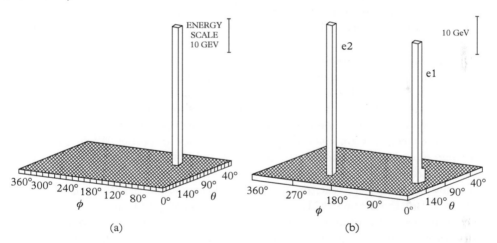

Fig. 22.5 The UA2 lego plots for $W \to e + \nu$ (a) and $Z \to e^+ + e^-$ (b). *(CERN, UA2 collaboration).*

to be compared with today's more precise values

$$m_W = 80.379 \pm 0.012 \text{ GeV} \quad \text{and} \quad m_Z = 91.1876 \pm 0.0021 \text{ GeV} \qquad (22.4)$$

This discovery gave Rubbia and van der Meer the 1984 Physics Nobel Prize.

22.2.3 The discovery of the heavy quarks and of the Brout–Englert–Higgs boson

As we mentioned earlier, the discovery of the charmed particles coincided with that of the τ lepton, which signalled the beginning of a third family. This initiated the search for the associated quarks b and t. We also recall that the existence of a third quark doublet was suggested back in 1973 by Kobayashi and Maskawa, as a natural way to accommodate CP violation. The first evidence for the b quark was found in 1977 at Fermilab by the E288 experiment led by L.M. Lederman. They found a very narrow resonance at 9.46 GeV, the analogue of the J/Ψ discovered by S. Ting and co-workers at Brookhaven. The analysis we presented in the first part of this section regarding the charmonium states applies also to the group of $b\bar{b}$ bound states known as *bottomonium*. They are shown as Υ resonances in Figure 21.5. We will describe the spectroscopy of hadrons containing a b quark in section 22.3.

The search for the top quark t took much longer. As we explained in Chapter 19, in the Standard Electroweak theory quark masses are generated through the Yukawa couplings of the quarks with the Brout–Englert–Higgs scalar and, as a result, they are arbitrary parameters of the theory. The requirement to suppress processes involving strangeness-violating neutral currents set upper bounds on the charm quark mass, but there is no similar phenomenological constraint for the quarks of the third family. Looking for the top quark became a central research goal in the agenda of every high energy accelerator in the final decades of the twentieth century. This applied, in particular, to LEP, the largest e^+–e^- collider ever built. It was commissioned in July 1989 with an initial centre-of-mass energy of 91 GeV, in order to study the region around the Z boson. In 1995, it was upgraded to higher energies, eventually reaching 209 GeV. It was dismantled in November 2000 to make room for the installation in the same tunnel of the LHC.

During its 11 years of operation LEP made the most complete study of Standard Model physics performing very accurate measurements of a large number of quantities, such as masses of particles, decay rates, cross sections and angular distributions. It is mostly through LEP results that the validity of the Standard Model at a very detailed level has been established. We give a collection of such results in section 22.7. In this section we concentrate on the top quark and the scalar boson. When LEP started operating, most of the parameters of the Standard Model had already been measured, some more accurately than others. There were only two whose values were still subject to speculation, the mass of the top quark m_t and the mass of the Brout–Englert–Higgs scalar m_H. But the Standard Model is a renormalisable quantum field theory and we can compute the value of any physical quantity A at any desired order of perturbation theory. In particular, a higher order diagram involving a top quark or a scalar boson loop will give A as a function of m_t and m_H. It follows that by measuring various As, LEP had indirect access to these two parameters.

We give an example. In equation (19.37) we introduced the quantity ρ and we noted that it characterises the nature of the symmetry breaking mechanism. At lowest order we found $\rho = 1$, but we can also compute the higher order corrections. To do that we must compute the corrections to m_W, m_Z and $\cos\theta_W$. In Figure 22.6 we show the 1-loop diagrams, which give corrections to m_W. They induce a correction $\Delta\rho_{\text{se}}$, ("se" stands for "self-energy") which, for $m_H > m_W$ and in the approximation in which m_t is much larger than any other fermion mass, is given by

$$\Delta\rho_{\text{se}} = \frac{3G_F m_W^2}{8\sqrt{2}\pi^2} \left[\frac{m_t^2}{m_W^2} - \frac{\sin^2\theta_W}{\cos^2\theta_W} \left(\ln\frac{m_H^2}{m_W^2} - \frac{5}{6} \right) + \cdots \right] \tag{22.5}$$

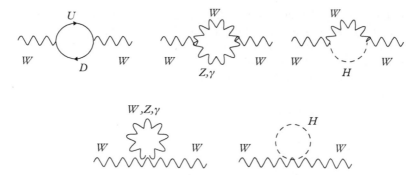

Fig. 22.6 The 1-loop diagrams which give corrections to m_W.

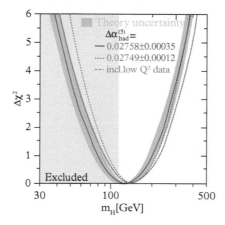

Fig. 22.7 $\Delta\chi^2$ as a function of m_H. *(Source: CERN, LEP collaborations).*

The important point is that the top quark corrections are quadratic in the quark mass, while the scalar boson ones are only logarithmic. The complete calculation, which includes also all diagrams contributing to the corrections of $\cos\theta_W$, is rather lengthy, but this dependence remains unchanged and applies also to all quantities that LEP could measure. It follows that combinations of many measurements could give very precise predictions for m_t and less precise ones for m_H. Indeed, in the early 1990s, LEP published the first estimate of $m_t \sim 170$ GeV, with errors less than 10%. In 1995, the Tevatron at Fermilab announced the direct observation of the top quark in full agreement with the LEP prediction. Today's value is $m_t = 173.0 \pm 0.4$ GeV, thus providing a further confirmation of the Standard Model.

Concerning m_H, LEP combined all measurements and produced the plot of Figure 22.7, which shows the likelihood χ^2, or *probability density function*, as a function of m_H. The yellow region corresponding to $m_H < 114$ GeV was excluded by direct searches at LEP and we see that the measurements favoured rather low values of m_H, lower than $\mathcal{O}(200$ GeV), with χ^2 increasing very fast for higher values.

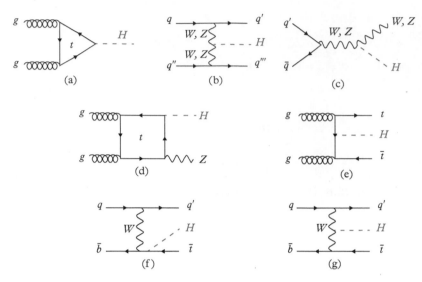

Fig. 22.8 Diagrams contributing to the production of a BEH scalar. *(Source: Particle Data Group).*

The situation remained unchanged until the operation of the LHC between 2009 and 2010. Even in its first configuration of 7 TeV total centre-of-mass energy, the LHC had more than enough energy to produce a particle with a mass between 100 and 200 GeV, but the same applied also to the Tevatron, which had been in operation since the early 1990s. The problem was not energy; it was production cross section and detection capabilities. None is easy to compute, but we can give some estimations.

The protons colliding at the LHC should be viewed as bags full of quarks, anti-quarks and gluons. We called them collectively "partons" in Chapter 21. In Figure 22.8, we draw a few Feynman diagrams representing processes producing a BEH boson. They are easy to compute but we need to know the probability of finding inside the proton a given parton carrying a fraction x of nucleon's total momentum. In Chapter 21 we called these probabilities "parton distribution functions" and we noted that they can be extracted from the deep inelastic scattering measurements at SLAC, extrapolated to the LHC energies using the renormalisation group equations of QCD and asymptotic freedom. It turns out that, numerically, at the LHC energies the most important contribution comes from the gluons, which carry more than one-half of proton's momentum. To first approximation a proton is a bag of gluons. As a result, the dominant mechanism for producing a BEH boson at the LHC is the process of gluon fusion, shown in Figure 22.8 (a). In principle, we should add all quark flavours in the loop, but it is clear that the top gives the most important contribution, because the scalar boson–fermion coupling is proportional to the fermion mass. This gives us an estimate of the production cross section $\sigma(p + p \to H + X) \sim \mathcal{O}(10^{-2})$ nb, to be compared with the proton–proton total cross section at these energies which is of order 10^8 nb. Production of the scalar boson is obviously necessary for the discovery, but it is not sufficient; one must also be able to detect it. Since its lifetime is very short, we

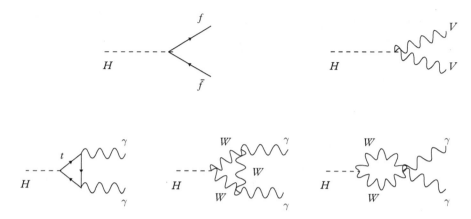

Fig. 22.9 Diagrams contributing to the decay of a BEH scalar. V stands for W or Z, which may be real, for a heavy H, or virtual giving lepton pairs.

must detect its decay products. Based on our knowledge of the couplings we found in Chapter 19, we present in Figure 22.9 various possible decay modes. Naturally, their relative importance depends on the mass of the scalar boson. If $m_H \geq 2m_W$ the two vector boson mediated decay modes dominate, because the corresponding coupling constants are very large, see equation (19.36). On the other hand, for a light scalar boson the leading mode is $H \to b\bar{b}$, because the b quark is the heaviest fermion to which the BEH boson can decay, followed by $H \to \tau^+\tau^-$. Figure 22.10 shows a graph with the estimated branching ratios as functions of the scalar boson mass. We see from this graph that a heavy boson would be easier to detect because it decays predominantly via WW or ZZ pairs, which can be detected through their leptonic modes. In the spring of 2011, the Tevatron collaborations announced that they could exclude the existence of a scalar boson with mass in the interval $156 < m_H < 177$ GeV and in the summer of the same year the LHC extended the excluded range to $145 < m_H < 460$ GeV. As a result, the allowed mass region was limited to $114 < m_H < 145$ GeV, where the $H \to b\bar{b}$ mode is expected to dominate.

However, as we noted already for the discovery of W and Z, it is not always the dominant decay modes which are the most promising. The order of magnitude of the cross sections we just mentioned shows that only one out of 10^{10} proton–proton collisions is expected to produce a BEH scalar, while a substantial part will produce b-hadrons. So, for every $b\bar{b}$ pair coming from an H decay, there are a billion others which do not. Digging the signal out of such huge background represents a formidable task. The $\tau^+\tau^-$ mode is also very difficult because in the decays of the taus we miss the two neutrinos.[6] It turned out that the most promising modes were very rare and this hindered the search considerably. In Figure 22.9 we show the diagrams, which contribute to the decay modes $H \to 2\gamma$ and $H \to l^+l^+l^-l^-$, with $l = e$ or μ. The first has an expected branching ratio (see Problem 22.2) $\Gamma(H \to 2\gamma)/\Gamma(H \to b\bar{b})$ of order

---

[6]Today, after several years of data taking and analysing, both these modes have been identified and measured, but they were not the ones which signed the discovery.

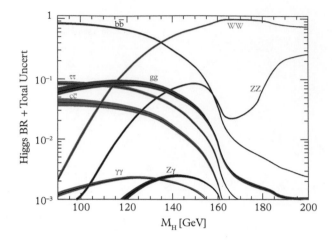

Fig. 22.10 The estimated branching ratios of the scalar boson decay modes as functions of its mass. The widths of the curves show the corresponding theoretical uncertainties. *(Source: arXiv: 1201.3084).*

10^{-3} and the others are smaller. Yet these modes were chosen to provide the signal for the discovery.

The background for the two-photon mode may have several origins. The first is direct production of two photons resulting from collisions among the proton constituents, quarks, anti-quarks or gluons. Using the parton distribution functions we can estimate the production rate. This is an *irreducible background,* in the sense that it has a distinct physical origin and cannot be reduced. Second is an instrumental background due to particle misidentification, for example a π^0 producing two photons, which are not resolved in the electromagnetic calorimeter. This is a *reducible background* because it can be reduced by improving the detector performance. The important point is that the pair of photons produced in either of these ways is expected to have an invariant mass[7] with a broad distribution, while the two photons with momenta k_1 and k_2, coming from a decay of a scalar boson will satisfy the equality $(k_1 + k_2)^2 = m_H^2$.

The four-lepton decay channel, coming from a ZZ intermediate state, is considered as a *gold-plated* channel. It is more rare, but it suffers from little background. Both LHC detectors had high efficiency in the detection of charged leptons and could measure their energy with a precision on the order of 1%.

The discovery was presented at a special conference held at CERN on 4 July 2012 by both ATLAS and CMS. Related figures are shown in Figure 22.11 and Figure 22.12. The hunting was over. The last missing particle of the Standard Model was finally discovered. It had taken almost fifty years, from the theoretical prediction made in 1964, until its experimental discovery in 2012.

---

[7] If k_1 and k_2 are the momenta of two photons their invariant mass is $(k_1 + k_2)^2$.

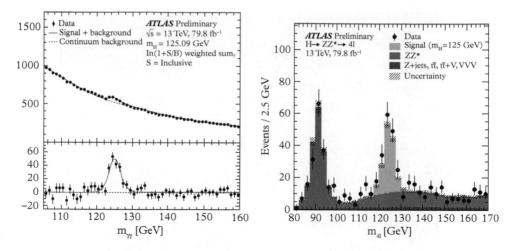

Fig. 22.11 The discovery of the BEH scalar in the decay modes 2γ (left) and $4l$ (right). The figures include the data of $\sqrt{s} = 13$ TeV. The plots present the number of events versus the two-photon (left) and the four-lepton invariant mass (right). On the left the upper curve shows the excess of events against a rapidly falling background. The lower curve shows the same data with the background subtracted. *(Source: CERN-ATLAS collaboration).*

22.3 Hadron Spectroscopy

Table 6.6 contains what we consider today to be the table of elementary particles. They are the 12 gauge bosons, the leptons and the quarks of the three families, and the BEH scalar. However, in the PDG lists we find hundreds of additional entries, which should be considered as bound states of quarks and gluons. A measure of our understanding of the fundamental forces is our ability to describe this rich spectroscopy. As we have seen already in this book, this understanding is only partial. We have had some notable successes but also several important failures.

The most important unsolved problem is that of confinement. We cannot prove that in an unbroken non-Abelian gauge theory the only possible asymptotic states are singlets under the gauge group. We know experimentally that this is true for QCD but, as we saw in Chapter 21, from the theoretical point of view we have only numerical evidence for it. The lattice simulations we presented in section 21.4.2 offer a remarkably good fit of the spectrum of light hadrons but they are limited to a small fraction of the observed states. In this section we give a very brief phenomenological description of the known hadronic states classified according to their assumed quark content.

Light quarks

We place in this category the quarks whose mass is smaller than the scale of the spontaneous breaking of chiral symmetry. The latter is of order a few hundred MeV so, looking at the Table 19.2, we see that this condition is certainly met by u and d,

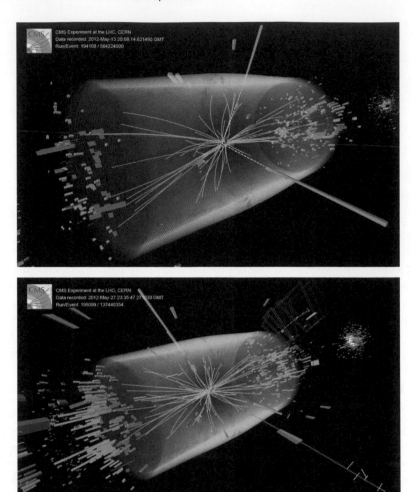

Fig. 22.12 Two beautiful events among those which established the discovery of the BEH particle. The upper figure shows a 2γ decay with the two photons shown as green tracks in the electromagnetic calorimeter. The lower figure shows an $e^+e^-\mu^+\mu^-$ decay with the electrons as green tracks in the e.m. calorimeter and the muons as red tracks in the muon chambers. *(Source: CERN-CMS collaboration).*

the quarks of the first family, and marginally by s. We have already pointed out an important property of the hadrons made exclusively out of these quarks:[8] their masses

---

[8]The term "exclusively" is not very precise for a relativistic theory, because we can always excite $q\bar{q}$ pairs of arbitrary flavour. However, we have seen that, at low energies, the probability of exciting a heavy quark flavour in a light hadron is very small. Since every quark carries a quantum number, which is conserved by strong and electromagnetic interactions, we could be tempted to use the quantum numbers to characterise the quark content of a hadron, but there exist mesons which are $c\bar{c}$ or $b\bar{b}$

are mainly due to the breaking of chiral symmetry; the masses of the constituents contribute very little.

The spectrum of the light states is accurately reproduced by QCD numerical simulations, see Figure 21.8. They are classified according to $SU(3)$ multiplets and, in the previous chapters, we have studied some of their properties. The mesons are $J^P = 0^-$ and 1^- $q\bar{q}$ bound states and form singlets and octets of $SU(3)$. The baryons are $\frac{1}{2}^+$ and $\frac{3}{2}^+$ qqq bound states and form an octet and a decouplet.

With the exception of the proton all these hadrons are unstable. As we have seen, some, like the neutron, decay via weak interactions, π^0 and Σ^0 decay via electromagnetic interactions, and the others via the strong interactions.

In a potential model for either $q\bar{q}$, or qqq we would expect the aforementioned hadrons to be the ground states. In addition, such a model would predict a series of excited states and, experimentally, we do observe a large number of hadrons, which could be candidates for such excited states. For example, in the PDG listings we find many pion–nucleon resonances with increasing spins and masses up to 2.5 GeV. Potential models, taking into account to a certain extent relativistic corrections, partly reproduce the low-lying part of the observed spectrum, but they provide only a phenomenological description of the experimental results and cannot be considered as calculations from first principles. On the other hand, QCD numerical simulations are not yet precise enough to describe excited states. As a result we cannot claim a deep understanding of the spectrum and the quantum numbers of these hadrons.

Heavy flavours

From the group theory point of view the Standard Model can be formulated with just one family. However, the presence of the other two opens up a vast domain of research. It gives rise to a rich hadron spectroscopy and offers new ways to study the underlying fundamental symmetries. We present only a brief summary.

With the successive discoveries of charm, bottom and top quarks the flavour group of particle physics extended from $SU(3)$ to $SU(4)$, $SU(5)$ and, finally, $SU(6)$. As we see in Table 19.2, these symmetries are very badly broken and this breaking seems to be due entirely to the very different couplings of the various quark species to the BEH scalar. Nevertheless, flavour symmetry can still be used as a guide to count and classify the new hadron states. The low lying states are expected to be 0^- and 1^- for $q\bar{q}$, and $\frac{1}{2}^+$ and $\frac{3}{2}^+$ for qqq bound states. As the symmetry breaking becomes stronger, we expect the mixing among the various states belonging to different representations to be more important. It is a simple exercise in group theory, see Problem 22.3, to find the dimensionalities of the representations, which generalise **8** and **10** for the higher groups. This gives a measure of the expansion in the number of hadron states. Not all the states predicted by the quark model have so far been identified experimentally and, even for those that have, the quantum numbers and partial decay widths have not always been measured.

---

bound states, which means that they are made out of heavy quarks and still do not carry any heavy quark quantum number.

Two experimental facts play an important role in the heavy flavour hadron spectroscopy: First, the quark masses grow larger and, as a result, we expect the corresponding new hadrons to grow heavier. Indeed, looking at the lightest 0^- states, we find π ($\simeq$140 MeV), K ($\simeq$ 490 MeV), D (with a c quark, $\simeq$ 1870 MeV) and B (with a b quark, $\simeq$ 5280 MeV). Hadrons containing the t quark have not been identified. With heavier masses we have a much larger phase space for available decays and we expect a large increase in the number of possible decay channels and, with other factors being approximately equal, shorter lifetimes. Indeed, it is what we observe, but a second fact complicates the picture. We find experimentally that the angles entering the V_{CKM} matrix are all very small: $s_{13} \ll s_{23} \ll s_{12} \ll 1$. This induces a hierarchy in various decays, as follows:

• *Cabibbo favoured decays.* These involve only cosine factors and, therefore, the initial and final quarks belong to the same family. Neutron β-decay is the obvious example. Among the available heavy flavours only the $c \to s$ transitions belong to this class.

• *Simply Cabibbo suppressed decays.* These involve one sine, which means that one quark changes family. Examples: $s \to u$, $c \to d$, or $b \to c$.

• *Doubly Cabibbo suppressed decays.* Two quarks change family, which means that the amplitude is proportional to the product of two sine factors. Example: $D^0 \to K^+\pi^-$, which, in quark language, means $(\bar{u}c) \to (\bar{s}u) + (\bar{u}d)$, or $\bar{u} \to \bar{s}$ and $c \to d$ with the creation of a $\bar{u}u$ pair.

This kind of hierarchy is observed in the lifetimes. In Problem 22.4 we ask the reader to go through the PDG tables and give qualitative explanations for particular features of the data.

• *M–$\overline{M}$ mixing and CP violation.* The neutral pseudoscalar D mesons exhibit the same phenomenon of D^0–$\overline{D}^0$ mixing we found for the K mesons. Therefore, the states which have well defined lifetimes are linear combinations D_1 and D_2. However, because the phase space for both of them is very large, their lifetimes are not very different. The mean lifetime is $\tau = (410.1 \pm 1.5) \times 10^{-15}$ s and we have $\Delta\tau/\tau \simeq 10^{-2}$ with $\Delta\tau = |\tau_1 - \tau_2|$. Similarly, the mass difference $\Delta m = |m_{D_1} - m_{D_2}|$ has been determined, with quite large errors, to be approximately 6×10^{-6} eV. We will discuss CP violation in section 22.4. The same is true for the neutral B mesons B^0 and B_s^0, which mix with their antiparticles.

• *$c\bar{c}$ and $b\bar{b}$ states.* Among the hadrons made out of heavy quarks, the $c\bar{c}$ and $b\bar{b}$ mesons play an important role: they are made out of heavy quarks although they do not carry any heavy quark quantum number. In section 22.2 we saw how charmed particles were discovered following the discovery and the study of the $c\bar{c}$ mesons. The same story was repeated for the b particles, i.e. hadrons containing at least one b quark.

If there is one sector of hadronic physics in which non-relativistic potential models could be considered trustworthy, it is precisely the study of $c\bar{c}$ and $b\bar{b}$ states. Because of the large mass of the constituent quarks the latter can be considered as two non-relativistic particles lying close to each other. The property of asymptotic freedom implies that the short distance part of the potential can be approximated by the one-gluon exchange force, therefore it is expected to be of the form $1/r$. Obviously this form breaks down at large distances because we need a rising potential to mimic the

Table 22.1 The first $c\bar{c}$ states

Particle	Spin^{PC}	M (MeV)	Γ (MeV)	τ (sec)
$\eta_c(1S)$	0^{-+}	2984	32	0.2×10^{-22}
$J/\Psi(1S)$	1^{--}	3097	9.3×10^{-2}	0.7×10^{-20}
$\chi_{c0}(1P)$	0^{++}	3415	10.3	0.6×10^{-22}
$\chi_{c1}(1P)$	1^{++}	3511	0.84	0.8×10^{-21}
$h_c(1P)$	1^{+-}	3525	0.7	0.8×10^{-21}
$\chi_{c2}(1P)$	2^{++}	3556	1.97	0.3×10^{-21}
$\eta_c'(2S)$	0^{-+}	3638	11.3	0.6×10^{-22}
$\Psi'(2S)$	1^{--}	3686	2.33×10^{-3}	0.3×10^{-18}
$\Psi(^3D_1)$	1^{--}	3770	27.2	0.2×10^{-22}

confining forces. These considerations lead to a simple parametrisation of the heavy quark–anti-quark potential of the form

$$V(r) = -\frac{\kappa}{r} + \sigma r \qquad (22.6)$$

with κ and σ two parameters to be determined by the data.[9] This potential is applied to both $c\bar{c}$ and $b\bar{b}$ states, although the non-relativistic approximation is better justified for $b\bar{b}$. So, the free parameters of this model are: the effective masses of the quarks m_c and m_b, the constants κ_c and κ_b which determine the $1/r$ part of the potential for charmonium and bottomonium respectively, and a common confining parameter σ, because confining forces are assumed to be flavour blind. This last assumption is also supported by lattice calculations. The Schrödinger equation is solved numerically and the low-lying states are compared with the experimental data. A rather good fit is obtained.

In Table 22.1 we show the first charmonium states. The notation is a mixture of the one used in atomic spectroscopy and the symbols assigned in particle physics.

All these states decay through strong interactions, and in section 22.2 we explained the long lifetimes of the first two Ψ states. Because of phase space, the number of final states is large. For J/Ψ the hadronic modes account for $(87.7\pm0.5)\%$ and among them we find 3, 5, 7 etc. pions. As the spectroscopic notation indicates, Ψ' is an excited state of J/Ψ and, indeed, its main decay mode is $J/\Psi + X$. The χ states decay mainly into an even number of pions. The $\Psi(3770)$ lies just above the threshold of $D\overline{D}$ and has a shorter lifetime. In fact, the first charmed particles were discovered by looking at the decay products of $\Psi(3770)$.

In section 22.2 we drew a parallel between the heavy quark–anti-quark states and the ones we know from positronium. Based on this analogy, theorists had predicted the existence of radiative transitions of the form $\Psi' \to \chi + \gamma$ and $\chi \to J/\Psi + \gamma$. Indeed,

---

[9]This is known as *the Cornell potential*. Other forms have also been used, for example replacing the linearly rising confining potential by a logarithmic one.

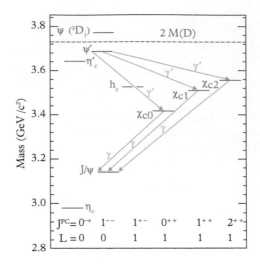

Fig. 22.13 The main radiative transitions in the charmonium system. *(Source: CLEO collaboration).*

such transitions have been observed and are shown in Figure 22.13. They offer a good confirmation of the charmonium idea.

An even richer structure is observed in the $b\bar{b}$, or bottomonium, system. As for charmonium, the lowest state is a 0^{-+} state, called $\eta_b(1S)$ with mass 9399 MeV and a lifetime of order 0.5×10^{-22} s. There exist three very narrow 1^{--} states, the analogues of the Ψ states, called Y states. They are $Y(1S)$ with mass 9460 MeV and lifetime $\simeq 0.6 \times 10^{-20}$ s, $Y(2S)$ with mass 10023 MeV and lifetime $\simeq 2 \times 10^{-20}$ s, and $Y(3S)$ with mass 10355 MeV and lifetime $\simeq 3 \times 10^{-20}$ s. The fourth excitation, called $Y(4S)$, has mass 10579 MeV and can decay into a $B\bar{B}$ pair. As a result it has a very short lifetime of order 3×10^{-23} s.[10] In addition, we have a large number of χ_b states, but their spin and parity quantum numbers have not been determined experimentally for all of them.

 • *t particles.* As we have said already, hadrons containing the t quark have not been found and we have good reasons to believe that they do not exist as particles with well defined masses and lifetimes. The reason is the enormous mass of the t quark. Being heavier than the W it can decay directly into a real W and a b quark: $t \to W + b$. This is a semi-weak, Cabibbo favoured process and the resulting lifetime, see Problem 22.5, is of order 0.5×10^{-24} s. Therefore, the t quark is expected to decay before it has time to form hadrons.[11] Another consequence of its large mass is that its coupling constant with the BEH scalar is large, which means that t quark loops dominate the corresponding higher order radiative corrections. We saw this effect in

---

[10]The so-called "*B*-factories" are $e^- e^+$ colliders tuned precisely at the mass of $Y(4S)$.

[11]As we have seen, the hadronisation scale is on the order of a few hundred MeV, which, in our system of units $\hbar = c = 1$, corresponds to $\mathcal{O}(10^{-23})$ s.

estimating the production cross section and the two-photon decay rate of the scalar boson.

$t\bar{t}$ pairs are produced in hadronic colliders, both at the Tevatron and the LHC. At the Tevatron the production cross section was on the order of 8 pb and it rises to almost 1 nb at the highest LHC energy of 14 TeV. These values are in agreement with theoretical estimations.

• *Gluon and multiquark bound states.* In the simple quark model mesons are $q\bar{q}$ and baryons qqq bound states. We saw that this picture describes quite well the majority of low-lying hadrons. However, in the PDG listings we find also states which do not seem to fit in this scheme. We call them collectively *exotic states* and, in fact, QCD does predict such states. Examples of exotic states are:

Glueballs. Even in the absence of quarks, QCD, which is an $SU(3)$ gauge theory, describes the self-interaction of an octet of gauge bosons, the gluons. Therefore it is natural to expect bound states containing only gluons, which we call "glueballs". This is confirmed by lattice computations, which predict a rich spectrum of such states. The ground state is predicted to have the quantum numbers 0^{++}, isospin 0 and mass around 1700 ± 100 MeV and the first excited state 2^{++} with a mass of about 2400 MeV. The calculations are not very accurate because they are mostly limited to the quenched approximation, i.e. they neglect the influence of the quarks in the dynamics among gluons. Experimentally there are three states in the mass region around 1500 MeV with 0^{++} quantum numbers: $f_0(1370)$, $f_0(1500)$ and $f_0(1710)$, while a simple quark model would predict only two. This is taken as a confirmation of the existence of glueballs, although the one-to-one correspondence is not meaningful because the glueballs are expected to mix with the $q\bar{q}$ bound states having the same quantum numbers. Such mixings cannot be computed in the quenched approximation, and lattice computations, have only recently gone beyond it.

Tetraquark mesons. Another example of exotic states are mesons ($qq\bar{q}\bar{q}$) made out of two quarks and two anti-quarks. They can be thought of as the analogue of two hydrogen atoms bound to form a hydrogen molecule. But independently of the particular dynamical mechanism, from a purely group theory point of view, we have several ways to combine the group indices of two quarks and two anti-quarks. In particular, we can imagine the scheme $(qq)\otimes(\bar{q}\bar{q})$. This has been suggested already for flavour $SU(3)$. We have: $\mathbf{3}\otimes\mathbf{3}=\mathbf{6}\oplus\bar{\mathbf{3}}$ and $\bar{\mathbf{3}}\otimes\bar{\mathbf{3}}=\bar{\mathbf{6}}\oplus\mathbf{3}$. Taking into account spin and colour indices and imposing full antisymmetry, people have argued that the lightest di-quark has spin 0 and is a colour $\bar{\mathbf{3}}$. Similarly the lightest di-anti-quark is a colour $\mathbf{3}$. Therefore, the lightest tetraquark states are predicted to form an $SU(3)$ nonet of scalar states: $\mathbf{3}\otimes\bar{\mathbf{3}}=\mathbf{1}\oplus\mathbf{8}$. The first candidate in the particle listings is the isosinglet $f_0(500)$ with quantum numbers 0^{++}. It is a very broad resonance with mass between 400 and 550 MeV and width between 400 and 700 MeV decaying mostly into two pions. Several states in the mass range of several hundred MeV could complete the nonet. More candidates for tetraquark states are found in the charmonium and bottomonium systems. For example, $X(3872)$, which is listed as an excited χ_{c1} 1^{++} state, is found to decay into $J/\Psi + \pi^+ + \pi^-$. Similarly, the charged $Z^{\pm}(4430)$ 1^{+-} state decays into $\Psi(2S) + \pi^{\pm}$. Such decays are not easy to understand if the initial states are pure $c\bar{c}$ and suggest rather a $(cq\bar{c}\bar{q})$ structure.

Pentaquark baryons. They are supposed to be $(qqqq\bar{q})$ bound states. Again, they can be viewed as baryon–meson molecules and there are several candidates of such states in the particle listings. Note, however, that their unambiguous experimental identification is not easy.

22.4   The Cabibbo–Kobayashi–Maskawa Matrix and *CP* Violation

The physics of the heavy flavours has become one of the most promising research fields in high energy physics. This is due to the fact that weak interactions do not conserve the individual family numbers and mix the various quark species. Many interesting phenomena appear to be related to these mixings. In the previous chapters we have studied the phenomenon of *CP* violation in the first three quark flavours. Two results are worth mentioning here. First, the phenomenon was observed only in *K*-decays. Second, it was a very small effect, on the order of a few per thousand. The simplest way to incorporate *CP* or *T*, violation in a Lagrangian field theory is to give a non-zero relative phase between two terms. In such a model this phase should be very small, of order 10^{-3}. In general, physicists don't like very small parameters, and this led Lincoln Wolfenstein in 1964 to formulate *the super-weak theory*, which postulated the existence of a fifth fundamental interaction, a thousand times weaker than ordinary weak interactions, maximally violating *CP*. In almost all cases the effects of such an interaction would be unobservable, buried under the background of the known interactions, with the exception of the neutral kaon decays. According to that theory, we had been lucky to observe *CP* violation in the first place; it was due to the particular structure of the K^0–$\overline{K^0}$ system, which, because of phase space, resulted in eigenstates having very different lifetimes.

The discovery of the heavy quark flavours changed the picture in many ways. First, it offered a richer hadronic spectroscopy, which we have briefly described in section 22.3. Second, the mixing among the quark flavours gave rise to a general 3×3 matrix, whose matrix elements should be determined by experiment. In an obvious notation, after diagonalising the quark mass matrix, the coupling of the charged weak vector bosons to the quarks is, see equation (19.30),

$$\frac{g}{\sqrt{2}}\overline{U_L}\gamma^\mu V_{\mathrm{CKM}}\widetilde{D}_L W_\mu^+ \ + \ \mathrm{h.c.} \tag{22.7}$$

where

$$U_L = \begin{pmatrix} u \\ c \\ t \end{pmatrix}_L , \quad \widetilde{D}_L = \begin{pmatrix} d \\ s \\ b \end{pmatrix}_L , \quad V_{\mathrm{CKM}} = \begin{pmatrix} V_{ud} & V_{us} & V_{ub} \\ V_{cd} & V_{cs} & V_{cb} \\ V_{td} & V_{ts} & V_{tb} \end{pmatrix} \tag{22.8}$$

In this section we study the properties of V_{CKM}.

22.4.1   The matrix elements

The matrix elements V_{ij} can be determined experimentally by studying various semi-leptonic weak decays. In the PDG tables we find:

 • $|V_{ud}| = 0.97420 \pm 0.00021$. This is the most accurately known from the measurement of a large number of $0^+ \rightarrow 0^+$ nuclear β-decay transitions.

• $|V_{us}| = 0.2243 \pm 0.0005$. This result is a weighted average from the measurements of neutral and charged kaon decays of the type $K \to \pi + e + \nu$, as well as hyperon β-decays. Note that in all cases the momentum transfer carried by the lepton pair is appreciable, on the order of the $K - \pi$ mass difference, so the interpretation of the results involves a theoretical extrapolation from this value to zero. In the most recent estimations this is done using the matrix elements of the weak current between hadronic states obtained from QCD lattice computations.

• $|V_{ub}| = (3.94 \pm 0.36) \times 10^{-3}$. The precision is rather poor due to both experimental and theoretical difficulties. On the experimental side the result is extracted from the study of exclusive and inclusive B-decays, which have significant backgrounds. The theoretical estimations involve large extrapolations, which increase the uncertainties.

• $|V_{cd}| = 0.218 \pm 0.004$. Obtained from the corresponding leptonic and semi-leptonic c-decays. It is similar to $|V_{us}|$ with larger errors, both experimental and theoretical.

• $|V_{cs}| = 0.997 \pm 0.017$. As we have said in Chapter 19, this corresponds to the dominant decay mode of the c quark and it is the analogue of $|V_{ud}|$.

• $|V_{cb}| = (42.2 \pm 0.8) \times 10^{-3}$. This is from exclusive and inclusive semi-leptonic decays of B mesons to charm particles.

• $|V_{td}|$, $|V_{ts}|$ and $|V_{tb}|$. In section 22.3 we pointed out that there are no t hadrons, i.e. hadrons containing a t quark among their constituents. So, these matrix elements are determined indirectly. For $|V_{tb}|$ we can look at the semi-weak decay $t \to W + b$, and for the others we can either measure the single t quark production cross section, or look at processes in which radiative corrections involving t quark loops are important. The results, as quoted by the PDG, are: $|V_{td}| = (8.1 \pm 0.5) \times 10^{-3}$, $|V_{ts}| = (39.4 \pm 2.3) \times 10^{-3}$ and $|V_{tb}| = 1.019 \pm 0.025$.

The phases of the matrix elements V_{ij} are determined by studying various CP violating processes, which we will present shortly.

22.4.2 Unitarity

In writing V_{CKM} in equation (19.25) we assumed that the matrix is unitary, which implies that there are no new, as yet unknown, states which mix with the quark flavours. In the Standard Model, for example, it means that there is no fourth family of quarks. This assumption made it possible to write V_{CKM} as a generalised rotation in a three-dimensional space. We adopted the phase convention according to which the rotations in the 1-2 and 2-3 planes are real, and this left us with a phase in the 1-3 plane rotation.

$$V_{\text{CKM}} = \begin{pmatrix} 1 & 0 & 0 \\ 0 & c_{23} & s_{23} \\ 0 & -s_{23} & c_{23} \end{pmatrix} \begin{pmatrix} c_{13} & 0 & s_{13}e^{-i\delta} \\ 0 & 1 & 0 \\ -s_{13}e^{i\delta} & 0 & c_{13} \end{pmatrix} \begin{pmatrix} c_{12} & s_{12} & 0 \\ -s_{12} & c_{12} & 0 \\ 0 & 0 & 1 \end{pmatrix} \qquad (22.9)$$

The three rotation angles and the phase δ are determined from the knowledge of the matrix elements V_{ij}, i.e.

$$s_{12} = \frac{|V_{us}|}{\sqrt{|V_{ud}|^2 + |V_{us}|^2}} \;, \quad s_{23} = s_{12}\frac{|V_{cb}|}{|V_{us}|} \;, \quad s_{13}e^{i\delta} = V_{ub}^* \qquad (22.10)$$

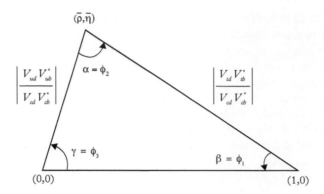

Fig. 22.14 An example of a unitarity triangle. The vertices are $(0,0)$, $(1,0)$ and $(\bar\rho, \bar\eta)$ with $\bar\rho = \rho(1 - \lambda^2/2)$ and $\bar\eta = \eta(1 - \lambda^2/2)$. *(Source: Particle Data Group).*

As we saw, the matrix elements obey some hierarchy relations, namely the diagonal elements are close to unity, which means that all three rotation angles are small, and their sines satisfy

$$s_{13} \ (\simeq 4 \cdot 10^{-3}) \ll s_{23} \ (\simeq 0.04) \ll s_{12} \ (\simeq 0.2) \ll 1 \tag{22.11}$$

In order to exhibit this hierarchy explicitly, Wolfenstein proposed another parametrisation of V_{CKM} by introducing the variables

$$s_{12} = \lambda \ , \quad s_{23} = A\lambda^2 \ , \quad s_{13} e^{i\delta} = A\lambda^3 (\rho + i\eta) \tag{22.12}$$

in terms of which V_{CKM}, in an expansion in powers of λ, takes the form

$$V_{\text{CKM}} = \begin{pmatrix} 1 - \lambda^2/2 & \lambda & A\lambda^3(\rho - i\eta) \\ -\lambda & 1 - \lambda^2/2 & A\lambda^2 \\ A\lambda^3(1 - \rho - i\eta) & -A\lambda^2 & 1 \end{pmatrix} + \mathcal{O}(\lambda^4) \tag{22.13}$$

The unitarity property implies that the matrix elements satisfy nine conditions: $\sum_i V_{ij} V_{ik}^* = \delta_{jk}$ and $\sum_j V_{ij} V_{kj}^* = \delta_{ik}$. Taking $j \neq k$ in the first sum and $i \neq k$ in the second, we obtain six relations of the form

$$V_{ud} V_{ub}^* + V_{cd} V_{cb}^* + V_{td} V_{tb}^* = 0 \tag{22.14}$$

Each of these relations can be represented as a triangle in the complex plane. In Problem 22.6 we ask the reader to prove that all six triangles have the same area, which measures the amount of CP violation. On the other hand, if, for some combination, the experimentally measured matrix elements do not form a triangle, this would imply a non-unitary matrix and the existence of new states. The most commonly used unitarity triangle is the one given by equation (22.14) normalised by dividing each side by the product $V_{cd} V_{cb}^*$, see Figure 22.14.

22.4.3 *CP* violation

In Chapters 6 and 18 we presented the experimental evidence for CP violation in the decays of strange particles. Here we want to complete this discussion by including the results from the decays of charm and b particles and present the phenomenological analysis based on our knowledge of the CKM matrix.

In the Standard Model CP violation is induced by the possibility of introducing a non-zero phase in the quark mixing matrix. As we have shown in Problem 19.1, this possibility requires the existence of at least three quark families. In equation (22.9) we adopted the convention of putting this phase in the rotation of the first and the third families, but we could have made a different choice and move the phase around in the matrix. Measurable quantities could not depend on such choices. We start by presenting a parametrisation independent measure of CP-violation, which was proposed in 1985 by C. Jarlskog.

Let us go back to the construction of the Electroweak Theory in section 19.4. In general, without assuming any particular mechanism for spontaneous symmetry breaking, we will have two 3×3 quark mass matrices, M_U for the up-type quarks u, c and t, and M_D for the down-type d, s and b. It is convenient to write them in a dimensionless form by dividing each one by its largest eigenvalue, m_t for M_U and m_b for M_D. Since these matrices are non-degenerate and do not have any zero eigenvalues, each one of them can be diagonalised by a unitary matrix

$$R_U M_U R_U^\dagger = \hat{M}_U = \text{diag}\left[\frac{m_u}{m_t}, \frac{m_c}{m_t}, 1\right]$$

$$R_D M_D R_D^\dagger = \hat{M}_D = \text{diag}\left[\frac{m_d}{m_b}, \frac{m_s}{m_b}, 1\right] \tag{22.15}$$

Diagonalising the mass matrices results in introducing in the charged weak current the CKM matrix which, in this notation, is given by

$$V_{\text{CKM}} = R_U R_D^\dagger \tag{22.16}$$

Jarlskog remarked that in general, M_U and M_D do not commute and $R_U \neq R_D$. The quark flavour mixing and CP violation are the results of this non-commutativity. Let us introduce the commutator

$$[M_U, M_D] = \text{i}C = R_U^\dagger [\hat{M}_U, V_{\text{CKM}} \hat{M}_D V_{\text{CKM}}^\dagger] R_U \tag{22.17}$$

with C a traceless Hermitian matrix. Equation (22.17) shows that, up to a unitary transformation, the matrix C is expressed in terms of measurable quantities, the quark masses and the elements of the CKM matrix. A useful quantity is the determinant of C. A direct calculation gives

$$\det C = -2F_U F_D J \tag{22.18}$$

where

$$F_U = (m_t - m_u)(m_t - m_c)(m_c - m_u)/m_t^3$$
$$F_D = (m_b - m_d)(m_b - m_s)(m_s - m_d)/m_b^3 \tag{22.19}$$

and J is defined through

$$\mathrm{Im}(V_{ij}V_{kl}V_{il}^*V_{kj}^*) = J \sum_{m,n=1}^{3} \epsilon_{ikm}\epsilon_{jln} \tag{22.20}$$

which, in terms of the parametrisations (19.25) or (22.13), gives

$$J = c_{12}c_{23}c_{13}^2 s_{12}s_{23}s_{13}\sin\delta \simeq \lambda^6 A^2 \eta \tag{22.21}$$

We see that the necessary conditions to have CP violation in the Standard Model are: (i) there should be no degeneracy in the mass spectrum, neither among the up, nor among the down quarks, (ii) all mixing angles should be different from 0 or $\pi/2$, (iii) δ should be different from 0 or π. Furthermore, equations (22.18), (22.19) and (22.21) show that the particular structure of the CKM matrix, in which all mixing angles as well as the factors F_U and F_D are small, implies that CP violation is very weak, even with large values of the angle δ.

We can now proceed with a phenomenological presentation of CP violation. Experimentally such violations have been observed in flavour violating decays of various hadrons. Let $|M\rangle$ denote an unstable hadron decaying into a final state $|f\rangle : M \to f$. We define also the CP transformed states:[12]

$$CP|M\rangle = |\overline{M}\rangle \ , \ \ CP|f\rangle = |\overline{f}\rangle \tag{22.22}$$

If $\mathcal{H}_W$ is the weak interaction Hamiltonian responsible for the transition $M \to f$, we can define the decay amplitudes

$$A_f = \langle f|\mathcal{H}_W|M\rangle, \ A_{\overline{f}} = \langle \overline{f}|\mathcal{H}_W|M\rangle, \ \overline{A}_f = \langle f|\mathcal{H}_W|\overline{M}\rangle, \ \overline{A}_{\overline{f}} = \langle \overline{f}|\mathcal{H}_W|\overline{M}\rangle \tag{22.23}$$

CP violation may manifest itself in several observables related to such decays. For historical, as well as pedagogical reasons, we distinguish three types of effects:

I. *CP violation in the decay.* It is clear that if CP is conserved, i.e. the operator CP commutes with the Hamiltonian, we have: $A_f = \overline{A}_{\overline{f}}$. Therefore any departure from this equality is direct evidence of CP violation. In fact, if the hadron M carries a non-zero conserved quantum number, such as electric charge or baryon number, this is the only possible manifestation of CP violation.

II. *CP violation in mixing.* If M is a neutral meson, such as K^0, D^0 or B^0, we expect to have a mixing between the states M and $\overline{M}$, like the one we described for K^0 and $\overline{K^0}$ in sections 6.10 and 18.11. The eigenstates of the Hamiltonian, which correspond to well defined values of mass and lifetime, are superpositions of the form[13]

---

[12]Since strong interactions conserve flavour, the relative phase between CP related hadronic states is arbitrary. Unfortunately, there is no generally accepted convention in the literature. Here we made the simplest choice of taking all phases equal to zero.

[13]For the neutral kaons we used the notation $|K_L^0\rangle$ and $|K_S^0\rangle$ to denote the long- and short-lived states because, as we have explained, the lifetimes were very different. This is no more the case for D and B mesons and, by convention, we label the states according to the values of their masses as $|M_L^0\rangle$ and $|M_H^0\rangle$ for *light* and *heavy*, respectively.

$$|M_L^0\rangle = \frac{p\,|M^0\rangle + q|\overline{M^0}\rangle}{\sqrt{|p|^2 + |q|^2}} \quad , \quad |M_H^0\rangle = \frac{p\,|M^0\rangle - q|\overline{M^0}\rangle}{\sqrt{|p|^2 + |q|^2}} \qquad (22.24)$$

If CP is conserved we must have $|p| = |q| = 1$ and $\langle M_L^0|M_H^0\rangle = 0$. So, any violation of these relations is evidence of CP non-conservation in the mixing of the two states.

III. *CP violation in interference between a decay without mixing, $M^0 \rightarrow f$, and a decay with mixing, $M^0 \rightarrow \overline{M^0} \rightarrow f$.*

Experimentally we have detected CP violation in K-decays, first established in 1964, in B-decays, since 2001, and in D-decays, observed only recently in 2019. At the time of writing, it is still a very active research programme and the situation may evolve in the near future. Here we give only a brief summary of the results.

We have already presented the evidence for CP violation in neutral K-decays in Chapter 18. They involve 2π decay modes ($\pi^+\pi^-$ and $\pi^0\pi^0$), 3π modes as well as the semi-leptonic modes ($\pi^+l^-\bar{\nu}$ and $\pi^-l^+\nu$, but also the K_{l4} modes, such as $\pi^0\pi^+l^-\bar{\nu}$ etc.) The resulting CP violating amplitudes are very small, for example the asymmetry in the semi-leptonic modes presented in equation (18.100) is at the level of 0.003. No measurable effect has been observed in charged kaon or strange baryon decays.

In the framework of the Standard Model the small values of CP violating parameters observed in kaon decays can be understood. As we have explained, we need the effective participation of all three families because a two-family model has no CP violation. The third family is much heavier than the others, so it suppresses the effects in the decays of light particles, such as kaons. A corollary of this argument is the expectation that in the decays of B particles, the heaviest available, CP violation could be stronger. As a result, in the last twenty years an intense experimental programme has been devoted to the study of CP violation in the B sector. Such effects have been detected in a plethora of decays and the values are quite large. We give here a few examples but a complete list can be found in the PDG tables.

• In $b \rightarrow c\bar{c}s$ or $b \rightarrow c\bar{c}d$ transitions, such as $B^0/\overline{B^0} \rightarrow J/\Psi + K^0/\overline{K^0}$ or $B^0/\overline{B^0} \rightarrow J/\Psi + \pi^0$, the parameter which describes the CP violating asymmetry is about 0.7.

• Equally significant CP violating effects have been found in $b \rightarrow c\bar{u}d$ transitions, such as $B^0/\overline{B^0} \rightarrow D^0(D^{*0}) + h^0$, with h^0 a neutral meson, for example a π^0.

• Type I CP violating effects have been observed, for example in the decay $B^+ \rightarrow (D^0)_+ + K^+$, where by $(D^0)_+$ we denote the combination of D^0 and $\overline{D^0}$ which is predominantly CP even.

• CP violation in the decays of baryons has not been found yet but, for b-baryons, theoretical estimations put it within reach of the forthcoming generation of experiments, LHCb at CERN and BELLE II in Japan.

CP violation in charmed meson decays has not been studied experimentally in great detail. The advantages of going to the higher mass B mesons were recognised quite early, and most dedicated experiments were designed primarily for B physics. In 2019, the LHCb collaboration announced the first observation of CP violation in D-decays.

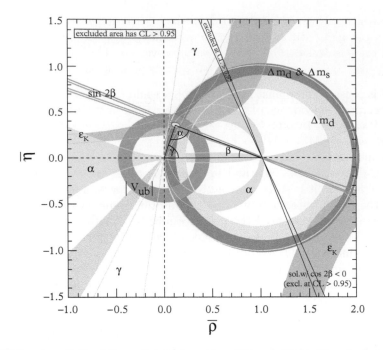

Fig. 22.15 The overall fit of the unitarity triangle of Figure 22.14. The shaded areas have 95% confidence level. *(Source: Particle Data Group).*

All these searches are very important because, as explained, CP violation may be a good portal to detect new physics beyond the Standard Model. In the latter this violation comes from the phase in the CKM matrix. Given the large number of independent measurements, the elements of this matrix are over-determined and they provide many sensitive tests of the consistency of the model. A good example is given by the various unitarity triangles we presented in equations (22.14). Let us take the one shown in Figure 22.14. The sides of the triangle are shown in the figure in terms of the absolute values of the transition matrix elements. The three angles are given in terms of the phases, which in the CKM matrix are functions of the phase δ. We find

$$\alpha \equiv \phi_2 \equiv \arg\left(-\frac{V_{td}V_{tb}^*}{V_{ud}V_{ub}^*}\right) \simeq \arg\left(-\frac{1-\rho-i\eta}{\rho+i\eta}\right)$$

$$\beta \equiv \phi_1 \equiv \arg\left(-\frac{V_{cd}V_{cb}^*}{V_{td}V_{tb}^*}\right) \simeq \arg\left(-\frac{1}{1-\rho-i\eta}\right)$$

$$\gamma \equiv \phi_3 \equiv \arg\left(-\frac{V_{ud}V_{ub}^*}{V_{cd}V_{cb}^*}\right) \simeq \arg(\rho+i\eta) \tag{22.25}$$

As we said, the triangle is over-determined. In Figure 22.15 we show the overall fit in the $\bar{\rho}, \bar{\eta}$ plane. The basis of the triangle, given by the normalisation we choose in Figure 22.14, is fixed with the two vertices at (0,0) and (1,0). The third vertex is determined by the intersection of two lines, determined by the angles β and γ, two circles, determined by the lengths of the two sides, and a third circle, determined by the angle α. All these determinations agree within errors. The agreement with the Standard Model is very good, but there is considerable margin for improvement. Upcoming measurements will either decrease considerably the error bars, or show evidence for new physics.

22.5 The Neutrino Matrix and *CP* Violation

When neutrino oscillations were first detected, there was hope that they would provide extra information which would help achieve a deeper understanding of the questions related to the existence of the three families in the Standard Model. Today, more than twenty years later, we have to admit that such expectations have not been fulfilled. In fact, our understanding of the lepton world is even poorer than that of the hadrons. Here we give a list of known results, but also a longer one with open questions.

• The absolute values of the neutrino masses. As we said in Chapter 20, we know the values of the two mass differences, but we have only upper bounds for the masses themselves. Upcoming experiments on the study of the end point spectrum in β-decay we presented in Chapter 20 will improve the bound but, if the masses are on the order of the differences, i.e. 10^{-3} or 10^{-5} eV, we do not expect to be able to measure them for quite some time.

• Even the ordering of the masses of the three neutrino states is not known. Current experiments are expected to settle the issue and distinguish between what we have called *normal* and *inverted hierarchy*.

• The number of neutrino states is not known. We do not know whether the three neutrinos oscillate entirely among themselves, or whether they also involve additional, often called *sterile*, neutrinos. This question may be answered in the near future thanks to a new generation of oscillation experiments using high energy neutrino beams.

• Do neutrinos and anti-neutrinos exhibit the same oscillation pattern? The question can be answered with currently planned dedicated experiments.

• Are neutrinos their own anti-particles; in other words, are they described by Majorana fields? Neutrinoless double beta decay experiments may provide the answer.

• The elements of the neutrino mixing matrix are not very well known. Assuming the 3×3 form of equation (20.9) and ignoring for the moment CP violating phases, we find with quite large error bars: $\tilde{s}_{12}^2 \simeq 0.3$, $\tilde{s}_{23}^2 \simeq 0.5$ and $\tilde{s}_{13}^2 \simeq 0.02$.[14] The important remark here is that the neutrino mixing matrix does not seem to follow a pattern similar to that of the CKM matrix. In the latter all three angles are small and they obey a hierarchy – nothing resembling the neutrino mixing matrix. The hope to relate and understand one from the other is fading away.

• No CP violating effects have ever been observed in the leptonic sector. However, the large values of the mixing angles and the possibility of having three CP violating phases, raises hopes of such effects being substantial. An intense experimental effort is dedicated to this question and we are looking forward to having new results soon. This is particularly important in astrophysics and cosmology. As already mentioned, CP violation is an essential ingredient in the effort to explain the matter–antimatter

---

[14]For several years it was thought that the third angle $\tilde{\theta}_{13}$ was zero. It was only in 2018 that an international collaboration, studying the oscillations of the anti-neutrinos coming from a complex of nuclear reactors in the Baya Bay area in SE China, established this non-zero value.

asymmetry in our universe. However, because of the smallness of the CKM mixing angles, the amount of CP violation coming from the quark sector seems to be insufficient to account for the observed asymmetry. If, however, CP violation turns out to be much stronger in leptons, we may have a scenario in which "baryogenesis", i.e. creation of baryons, is triggered through "leptogenesis", i.e. creation of leptons.

All these unanswered questions make neutrinos a very active and fascinating subject in experimental high energy physics.

22.6   Questions of Flavour

We encountered the first "question of flavour" in section 6.7 in the form of Rabi's question concerning the muon: "Who ordered that?" Today, more than seventy years later, we have to admit that we still do not know the answer. Instead, although flavour physics has grown into a rich and active field of research, the fundamental questions remain unanswered. In this section we present a brief history of the subject and indicate why the study of flavour could pave the way to physics beyond the Standard Model.

22.6.1   A brief history of flavour

In Chapter 6 we presented the table of the elementary particles known in 1932. It contains the nucleons, proton and neutron, the leptons, neutrino and electron and the photon. Based on this we established three rules: (i) Matter constituents have spin $\frac{1}{2}$, the quantum of radiation has spin 1. (ii) There exists a symmetry between hadrons and leptons. (iii) The role of all these particles in the structure of matter is clear. The world appeared to be simple with no mysteries!

The appearance of the muon shattered this picture of simplicity and understanding. It was, and still is, a particle nobody was looking for, nobody wanted and nobody knew what to do with.[15] In Chapter 6 we described the confusion which prevailed during the early years of accelerators, the years of chaotic inflation in the number and variety of "elementary" particles. All three simple rules were violated. The very distinction between matter constituents and mediators of forces was lost, not even a shadow of a lepton–hadron symmetry, and the physicists did not even dare to ask the question of the role of all these particles.

We will not repeat here the story we presented in Chapter 6, we just highlight the main events:

• 1944: The first evidence for "strange" particles in cosmic rays.

• 1953: M. Gell-Mann and K. Nishijima introduce strangeness as a new quantum number. The first concept of "flavour".

• 1960: M. Gell-Mann and S.L. Glashow note a peculiar behaviour in the weak decays of strange particles: they do not decay through strangeness changing neutral currents. More about that later.

• 1962: The experimental discovery of the muon neutrino. The concept of flavour is extended to neutrinos.

• 1964: The quarks (M. Gell-Mann and G. Zweig). The missing triplet of $SU(3)$.

• 1964: The Cabibbo angle, the precursor of the CKM matrix.

---

[15] A. Pais, in his book *Inward Bound*, calls the muon "the divine laughter".

In 1964 the world was not simple. The known elementary particles were three quarks (u, d, s) and four leptons $(e^-, \nu_e, \mu^-, \nu_\mu)$ to which we should add the mediators of forces. Only the photon was known, although the possible existence of intermediate vector bosons for the weak interactions and one vector "gluon" for the strong ones had been conjectured. Out of the three simple 1932 rules only the first one could be considered as valid.

- 1970: The GIM mechanism. The existence of the charm quark is conjectured. Lepton-hadron symmetry is restored.

- 1972: Families must be complete. Lepton–hadron symmetry becomes a fundamental law of nature. We can prove that this is a necessary condition for the mathematical consistency of the theory.

- 1975: The third family. The existence of t and b was conjectured by M. Kobayashi and T. Maskawa as a natural source of CP violation. That of the accompanying leptons τ and ν_τ was conjectured in order to complete the family.

Today, after all these discoveries, the world is still not very simple. Looking at the present Table 6.7 of elementary particles we see that matter constituents have spin $\frac{1}{2}$ and radiation quanta spin 1, but we have also the BEH scalar, which has spin zero. There is a well established and understood lepton–hadron symmetry but the role of flavour is still elusive. Why three families? This brings us back to Rabi's question: Who ordered flavour?

22.6.2 Rare flavour decays

Even if we do not know why they are there, heavy flavours have opened vast fields of research. We have already seen many of them. In this section we will concentrate on a particular one, namely the study of rare events as signs of new physics.

By "rare events" we mean processes which are either forbidden or highly suppressed in the framework of the Standard Model. Therefore they are sensitive to possible deviations, implying new physics. The simplest examples are decays which violate the conservation of quantum numbers, such as baryon or lepton number. We saw in section 20.2.4 the neutrinoless double β-decay, which violates the conservation of lepton number by two units. Another example is proton decay. Both processes are forbidden in the Standard Model, so detection of any of these will be proof of the existence of physics beyond this model.

Another interesting case is decays which are highly suppressed, although not absolutely forbidden. We will present here a few processes, which require the existence of flavour-changing neutral currents (FCNC). They are studied theoretically and actively searched for experimentally in all three flavours, s, c and b.

In recent years, experiments have reached the sensitivity to measure some FCNC processes in the decays of strange particles. The most common examples are $K_L^0 \rightarrow l^+ + l^-$ or $K \rightarrow \pi + l + \bar{l}$, where in the first example l may be an electron or a muon and in the second a charged lepton or a neutrino. The same type of decays we can have for D^0 or B^0. At the quark level they all proceed through the virtual process $q_i \rightarrow q_j + \ldots$, where dots stand for a lepton pair and possible $q\bar{q}$ states. For K^0 the transitions are $s \rightarrow d$, for D^0 $c \rightarrow u$ and for B^0 $b \rightarrow d$. At lowest order in the weak interactions these decays are forbidden, because the couplings of the neutral

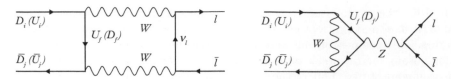

Fig. 22.16 Diagrams contributing to FCNC transitions. $U(D)$ denote quarks of up (down) type.

vector boson Z are diagonal in flavour space, see equation (19.33). At 1-loop we can draw several diagrams, depending on the process we are looking at. Examples are in Figure 22.16. Since they are second order in the weak interactions, we would expect, naïvely, the amplitude to be of order αG_F, which means a suppression factor in the branching ratio on the order of 10^{-5}–10^{-4}. As we explained in section 19.4, this is not enough; the experimental limits are several orders of magnitude smaller. In Problem 22.7 we ask the reader to give a rough estimation of these diagrams. The result can be expressed as follows: Let us look, for example, at the box diagram of Figure 22.16. We call $A_U^{ij}(D_i \to D_j + \dots)$, $U = u, c, t$, the value of this diagram when the up-type quark exchanged in the internal line is u, c or t, and $A_D^{ij}(U_i \to U_j + \dots)$, $D = d, s, b$ when it is a d, s or b. In Problem 22.7 we find

$$\sum_{U=u,c,t} A_U(s \to d + \dots) \simeq \frac{m_t^2}{m_W^2}\lambda_t + \frac{m_c^2}{m_W^2}\lambda_c + \frac{m_u^2}{m_W^2}\lambda_u + \dots \tag{22.26}$$

where λ_U is given by the product of the CKM matrix elements $V_{Ud}V_{Us}^*$, and the dots stand for subdominant terms. Similar expressions are obtained for transitions $s \to b$, $d \to b$ or $u \to c$. Several conclusions follow from this calculation: (i) As a consequence of the unitarity of the CKM matrix, if all up-type quarks had the same mass, the sum would vanish. (ii) Every single term in (22.26) is highly suppressed: the u term because of the very small mass of the up quark, the c term because $m_c/m_W \sim 0.02$ and $\lambda_c \sim \mathcal{O}(\lambda)$, and the t term because $\lambda_t \sim \mathcal{O}(\lambda^5)$, with λ the Wolfenstein expansion parameter of equation (22.13). In fact, it was this argument which was used in 1970, when the only known quarks were u, d and s, in order to predict the existence of the c quark and set an upper bound on its mass. Note also that the t quark term contains the phase of CP violation. (iii) The t quark exchange gives the dominant contribution, therefore we expect such processes to be more important for K^0 and B^0 decays, rather than D^0 ones.

This was only an order of magnitude estimation. For a realistic computation we must go from quarks to hadrons, and this cannot be done in perturbation theory. We must either use some model for hadronisation, or appeal to lattice QCD calculations. This has been done for most interesting processes.

On the experimental side we have results for the following branching ratios: $\text{BR}(K^\pm \to \pi^\pm + e^+ + e^-) \simeq (3 \pm 0.09) \times 10^{-7}$, $\text{BR}(K^\pm \to \pi^\pm + \mu^+ + \mu^-) \simeq (9.4 \pm 0.6) \times 10^{-8}$, and $\text{BR}(K^\pm \to \pi^\pm + \nu + \bar{\nu}) \simeq (1.7 \pm 1.1) \times 10^{-10}$. The last one, which is very sensitive to new physics, is estimated in the Standard Model to be $(0.78 \pm 0.08) \times 10^{-10}$, in agreement with the measurement. In an upcoming run, the CERN NA62 experiment plans to collect around 100 events of this type, greatly increasing the precision. On the other hand, for the ratios of $K_L^0 \to \pi^0 + l + \bar{l}$ we have only upper bounds substantially larger than the theoretical estimations.[16]

The $K_L^0 \to l^+ + l^-$ branching ratios have also been measured: $\mathrm{BR}(K_L^0 \to \mu^+ + \mu^-) \simeq (6.84 \pm 0.11) \times 10^{-9}$ and $\mathrm{BR}(K_L^0 \to e^+ + e^-) \simeq 9_{-4}^{+6} \times 10^{-12}$, the smallest branching ratio ever measured. These values are in agreement with the Standard Model estimations.

In charm decays we have a bound for the branching ratio $\mathrm{BR}(D^0 \to \mu^+ + \mu^-) < 6.2 \times 10^{-9}$, and in B decays both LHCb and BELLE have reached the required sensitivity and have obtained precise results. The measurements show some tension with the Standard Model predictions and will be briefly discussed in the following section.

22.7   An Overall Fit

With the discovery of the BEH boson, the Standard Model is complete. All its parameters have been measured independently. In Table 19.1 we show the values for some of them. In addition, a large number of quantities have been computed theoretically and measured experimentally. The agreement between theory and experiment is impressive. In this section we want to present this agreement, but also to point out a few cases for which some tension between theoretical predictions and experimental measurements persists.

Figure 22.17 presents, in a pictorial way, the agreement between theory and experiment for a collection of observables. They include all those for which a potential problem exists. As we see, the discrepancies are in general very small and they never reach the threshold of three standard deviations, above which they are considered significant.

In addition to the observables shown in Figure 22.17, we have singled out two for which the difference between the theoretical predictions and the experimental measurements exceeds the three σ level. The first is the muon anomalous magnetic moment and the other is related to decay properties of hadrons containing a b quark.

• *Muon g-2.* The anomalous magnetic moment of the muon can be computed along the lines we presented in Chapter 16 for the anomalous magnetic moment of the electron. We can decompose the theoretical calculation into four parts, see Table 22.2. Part I is the pure QED contribution, which differs from the corresponding calculation for the electron only by the $e - \mu$ mass difference. We can assume that it is known with the required precision as shown in the table. Part II contains the weak interaction contributions with the Z boson replacing the photon. Again, we do not expect any uncertainties in the calculation, and we can assume that the value given in the Table is correct with very small errors. Part III contains the hadronic corrections. The most important part seems to be given by the hadronic corrections to the photon propagator. In the table we give the results for the leading order (LO) calculation, as well as the next-to-leading order (NLO) and the next-to-next-to-leading order (NNLO) corrections. The LO is very large, but it has been computed by two groups independently.

---

[16]However, recently the KOTO collaboration in Japan announced the observation of four candidate events of the decay $K_L^0 \to \pi^0 + \nu + \bar{\nu}$. The expectation from Standard Model calculations was 0.1.

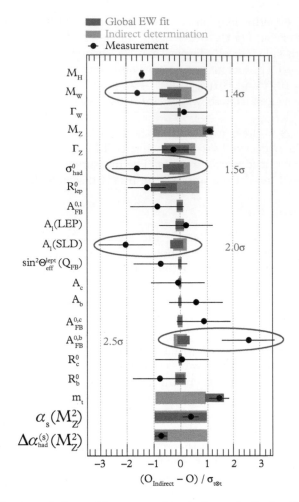

Fig. 22.17 A collection of observables, as predicted theoretically by the Standard Model and as measured experimentally. The horizontal axis shows the deviations between theory and experiment. The observables include masses and widths of various particles, coupling constants and parameters such as forward–backward and left–right asymmetries in various decays. *(Source: Particle Data Group).*

Finally, Part IV contains the hadronic contribution to the scattering of light by light. It has been estimated in model calculations, as well as in lattice simulations. The combined result is shown in the table, together with the 2004 Brookhaven measurement. As we can see, the discrepancy is of order 3.6σ. A new experiment is in progress at Fermilab with the goal to reduce the experimental error by a factor of 4. It should be pointed out, though, that it is hard to evaluate the significance of the discrepancy in view of the size of the LO hadronic corrections.

Table 22.2 Comparison between theory and experiment for the muon $g-2$

MUON $g-2$		
Contribution	Value$\times 10^{10}$	Uncertainty$\times 10^{10}$
QED	11658471.895	0.008
EW Corrections	15.4	0.1
HVP (LO)	693.5	4.3
HVP (NLO)	-9.84	0.06
HVP (NNLO)	1.24	0.01
HLbL	10.5	2.6
Total SM Th	11659183	5.0
BNL Exp	11659209.1	6.3
Fermilab target		1.6

• *Lepton universality in B-meson decays.* Three experimental collaborations, BaBar at SLAC, BELLE in Japan and LHCb at CERN, have measured the decay properties of B-mesons, i.e. mesons containing one b quark. Among the physics questions they address is that of lepton universality. A basic assumption of the Standard Model is that the three charged leptons, to wit the electron, the muon and the tau, have identical properties, and their only difference is their mass. The three experiments attempted to test this assumption in the semi-leptonic decays of B mesons and they measured ratios of branching ratios of the form

$$R(D) = \frac{BR(\overline{B} \to D + \tau + \bar{\nu}_\tau)}{BR(\overline{B} \to D + l + \bar{\nu}_l)} \ , \quad R(D^*) = \frac{BR(\overline{B} \to D^* + \tau + \bar{\nu}_\tau)}{BR(\overline{B} \to D^* + l + \bar{\nu}_l)} \tag{22.27}$$

where l stands for either μ, or e, and D (D^*) is the 0^- (1^-) charmed meson. From the experimental point of view, forming ratios such as those shown in (22.27) reduces some uncertainties related to detection efficiencies and calibrations. Theoretically these ratios are expected to be different from 1, because the tau is much heavier than the other two leptons. This has two effects: a trivial kinematic effect due to a different phase space, and a more subtle dynamic one, because the matrix elements of the hadronic weak current $\langle D | h_\lambda | \overline{B} \rangle$ should be computed at different values of the momentum transfer. This last effect has been estimated using a combination of model calculations and lattice simulations. The results are

$$R(D)^{\text{Th}} = 0.299 \pm 0.003 \ , \quad R(D^*)^{\text{Th}} = 0.252 \pm 0.003 \tag{22.28}$$

The comparison with the measured values is given in Figure 22.18. Since the experimental errors in the two measurements are correlated, it is more meaningful to

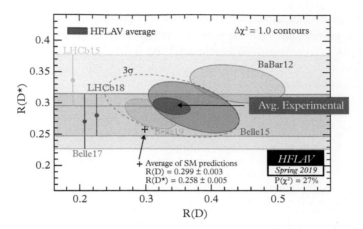

Fig. 22.18 $R(D)$ plotted against $R(D^*)$ showing the experimental measurements and the Standard Model prediction. *(Source: CERN, LHCb collaboration).*

plot the two ratios one against the other. We see that there is a discrepancy between theory and experiment of order 3σ.

The LHC experiments have recently reached the sensitivity predicted by the Standard Model in the FCNC decays of B^0:

$\text{BR}(B_d^0 \to \mu^+ + \mu^-) < 3.4 \times 10^{-10}$, with 95% confidence level,

$\text{BR}(B_s^0 \to \mu^+ + \mu^-) = (3.0 \pm 0.4) \times 10^{-9}$

Several collaborations (BELLE, LHCb, ATLAS, CMS) have measured the decay $B \to K(K^*) + l^+ + l^-$ where some discrepancies with respect to Standard Model predictions have been found. In particular, LHCb finds that the ratio of produced muons versus electrons is too small, especially if you plot the number of events as a function of the lepton pair invariant mass. If we combine all experiments, the discrepancy reaches several σ, but this may not be the right thing to do. This subject is right now a very active field of research and it is probable that the situation will evolve considerably in the near future. For the moment, no clear theoretical conclusion can be drawn from these effects in view of the uncertainties in the theoretical calculations and the difficulty to evaluate very precisely possible experimental biases.

Note added in proof: In April 2021 Fermilab announced the results from Run 1 of their muon $g-2$ measurement (*arXiv:2104.03281v1*) which are in agreement with those of the 2004 Brookhaven experiment. The uncertainties in both measurements are dominated by statistics and can be combined together. This changes only slightly the value given in Table 22.2 but reduces the errors to $a_\mu(\text{Exp})=116592061(41) \times 10^{-11}$. The difference with the theoretical estimation is now $a_\mu(\text{Exp})-a_\mu(\text{SM})=(251 \pm 59) \times 10^{-11}$ which brings the discrepancy to 4.2σ. However a new theoretical estimation of the leading order hadronic vacuum polarisation contribution (HVP-LO) was published (*arXiv:2002.12347v3*), based entirely on lattice QCD simulations, with the result $a_\mu(\text{HVP-LO})=(707 \pm 5.5) \times 10^{-10}$. If this new value is considered the discrepancy between theory and experiment becomes essentially insignificant. So, right now the situation is unclear. During the next two years Fermilab will analyse and publish the results of Runs 2, 3 and 4 of their experiment reducing considerably the experimental errors. On the theoretical side the lattice calculation should be checked by independent groups and, if the result is confirmed, the origin of the difference between the two theoretical estimations should be understood. In addition, there are proposals for dedicated experiments to settle the issue. Notice also that the HVP value is the hadronic contribution to the photon propagator which controls the running of α_{QED}. Therefore it affects many other calculations and the knowledge of its precise value is very important.

22.8 Problems

Problem 22.1 This problem aims at an order of magnitude estimation of the required energy and luminosity of a $p\text{-}\bar{p}$ collider capable of making the production and detection of the weak vector bosons possible.

   1. *Energy.* Estimate the minimum centre-of-mass energy of a collider capable of producing a vector boson of $\sim 10^2$ GeV mass. *Hint: The boson is expected to be produced in $q\text{-}\bar{q}$ collisions and roughly half of a proton momentum is carried by gluons.*

   2. *Luminosity.* Assume a cross section for inclusive Z production on the order of 1.5 nb. A promising channel for its detection is the decay $Z \to e^+ + e^-$. What is the minimum luminosity of the machine if we want to have at least one such event per day?

Problem 22.2 Estimate the branching ratio $\Gamma(H \to 2\gamma)/\Gamma(H \to b\bar{b})$.

Problem 22.3 With the discovery of the heavy quarks c and b the $SU(3)$ flavour symmetry, which was used to classify light hadrons, was extended to $SU(4)$ and $SU(5)$. Find in each case the dimensionalities of the representations which generalise the octet and the decouplet.

Problem 22.4 Go through the PDG tables and give qualitative explanations for the following facts about the lifetimes of various particles:

   1. For the charmed meson D^+ the branching ratio $\Gamma(D^+ \to \pi^+ + \pi^0)/\Gamma(D^+ \to all)$ has been measured and is equal to $(1.24 \pm 0.06) \times 10^{-3}$, even though the phase space is very large. Explain why.

   2. (i) The charmed state D^{*+} has a lifetime equal to 0.8×10^{-21} s, while that of D^+ is 1.08×10^{-12} s. (ii) The b-states $\Sigma_b^\pm$ and $\Sigma_b^{*\pm}$ have lifetimes $\sim 10^{-22}$ s to be compared with those of other b particles which are $\sim 10^{-12}$ s. Explain why.

   3. The B mesons have a typical lifetime of order 10^{-12} s, comparable to those of the charmed particles, although they are much heavier and the phase space is larger. Explain why.

   4. The lifetime of the charmed D^{*0} meson has not been accurately measured. Explain why.

Problem 22.5 Estimate the lifetime of a t quark decaying through $t \to W + b$.

Problem 22.6 Prove that all six unitarity triangles we can construct from the CKM matrix have the same area. Prove that it gives a measure of CP violation.

Problem 22.7 Estimate the amplitudes for flavour changing neutral currents.

1. Prove that at the limit of exact flavour symmetry all flavour changing transitions vanish.

2. Give an estimate of the diagrams of Figure 22.16 in terms of the quark masses and mixing angles.

23
Beyond the Standard Model

The construction of the Standard Model has been a great triumph of abstract theoretical thought. We have been extremely fortunate to build a theory in remarkable agreement with experiment. Given this enormous success, what does "beyond" mean? In other words, what is wrong with the Standard Model and why do we wish to go beyond it? In this chapter we will address these questions. However, in order to avoid any misunderstanding, let us first state clearly that so far there is no concrete theoretical scheme which could replace, or complete, the Standard Model and which has received any experimental support. Therefore this chapter will follow a different approach from the one which prevailed up to now in this book. So far, all concepts and ideas we have presented, even the most abstract ones, were solidly anchored in experimental results. The mathematical methods we used were chosen in order to serve a well defined phenomenological purpose. In this chapter we shall break away from this approach and present instead a panorama of speculations. The choice is ours and reflects our own prejudices. With no experiments to guide us, the speculations will follow a variety of routes, not always mutually compatible. We shall ask many questions, but we shall give very few answers.

23.1 Many Questions

In this section we present a list of questions which we expect a fundamental theory to be able to answer and which the Standard Model does not. Some of them have already been mentioned in this book.

- *Unification.* The Standard Model describes three fundamental interactions among elementary particles in the unified framework of gauge quantum field theories. Nevertheless, it is *not* a unified theory because it contains three independent gauge coupling constants.

- *Quantisation of the electric charge.* We have shown in Chapter 13 that only gauge theories based on non-Abelian groups exhibit charge quantisation. In the Standard Model the presence of the $U(1)$ factor which intersects the electromagnetic $U(1)_{\mathrm{em}}$, implies, for example, that the equality of the electric charges of the electron and the muon must be imposed by hand.

- *The three families.* This is part of the general problem of flavour which we presented in section 22.6.

- *The large number of arbitrary constants.* In addition to the three gauge coupling constants, the Standard Model contains a large number of masses, mixing angles and couplings which, at the level of our understanding, are arbitrary parameters, which

Elementary Particle Physics. John Iliopoulos and Theodore N. Tomaras, Oxford University Press.
© John Iliopoulos and Theodore N. Tomaras (2021). DOI: 10.1093/oso/9780192844200.003.0023

are measured experimentally. Their numerical values, especially if we include the neutrino masses, are spread over many orders of magnitude. We would expect a really fundamental theory to be able to predict the values of all its dimensionless constants, or at least to reduce this number considerably.

• *Questions related to astrophysics and cosmology.* Examples are the absence of a dark matter candidate, or an explanation for the observed matter–antimatter asymmetry in the universe.

• *The gravitational interactions are not included.*

• *The particular value of the* BEH *boson mass.* Let us consider a quantum field theory described by a Lagrangian density $\mathcal{L}$ written in terms of a set of fields $\Psi_i(x)$, $i = 1, \ldots, N$. Even if $\mathcal{L}$ is in agreement with all known data, we want to argue[1] that $\mathcal{L}$ should be viewed as an effective theory valid in a certain energy range. It contains the degrees of freedom which are relevant in that range. No theory can be considered as valid at all energies, because we cannot claim that we know the physics, and the relevant degrees of freedom, at arbitrarily high energies. For the Standard Model today this energy range is up to a scale M on the order of a few TeV. Following Wilson, let us integrate over all degrees of freedom which are heavier than M. The result will be an effective theory $\mathcal{L}_{\text{eff}}$ written in terms of the degrees of freedom ψ_j, $j = 1, \ldots, N'$ which are lighter than M. Let us write the most general form of $\mathcal{L}_{\text{eff}}$. It will be an infinite sum of monomials made out of the fields ψ_j and their derivatives restricted only by symmetry requirements. Since integrating over the heavy degrees of freedom does not break any symmetry, every term in $\mathcal{L}_{\text{eff}}$ will be invariant under whichever symmetries we had assumed in $\mathcal{L}$.

$$\mathcal{L}_{\text{eff}} = \sum_i C_i(g, M)\, O_i \tag{23.1}$$

where O_i are monomials in the fields ψ and $C_i(g, M)$ are numerical coefficients which depend on the parameters of the original theory, denoted collectively by g, and the scale M. Even if we cannot compute the C_is, we can find their dependence on M from dimensional analysis. Let d_i be the dimension of the operator O_i. Then, assuming M to be much larger than any other mass parameter in the theory, we have that $C_i \sim M^{4-d_i}$. We thus obtain three classes of operators O_i:

(i) An infinite class of operators with $d_i > 4$. The corresponding coefficients behave like negative powers of M and can be neglected for large M. We call them *irrelevant operators.*

(ii) Operators having $d_i = 4$. The coefficient functions are independent[2] of M. We call them *marginal operators.*

(iii) Operators with dimension smaller than 4. The coefficient functions behave like positive powers of M and we call them *dominant operators.*

Let us apply this reasoning starting from the Lagrangian of the Standard Model. We must choose M to be larger than a few TeV, the scale already explored by LHC, but otherwise the choice is arbitrary. In the spirit of Wilson, M represents the scale

---

[1] The argument is due to Ken Wilson.

[2] This argument neglects any logarithmic dependence of the coefficient functions on M.

up to which the Standard Model can be trusted. If M is large enough the irrelevant operators die away and can be neglected. The marginal operators are precisely the dimension four terms in the Lagrangian of the Standard Model. There are only two dominant operators: the unit operator **1** with dimension zero and the mass operator of the BEH scalar ϕ^2 with dimension two. The coefficient of the first behaves like M^4 and that of the second like M^2. The first gives a constant term in the Lagrangian which, in the absence of gravitation, can be dropped.[3] The second gives a correction to m_H^2 which grows like M^2. The coefficient C_{ϕ^2} contains coupling constants and, generically, it is expected to be of order $\alpha \sim 10^{-2}$.[4] This would mean that M cannot exceed the TeV scale by much. This is the puzzle: what keeps the mass of the scalar boson at 125 GeV? Why has LHC seen no signs of new physics? This problem is known in the literature as the problem of *naturalness* and it is common to all theories which contain elementary scalar fields. Unless specifically tuned, a quantum field theory is unable to sustain naturally a light scalar. By "specifically tuned" we mean theories for which the parameters satisfy certain relations ensuring that C_{ϕ^2} turns out to be very small, or even zero. We shall see examples later on.

Although this list is not exhaustive, it shows that we have good reasons to look for a more fundamental theory beyond the Standard Model. Here "beyond" does not mean "to abolish" but "to fulfil". As we have said, we have no candidate theory which could answer convincingly all these questions and the search is still open.

23.2 Beyond, but How?

23.2.1 Grand unified theories (GUTs)

The gauge group of the Standard Model is $SU(3) \times SU(2) \times U(1)$, but at low energies an observer will see only $SU(3) \times U(1)_{\text{em}}$, the part which remains unbroken. The idea of grand unification is to assume that the same story is repeated at much higher energies. More precisely the assumption is that $SU(3) \times SU(2) \times U(1)$ is the remnant of a larger, simple, or semi-simple, group G, which is spontaneously broken at very high energies. The scheme, first proposed by H. Georgi and S.L. Glashow, looks like

$$G \xrightarrow{M} SU(3) \times SU(2) \times U(1) \xrightarrow{m_W} SU(3) \times U(1)_{\text{em}} \qquad (23.2)$$

where the breaking of G may be a multistage one and M is one (or several) characteristic mass scale(s).

We want G to be simple, or semi-simple, because we want to satisfy naturally the electric charge quantisation. In order to simplify the search for the group G let us make the assumption that the 16 chiral fermion fields which form a family of the Standard Model fill a representation (not necessarily irreducible) of G; in other words, we assume that there are no other, as yet unobserved, particles which belong to the same representation. Since G contains the group of the Standard Model, the electric charge operator Q must be one of the generators of the Lie algebra of G and, since G

---

[3]If we include gravitational interactions this term is known as *the cosmological term*.

[4]In the Standard Model C_{ϕ^2} turns out to be larger because of the large value of the coupling constant between the BEH scalar and the t quark.

is semi-simple, all its generators are represented by traceless matrices. It follows that, in any representation of G, we must have

$$\mathrm{Tr}\, Q = 0 \tag{23.3}$$

in other words, the sum of the electric charges of all particles in a given representation vanishes. This has a very important consequence: As we have remarked, the members of a family satisfy (23.3) because the sum of their charges vanishes. This, however, is not true if we consider leptons or quarks separately. Therefore each irreducible representation of G will contain both leptons and quarks. This means that there exist gauge bosons of G, which can convert a lepton into a quark and vice versa. We conclude that a grand unified theory, which satisfies our assumption, cannot conserve baryon and lepton numbers separately. This sounds like both bad news and good news. Bad news because we must make sure that in such a theory protons do not decay too fast and good news because we have a chance to explain the matter–anti-matter asymmetry of the universe.

The amplitude for proton decay is given by the exchange of the corresponding gauge boson whose mass is of order M, the scale of the breaking of G. The resulting proton lifetime will be of order

$$\tau_p \sim \frac{M^4}{m_p{}^5} \tag{23.4}$$

Using the experimental bound (for particular decay modes) of a few times 10^{33} years, we can put a lower bound on M, namely

$$M \geq 10^{16} \mathrm{GeV} \tag{23.5}$$

Grand unification is not a low energy phenomenon!

If G is simple, we have a second, model independent prediction noticed by H. Georgi, H. Quinn and S. Weinberg. At scales well above M the gauge theory must have one gauge coupling constant g, as we explained in section 13.2. It must be the same with the coupling constants of the Standard Model, extrapolated through the renormalisation group equations to that scale.[5] So, the prediction is that at $\mu > 10^{16}$ GeV the three coupling constants come together. The results of the calculation are shown in Figure 23.1.

Models of GUTs have been studied for several groups. The simplest ones are: (i) $SU(5)$, in which the 16 Weyl spinors of a Standard Model family fill a $\bar{\mathbf{5}} \oplus \mathbf{10} \oplus \mathbf{1}$ representation and (ii) $SO(10)$, which has a 16-dimensional irreducible representation appropriate to accommodate all 16 members of the family. More exotic choices include the exceptional groups E_6 or E_8.

23.2.2   The trial of scalars

The Standard Model contains three independent worlds. The radiation world of the gauge bosons, the matter world of the fermions and, finally, in our present understanding, the world of BEH scalars. In the framework of gauge theories these worlds are

---

[5]For the $U(1)$ coupling constant, which we called g' in Chapter 19, we must normalise the corresponding generator in order to embed it into the algebra of G, see Problem 19.6.

largely unrelated to each other. Given a group G, the world of radiation is completely determined, but we have no way to know a priori which and how many fermion representations should be introduced; the world of matter is, to a great extent, arbitrary.

This arbitrariness is even more disturbing if one considers the world of BEH scalars. Not only their number and their representations are undetermined, but their mere presence introduces a large number of arbitrary parameters into the theory. What makes things worse is that these arbitrary parameters appear with a wide range of values.

One possible remedy is to throw away the scalars as fundamental elementary particles. After all, their sole purpose is to induce the spontaneous symmetry breaking through their non-vanishing vacuum expectation values. This phenomenon is known to occur with or without the presence of gauge interactions, with the role of the scalar fields played by fermion bilinears. Well known examples are the spontaneous breaking of chiral symmetry in QCD, as well as many phenomena in non-relativistic condensed matter physics.

This idea of dynamical symmetry breaking has been studied extensively, especially under the name of "technicolor" and variations thereof. It is a model which mimics QCD at higher energies. It has many attractive features but suffers from several important difficulties. First, the available field theory technology does not allow for any precise quantitative computation of bound state effects and everything has to be based on analogy with the chiral symmetry breaking in QCD. Second, nobody has succeeded in producing an entirely satisfactory phenomenological model. In particular, the simplest models have large flavour changing neutral current effects. Third, and most important, we expect to have an entire spectroscopy of bound states, the analogues of ordinary hadrons, and the LHC does not see any of them.

23.2.3 Supersymmetry, or the defence of scalars

Supersymmetry is a rather unfortunate name for a symmetry which mixes fermions with bosons. In a four-dimensional space-time it was proposed by J. Wess and B. Zumino. Rather than eliminating scalars, supersymmetry aims at a fundamental unification of all three worlds of the Standard Model, to wit the gauge bosons, the fermions and the BEH scalars. In the usual notation, an infinitesimal supersymmetry transformation has the form

$$\delta\phi^i(x) = \epsilon^a (T_a)^i_j \phi^j(x) \tag{23.6}$$

where $a = 1, \ldots, n$ with n denoting the number of generators of the Lie algebra of the group we are considering. T_a is the matrix of the representation of the fields and ϵ^a are n infinitesimal parameters. Since we want to mix fermions and bosons, the ϵs must be anti-commuting spinors. We shall call such transformations "supersymmetry transformations" and we see that a given irreducible representation will contain both fermions and bosons. It is not a priori obvious that such supersymmetries can be implemented consistently in particle physics, but in fact they can. The corresponding algebraic scheme closes using both commutators and anti-commutators. We proceed as follows: we start with the known symmetries, namely the Poincaré group, with generators $\mathcal{J}^{\mu\nu}$ and P^μ which satisfy the algebra (5.181), times some internal symmetry group G with generators T_a satisfying the algebra (5.26). Then we consider a set of operators Q_b^r, $b = 1, \ldots N$, which transform according to a finite-dimensional representation of the sum $\mathfrak{g} \oplus \mathfrak{P}$. In all our applications we shall choose the Qs to be translationally invariant Majorana (or Weyl) spinors belonging to an N-dimensional representation of G, so, in addition to the commutation relations (5.181) and (5.26), we shall have in an obvious notation

$$[P_\mu, Q_b^r] = 0 \;, \quad [Q_b^r, \mathcal{J}_{\mu\nu}] = \mathrm{i}\gamma_{\mu\nu}^{rs} Q_b^s \;, \quad [T_a, Q_b^r] = s_{abc} Q_c^r \qquad (23.7)$$

Let us call collectively A_i the generators of $\mathfrak{g} \oplus \mathfrak{P}$ and Q^m the operators Q. We shall say that the set of operators A and Q form a *graded Poincaré superalgebra* if we can find constants h_i^{mn} such that

$$\{Q^m, Q^n\} = Q^m Q^n + Q^n Q^m = h_i^{mn} A^i \qquad (23.8)$$

The constraint on the constants h is that they must satisfy the corresponding Jacobi identities for the set of equations (23.7) and (23.8) to be consistent. There exist theorems which give a classification of graded superalgebras analogous to the Cartan classification of Lie algebras, but we shall not need them here. The only superalgebras we shall use are defined through

$$\{Q_a^r, Q_b^s\} = 0 \;, \quad \{Q_a^r, \overline{Q}_b^s\} = -2\delta_{ab}\gamma_\mu^{rs} P^\mu \qquad (23.9)$$

where $\overline{Q} = Q^\dagger \gamma^0$. This algebra admits $SL(2, C) \times G$ as a group of automorphisms. Supersymmetry has been extensively studied during the last fifty years and there exists a vast literature on the subject. Here we only mention a few results.

• *Representations in terms of one particle states.* We can construct them following the same method we used in section 5.6.2 for the representations of the Poincaré algebra. We start by observing that the spinorial charges commute with P_μ and therefore they do not change the momentum of the one-particle state. Furthermore, the operator P^2 commutes with all the operators of the algebra, which implies that all the members of a supermultiplet will have the same mass. Furthermore, every supermultiplet will contain the same number of fermionic and bosonic states. Obviously, this number depends on N, the number of spinorial generators. We list here some low-spin multiplets for $N = 1$, which may turn out to be relevant for particle physics.[6]
–Chiral multiplet, any mass: $(0^+, 0^-, \frac{1}{2})$
–Massive vector multiplet: $(0^+, 1^-, 2 \times \frac{1}{2})$ or $(0^-, 1^+, 2 \times \frac{1}{2})$
–Massless vector multiplet: $(1^-, \frac{1}{2})$
–Massless tensor multiplet: $(2^+, \frac{3}{2})$

• *Supersymmetric quantum field theory.* We can write quantum field theory models whose equations of motion are invariant under supersymmetry and study their properties. The most important result, which made supersymmetry a central element in most attempts to go beyond the Standard Model, is that these theories have remarkable convergence properties. We can understand them intuitively by the following argument. The divergences in a perturbation expansion are due to integrations over loop

---

[6]The spinors are Majorana or Weyl.

momenta. In supersymmetry for every fermion loop we have a corresponding boson loop, therefore we expect certain divergences to cancel. A list of such cancelations includes:

−*Vacuum energy.* In a supersymmetric theory the vacuum energy vanishes identically. No normal ordering of the Hamiltonian is necessary. In the expansion (23.1) the coefficient C_1 of the unit operator is zero.

−*Scalar mass.* The same is true for the coefficient C_{ϕ^2} if the scalar field is a member of a chiral multiplet. Since massive vector multiplets (or higher) lead to non-renormalisable interactions, it follows that a renormalisable supersymmetric theory has no naturalness problem.

−*Scale invariance of supersymmetric gauge theories with $N > 1$.* In a supersymmetric gauge theory with $N = 4$ spinorial generators the renormalisation group β-function vanishes identically. It follows that Green functions do not depend on any mass parameter and exhibit scale invariance. For a theory with $N = 2$ the β-function is non-zero only at 1-loop order, therefore the violations of scale invariance are exactly known.

• *Supersymmetry and particle physics.* The spectrum of elementary particles shows no trace of supersymmetry. Leptons and quarks do not have any bosonic partners with the same mass. If supersymmetry is part of the symmetries of the final theory of particle physics, it must be badly broken. Consequently, the cancellations mentioned above will be partial, with characteristic scale determined by that of supersymmetry breaking. Furthermore, we have 96 fermionic degrees of freedom and only 28 bosonic ones, and gauge bosons belong to the adjoint representation of the group, while fermions belong to the fundamental one. Two conclusions follow from these considerations regarding a possible supersymmetric extension of the Standard Model. First, it will necessarily introduce new particles and, second, it will associate known bosons with unknown fermions and known fermions with unknown bosons. In Table 23.1 we present the particle content in the supersymmetric standard model proposed by P. Fayet. Although the mass values of these particles are model dependent, their very existence is a crucial test of supersymmetry.

We end with a figure which shows the possible influence of supersymmetry in grand unified theories. We mentioned that a prediction of these theories is that the three gauge coupling constants of the Standard Model should come together at a scale on the order of 10^{16} GeV. Supersymmetry affects this extrapolation because it changes the values of the β-functions as a result of the introduction of new particles. Figure 23.1 shows this evolution without and with supersymmetry. Only in the second case do the three curves appear to come together, an observation which could be thought of as an argument in favour of supersymmetry.

• *Local supersymmetry or supergravity.* Since the supersymmetry algebra (23.9) includes the generators of translations, it follows that a local version of supersymmetry will include general relativity. In fact, there exists a remarkable result: if we write the interaction of the metric tensor $g_{\mu\nu}$ with a massless spin-$\frac{3}{2}$ Majorana fermion, which is invariant under general relativity, it is automatically supersymmetric.[7] Unfortunately, this does not seem to solve the renormalisation problems of gravity.

---

[7]This result was first obtained by S. Ferrara, D.Z. Freedman and P. van Nieuwenhuizen.

Table 23.1 The particle content of the supersymmetric Standard Model. By convention, the fermionic partners of known bosons are denoted by the ending "-ino". The assignment is conventional. In any particular model the physical particles may be linear combinations of those appearing in the table

SPIN-1	SPIN-$\frac{1}{2}$	SPIN-0		
Gluons	Gluinos	no partner		
Photon	Photino	no partner		
$W^\pm$	2 Dirac Winos	$w^\pm$	B	
Z	2 Majorana Zinos	z	E	b
		standard ϕ^0	H	o
	1 Majorana Higgsino			s
		pseudoscalar $\phi^{0\prime}$		o
				n
				s
	Leptons	Spin-0 leptons		
	Quarks	Spin-0 quarks		

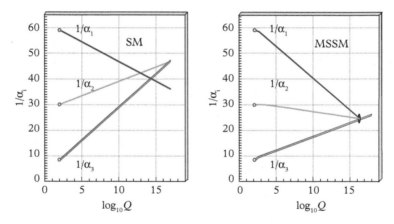

Fig. 23.1 The renormalisation group evolution of the inverse coupling constants $\alpha_i^{-1}, i = 1, 2, 3$ of the three Standard Model groups $U(1)$, $SU(2)$ and $SU(3)$ without supersymmetry (left), and with supersymmetry (right). The uncertainties are also shown. The calculations include two loop effects. *(Source: Particle Data Group).*

We can go further and increase the number of supersymmetry generators. It is easy to see that if we do not want to introduce fields with spin higher than 2, the maximum number is $N = 8$. The multiplet is formed by 1 spin-2 graviton, 8 spin-$\frac{3}{2}$ Majorana fermions, 28 spin-1 vector bosons, 56 spin-$\frac{1}{2}$ Majorana fermions and 70 spin-0 bosons. As was shown by E. Cremmer and B. Julia, the action which involves all these fields is invariant under eight local supersymmetry transformations. The hope was that the large number of supersymmetries would result in a cancellation of the divergences of perturbation theory so as to make the theory finite, but nobody has yet found a convincing proof of such a property.

In some sense $N = 8$ supergravity can be viewed as the end of a road, the road of local quantum field theory. Usually, the response of physicists whenever faced with a new problem was to seek the solution in an increase of symmetry. This quest for larger and larger symmetry led us to the Standard Model, to grand unified theories and then to supersymmetry, to supergravity and, finally, to the largest possible supergravity, that with $N = 8$. In the traditional framework we are working, that of local quantum field theory, there exists no known larger symmetry scheme. The next step had to be a very radical one. The very concept of point particle, which had successfully passed all previous tests, was abandoned. During the last decades the theoretical investigations have moved towards the theory of interacting extended objects, such as open and closed strings or membranes. Especially the superstring theory has attracted considerable attention and its relevance as a theory of high energy physics is still under investigation.

23.3 Closing Remarks

As announced, this chapter does not present a generally accepted doctrine. It contains exploratory ideas and research which is still in progress. The reader should be particularly critical. It is true that no model among those we have studied so far imposes itself on the grounds of experimental support or aesthetic beauty, which means that, most probably, none is the right one. So far, they offer at best only partial answers to the questions we formulated in section 23.1. However, they all contain interesting and intriguing ideas and we believe that, embedded in a future scheme, they may be part of the actual theory beyond the Standard Model. The really puzzling fact is that LHC sees no sign of new physics, despite the fact that all the arguments we have presented indicate that such signs should already have been visible.

Appendix A
A Collection of Useful Formulae

The numbers in parenthesis refer to the sections in the book in which these concepts are presented.

A.1 Units and Notations

- The standard unit system of high energy physics is $c = \hbar = 1$. A useful conversion formula is $(200 \ \text{MeV})^{-1} \simeq 10^{-13}$ cm $\simeq 3.3 \times 10^{-24}$ s.
- The Minkowski metric is denoted by $\eta_{\mu\nu}$ and it is given by $\eta_{00} = 1$, $\eta_{ii} = -1$, $\eta_{\mu\nu} = 0$ for $\mu \neq \nu$.

The scalar product of two four vectors p and q is denoted by $p \cdot q = p^0 q^0 - \boldsymbol{p} \cdot \boldsymbol{q}$

The mass shell condition for a particle of mass $m \geq 0$ is $p^2 = m^2$, $E_p = \sqrt{\boldsymbol{p}^2 + m^2}$.

The invariant measure on the positive energy branch of the mass hyperboloid $p^2 = m^2$ is given by equation (2.19)

$$d\Omega_m = \frac{d^4 p}{(2\pi)^4}(2\pi)\delta(p^2 - m^2)\theta(p^0) = \frac{d^3 p}{2(2\pi)^3 E_p} \tag{A.1}$$

- The Fourier transform of a function $f(x)$ in d dimensions is defined by

$$\tilde{f}(p) = \frac{1}{(2\pi)^d} \int d^d x \ e^{-ip\cdot x} f(x) \tag{A.2}$$

- The $SU(2)$ Pauli matrices are denoted by either $\boldsymbol{\sigma}$ or $\boldsymbol{\tau}$ and are given by

$$\sigma_1 = \begin{pmatrix} 0 & 1 \\ 1 & 0 \end{pmatrix}, \quad \sigma_2 = \begin{pmatrix} 0 & -i \\ i & 0 \end{pmatrix}, \quad \sigma_3 = \begin{pmatrix} 1 & 0 \\ 0 & -1 \end{pmatrix} \tag{A.3}$$

- The $SU(3)$ Gell-Mann matrices are denoted by λ_a, $a = 1, ..., 8$ and are given by

$$\lambda_1 = \begin{pmatrix} 0 & 1 & 0 \\ 1 & 0 & 0 \\ 0 & 0 & 0 \end{pmatrix}, \quad \lambda_2 = \begin{pmatrix} 0 & -i & 0 \\ i & 0 & 0 \\ 0 & 0 & 0 \end{pmatrix}, \quad \lambda_3 = \begin{pmatrix} 1 & 0 & 0 \\ 0 & -1 & 0 \\ 0 & 0 & 0 \end{pmatrix}$$

$$\lambda_4 = \begin{pmatrix} 0 & 0 & 1 \\ 0 & 0 & 0 \\ 1 & 0 & 0 \end{pmatrix}, \quad \lambda_5 = \begin{pmatrix} 0 & 0 & -i \\ 0 & 0 & 0 \\ i & 0 & 0 \end{pmatrix}, \tag{A.4}$$

$$\lambda_6 = \begin{pmatrix} 0 & 0 & 0 \\ 0 & 0 & 1 \\ 0 & 1 & 0 \end{pmatrix}, \quad \lambda_7 = \begin{pmatrix} 0 & 0 & 0 \\ 0 & 0 & -i \\ 0 & i & 0 \end{pmatrix}, \quad \lambda_8 = \frac{1}{\sqrt{3}} \begin{pmatrix} 1 & 0 & 0 \\ 0 & 1 & 0 \\ 0 & 0 & -2 \end{pmatrix}$$

Elementary Particle Physics. John Iliopoulos and Theodore N. Tomaras, Oxford University Press.
© John Iliopoulos and Theodore N. Tomaras (2021). DOI: 10.1093/oso/9780192844200.004.0001

- The γ matrices satisfy the following, representation-independent, relations, see section 7.3.4:

$$\{\gamma^\mu, \gamma^\nu\} = \gamma^\mu\gamma^\nu + \gamma^\nu\gamma^\mu = 2\,\eta^{\mu\nu}\,\mathbf{1}\;,\;\;(\gamma^0)^\dagger = \gamma^0,\;\gamma^{i\dagger} = -\gamma^i,\;\gamma^{\mu\dagger} = \gamma^0\,\gamma^\mu\,\gamma^0$$

$$\gamma^5 = \gamma_5 = \mathrm{i}\gamma^0\gamma^1\gamma^2\gamma^3 = \frac{-\mathrm{i}}{4!}\,\varepsilon_{\mu\nu\varrho\sigma}\,\gamma^\mu\gamma^\nu\gamma^\varrho\gamma^\sigma\;,\;\;(\gamma_5)^2 = 1\;,\;\;\{\gamma_5,\gamma^\mu\} = 0$$

If we call $T^{\mu_1\cdots\mu_n} \equiv \mathrm{Tr}(\gamma^{\mu_1}\gamma^{\mu_2}\ldots\gamma^{\mu_n})$ and $T_5^{\mu_1\cdots\mu_n} \equiv \mathrm{Tr}(\gamma_5\gamma^{\mu_1}\gamma^{\mu_2}\ldots\gamma^{\mu_n})$, we have

$$T^{\mu_1\cdots\mu_{2n+1}} = T_5^{\mu_1\cdots\mu_{2n+1}} = 0 \quad\text{and}\quad T_5^{\mu_1\cdots\mu_n} = 0 \quad\text{for}\;\; n < 4$$

$$T^{\mu_1\mu_2} = 4\,\eta^{\mu_1\mu_2}\;,\quad T^{\mu_1\cdots\mu_4} = 4\left(\eta^{\mu_1\mu_2}\eta^{\mu_3\mu_4} - \eta^{\mu_1\mu_3}\eta^{\mu_2\mu_4} + \eta^{\mu_1\mu_4}\eta^{\mu_2\mu_3}\right)$$

$$T_5^{\mu_1\cdots\mu_4} = -4\mathrm{i}\epsilon^{\mu_1\cdots\mu_4}\;,\quad T^{\mu_1\cdots\mu_{2n}} = \sum_{k=1}^{2n-1}(-)^k\eta^{\mu_1\mu_k}T^{\mu_2\cdots\mu_{k-1}\mu_{k+1}\cdots\mu_{2n}}$$

- The γ matrices in the standard, or Dirac representation are given by

$$\gamma^i{}_{\mathrm{D}} = \begin{pmatrix} 0 & \sigma^i \\ -\sigma^i & 0 \end{pmatrix},\quad \gamma^0{}_{\mathrm{D}} = \begin{pmatrix} 1 & 0 \\ 0 & -1 \end{pmatrix},\quad \gamma^5{}_{\mathrm{D}} = \begin{pmatrix} 0 & 1 \\ 1 & 0 \end{pmatrix}$$

and in the spinorial or Weyl representation are given by

$$\gamma^i{}_{\mathrm{W}} = \begin{pmatrix} 0 & -\sigma^i \\ \sigma^i & 0 \end{pmatrix},\quad \gamma^0{}_{\mathrm{W}} = \begin{pmatrix} 0 & 1 \\ 1 & 0 \end{pmatrix},\quad \gamma^5{}_{\mathrm{W}} = \begin{pmatrix} 1 & 0 \\ 0 & -1 \end{pmatrix}$$

A.2 Free Fields

Real scalar field
- Lagrangian density: (section 7.2) $\mathcal{L} = \frac{1}{2}\left((\partial\phi)^2 - m^2\phi^2\right)$
- Equation of motion: (section 7.2) $(\Box + m^2)\phi = 0$
- Feynman propagator: (section 7.2) $\mathrm{i}/(p^2 - m^2 + \mathrm{i}\epsilon)$
- Expansion in plane waves: (section 7.2) $\phi(x) = \int \mathrm{d}\Omega_m\left[a(\boldsymbol{p})\mathrm{e}^{-\mathrm{i}p\cdot x} + a^*(\boldsymbol{p})\mathrm{e}^{\mathrm{i}p\cdot x}\right]$
- Canonical commutation relations: (section 10.2) $\left[a(\boldsymbol{p}), a^\dagger(\boldsymbol{p'})\right] = (2\pi)^3 2E_p\delta^3(\boldsymbol{p}-\boldsymbol{p'})$

Complex scalar field
- Lagrangian density: (section 7.2) $\mathcal{L} = \partial_\mu\phi^*\partial^\mu\phi - m^2\phi^*\phi$
- Expansion in plane waves: (section 7.2) $\phi(x) = \int \mathrm{d}\Omega_m\left[a(\boldsymbol{p})\mathrm{e}^{-\mathrm{i}p\cdot x} + b^*(\boldsymbol{p})\mathrm{e}^{\mathrm{i}p\cdot x}\right]$
- Canonical commutation relations: (section 10.2)

$$\left[a(p), a^\dagger(p')\right] = \left[b(p), b^\dagger(p')\right] = (2\pi)^3 2E_p\delta^3(\boldsymbol{p}-\boldsymbol{p'})$$

Dirac field
- Lagrangian density: (section 7.3) $\mathcal{L} = \bar{\psi}(\mathrm{i}\partial\!\!\!/ - m)\psi$
- Equation of motion: (section 7.3) $(\mathrm{i}\partial\!\!\!/ - m)\psi = 0$

- Feynman propagator: (section 7.3) $i/(\not{p} - m + i\epsilon)$
- Expansion in plane waves: (section 7.3) $\psi(x) = \int d\Omega_m \sum_{\alpha=1}^{2}$
$\left[a_\alpha(\boldsymbol{p})u^{(\alpha)}(\boldsymbol{p})e^{-ip\cdot x} + b_\alpha^*(\boldsymbol{p})v^{(\alpha)}(\boldsymbol{p})e^{ip\cdot x} \right]$
- Canonical anti-commutation relations: (section 10.3)

$$\{a_s(\boldsymbol{p}), a_{s'}^\dagger(\boldsymbol{p}')\} = (2\pi)^3 2E_p \delta^3(\boldsymbol{p} - \boldsymbol{p}')\,\delta_{ss'} \;,\quad \{b_s(\boldsymbol{p}), b_{s'}^\dagger(\boldsymbol{p}')\} = (2\pi)^3 2E_p \delta^3(\boldsymbol{p} - \boldsymbol{p}')\,\delta_{ss'}$$

Massless vector field in a covariant gauge
- Lagrangian density: (section 7.4)

$$\mathcal{L} = -\frac{1}{4}F_{\mu\nu}F^{\mu\nu} + \frac{1}{2\lambda}(\partial_\mu A^\mu)^2\,, \qquad F_{\mu\nu} = \partial_\mu A_\nu - \partial_\nu A_\mu$$

- Equation of motion: (section 7.4) $\partial^\mu F_{\mu\nu} + \frac{1}{\lambda}\partial_\nu \partial_\mu A^\mu = 0$
- Feynman propagator: (section 7.4)

$$\frac{-i}{p^2 + i\epsilon}\left[\eta^{\mu\nu} - \frac{p^\mu p^\nu}{p^2(1-\lambda)}\right]$$

- Expansion in plane waves: (section 7.4)

$$A_\mu(x) = \int d\Omega_0 \sum_{\lambda=0}^{3}\left[a^{(\lambda)}(\boldsymbol{p})\epsilon_\mu^{(\lambda)}(\boldsymbol{p})e^{-ip\cdot x} + a^{(\lambda)*}(\boldsymbol{p})\epsilon_\mu^{(\lambda)*}(\boldsymbol{p})e^{ip\cdot x}\right]$$

- Canonical commutation relations: (section 10.4)

$$\left[a^{(\lambda)}(\boldsymbol{p}), a^{(\lambda')\dagger}(\boldsymbol{p}')\right] = -(2\pi)^3 2E_p \delta^3(\boldsymbol{p} - \boldsymbol{p}')\eta^{\lambda\lambda'}$$

Massive vector field
- Lagrangian density: (section 7.4) $\mathcal{L} = -\frac{1}{4}F^2 + \frac{1}{2}m^2 A^2$
- Feynman propagator: (section 7.4)

$$\frac{-i}{p^2 - m^2 + i\epsilon}\left[\eta^{\mu\nu} - \frac{p^\mu p^\nu}{m^2}\right]$$

A.3 Feynman Rules for Scattering Amplitudes

A scattering amplitude relating a set of initial to a set of final particles is given, order by order in perturbation theory, by the sum of all connected and amputated Feynman diagrams to that order in the perturbation expansion, with external lines corresponding to the initial and final particles of the amplitude. The rules for calculating these diagrams are the following (section 12.3).

- **External lines**
  - All external momenta are put on the mass shell of the corresponding particle: $p^2 = m^2$.

A scalar line   — — —◯   1

Fermion lines

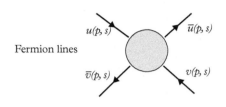

The factors applied to every external scalar, spinor or vector line in a Feynman diagram for a scattering amplitude. The momenta and spins are those of the external particle. Possible internal symmetry indices heve been suppressed. For the fermion lines the two on the left are supposed to be incoming and the two on the right outgoing.

A spin – 1 line   ∿∿∿◯$^{\mu}$ $\epsilon_{\mu}(p)$

Fig. A.1

   –All polarisation vectors correspond to physical particles. In particular, photon polarisations are transverse.

   –A factor, corresponding to the wave function of the external particle, multiplies each external line. They are given in Figure A.1.

- **Internal lines**
   –To every internal line there corresponds a Feynman propagator, according to the spin of the line. In general, the propagators are matrices carrying Lorentz and internal symmetry indices. The expressions are given in section A.2.

   –To every internal line of momentum p there is an integration $\int \mathrm{d}^4p/(2\pi)^4$

- **Vertices** (sections 12.3 and 13.2)
   We give a list of commonly used interactions.
– ϕ^4 (Figure A.2 (a)): The interaction Lagrangian density is $\mathcal{L}_I = (-\lambda/4!)\phi^4(x)$. The vertex is $-i\lambda/4!$
–Yukawa (Figure A.2 (b)): The interaction Lagrangian density is
$\mathcal{L}_I = g\overline{\psi}(x)\psi(x)\phi(x)$ and the vertex is ig.
–$SU(2)$ pseudo-scalar Yukawa theory with a doublet of fermions and a triplet of bosons (Figure A.2 (c))
$\mathcal{L}_I = g\overline{\Psi}(x)\gamma_5 \boldsymbol{\tau}\Psi(x) \cdot \boldsymbol{\phi}(x)$ and the vertex is $ig(\tau^k)_{ij}(\gamma_5)_{\alpha\beta}$.
–QED (Figure A.2 (d)): $\mathcal{L}_I = -e\overline{\psi}(x)\gamma_\mu\psi(x)A^\mu(x)$ and the vertex is $-ie(\gamma_\mu)_{\alpha\beta}$.

–Scalar QED (Figure A.2 (e) and (f)): There are two interaction terms.

$$\mathcal{L}_I^{(1)} = ieA_\mu(\phi\partial^\mu\phi^* - \phi^*\partial^\mu\phi) \text{ with vertex } -ie(p+p')_\mu.$$
$$\mathcal{L}_I^{(2)} = e^2 A_\mu A^\mu \phi\phi^* \text{ with vertex } 2ie^2\eta_{\mu\nu}.$$

–Yang–Mills (Figure A.2 (g), (h) and (i)):
$\mathcal{L}_I^{(1)} = -\frac{1}{2}gf_{abc}(\partial_\mu A_\nu^a - \partial_\nu A_\mu^a)A^{b\mu}A^{c\nu}$ with vertex
$.gf_{abc}[\eta_{\mu\nu}(p_1 - p_2)_\rho + \eta_{\nu\rho}(p_2 - p_3)_\mu + \eta_{\rho\mu}(p_3 - p_1)_\nu]$

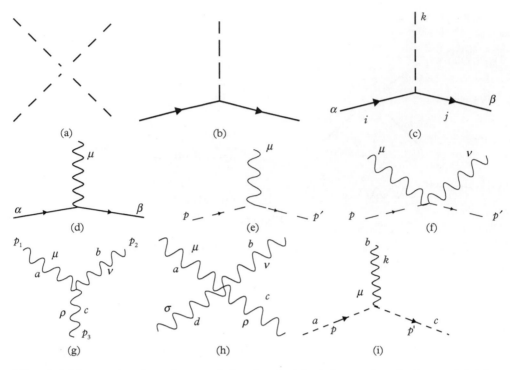

Fig. A.2 The vertices for a few usual theories. In (c) α, β are spinor indices and i, j, k are isospin indices.

$\mathcal{L}_I^{(2)} = -\frac{1}{4}g^2 f_{abc} f_{ab'c'} A_\mu^b A_\nu^c A^{b'\,\mu} A^{c'\nu}$ with vertex

$-ig^2 \left[f_{eab} f_{ecd} (\eta_{\mu\rho}\eta_{\nu\sigma} - \eta_{\mu\sigma}\eta_{\nu\rho}) + f_{eac} f_{edb} (\eta_{\mu\sigma}\eta_{\rho\nu} - \eta_{\mu\nu}\eta_{\rho\sigma}) + f_{ead} f_{ebc} (\eta_{\mu\nu}\eta_{\rho\sigma} - \eta_{\mu\rho}\eta_{\nu\sigma}) \right]$

$\mathcal{L}_I^{(3)} = g f_{abc} (\partial^\mu \overline{\chi}^a) A_\mu^b \chi^c$ with vertex $g p^\mu f_{abc}$

- **Cross section** (section 12.4)

Consider the reaction $A(q_1) + B(q_2) \to C_1(p_1) + \cdots + C_m(p_m)$. The cross section is given by

$$\sigma = \frac{1}{4\sqrt{(q_1 \cdot q_2)^2 - m_1^2 m_2^2}} \int (2\pi)^4 \, \delta^{(4)} \left(q_1 + q_2 - \sum p_j \right) |\mathcal{M}_{fi}|^2 \mathrm{d}\Omega_1 \ldots \mathrm{d}\Omega_m$$

- **Decay rate** (section 12.4)

$$w = \frac{1}{\tau} = \frac{1}{2M} \int (2\pi)^4 \delta \left(M - \sum p_j^0 \right) \delta^{(3)} \left(\sum \boldsymbol{p}_j \right) |\mathcal{M}_{fi}|^2 \mathrm{d}\Omega_1 \ldots \mathrm{d}\Omega_m$$

Index

John Iliopoulos is a Director of Research Emeritus at the École Normale Supérieure in Paris. He has taught on many introductory courses in Theoretical Physics, including Quantum Field Theory and the Theory of Elementary Particles, at the École Normale Supérieure and the École Polytechnique as well as in various Schools and Universities. In 1970, in collaboration with Sheldon Glashow and Luciano Maiani, he predicted the existence of the charm quark and proposed the GIM mechanism, an important step in the construction of the Standard Model. He also contributed to the development of supersymmetry, with Bruno Zumino and Pierre Fayet. He has received many awards, including the Ricard Prize of the French Physical Society, the Sakurai Prize of the American Physical Society, the High Energy Physics Prize of the European Physical Society and the Dirac Medal.

Following his PhD at Harvard University, **Theodore Tomaras** worked as research associate at CalTech and junior faculty at Rockefeller University, before joining the University of Crete, Greece, where he is now Professor of Physics Emeritus. He has taught many undergraduate and postgraduate courses, on elementary particle physics, quantum field theory, and gravitation and cosmology. He has contributed to the study of magnetic monopoles in GUT models, to the physics beyond the Standard Model, to the study of solitons in High Energy and Condensed Matter Physics, and to astroparticle physics. He has served as Head of the Department of Physics and of the Institute of Theoretical and Computational Physics of the University of Crete for several years, and was recently honoured with the 'S. Pihorides Award for Exceptional University Teaching'.

Note from the publisher: At the point of sending the final book to press, it is with regret that we have been informed of the passing of Theodore Tomaras in August 2021.